T0345918

Education, Skills, and Technical Change

Studies in Income and Wealth
Volume 77

National Bureau of Economic Research
Conference on Research in Income and Wealth

Education, Skills, and Technical Change
Implications for Future US GDP Growth

Edited by **Charles R. Hulten
and Valerie A. Ramey**

The University of Chicago Press

Chicago and London

The University of Chicago Press, Chicago 60637
The University of Chicago Press, Ltd., London
© 2019 by the National Bureau of Economic Research
All rights reserved. No part of this book may be used or reproduced
in any manner whatsoever without written permission, except in the
case of brief quotations in critical articles and reviews. For more
information, contact the University of Chicago Press, 1427 E. 60th St.,
Chicago, IL 60637.
Published 2019
Printed in the United States of America

28 27 26 25 24 23 22 21 20 19   1 2 3 4 5

ISBN-13: 978-0-226-56780-8 (cloth)
ISBN-13: 978-0-226-56794-5 (e-book)
DOI: https://doi.org/10.7208/chicago/9780226567945.001.0001

Library of Congress Cataloging-in-Publication Data

Names: Education, Skills, and Technical Change: Implications
    for Future U.S. GDP Growth (Conference) (2015 : Bethesda,
    Maryland) | Hulten, Charles R., editor. | Ramey, Valerie A. (Valerie
    Ann), editor.
Title: Education, skills, and technical change : implications for future
    US GDP growth / edited by Charles R. Hulten and Valerie A.
    Ramey.
Other titles: Studies in income and wealth ; v. 77.
Description: Chicago : The University of Chicago Press, 2019. | Series:
    Studies in income and wealth ; v. 77 | "This volume contains revised
    versions of the papers presented at the Conference on Research in
    Income and Wealth titled "Education, Skills, and Technical Change:
    Implications for Future U.S. GDP Growth," held in Bethesda,
    Maryland, on October 16–17, 2015"—Publisher info. | Includes
    bibliographical references and index.
Identifiers: LCCN 2018013221 | ISBN 9780226567808 (cloth : alk.
    paper) | ISBN 9780226567945 (e-book)
Subjects: LCSH: Labor supply—Effect of education on—United
    States—Congresses. | Labor supply—Effect of technological
    innovations on—United States—Congresses. | Education—Effect of
    technological innovations on—United States—Congresses. | Gross
    domestic product—Social aspects—United States—Congresses. |
    Human capital—United States—Congresses.
Classification: LCC HD5724 .E28 2019 | DDC 338.973—dc23
LC record available at https://lccn.loc.gov/2018013221

♾ This paper meets the requirements of ANSI/NISO Z39.48-1992
    (Permanence of Paper).

Relation of the Directors to the
Work and Publications of the
National Bureau of Economic Research

1. The object of the NBER is to ascertain and present to the economics profession, and to the public more generally, important economic facts and their interpretation in a scientific manner without policy recommendations. The Board of Directors is charged with the responsibility of ensuring that the work of the NBER is carried on in strict conformity with this object.

2. The President shall establish an internal review process to ensure that book manuscripts proposed for publication DO NOT contain policy recommendations. This shall apply both to the proceedings of conferences and to manuscripts by a single author or by one or more co-authors but shall not apply to authors of comments at NBER conferences who are not NBER affiliates.

3. No book manuscript reporting research shall be published by the NBER until the President has sent to each member of the Board a notice that a manuscript is recommended for publication and that in the President's opinion it is suitable for publication in accordance with the above principles of the NBER. Such notification will include a table of contents and an abstract or summary of the manuscript's content, a list of contributors if applicable, and a response form for use by Directors who desire a copy of the manuscript for review. Each manuscript shall contain a summary drawing attention to the nature and treatment of the problem studied and the main conclusions reached.

4. No volume shall be published until forty-five days have elapsed from the above notification of intention to publish it. During this period a copy shall be sent to any Director requesting it, and if any Director objects to publication on the grounds that the manuscript contains policy recommendations, the objection will be presented to the author(s) or editor(s). In case of dispute, all members of the Board shall be notified, and the President shall appoint an ad hoc committee of the Board to decide the matter; thirty days additional shall be granted for this purpose.

5. The President shall present annually to the Board a report describing the internal manuscript review process, any objections made by Directors before publication or by anyone after publication, any disputes about such matters, and how they were handled.

6. Publications of the NBER issued for informational purposes concerning the work of the Bureau, or issued to inform the public of the activities at the Bureau, including but not limited to the NBER Digest and Reporter, shall be consistent with the object stated in paragraph 1. They shall contain a specific disclaimer noting that they have not passed through the review procedures required in this resolution. The Executive Committee of the Board is charged with the review of all such publications from time to time.

7. NBER working papers and manuscripts distributed on the Bureau's web site are not deemed to be publications for the purpose of this resolution, but they shall be consistent with the object stated in paragraph 1. Working papers shall contain a specific disclaimer noting that they have not passed through the review procedures required in this resolution. The NBER's web site shall contain a similar disclaimer. The President shall establish an internal review process to ensure that the working papers and the web site do not contain policy recommendations, and shall report annually to the Board on this process and any concerns raised in connection with it.

8. Unless otherwise determined by the Board or exempted by the terms of paragraphs 6 and 7, a copy of this resolution shall be printed in each NBER publication as described in paragraph 2 above.

Contents

Prefatory Note

This volume contains revised versions of the papers presented at the Conference on Research in Income and Wealth titled "Education, Skills, and Technical Change: Implications for Future U.S. GDP Growth," held in Bethesda, Maryland, on October 16–17, 2015.

We gratefully acknowledge the financial support for this conference provided by the Bureau of Economic Analysis. Support for the general activities of the Conference on Research in Income and Wealth is provided by the following agencies: Bureau of Economic Analysis, Bureau of Labor Statistics, Bureau of the Census, Board of Governors of the Federal Reserve System, Statistics of Income/Internal Revenue Service, and Statistics Canada.

We thank Charles R. Hulten and Valerie A. Ramey, who served as conference organizers and as editors of the volume.

Executive Committee, December 2016

Introduction

Charles R. Hulten and Valerie A. Ramey

Overview

The growth in future living standards in the United States will likely depend to a significant degree on the continued evolution in the "knowledge" segments of the economy. These are the high-value-added sectors where product and organizational innovation generates high levels of productivity and creates new goods and markets. They are also the sectors that are the least vulnerable to global competition from low-wage manufacturing economies. Technology has already transformed many sectors with innovations like mobile communication devices, e-commerce, global supply-chain management, customization of manufacturing products, and GPS-based transportation management, and there is likely more to come with big data, the evolution of automated "workerless" factories and driverless vehicles, and developments in the areas of artificial intelligence, 3-D printing, nano-technology, and genomics. Evidence suggests that such innovations often require a parallel transformation in worker skills in order to implement and operate the new technology and business models. A workforce that cannot play this role may limit the rate of innovation and may slow the growth in living standards.

A century ago the United States became a world leader in the expansion of secondary and tertiary education, a development that helped propel US

Charles R. Hulten is professor of economics emeritus at the University of Maryland and a research associate of the National Bureau of Economic Research. Valerie A. Ramey is professor of economics at the University of California, San Diego, and a research associate of the National Bureau of Economic Research.

For acknowledgments, sources of research support, and disclosure of the authors' material financial relationships, if any, please see http://www.nber.org/chapters/c13693.ack.

productivity growth for decades, a thesis advanced in the 2010 study by Goldin and Katz. However, recent macroeconomic evidence suggests that the contribution of human capital accumulation to US growth has slowed in recent decades and the slowdown may last into the future. Moreover, the long-standing problem of the *quality* of the US primary and secondary education system has continued to be a source of concern, despite decades of efforts to improve the US education system. According to the Organisation for Economic Co-operation and Development (OECD)'s 2015 PISA survey of fifteen-year-olds, the US math performance was significantly below the mean OECD performance.[1]

The 2013 Programme of International Assessment of Adult Competencies (PIAAC) tells a similar story in its survey of the skill distribution of adults age sixteen to sixty-five in twenty-four countries. The literacy results for the US population are slightly below those of the OECD as a whole, but are considerably below the OECD in numeracy. Indeed, only a third of US respondents scored at the upper levels in math compared to around a half of OECD respondents.[2] This is all too consistent with the results of the recent "Nation's Report Card" (NAEP 2015) from the US Department of Education. This survey of American 12th graders found that only one in four were proficient or higher in mathematics and only two in five in reading ability. The study also found that the literacy and numeracy skills of 12th graders have been stagnant in recent years.

The implications of the trend in human capital formation and its interaction with technology for the future of US growth are the subject of the Conference on Research in Income and Wealth conference "Education, Skills, and Technical Change: Implications for Future US GDP Growth," held in Bethesda, Maryland, October 16–17, 2015. This conference volume contains twelve chapters exploring various aspects of this question, with discussant comments for many of the chapters. The contributors span an unusually broad range of expertise, including experts on aggregate productivity growth, cross-country comparisons of test scores and skill levels, the skill and task requirements of jobs, broader concepts of labor skills such as "noncognitive skills," alternatives to traditional education such as on-the-job training and online education, the role of immigration in skill supply, and the structure of the higher education sector.

We begin this introduction with some general observations about the way human capital affects economic growth and review the channels through which the skills and education of the labor force impact gross domestic product (GDP) growth. We then offer our own summary assessments of many of the salient issues before providing a brief summary of the chapters themselves.

---

1. OECD (2016, *Snapshot Table*, 5).
2. OECD (2013, tables A2.1 and A2.5).

Human Capital's Contribution to GDP Growth

Virtually every aspect of economic activity involves human agency of some sort, whether it involves decisions about business models and management procedures, innovation, capital investment, and, perhaps most important of all, the skills and motivation that workers bring to their jobs. The quantity and quality of this agency matter, and this is where education comes into play. While formal education is not the only way that human capital is built, it provides the foundational infrastructure of literacy, numeracy, and general information that informs the functioning of an advanced society, including its economy. It also provides important vocational and professional skills.

How important is education and the knowledge it imparts compared to other factors that affect economic activity? Economic historians and economists specializing in the field of education generally see educational attainment and human capital development as critical factors in the process of economic growth. Hanushek and Woessmann (2015, 1) start their book, *The Knowledge Capital of Nations*, with the statement that "knowledge is the key to economic development. Nations that ignore this fact suffer, while those that recognize it flourish." Moreover, it is not just the average level of education that matters. Economic historian Joel Mokyr argued in 2005 that it was those in the upper tail of the knowledge distribution that were responsible for much of the technological development that drove the Industrial Revolution. David Landes (1998, 276), in his appraisal of the factors that determine the *Wealth and Poverty of Nations*, sums up with the following observation: "Institutions and culture first; money next; but from the beginning and increasingly, the payoff was to knowledge."

The importance of acquiring knowledge is well understood by the population at large, if historical statistics on educational attainment are any indication. The proportion of persons older than age twenty-five with college degrees increased from around 5 percent in 1950 to 30 percent in 2010, and two-thirds of high school graduates went on to some form of tertiary education in 2012, up from 50 percent in 1975.[3] This increase was driven, in part, by the growing wage premium for a college education documented in the work of Goldin and Katz (2010), and by Valetta writing in chapter 9 of this volume. The dramatic increase in schooling was matched by a large increase in the national commitment to education. Annual real expenditures per student rose over the period 1960 to 2011, from around $3,000 to $11,000, and when private spending is added to public outlays, the combined direct investment rate in education in the United States in 2011 was nearly 7 percent of GDP.[4]

3. US Census Bureau (2015).
4. These estimates are from table 236.55 of the *2013 Digest of Educational Statistics* (NCES 2013).

This is an impressive record. There is, however, another important question: Does more education necessarily lead to more economic growth? Are past results indicative of future returns? On the one hand, the demand for college graduates may have decreased, and, as noted, the macroeconomic contribution of education to aggregate output growth seems also to have slowed. On the other hand, the underlying factors that have propelled the demand for higher education and more complex skills—skill-biased and labor-saving technical change and the globalization of the world economy— proceed apace (for now), and the demand for college-educated workers is increasingly a demand for postgraduate and professional education. These are issues that high-income societies like the United States face today in their efforts to sustain the economic growth needed to improve living standards for a broad range of the population, and not just for those with college degrees.

The Channels through Which Human Capital Affects GDP Growth

Economic growth is a complex process influenced by many factors, and education is a multifaceted process that affects growth through multiple channels. As a backdrop for the material presented in the various chapters of this volume, we identify and comment on five of these channels:

1. **Worker Productivity.** Education operates directly by raising the marginal productivity of workers. The Mincer wage equation is a staple of labor economics, linking education, cognitive skills, and other individual characteristics to wage rates, which are in turn linked to the value of the marginal product of labor. When these individual productivity effects are aggregated, they constitute a potentially important source of growth in real GDP. The size of and relative importance of this effect can be estimated using the growth-accounting method pioneered by Jorgenson and Griliches (1967) in their pathbreaking paper and employed by the Bureau of Labor Statistics (BLS 1983) in their Multifactor Productivity program. The chapters by Jorgenson, Ho, and Samuels, and by Bosler, Daly, Fernald, and Hobijn in this volume provide estimates based on this method, which suggest that education may make a relatively smaller contribution to growth than in the past.

2. **Skill-Biased Technical Change.** Changes in the nature of technology in recent decades have shifted the demand for labor skills in favor of those involving nonroutine cognitive activities. Education is one factor that accommodates this skill-biased technical change, which can affect output growth above and beyond the direct marginal product effect, as set out in the important 2011 and 2012 contributions by Acemoglu and Autor. Moreover, shifts in the microstructure of production activities have tended to involve workers with advanced skills that are strong complements with the more sophisticated types of capital and technology, and are thus necessary inputs whose absence can limit growth (Hulten, chapter 3, this volume). This

demand for these "necessary" workers is one factor driving the growth of the college wage premium.

3. Innovation. The education sector is a prime source of the new ideas and perspectives that lead to technical innovation, and education is important for the adoption and diffusion of technology, as Nelson and Phelps (1966) emphasize in their contribution. Other research suggests that technologies diffuse more quickly when basic literacy and numeracy are more widespread.[5] In other words, innovation is an endogenous process that depends in part on education, both for its development and diffusion.

4. Knowledge Spillovers. The development and transmission of knowledge involves spillover externalities in which the social return to investments in both education and research and development (R&D) exceed the private return. In the case of education, the spillover occurs because educated people interact in ways that are not mediated by a labor market return (Lucas 1988). With R&D, the knowledge spillover arises from the inability of innovators to completely protect their property rights against diffusion to other users (Romer 1986, 1990).

5. Social Capital. Education is part of the foundational infrastructure that sustains social, political, and economic institutions. This mechanism is perhaps not so much a specific channel as it is an infrastructural investment in building or maintaining social capital. It involves the Landes emphasis on institutions and culture as sources of national prosperity, but the following quote, attributed to Thomas Jefferson, perhaps says it best: "If the children are untaught, their ignorance and vices will in future life cost us much dearer in their consequences than it would have done in their correction by a good education."

The chapters in the volume are focused largely on various aspects of the first two channels. This focus should be kept in mind when assessing the impact and value of education, since a great deal of education's overall value is created through the other channels.

The Supply and Demand for Skills and Education: An Overview

Individual chapters are summarized briefly in the next section, but, before going there, we offer a summary assessment of what we see as the main points. They reflect our reading of the chapters, as well as our own research and understanding of the issues, and they should not necessarily be attributed to any individual author or discussant whose work appears in the volume.

1. A strong education system is essential for the proper functioning of modern economies, and is the hallmark of an advanced society. Evidence suggests that those societies with the highest income per capita are also

5. See, for example, Benhabib and Spiegel (2005).

those with the greatest educational attainment. Education played a particularly key role in the transition over the last half century to a globalized "knowledge economy" by helping provide the requisite nonroutine cognitive and noncognitive skills. Without the appropriate supply response to the changing demand for skills, it is hard to see how this revolution could have occurred in its current form.

2. More is involved in skill development and learning than formal education alone. Home environment is an important determinant of skill formation, with the cognitive and noncognitive skills developed in early childhood playing a fundamental role in a child's ability to learn. The socioeconomic status of the family also matters (see, e.g., Ramey and Ramey 2010), as do idiosyncratic factors like ability. Moreover, skill development does not stop at graduation. Research at the BLS reported in the Gittleman, Monaco, and Nestoriak chapter in this volume has found that the formal school preparation placed third behind training and job experience as a source of skill development. On the other hand, education does provide the general skills of literacy and numeracy needed for the further development of many task-related skills, and is the main *systematic* way that children are prepared for adult life and the world of work. It also provides vocational training and preparation for various professions, and educational attainment has been found to be positively correlated with employment in jobs requiring more complex cognitive and noncognitive skills.

3. Much of the recent focus on the demand side of skill development has been on the higher-order cognitive and noncognitive skills needed for the growing complexities of the technology revolution. This is appropriate, given that these skills are an important enabler of that revolution and the income growth it has created. However, it is also true that only a fraction of all jobs involve complex tasks (around 15 percent, according to the BLS study in this volume), and only a quarter of all jobs require a college degree. Any discussion of the demand for skills must acknowledge the fact that the education system needs to prepare students for a broad range of skills and vocations, not just those at the top ends of the skill and educational attainment scales. This is all the more important because the requirements of many "routine" skills have shifted as a result of sectoral changes in the structure of the economy and the growing presence of information technology.

4. Much of the initial focus on the demand for skills was on higher-order cognitive skills, but the importance of noncognitive "soft" skills has been increasingly appreciated. These soft-skill traits include self-discipline, conscientiousness, and the ability to get along with others. These traits are hard to pin down analytically, but studies suggest that they are rewarded in the labor market (see the study by Lundberg and the discussion by Deming in this volume). They are important for the full spectrum of jobs, but are particularly important for jobs that involve less direct supervision.

5. Increased college-participation rates are not a panacea for address-

ing income equity and prompting more rapid economic growth. Not only are there limits on the demand for the skills of college-educated workers, there are supply-side issues as well. Research by James Heckman and colleagues has emphasized the importance of "college readiness" and the limits it imposes on individual higher education outcomes.[6] While the average college wage premium is still large, not everyone receives this premium. A study by Abel and Deitz (2014) finds that the lowest quartile incomes of college graduates only marginally outperformed the median incomes of high school graduates.

6. At the other end of the wage premium spectrum, the United States stands out in the PIAAC international comparison in its propensity to reward those with the highest skills (Broecke, Quintini, and Vandeweyer, chapter 7, this volume). This is significant in view of the Mokyr hypothesis that those in the upper-tail knowledge of the distribution play a key role in technological development. They are prominent in the research labs of universities and companies, the C suites of corporations, and software development divisions of technology companies.

7. Education is a process that unfolds over time for any given individual and is fraught with uncertainty and institutional problems and rigidities. Thus, the adjustment of the supply of new graduates to a change in demand for a skill or occupation cannot occur immediately, leading to periods in which demand growth may outstrip supply. Goldin and Katz argue that this phenomenon occurred as the information revolution increased the demand for complex skills and higher education, and a lagging supply response led to a college wage premium as the natural market outcome. Some have interpreted this as a worrisome "skills gap," but standard economic logic sees it as a period of labor market adjustment. Indeed, recent evidence suggests that the uptake of college graduates may be slowing, along with the wage premium for college (see Beaudry, Green, and Sand [2016] and chapter 9 in this volume by Valletta, as well as the comment on chapter 9 by Autor).

8. Immigration is an important source of the supply of highly skilled and educated workers, and is particularly important in the science, technology, engineering, and mathematics (STEM) areas. Hanson and Slaughter, writing in chapter 12 of this volume, report that the foreign-born share of STEM employment in 2013 was approximately 20 percent among those with bachelor's degrees, 40 percent among those with master's degrees, and 55 percent among PhDs. Expressed in terms of hours among prime-age workers (those thirty to forty-five years of age) with an advanced degree, the foreign born accounted for nearly one-half of total hours worked in STEM occupations in 2013, up from around one-quarter in the 1990s and one-fifth in the 1980s. These estimates refer to STEM workers. Immigration has also

---

6. See Heckman, Stixrud, and Urzua (2006) and Heckman, Humphries, and Veramendi (2016).

been an important source of entrepreneurship, according to the study by Kerr and Kerr (2017).

9. The quality of education matters as well as the quantity. In this regard, the success of the US education system in preparing students with the skills needed for the economy of the twenty-first century gets a mixed report card. According to Current Population Survey (CPS) data, most students today finish high school (some 90 percent), and two-thirds go on to some form of tertiary education.[7] Not all succeed in obtaining a four-year college degree, as only around one-third of the population end up with a four-year college degree or more (though Abel and Deitz, in chapter 4 of this volume, show that many of those who do not find jobs requiring a college degree end up in fairly well-compensated employment). The quality of US higher education is very high in international comparisons, but there are still problems facing college students: rising tuition (see chapter 10 in this volume by Gordon and Hedlund), the growing burden of student debt, and retention and lengthy time-to-graduation are issues. The college "industry" is also undergoing changes in the technology of teaching made possible by the digital revolution, not the least of which is the rise of online education (Hoxby, chapter 11, this volume). On the other hand, the educational outcomes at the K–12 level revealed by the National Assessment of Educational Progress (NAEP 2015) and by international comparisons point to deeper and more persistent problems.[8] However, the K–12 results cannot be attributed to the quality of schooling alone. Research suggests that the cognitive and noncognitive skills developed by age three have fundamental effects on the ability to learn. Thus, K–12 schools have little control over a key input into their production functions.

10. Combined with those students who do not finish high school, the test-score results suggest that a substantial portion of US youth is not being prepared for the needs of the knowledge economy and the affluence it conveys, or for the remaining medium-skill jobs that in the past have provided middle-class affluence. While higher education, with its large wage premium, is a pathway to higher incomes for some, many others are left behind. Finding an answer to this equity versus growth conundrum is one of the great educational and economic challenges of the years ahead.

We emphasize, again, that these points reflect our own views and understanding of the subjects covered and should not be attributed to any individual author or discussant.

---

7. US Census Bureau (2015).

8. One NAEP result is particularly noteworthy in this regard. More than a third of the 12th grade students surveyed scored in the below basic category in reading and almost 40 percent in mathematics. These deficits have persisted over time and they do not bode well for future employment in an increasingly technological world economy.

Summary of the Chapters in the Volume

The chapters in this volume touch on one or more of the issues raised in the preceding section. We turn now to a brief summary of these chapters and discuss how they help address those issues.

The Macroeconomic Link between Education and Real GDP Growth

The volume begins with three chapters that use a growth-accounting model to measure the contribution of labor quality to GDP growth. These are the chapters by Jorgenson, Ho, and Samuels; Bosler, Daly, Fernald, and Hobijn; and Hulten. The first two chapters are followed by a general discussion of the issues by Douglas W. Elmendorf, whose perspective as former head of the Congressional Budget Office (CBO) illustrates the policy relevance of the questions being asked.

The first two chapters use the Jorgenson and Griliches (1967) extension of the Solow (1957) growth-accounting framework as a starting point. The great advantage of the Solow framework is its ability to sort out the contributions of the three general factors responsible for growth: labor, capital, and technology. Jorgenson and Griliches took this a step further by adding the labor "quality" to this list, defining it as the shift in the composition of labor force characteristics (including education) to those with higher or lower marginal products. This framework disaggregates labor into its various characteristics and assumes that wage rates accurately reflect the corresponding marginal products. It then resolves the results into indexes of the quantity of labor input and its composition/quality.

The chapter "Educational Attainment and the Revival of US Economic Growth" by Jorgenson, Ho, and Samuels analyzes the recent past and projected future of labor-quality growth and overall GDP growth using a newly constructed KLEMS (capital, labor, energy, materials, and purchased services) sixty-five-industry data set from 1947 through 2014. Despite an overall slowdown in educational attainment of the population, Jorgenson et al.'s labor-quality series shows a continuing significant contribution of educational attainment to labor quality from 2007 through 2014. The source of this discrepancy is the decline in employment participation of the less educated, so the average educational attainment of the employed continued to rise. Looking forward, Jorgenson et al. project that labor-quality growth will contribute essentially nothing to growth from 2014 to 2024 if the recent decline in the employment participation rate of the less educated is reversed.

An empirical challenge facing users of the Jorgenson-Griliches framework is the construction of the labor-quality index, since it is not directly observable. The chapter "The Outlook for US Labor-Quality Growth" by Bosler, Daly, Fernald, and Hobijn begins by addressing this problem. The standard way to estimate labor quality is to invoke the assumption of competitive factor markets and use wages as a measure of marginal product. One approach used in the labor economics literature regresses the wages of individual workers on their observable characteristics such as education level, gender, experience, and so forth, and then uses the estimated coefficients to derive weights in order to construct a labor-quality index. As Bosler et al. explain, researchers face a trade-off: adding more detailed characteristics explains more of the variation of wages across workers, but at the same time reduces the precision of the marginal product estimates because the number of workers in each cell falls. Bosler et al. explicitly show the trade-off across almost 2,000 specifications that vary in the number of worker characteristics included, how finely these characteristics are disaggregated, and the functional form. The authors then construct an index of labor quality for their preferred specification, as well as several of the leading alternatives.

Bosler et al.'s analysis confirms Jorgenson et al.'s findings that the much-discussed decline in the employment-population ratios of the less educated has contributed to labor-quality growth through a composition effect on the employed. These same employment-population movements create uncertainty about the future growth rate of labor quality, however. If the employment of the less educated recovers, the labor force will grow faster than otherwise expected, but labor-quality growth will be slower. Bosler et al. also offer several projections of future labor-quality growth. Their preferred projections are for labor quality to grow relatively slowly, from 0.1 to 0.25 percent per year, for the longer run reaching 2025. If these projections are borne out, they mean that labor-quality growth will be a less important part of GDP growth in the future than it has been in the past. In other words, the slowdown in educational attainment in the United States will finally start showing up in aggregate labor-quality growth.

The chapter by Hulten, "The Importance of Education and Skill Development for Economic Growth in the Information Era," is the third of the chapters in the volume that deal with growth accounting. Where the methodology of Jorgenson et al. essentially follows the approach of Jorgenson and Griliches (1967), and Bosler et al. explore alternative ways of measuring the labor-composition term of that model, the Hulten chapter proposes an alternative way of looking at the technology that underpins the growth-accounting framework. This alternative approach is motivated, in part, by the view that education plays a more fundamental role in enabling economic activity than is implied by the labor-composition effect, and that this might help explain the relatively small *measured* role in output growth over the course of the information revolution. Hulten builds on the Acemoglu and Autor (2012) insights about task-skill links, but develops them in the context of a disaggregated activity-analysis technology. In this framework, the business model of a firm specifies the kinds of goods to be produced and how they are marketed, and the execution of these decisions is broken down into various activities within the firm. In the strict version of this model, each activity uses inputs in a fixed proportion, meaning that each type of skilled labor and capital is a necessary input. This provides a mechanism through which the more complex forms of capital, both tangible and intangible, are linked to the higher-order labor skills needed to operate that capital. This "necessary input" model contrasts with the conventional aggregate production function approach to growth accounting, which groups input into capital and labor aggregates and assumes a high degree of substitutability between them.

One goal is to examine the implications of this "necessary input" feature of the activity-analysis model for conventional aggregate sources-of-growth estimates. This leads to the salient result that the empirical sources-of-growth results reported by BLS *could equally have been generated by the activity-analysis model.* This enables these results to be interpreted in a very different way than under the standard Solow aggregate production function interpretation, one that assigns a greater importance to labor skills and education.

Jobs and Skills Requirements

Preparing students for jobs is not just a matter of inducing them to attend school for a certain number of years, since there is no guarantee that the skills students learn in school will match those demanded by employers. The two chapters in this section shed light on the issue of this match and the demand for skills. The first chapter studies the outcomes of recent college graduates, and the second surveys the skill requirements of jobs.

"Underemployment in the Early Careers of College Graduates following the Great Recession" by Abel and Deitz studies an issue that has received much attention from the press: Are recent college graduates finding jobs that match their education level? Following the Great Recession, newspapers published a number of stories about recent college graduates who ended up working as baristas in coffee shops. Abel and Deitz study the validity of this picture by constructing and analyzing detailed data on the unemployment and underemployment experiences for recent graduates. Unemployment rates by education are readily available, but *underemployment* rates are not part of the standard government statistics. The authors construct new series on underemployment rates of recent graduates using information from the Department of Labor's O*NET database, which contains information on the characteristics of hundreds of occupations based on interviews of incumbent workers and occupational specialists. They discover that underemployment of this group is not a new phenomenon. In fact, their series shows a rough V-shape since 1990. The current level of 45 percent underemployment of recent college graduates still lies below the level that prevailed in the first half of the 1990s.

A question that arises is, What sort of jobs do the underemployed recent college graduates take? The Abel-Deitz results show that most underemployed recent graduates did not end up working in low-paid service jobs (e.g., baristas). Rather, nearly half ended up in relatively high-paying occupations, such as information processing and office and administrative support. Only 9 percent of all recent college graduates began their careers in low-paying service jobs. Thus, even if a college degree did not guarantee an initial placement in an occupation requiring a college degree, it did give individuals a competitive advantage in the occupations that did not require a college degree.

"The Requirements of Jobs: Evidence from a Nationally Representative Survey" by Gittleman, Monaco, and Nestoriak describes a new survey conducted by the Bureau of Labor Statistics (BLS) and reports findings from the preproduction test survey. The BLS launched the Occupational Requirements Survey (ORS) in collaboration with the Social Security Administration as a data source in disability adjudication. The rich information from the survey can be used to answer a number of other economic questions, including the demand for and returns to education and skills in occupations.

Gittleman et al. use these data to study the requirements of jobs. An important finding is that fewer than 25 percent of jobs require a college degree or higher degree, somewhat less than reported in the O*NET data (around 27 percent). This relatively small fraction stands in contrast to the common assertion that earning a college degree has become de rigueur for employment in the twenty-first-century US economy. The bottom line is that three-quarters of all current jobs do not *require* a four-year college degree.

Additional results suggest that there are many jobs that do not require complex tasks, or that allow only loose control. Any policy aimed at significantly increasing college enrollments should take note of these findings. However, it is also important to note that these results do not diminish the importance of a higher education for those jobs for which it is needed. Moreover, Gittleman et al.'s analysis of average wages by job characteristic reflect large premiums for education. Thus, the more nuanced interpretation of the Gittleman et al. results is that while there are many jobs available for individuals with low education and skill levels, those jobs pay much less than those with higher education and skill levels.[9]

Skills, Inequality, and Polarization

The chapters in the last sections go beyond the standard practice of equating labor quality or skill with years of education. The chapters in this section consider additional dimensions. One chapter branches out to consider

9. We emphasize that these wage outcomes should not be interpreted as a type of "demand" for skills indicator irrespective of supply. The creation of a job or occupation is the outcome of the interaction of particular demands in the face of a supply of skills in an economy. Thus, firms facing a badly educated workforce would be expected to adapt by fashioning their job requirements around the supply of skills, and using technology in ways that overcome gaps in skill supply.

noncognitive skills, and the other three consider the distribution of skills rather than just the average.

"Noncognitive Skills as Human Capital" by Lundberg discusses both what we know about the importance of noncognitive skills in individuals' outcomes and the measurement challenges for quantifying these types of skills. The standard measures of human capital include years of education, cognitive test scores, and/or IQ-related measures (such as the Armed Forces Qualifying Test [AFQT]). A literature that emerged in the first decade of the twenty-first century showed that it might be valuable for economists to broaden their concept of human capital to include "noncognitive skills" in the form of personality traits. As Lundberg points out, however, measures of noncognitive skills are not always reliable in all applications. She cites a lack of consensus on what noncognitive skills really are, as well as a lack of a consistent set of metrics across studies. Part of her chapter points out the current gaps and what would be needed to consider the role of noncognitive skills in economic growth. Among the challenges are establishing a *causal* channel based on estimated relationships in which unobserved factors may be playing a role and evidence on the heterogeneity of returns to noncognitive skills across different environments.

To illustrate the issues involved, Lundberg uses the NLSY97 and the Add Health surveys to estimate the relationships between noncognitive skills and outcomes. A number of interesting results emerge that show the difficulty of interpreting results. First, the correlation between various measures of noncognitive skills is surprisingly low. Second, the important and statistically significant effects of many of the noncognitive skill measures on wages and employment often disappear once educational attainment is included in the regressions. These results suggest that a key channel of influence of noncognitive skills on labor market outcomes might be through educational attainment and not through the direct channel of on-the-job performance. Third, the importance of certain measures of noncognitive skills in predicting outcomes such as crime are not necessarily robust to adding other measures of noncognitive skills.

Overall, Lundberg's chapter highlights the fact that noncognitive skills are potentially very important for thinking about human capital and productivity more broadly. There are still many problems to be solved in making this analysis more concrete and filling in the causal steps. Lundberg's chapter is very useful for pointing out the key gaps that need to be filled in the literature.

The next chapter in the section, by Broecke, Quintini, and Vandeweyer, uses data from the latest survey of the Programme for the International Assessment of Adult Competencies (PIAAC) to determine how much of the differences in wage inequality across countries can be explained by differences in the endowments of and return to skills across countries. Their chapter contributes to a debate about whether a difference in skill distributions or institutions can best explain differences in inequality across countries.

Broecke et al. begin by comparing the distribution of skills—they concentrate on numeracy in particular—and the distribution of wages within a number of countries. They find that the United States has one of the lowest average levels of adult skills, but also one of the highest dispersions of skills. Moreover, the United States has the highest returns to skills, is among the countries with the highest average levels of wages, and is near the top in wage inequality.

Broecke et al. conduct accounting exercises in order to analyze the extent to which the endowment of skills and the return to skills can explain wage inequality differences across countries. They find that differences in the *returns to skills* in the United States are much more important than differences in the *endowment of skills* in accounting for the inequality of wages in the United States relative to other countries. Overall, this chapter shows how concrete measures of skills and their returns can help explain differences in inequality across countries. An additional outcome of their study is the clear demonstration that the average skill level of American adults lags behind many other OECD countries. It is also apparent, however, that the demand for skills in the United States remains high, as evidenced by the high-skill premium.

Erik Hanushek's chapter, "Education and the Growth-Equity Trade-Off," considers a number of important issues concerning the link between cognitive skills, growth, and inequality. He first considers the role of human capital in growth models. As he points out, in neoclassical models, a rise in human capital will raise the *level* of output, but not the steady-state growth rate of output. In contrast, in endogenous growth models, a rise in human capital can potentially raise the steady-state growth rate of output. The second point he makes is how years of educational attainment is a poor measure of human capital. Hanushek notes that the quality of educational systems differs dramatically across countries, and even possibly across time. Illustrating the findings from his earlier work with coauthors, he shows that in a cross-section regression of long-run growth rates, average years of education performs poorly relative to his preferred measures that use the results of international assessments of test scores and similar metrics.

Robert Valletta's chapter, "Recent Flattening in the Higher Education Wage Premium: Polarization, Skill Downgrading, or Both?," focuses on trends in wage premiums. He particularly studies possible sources for the documented flattening in the returns to education. Since 1980, educational wage premiums have increased, but they have done so at a decreasing rate. The premium for college only (i.e., four-year college degree, but no graduate school) over high school rose the fastest in the 1980s, slightly less fast in the 1990s, and then stalled since 2000. The premium for graduate degrees rose more robustly during most decades, but appears to have stalled since 2010. Valletta then considers the extent to which two possible hypotheses can explain these trends. First is the job polarization hypothesis (e.g., Autor, Katz, and Kearney 2008; Acemoglu and Autor 2011), which argues that skill-biased technological change has reduced the demand for routine jobs that can be computerized. In this hypothesis, the middle-educated (e.g., some college or college only) lose their jobs and are forced to move down to nonroutine, noncognitive jobs that pay much less. A second hypothesis, which expands on the polarization hypothesis, is "skill downgrading" by Beaudry, Green, and Sand (2016). They argue that the rise in educational premiums was in part a transitional effect of moving to a higher level of intangible organizational capital. Demand for cognitive skills was high when investment in information technology (IT) was high during the transition to the new steady state, but once the new state was reached, there was less demand for those types of cognitive skills. To shed some light on the forces at play, Valletta analyzes changes in premiums within and between broad occupation categories as well as shares of workers by education in those groups. Valletta interprets his results as suggesting rising competition among educated workers for high-paying jobs that are becoming more scarce. He argues that even if the social return to higher education might be slowing down, the private returns are still large because it enables workers to compete for the best-paying jobs.

The Supply of Skills

Our opening comments describe some of the frictions arising in the formal education sector in the United States that tend to slow the supply response of skills to shifts in demand. In the same vein, this section begins with a chapter that examines the sources of the rise in college tuition in the United States and then moves on to consider some nontraditional means for increasing the supply of educated workers.

A potentially important impediment to the growth in educational attainment of the US population is the dramatic rise in college tuition. Tuition and fees, even net of institutional aid, grew by 100 percent between 1987 and 2010. This rise dwarfs even the rise in health care costs. In "Accounting for the Rise in College Tuition," Gordon and Hedlund seek to understand the sources of this rise since 1987.

Assessing the importance of the leading factors would be difficult to do with purely empirical methods, since tuition and many of the candidate factors are all trending up together. To answer the question, Gordon and Hedlund thus turn to quantitative methods. In particular, they specify a theoretical model that embeds a college sector in an open-economy model. They then calibrate the model to match key data moments since 1987 and use it to assess the sources of the rise in college tuition between 1987 and 2010. They find that demand changes due to changes in financial aid can account for virtually all of the rise in tuition. The rise in the college wage premium (due to skill-biased technological change) alone can account for 20 percent of the rise. In contrast, they find a *negative* role for Baumol's cost disease. This surprising result becomes clearer once one considers equilibrium effects: while the cost disease might explain tuition increases at a *given* university, in equilibrium students are substituting into cheaper universities, so this factor does not raise *overall* tuition.

The Gordon and Hedlund chapter represents a serious first step in using quantitative models to study the sources of the rise in college tuition. As they acknowledge, however, the model is very stylized in some dimensions and misses some potentially important features. Thus, the results are only suggestive at this point. However, their analysis is a good foundation for future research using quantitative methods.

The role of education in innovation and the production of output has been a general theme of this conference. "Online Postsecondary Education and Labor Productivity" by Caroline M. Hoxby turns this question around and looks at one of the most notable innovations in higher education itself. Enrollment in online education has experienced explosive growth in recent years and the online postsecondary education sector (OLE) has been hailed as the wave of the future by its enthusiasts. Hoxby takes a close look at the evidence, examining both its pros and cons in comparison with traditional "in-person" brick-and-mortar institutions (B&M), including those that are less "competitive" and also have an online presence. Hoxby uses longitudinal data from the IRS on nearly every person who engaged substantially in online postsecondary education between 1999 and 2014 (supplemented, in places, by National Center for Education Statistics [NCES] data). Her basic objective is to calculate the return on investment (ROI) to see if students recoup enough in additional discounted lifetime wages to cover the cost of the OLE, inclusive of the opportunity cost of time. In addition, the study computes a social return that includes the cost of public subsidies.

This first in-depth study of the returns to online education uncovers many interesting, and sometimes surprising, dimensions of online education. For example, she finds that the undergraduate tuition paid by the OLE students is actually higher than that paid by those in *nonselective* brick-and-mortar institutions. Yet, the resources devoted to students in OLE are lower. Estimates of ROIs suggest that the earnings of most online students do not increase by enough to cover even their private costs, though there are exceptions. Moreover, while online enrollment episodes do usually raise students' earnings, it is almost never by an amount that covers the *social* cost of their education.

Last, but by no means least, in the topic of skill supply is the important issue of immigration as a source of supply for the skills needed in high-technology employments. The chapter "High-Skilled Immigration and the Rise of STEM Occupations in US Employment" by Hanson and Slaughter explores the contribution of immigrants to employment in US STEM fields. The STEM workers overall tend to have much higher formal education than the average worker. Moreover, as previously noted, Hanson and Slaughter show that the immigrant share of hours worked in the STEM occupations has increased to the point that prime-age workers with advanced degrees now account for almost half the total hours worked, more than double the proportion of the hours worked in 1980.

The foreign-born share of STEM employment is higher than for non-STEM employment. Hanson and Slaughter consider possible explanations for the foreign-born comparative advantage in STEM fields. The hypothesis with the most support is that it is relatively more difficult for foreign-born higher-educated workers to gain entry into nontechnical occupations because many of those occupations require elevated knowledge of the subtleties of US culture that are important for face-to-face communication with customers. The authors compare wages and find that, while the foreign born have significantly lower wages than natives in the nontechnical occupations, the foreign born have similar wages to natives in the STEM occupations. Hanson and Slaughter's findings suggest that, to the extent that STEM occupations are important for technological change and growth in the United States, then immigrants with college and advanced degrees have played an important role in US growth.

We also recommend the comments made by discussants of the various chapters. The discussants are eminent experts and their discussions are well worth reading as contributions in their own right.

Conclusion

The chapters in this volume cover a wide range of issues drawn from different literatures within the field of economics. The goal was to bring together a mix of researchers in order to address an important question that spans these literatures: How will current trends in human capital formation affect future US growth? The macroeconomic literature on the sources of growth has long recognized the potential importance of human capital accumulation for growth but has only begun to study the microeconomic mechanisms of that accumulation. On the other hand, the microeconomic literature on education and human capital formation studies many detailed aspects of skill supply and demand at the microeconomic level but seldom draws out the implications for the future of macroeconomic growth. While there is still considerable debate over many of the issues touched on in this volume, we believe that the research presented is a significant step toward linking these research areas in a way that informs the larger questions of how well students are being prepared for the current and future world of work, and whether this preparation will sustain the growth of an increasingly knowledge-based economy.

References

Abel, Jaison R., and Richard Deitz. 2014. "College May Not Pay Off for Everyone." Liberty Street Economics, Federal Reserve Bank of New York, September. http://libertystreeteconomics.newyorkfed.org/2014/09/college-may-not-pay-off -for-everyone.html.

Acemoglu, Daron, and David Autor. 2011. "Skills, Tasks and Technologies: Implications for Employment and Earnings." *Handbook of Labor Economics* 4:1043–171.

———. 2012. "What Does Human Capital Do? A Review of Goldin and Katz's *The Race between Education and Technology.*" *Journal of Economic Literature* 50 (2): 426–63.

Autor, David H., Lawrence F. Katz, and Melissa S. Kearney. 2008. "Trends in U.S. Wage Inequality: Revising the Revisionists." *Review of Economics and Statistics* 90 (2): 300–323.

Beaudry, Paul, David A. Green, and Benjamin M. Sand. 2016. "The Great Reversal in the Demand for Skill and Cognitive Tasks." *Journal of Labor Economics* 34 (S1): S199–247.

Benhabib, Jess, and Mark M. Spiegel. 2005. "Human Capital and Technology Diffusion." *Handbook of Economic Growth* 1:935–66.

Bureau of Labor Statistics (BLS). 1983. *Trends in Multifactor Productivity, 1948–81,* vol. 2178 of *Bulletin of the United States Bureau of Labor Statistics.* Washington, DC: United States Government Printing Office.

Goldin, Claudia, and Lawrence F. Katz. 2010. *The Race between Education and Technology.* Cambridge, MA: Belknap Press.

Hanushek, Eric A., and Ludger Woessmann. 2015. *The Knowledge Capital of Nations: Education and the Economics of Growth.* Cambridge, MA: MIT Press.

Heckman, James J., John Eric Humphries, and Gregory Veramendi. 2016. "Returns to Education: The Causal Effects of Education on Earnings, Health and Smoking." NBER Working Paper no. 22291, Cambridge, MA.

Heckman, James J., Jora Stixrud, and Sergio Urzua. 2006. "The Effects of Cognitive and Noncognitive Abilities on Labor Market Outcomes and Social Behavior." *Journal of Labor Economics* 24 (3): 411–82.

Jorgenson, Dale W., and Zvi Griliches. 1967. "The Explanation of Productivity Change." *Review of Economic Studies* 34:349–83.

Kerr, Sari Pekkala, and William R. Kerr. 2017. "Immigrant Entrepreneurship." In *Measuring Entrepreneurial Business: Current Knowledge and Challenges,* edited by John Haltiwanger, Erik Hurst, Javier Miranda, and Antoinette Schoar, 187–249. Chicago: University of Chicago Press.

Landes, David S. 1998. *The Wealth and Poverty of Nations.* New York: W. W. Norton & Company

Lucas, Robert E. Jr. 1988. "On the Mechanics of Economic Development." *Journal of Monetary Economics* 22:3–42.

Mokyr, Joel. 2005. "Long-Term Economic Growth and the History of Technology." *Handbook of Economic Growth* 1:1113–80.

National Center for Education Statistics (NCES). 2013. *Digest of Educational Statistics, 2013.* http://nces.ed.gov/programs/ digest/d13/tables/dt13_236.55.asp.

Nation's Report Card (NAEP). 2015. *Math and Reading Scores at Grade 12.* National Assessment of Educational Progress. https://www.nationsreportcard.gov/reading _math_g12_2015/#.

Nelson, Richard R., and Edmund S. Phelps. 1966. "Investment in Humans, Technological Diffusion, and Economic Growth." *American Economic Review* 56 (1/2): 69–75.

Organisation for Economic Co-operation and Development (OECD). 2013. *OECD Skills Outlook 2013: First Results from the Survey of Adult Skills.* Paris: OECD Publishing. http://dx.doi.org/10.1787/9789264204256-en.

———. 2016. *PISA 2015: PISA Results in Focus.* Paris: OECD Publishing. http://www.oecd.org/pisa/pisa-2015-results-in-focus.pdf.

Ramey, Garey, and Valerie A. Ramey. 2010. "The Rug Rat Race." *Brookings Papers on Economic Activity* Spring:129–76.

Romer, Paul M. 1986. "Increasing Returns and Long-Run Growth." *Journal of Political Economy* 94 (5): 1002–37.

———. 1990. "Endogenous Technological Change." *Journal of Political Economy* 98 (5, pt. 2): S71–102.

Solow, Robert M. 1957. "Technical Change and the Aggregate Production Function." *Review of Economics and Statistics* 39:312–20.

US Census Bureau. 2015. *CPS Historical Time Series Tables.* Table A.1 "Years of School Completed by People 25 Years and over, by Age and Sex: Selected Years 1940 to 2015." https://www.census.gov/data/tables/time-series/demo/educational-attainment/cps-historical-time-series.html.

I

The Macroeconomic Link between Education and Real GDP Growth

1

Educational Attainment and the Revival of US Economic Growth

Dale W. Jorgenson, Mun S. Ho, and Jon D. Samuels

1.1 Introduction

Labor-quality growth captures the upgrading of the labor force through higher educational attainment and greater experience. While much attention has been devoted to the aging of the labor force, the implications of the coming plateau in educational attainment have been neglected.[1] Average levels of educational attainment remain high for people entering the labor force, but will no longer increase. Rising average educational attainment will gradually disappear as a source of US economic growth.

We define the employment rate as the number employed as a proportion of the corresponding population. We find that the employment rate for each age-gender category increases with educational attainment. The investment boom of 1995–2000 drew many younger and less educated workers into employment. After attaining a peak in 2000, the employment rates for these workers declined during the recovery of 2000–2007 and dropped further during the Great Recession of 2007–2009. The employment rates for the highly educated groups also fell during the Great Recession, but by 2015 they had recovered more than the employment rates of the less educated groups.

Dale W. Jorgenson is the Samuel W. Morris University Professor at Harvard University. Mun S. Ho is a visiting scholar at Resources for the Future. Jon D. Samuels is a research economist at the Bureau of Economic Analysis.

The views expressed in this chapter are solely those of the authors and not necessarily those of the US Bureau of Economic Analysis or the US Department of Commerce. We are grateful to Douglas Elmendorf and the editors, Charles Hulten and Valerie Ramey, for their comments on an earlier version of our manuscript. We are also grateful to the BEA for sharing their labor-input estimates. For acknowledgments, sources of research support, and disclosure of the authors' material financial relationships, if any, please see http://www.nber.org/chapters/c13695.ack.

1. See Aaronson, Hu, et al. (2014) and Aaronson, Cajner, et al. (2014).

In order to assess the prospects for US economic growth in more detail, we present a new data set on US output and productivity growth by industry for the postwar period, 1947–2014. This includes outputs for the sixty-five industries represented in the US National Income and Product Accounts (NIPAs). The new data set also includes inputs of capital (K), labor (L), energy (E), materials (M), and services (S), hence the acronym KLEMS. The rate of growth of productivity is the key indicator of innovation, where productivity is defined as the ratio of output to input for each industry.

A distinctive feature of our new US data set is detailed information on employment for the US labor force. This covers the period 1947–2014 and enables us to characterize the relationship between employment and the age, gender, and educational attainment of workers over more than six decades. Since the revival of US economic growth depends critically on the recovery of US employment rates, we utilize this new information on employment in assessing the prospects for a US growth revival.

Are the lower employment rates of the less educated workers a "new normal" for the US labor force that will persist for some time? Or, will the continuing economic recovery enable these workers to resume the higher employment rates that preceded the Great Recession? The answers to these questions are critical for the future growth of the US economy. In order to assess the prospects for recovery of employment as a potential source for the revival in US economic growth, we account for the employment rate of each age-gender-education group.

We build on the work of Jorgenson, Ho, and Stiroh (2005), who presented an industry-level data set for outputs, inputs, and productivity for the US economy for the period 1977–2000. For the earlier period 1947–1977, our new data set captures the postwar recovery of the US economy, ending with the energy crisis of 1973. For the recent period 2000–2014, our new data set highlights the slowdown in productivity growth after 2007, the fall in investment during the Great Recession of 2007–2009, and the slow recovery since 2009.

Paul Schreyer's (2001) Organisation for Economic Co-operation and Development (OECD) manual, *Measuring Productivity*, established international standards for economy-wide and industry-level productivity measurement. These standards are based on the production account for the US economy presented by Jorgenson, Gollop, and Fraumeni (1987) in their book, *Productivity and U.S. Economic Growth*. This was recommended by the Statistical Working Party of the OECD Industry Committee (2001). The Statistical Working Party was chaired by Edwin Dean, former Associate Commissioner for Productivity of the US Bureau of Labor Statistics (BLS).

We present a prototype production account within the framework of the US national accounts. This production account includes newly available estimates for the growth of outputs and intermediate inputs for the period 1998–2014 from the Bureau of Economic Analysis (BEA). We combine these estimates with data from the production account for the United States for the period 1947–2012 that we presented in Jorgenson, Ho, and Samuels (2016). We aggregate industries by means of the production possibility frontier employed by Jorgenson, Ho, and Stiroh (2005) and Jorgenson and Schreyer (2013). This links industry-level data on US growth and productivity to the economy-wide data from the US national accounts presented by Harper et al. (2009).

The first application of our new industry-level data set on outputs, inputs, and productivity is to analyze the sources of postwar US economic growth. We divide the Postwar Period, 1947–2014, into three subperiods—the Postwar Recovery, 1947–1973; the Long Slump following the 1973 energy crisis, 1973–1995; and the recent period of Growth and Recession, 1995–2014. We provide more detail on the period of Growth and Recession by considering the subperiods of the Investment Boom, 1995–2000; the Jobless Recovery, 2000–2007; and the Great Recession, 2007–2014.

We show that nearly 80 percent of US economic growth since 1947 is due to the growth of capital and labor inputs. This reflects the expansion and upgrading of the labor force and investments in plant equipment, and intangible assets like research and development and software. Only 20 percent of US growth is due to growth in productivity, output per unit of input, which captures innovation. Of course, economic growth involves both accumulation of capital and labor inputs and the introduction of new technologies, but factor accumulation greatly predominates as a source of US economic growth.

Our finding on the relative unimportance of innovation is the reverse of the well-known conclusions of Robert M. Solow (1957) and Simon Kuznets (1971). Solow and Kuznets found that innovation, represented by productivity growth, accounts for 80 percent of US economic growth, while accumulation of capital and labor inputs, the primary factors of production, accounts for only 20 percent. The sharp reversal of this conclusion is the most important empirical finding from several decades of research on productivity growth summarized by Jorgenson (2009) and Jorgenson, Fukao, and Timmer (2016).[2]

The reversal of the key empirical findings from the research of Solow (1957) and Kuznets (1971) can be traced to the critically important changes in methodology introduced by Jorgenson, Gollop, and Fraumeni (1987) and Jorgenson, Ho, and Stiroh (2005). These changes are summarized by Schreyer's (2001, 2009) OECD manuals. The new methodology for measuring productivity has had an enormous impact on the practice of pro-

---

2. An industry-level production account for the United States for the period 1947–2012 is presented in our paper, Jorgenson, Ho, and Samuels (2016). The official industry-level production account for the period 1998–2012 is presented by Rosenthal et al. (2016) and in Jorgenson, Fukao, and Timmer (2016, chapter 11, 377–428).

ductivity measurement. More than forty countries have employed the new methodology for productivity measurement and more than a dozen of these countries, including the United States, use this methodology to generate official estimates of productivity growth within the framework of the national accounts.

The predominant role of growth in capital and labor inputs in US economic growth is crucial for the formulation of economic policy. During the prolonged recovery from the Great Recession of 2007–2009, economic policy must focus on reviving investment and reestablishing the prerecession employment rates of the labor force. Policies for stimulating innovation would have a very limited impact.

The second application of our new data set is to project the future growth of the US economy. For this purpose we employ the methodology of Jorgenson, Ho, and Stiroh (2008).[3] We aggregate over industries to obtain data for the US economy as a whole. We project the future growth of labor input and productivity. We then determine the future growth of output consistent with the assumption that output and capital must grow at the same rate. This assumption eliminates the transitional dynamics associated with the accumulation of capital. We discuss the methodology for projecting future US economic growth in more detail in the appendix to this chapter.

We first consider the growth of labor input as a determinant of US economic growth. We project the size of the labor force from the growth and composition of the population. We then project the future growth of labor quality from the educational attainment of age cohorts of the population as they enter the labor force and the increase in experience as these cohorts age. Finally, we account for the employment rates for each age-gender-education category of the labor force, projecting them from 2014 levels.

We next consider productivity growth as a determinant of future US economic growth. To characterize the uncertainty that characterizes future trends, we construct a Base Case projection based on productivity growth for the period of Growth and Recession, 1995–2014. We then develop a Low Growth Case that also incorporates productivity trends for 1973–2014, including the Long Slump of 1973–1995, as well as the period 1995–2014. Finally, we present a High Growth Case based on productivity growth during the Investment Boom of 1995–2000 and the Jobless Recovery of 2000–2007. This excludes the Recession and Recovery of 2007–2014. We find that US economic growth will continue to recover from the Great Recession of 2007–2009 through the resumption of growth in productivity and labor input. However, the growth rate of the US economy in the next decade will depend critically on the revival of employment rates that prevailed before the Great Recession. We compare our results with the

---

3. Jorgenson and Vu (2017) employ this methodology to project the growth of the United States and the world economy.

projections by the Congressional Budget Office (2016) and John Fernald (2014, 2016). The final section of the chapter presents our conclusions.

1.2   A Prototype Industry-Level Production Account for the United States, 1947–2014

Our first objective is to construct a new data set for growth and productivity of the US economy at the industry level. This is greatly facilitated by the progress of the Bureau of Economic Analysis (BEA) in developing a system of industry accounts within the framework of the US National Income and Product Accounts. The BEA has successfully integrated three separate industry programs—benchmark input-output tables, released every five years; annual input-output tables; and annual estimates of gross domestic product by industry. The BEA's system of industry accounts is described by Mayerhauser and Strassner (2010).

McCulla, Holdren, and Smith (2013) summarize the 2013 benchmark revision of the NIPAs. A particularly significant innovation is the addition of intellectual property products such as research and development and entertainment, artistic, and literary originals. Intellectual property products are treated symmetrically with other capital expenditures. Investments in intellectual property are included in the gross domestic product (GDP), and the capital services generated by these investments are included in the national income.

Kim, Strassner, and Wasshausen (2014) describe the 2014 benchmark revision of BEA's industry accounts. These accounts include annual input-output tables and gross domestic product by industry and cover the period 1997–2012. The BEA's industry data are consistent with the 2013 benchmark revision of the NIPAs and the benchmark input-output table for 2007. The industry accounts and the annual input-output tables have been updated to 2013 and 2014 by BEA.

Lyndaker et al. (2016) have extended BEA's estimates of output and intermediate inputs to the period 1947–1996. This extension incorporates earlier benchmark input-output tables for the United States, including the first benchmark table for 1947. The BEA has linked these benchmark input-output tables to the annual input-output tables and industry accounts for 1997–2014. The BEA industry data are available for forty-six industries for 1947–1962, and sixty-five industries for 1963–2014. The BEA's historical data set includes estimates of output and intermediate input in current and constant prices. We incorporate these estimates into our prototype industry-level production account.[4]

---

4. For the period before 1998, BEA uses the industry, commodity, and import prices developed in Jorgenson, Ho, and Samuels (2016) to estimate constant-price industry output and intermediate input. For the 1963–2014 period, we use the BEA estimates in current and constant prices. For the 1947–1962 period, we scale the sixty-five-sector estimates developed by Jorgenson, Ho, and Samuels (2016) to the forty-six industries published by the BEA.

The BEA has prepared estimates of capital and labor inputs for the period 1998–2014. Our labor-input estimates are taken from Jorgenson, Ho, and Samuels (2016) for 1947–2012. We extrapolate these estimates to 2014, using the version of our labor data set maintained by BEA. This labor data set is used to generate an integrated industry-level production account beginning in 1998 by Steven Rosenthal and Lisa Usher of BLS and Matthew Russell, Samuels, and Strassner of BEA (2016).

Similarly, our estimates of capital input for 1947–2012 are taken from Jorgenson, Ho, and Samuels (2016) and updated to 2014, using capital input estimates in the BEA-BLS integrated industry-level production account. Combining the estimates of labor and capital inputs with estimates of output and intermediate inputs, we obtain an industry-level production account for the United States. This prototype production account covers the period of 1947–2014 in current and constant prices for all sixty-five industries included in the US national accounts. Jorgenson and Schreyer (2013) show how to integrate our prototype industry-level production account into the United Nations'(2009) *System of National Accounts 2008*.

Our new KLEMS-type data set for the United States is the culmination of our previous research on industry-level outputs, inputs, and productivity for the postwar period. This data set is consistent with BEA's industry accounts and annual input-output tables for 1947–2014 and provides greater industry detail for 1947–1962. The BEA/BLS integrated industry-level production account for 1998–2014, released on January 13, 2017, uses similar methodology. However, our industry-level production account covers the entire postwar period, beginning in 1947.

1.2.1   Changing Structure of Capital Input

Swiftly falling information technology (IT) prices have provided powerful economic incentives for the rapid diffusion of IT through investment in hardware and software. A substantial acceleration in the IT price decline occurred in 1995, triggered by a much sharper acceleration in the price decline for semiconductors. The IT price decline after 1995 signaled even faster innovation in the main IT-producing industries—semiconductors, computers, communications equipment, and software—and ignited a boom in IT investment. Figure 1.1 presents price indices for 1973–2014 for asset categories included in our measures of capital input—equipment, computers, software, research and development, artistic originals, and residential structures.

The price of an asset is transformed into the price of the corresponding capital input by multiplying the asset price by the *cost of capital* introduced by Jorgenson (1963). The cost of capital includes the nominal rate of return, the asset-specific rate of depreciation, and the rate of capital loss due to declining prices. The distinctive characteristics of IT prices—high rates of price decline and rates of depreciation—imply that cost of capital for the

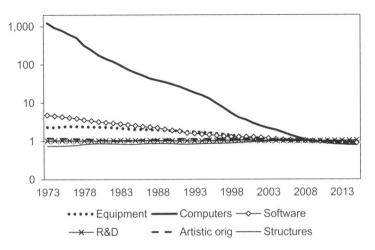

Fig. 1.1   Price of investment relative to GDP deflator (log scale)

price of IT capital input is very high, relative to the cost of capital for the price of non-IT capital input.

Schreyer's (2009) OECD manual provides detailed recommendations for the construction of prices and quantities of capital services. Incorporation of data on labor and capital inputs in constant prices into the national accounts is described in the *2008 System of National Accounts* (United Nations 2009, chapters 19 and 20). In chapter 20 of *2008 SNA*, estimates of capital services are described as follows: "By associating these estimates with the standard breakdown of value added, the contribution of labor and capital to production can be portrayed in a form ready for use in the analysis of productivity in a way entirely consistent with the accounts of the System" (United Nations 2009, 415).

To capture the impact of the rapid decline in IT equipment prices and the high depreciation rates for IT equipment, we distinguish between the flow of capital services and the stock of capital. Capital quality is defined as the ratio of the flow of capital services to capital stock. Figure 1.2 gives the share of IT in the value of total capital stock, the share of IT capital services in total capital input, and the share of IT services in total output. The IT stock share rose from 1960 to 1995—on the eve of the IT boom—and reached a high in 2001 after the dot-com bubble. This share fell during the Jobless Recovery with the plunge in IT investment. The share of the IT service flow in the value of total capital input is much higher than the IT share in total capital stock. This reflects the rapid decline in IT prices and the high depreciation rates of IT equipment that enter the formula for the cost of capital associated with the IT service flow. The share of the IT service flow was fairly stable during the period 1960–1980 and then began to rise, reaching a peak

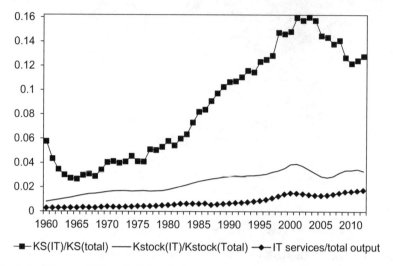

−■−KS(IT)/KS(total) ——Kstock(IT)/Kstock(Total) −◆−IT services/total output

Fig. 1.2   Shares of IT stock, IT capital services, and IT service output in total economy
Note: IT services = (information and data processing, computer system design).

in 2000. The IT service flow then declined and ended with a sharp plunge during the Great Recession.

The IT service industries, information and data processing, and computer system design have shown persistent growth. The share of the output of these two industries in the value of the GDP, shown in figure 1.2, declined slightly from 2000 to 2005 and then continued to rise, reaching a high in 2014. This reflects the displacement of IT hardware and software by the growth of IT services like cloud computing. Investment in intellectual property (IP) products since 1973 is shown as a proportion of the GDP in figure 1.3. This proportion grew during the Investment Boom of 1995–2000 and has declined only slightly since the peak around 2000. Investment in research and development also peaked around 2000, but has remained close to this level through the Great Recession.

1.2.2   Changing Structure of Labor Input

Our measure of labor input recognizes differences in labor compensation for workers of different ages, educational attainment, and gender, as described by Jorgenson, Ho, and Stiroh (2005, chapter 6). The rate of labor-quality growth is the difference between the growth rate of labor input and the growth rate of hours worked. For example, a shift in the composition of labor input toward more highly educated workers, who receive higher wages, contributes to the growth of labor quality. Figure 1.4 shows the decomposition of changes in labor quality into age, education, and gender components.

During the Postwar Recovery of 1947–1973, the massive entry of young,

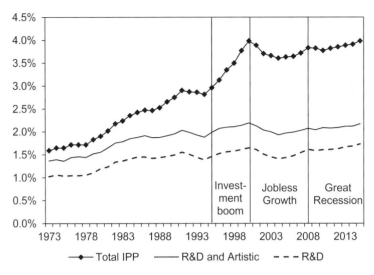

Fig. 1.3   Share of intellectual property investment in GDP (percentage)

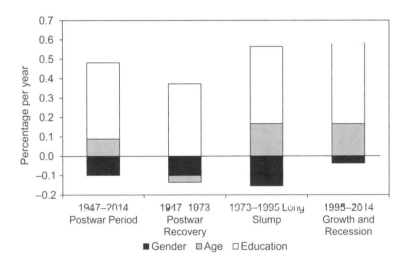

Fig. 1.4   Contribution of education, age, and gender to labor-quality growth

lower-wage workers contributed negatively to labor-quality growth. The rapidly increasing female labor force also contributed negatively, reflecting the lower average labor compensation of female workers. Rising educational attainment generated substantial growth in labor quality. During the Long Slump of 1973–1995, the increase in employment of female workers accelerated and the contribution of the gender composition became more negative. The aging of the labor force contributed positively to labor quality through

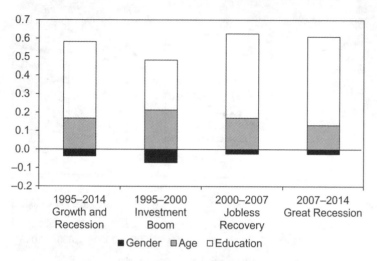

Fig. 1.5 **Contribution of education, age, and gender to labor-quality growth, 1995–2014**

increased experience, while educational attainment continued to rise and the growth of labor quality became more rapid. The negative impact of increased female employment diminished and labor quality continued to grow as workers gained experience. Considering the period of Growth and Recession in more detail in figure 1.5, we see that labor quality rose steadily throughout the period. The growth rate declined slightly in 1995–2000, relative to the Long Slump of 1973–1995, as a consequence of a jump in employment by younger and less educated workers. The less negative gender contributions during the Jobless Recovery of 2000–2007 and the Great Recession of 2007–2014 reflect the fact that unemployment rates rose much more sharply for men than for women.

The level of educational attainment of US workers is shown in figure 1.6. In 1947 only a modest proportion of the US workforce had four or more years of college. By 1973 the proportion of college-educated workers had risen dramatically, and this proportion has continued to grow. There was a change in classification in 1992 from years enrolled in school to years of schooling completed. By 2014 almost a third of US workers had completed a BA degree or higher. The fall in the share of workers with lower educational attainment accelerated during the Great Recession.

Figure 1.7 shows that educational attainment of the twenty-five to thirty-four age group improved substantially during the Postwar Recovery from 1947 to 1973, followed by a pause during the Long Slump of 1973–1995. Gains in educational attainment resumed during the Investment Boom of 1995–2000, and have continued to the present. During the Great Recession, less educated workers had much higher unemployment rates and the average educational attainment rose for workers.

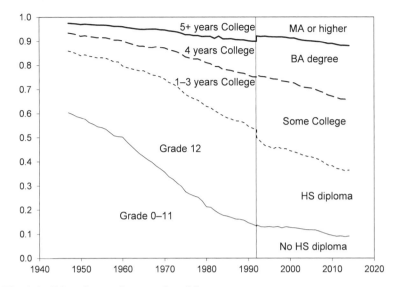

Fig. 1.6   Education attainment of workforce

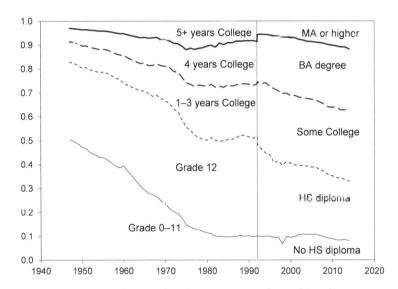

Fig. 1.7   Education attainment of workers age twenty-five to thirty-four

Figure 1.8 gives employment rates of males and females for three age groups—twenty-five to thirty-four, thirty-five to forty-four, and forty-five to fifty-four years old. Better-educated workers are much more likely to be employed for both genders and all three age groups. Male workers with BA degrees have very high employment rates for all years except the recessions. Employment rates for males with high school diplomas are substantially lower. The Investment Boom of 1995–2000 drew in many less educated and

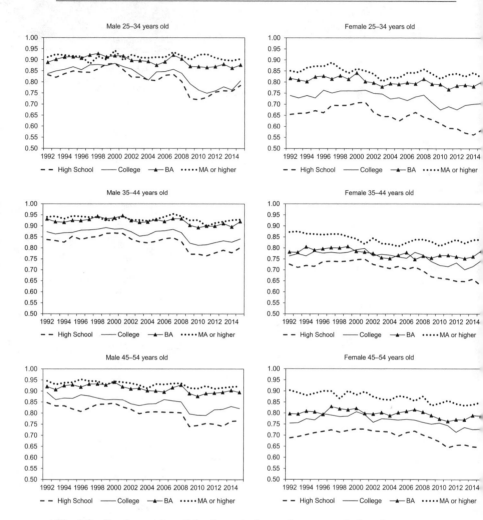

Fig. 1.8   Employment participation rates by gender, age, and education

younger workers, raising their employment rates. The employment rates have fallen since 2000 for the less educated. These rates declined further during the Great Recession.

Although the decline in employment is widely discussed, employment rates by gender, age, and educational attainment, like those presented in figure 1.8, have not been considered until now. A model of employment and unemployment is presented by Kroft et al. (2016). This model has been elaborated by Krueger, Cramer, and Cho (2014).

The modeling of employment and unemployment could be extended to a more detailed breakdown of alternatives to employment for members of the working-age population. These would include disability status and increased participation in welfare programs. Both of these increased as a proportion of

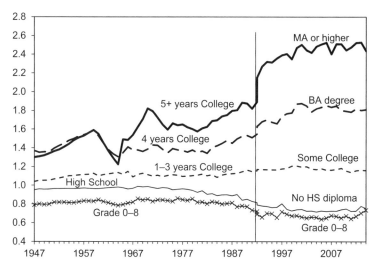

Fig. 1.9   Compensation by education attainment (relative to those with high school diploma)

the working-age population during the Great Recession with relaxation of requirements for eligibility. Employment may have been adversely affected by extended benefit periods for the unemployed, now expired, and lower income requirements for food stamps.[5]

The increase in the "college premium," the difference between wages earned by workers with college degrees and wages of those without degrees, has been widely noted. In figure 1.9 we plot the compensation of workers by educational attainment, relative to those with a high school diploma (four years of high school). We see that the four-year college premium was stable in the 1960s and 1970s, but rose during the 1980s and 1990s. The college premium stalled throughout the first decade of the twenty-first century. The master's-and-higher degree premium rose even faster than the BA premium between 1980 and 2000 and continued to rise through the middle of the first decade of the twenty-first century.

A possible explanation for the rise in relative wages for college-educated workers with a rising share of these workers in the labor force is that their labor services are complementary to the use of information technology.[6] The most rapid growth of the college premium occurred during the 1995–2000 boom when IT capital made its highest contribution to GDP growth. Our industry-level view of postwar US economic history allows us to consider the role of changing industry composition in determining relative wages. Table 1.1 gives characteristics of the workforce for each industry for 2010.

5. The long-term decline in labor force participation for prime-age males is analyzed along these lines by the Council of Economic Advisers (2016).
6. See Goldin and Katz (2008) for more details and historical background.

Table 1.1   Labor characteristics by industry, year 2010

	Percentage workers college educated	Compensation ($/hour)	Percentage total hours (age 16–35)	Percentage total hours (females)	Percentage females college educated	Percentage males college educated
1 Farms	15.1	19.5	20.3	14.7	18.6	14.3
2 Forestry, fishing, and related activities	16.4	16.6	30.8	15.3	30.6	13.3
3 Oil and gas extraction	38.6	79.5	14.6	22.2	44.4	36.7
4 Mining, except oil and gas	11.8	39.2	20.2	8.8	28.0	10.1
5 Support activities for mining	26.0	37.6	25.8	13.8	39.4	23.4
6 Utilities	24.0	64.0	22.0	23.4	28.6	22.6
7 Construction	14.0	31.6	33.9	8.9	24.8	12.8
8 Wood products	12.2	26.0	32.9	15.1	17.3	11.1
9 Nonmetallic mineral products	18.1	32.3	26.0	19.7	21.6	17.2
10 Primary metals	17.7	39.7	26.0	13.5	24.8	16.6
11 Fabricated metal products	15.2	32.2	27.7	17.7	15.9	15.0
12 Machinery	24.5	38.7	25.6	19.4	24.2	24.6
13 Computer and electronic products	62.3	56.7	31.0	30.3	54.2	66.0
14 Electrical equipment appliances	44.2	52.5	26.4	30.9	33.6	49.2
15 Motor vehicle bodies and parts	23.6	37.9	28.4	21.8	20.8	24.4
16 Other transportation equipment	31.4	50.6	22.6	17.3	30.9	31.7
17 Furniture and related products	15.6	26.3	31.5	24.3	17.4	15.0
18 Miscellaneous manufacturing	32.1	40.7	26.8	35.6	26.3	35.7
19 Food, beverage, and tobacco	23.8	27.2	24.3	31.5	23.2	24.2
20 Textile mills and textile-product mills	14.0	25.6	26.5	45.2	11.9	15.8
21 Apparel and leather products	17.6	27.0	27.4	55.9	15.4	20.9
22 Paper products	18.8	37.3	23.9	20.7	18.5	18.9
23 Printing and related support activities	22.0	29.5	28.7	32.2	23.1	21.4
24 Petroleum and coal products	32.9	81.5	17.7	17.4	45.2	30.0
25 Chemical products	49.5	54.1	27.4	35.2	49.1	50.3
26 Plastics and rubber products	16.4	30.7	30.2	28.5	11.4	18.5
27 Wholesale trade	32.0	41.2	29.1	26.0	32.6	31.7
28 Retail trade	15.8	23.0	35.4	42.0	14.4	17.3
29 Air transportation	38.2	49.5	28.6	35.9	36.7	39.1
30 Rail transportation	13.2	50.7	14.0	8.3	28.7	11.7

32	Truck transportation	8.6	28.0	24.6	11.1	14.4	7.8
33	Transit, ground passenger transportation	16.3	22.8	18.4	23.5	11.5	18.1
34	Pipeline transportation	32.8	65.6	17.5	18.4	45.6	29.6
35	Other transportation and support	19.7	33.5	34.1	20.7	22.3	19.0
36	Warehousing and storage	12.6	29.2	35.6	26.3	13.2	12.4
37	Publishing industries (includes software)	60.2	52.5	38.1	42.8	59.7	60.5
38	Motion picture and sound recording	45.9	46.4	47.9	31.6	48.8	44.3
39	Broadcasting and telecommunications	39.5	46.7	37.9	39.0	42.4	37.7
40	Information and data-processing services	55.4	55.0	47.7	40.8	50.8	59.1
41	Federal Reserve banks, credit intermediation	42.4	42.1	36.5	60.1	30.3	62.8
42	Securities, commodity contracts	71.9	120.6	38.3	35.2	58.0	80.7
43	Insurance carriers	46.6	43.7	28.5	56.4	33.9	65.0
44	Funds, trusts, and other financial vehicles	71.0	99.4	40.7	37.3	57.1	80.4
45	Real estate	40.6	31.1	18.6	46.6	36.1	45.1
46	Rental, leasing, and lessors of intangibles	25.4	31.1	45.0	28.8	24.1	26.0
47	Legal services	65.5	57.5	29.0	53.1	46.3	90.6
48	Computer systems design	68.6	56.7	41.1	28.5	67.0	69.3
49	Miscellaneous professional and technical services	65.3	46.9	31.1	42.3	58.9	70.6
50	Management of companies	53.4	62.2	28.9	51.4	39.8	69.4
51	Administrative and support services	20.1	24.8	37.7	40.4	23.2	17.9
52	Waste management	10.2	32.5	33.9	14.3	16.1	9.2
53	Educational services	64.2	28.8	27.5	65.9	64.2	64.2
54	Ambulatory health care services	38.8	39.2	27.5	74.2	30.8	66.6
55	Hospitals, nursing, and residential care	30.4	28.4	28.1	79.5	29.4	34.4
56	Social assistance	30.0	18.8	36.1	86.7	28.9	37.4
57	Performing arts, spectator sports	48.7	53.8	29.1	43.8	55.1	43.1
58	Amusements, gambling, and recreation	21.7	20.1	39.4	41.0	22.2	21.4
59	Accommodation	18.6	22.1	35.8	52.7	16.3	21.3
60	Food services and drinking places	11.1	14.8	53.5	47.9	9.9	12.2
61	Other services except government	17.9	25.7	26.7	64.8	18.8	19.3
62	Federal general government	52.0	63.3	19.5	54.6	49.6	54.9
63	Federal government enterprises	19.6	42.0	14.5	34.6	20.0	19.3
64	S&L government enterprises	29.9	40.9	25.4	40.2	28.9	30.7
65	S&L general government	48.6	36.3	23.5	61.2	48.6	50.6

Note: "College-educated" workers are those with BA or BA and higher degree.

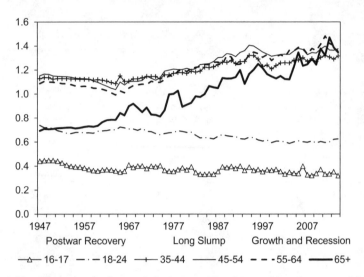

Fig. 1.10   Compensation by age (relative to twenty-five- to thirty-four-year-olds)

The industries with the higher share of college-educated workers include the IT-producing industries—computer and electronic products, publishing (including software), information and data processing, and computer systems design. The industries with higher shares of college-educated workers also include those that use IT products and services intensively—securities and commodity contracts, legal services, professional and technical services, and educational services.

After educational attainment, the most important determinant of labor quality is experience, captured by the age of the worker. We have noted that the entry of the baby boomers into the labor force contributed negatively to labor-quality growth during 1947–1973, and that the aging of these workers contributed positively after 1973. We show the wages of different age groups, relative to the wages of workers age twenty-five to thirty-four, in figure 1.10. The wages of the prime age group, forty-five to fifty-four, rose steadily relative to the young from 1970 to 1994. During the Investment Boom of 1995–2000, the wages of the younger workers surged and the prime-age premium fell.

The wage premium of the thirty-five to forty-four and fifty-five to sixty-four age groups shows the same pattern as the premium of prime-age workers, first rising relative to the twenty-five- to thirty-four-year-olds, then falling or flattening out during the Investment Boom. The wage premium of the oldest workers is the most volatile, but showed a general upward trend throughout the Postwar Period, 1947–2014. The share of workers age sixty-five and older has been rising steadily since the mid-1990s, during a period of large swings in the wage premium. The relative wages of the very young, eighteen to twenty-four, has been falling steadily since 1970, reflecting the rising demand for education and experience.

Our new industry-level data set provides detailed information for the period 1947–2014 on the growth of outputs, capital, labor, energy, materials and services inputs, and productivity for the sixty-five industries that make up the US economy. We present new information on educational attainment and the relationship between employment and educational attainment. We also provide detailed information on labor compensation by age and educational attainment. We next consider the application of our new data set to an analysis of the sources of US economic growth. This will be followed by the application of this data set to the projection of the future growth of the US economy.

1.3   Sources of US Economic Growth

In analyzing the sources of US economic growth, we first consider the contributions of three major industry groups to the growth of aggregate output. These are the IT-producing industries, the IT-using industries, and non-IT industries, defined more precisely below. We then consider the contributions of these industry groups to aggregate productivity growth, defined as the difference between the growth rates of output and input. Although the IT-producing industries account for a relatively small proportion of the value of US output, they generate a much larger share of productivity growth.

Finally, we consider the growth of capital and labor inputs, as well as productivity growth, as sources of US economic growth. We divide the growth of capital input among IT equipment and software, intellectual property, and all other capital inputs. In order to emphasize the role of the dramatic increases in educational attainment, we divide the growth of labor input between college and noncollege labor inputs. We find that the growth of capital and labor inputs greatly predominates over productivity growth as a source of US economic growth for the Postwar Period, 1947–2014, as well as for the subperiods we consider.

In *Information Technology and the American Growth Resurgence*, Jorgenson, Ho, and Stiroh (2005) analyze the economic impact of IT at the aggregate level for 1948–2002 and the industry level for 1977–2000. They also provide a concise history of the main technological innovations in information technology during the Postwar Period, beginning with the invention of the transistor in 1947. Jorgenson, Ho, and Samuels (2012) convert the industrial classification to the North American Industry Classification System (NAICS). They update and extend the data to cover seventy industries for the period 1960–2007.

The NAICS industry classification includes the industries identified by Jorgenson, Ho, and Samuels (2012) as IT-producing industries, namely,

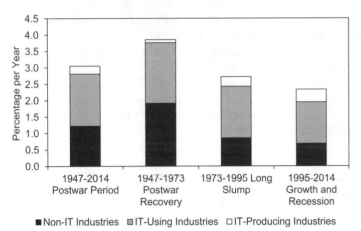

■ Non-IT Industries   ◪ IT-Using Industries   □ IT-Producing Industries

Fig. 1.11   Contributions of industry groups to value-added growth, 1947–2014

computers, electronic products, and software, and the two IT service industries, information and data processing and computer systems design. Jorgenson, Ho, and Samuels (2012) define an IT intensity index as the ratio of the sum of IT capital input and IT services to the sum of all capital input and IT services. They classify industries as IT-using if the IT intensity index is greater than the median for all US industries that do not produce IT equipment, software, and services. We classify all other industries as non-IT.

Value added in the IT-producing industries during 1947–2014 is only 2.5 percent of the US economy, while value added in the IT-using industries is 47.5 percent with value added in the non-IT industries accounting for the remaining 50 percent. The IT-using industries are mainly in trade and services. Most manufacturing industries are in the non-IT sector. The NAICS industry classification provides much more detail on services and trade, especially the industries that are intensive users of IT. We begin by discussing the results for the IT-producing sectors, now defined to include the two IT service sectors.

Figure 1.11 reveals a steady increase in the share of IT-producing industries in the growth of value added since 1947. This corresponds to a decline in the contribution of the non-IT industries, while the share of IT-using industries remains relatively constant. Figure 1.12 decomposes the growth of value added for the period 1995–2014. The contributions of the IT-producing and IT-using industries peaked during the Investment Boom of 1995–2000 and have declined since then. However, the contribution of the non-IT industries also revived during the Investment Boom and declined substantially during the Jobless Recovery and the Great Recession. Figure 1.13 gives the contributions to value added for the sixty-five individual industries over the

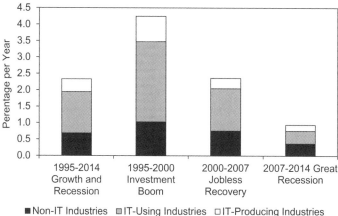

Fig. 1.12   Contributions of industry groups to value-added growth, 1995–2014

period 1947–2014. The leading contributors are real estate, wholesale and retail trade, and computer and electronic products.

In order to assess the relative importance of productivity growth at the industry level as a source of US economic growth, we express the growth rate of aggregate productivity as a weighted average of industry productivity growth rates, using the ingenious weighting scheme of Evsey Domar (1961).[7] The Domar weight is the ratio of the industry's gross output to aggregate value added. The Domar weights for all industries sum to more than one. This reflects the fact that an increase in the rate of growth of the industry's productivity has a direct effect on the industry's output and an indirect effect via the output delivered to other industries as intermediate inputs.

The rate of growth of aggregate productivity also depends on the reallocations of capital and labor inputs among industries. The rate of aggregate productivity growth exceeds the weighted sum of industry productivity growth rates when these reallocations are positive. This occurs when capital and labor inputs are paid different prices in different industries and industries with higher prices have more rapid input growth rates. Aggregate capital and labor inputs then grow more rapidly than weighted averages of industry-capital and labor-input growth rates, therefore the reallocations are positive. When industries with lower prices for inputs grow more rapidly, the reallocations are negative.

Figure 1.14 shows that the contributions of IT-producing, IT-using, and non-IT industries to aggregate productivity growth are similar in magnitude for the period 1947–2014. The non-IT industries contributed substantially

7. The formula is given in Jorgenson, Ho, and Stiroh (2005, equation 8.34).

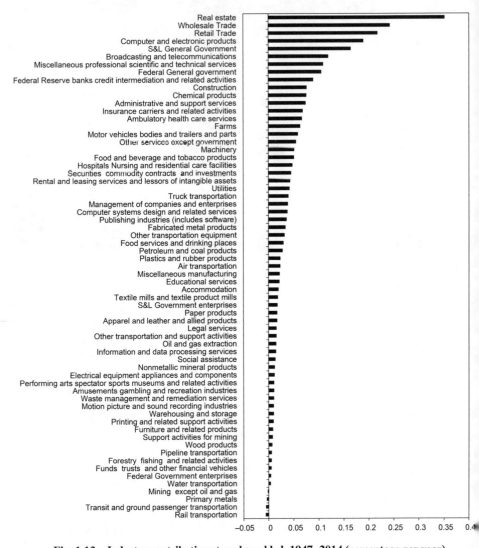

Fig. 1.13   Industry contributions to value added, 1947–2014 (percentage per year)

to productivity growth during the Postwar Recovery, 1947–1973, but this contribution became negative during the Long Slump, 1973–1995. The contribution of IT-producing industries was very small during the Postwar Recovery, but became the predominant source of US productivity growth during the Long Slump, 1973–1995. The contribution of IT-producing industries increased considerably during the period of Growth and Recession, 1995–2014.

The IT-using industries contributed substantially to US productivity

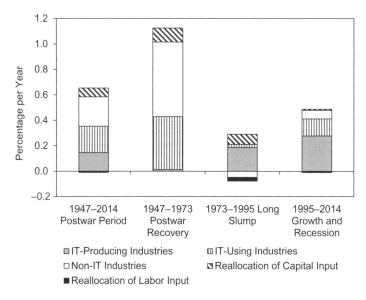

Fig. 1.14   **Contribution of industry groups to aggregate productivity growth, 1947–2014**

growth during the Postwar Recovery, but this contribution nearly disappeared during the Long Slump, 1973–1995, before reviving after 1995. The reallocation of capital input made a small but positive contribution to productivity growth during the Postwar Period, 1947–2014, and each of the subperiods. The contribution of reallocation of labor input was negligible for the period as a whole. During the Long Slump and the period of Growth and Recession, the contribution of the reallocation of labor input was slightly negative.

Considering the period of Growth and Recession in more detail in figure 1.15, all three industry groups contributed to aggregate productivity growth during the period as a whole. However, the IT-producing industries predominated as a source of productivity growth during the period as a whole and the three subperiods. The contribution of these industries remained substantial during each of the subperiods (1995–2000, 2000–2007, and 2007–2014) despite the sharp contraction of economic activity during the Great Recession of 2007–2009.

The contribution of the IT-using industries was considerable during the Investment Boom of 1995–2000, remained substantial in the Jobless Recovery of 2000–2007, but became slightly negative during the Great Recession of 2007–2014. The non-IT industries contributed positively to productivity growth during the Investment Boom. This contribution rose during the Jobless Recovery and then became negative during the Great Recession.

Figure 1.16 gives the contributions of each of the sixty-five industries to

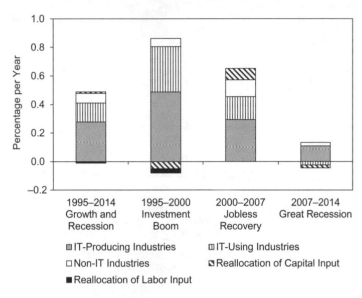

Fig. 1.15   Contribution of industry groups to aggregate productivity growth, 1995–2014

productivity growth for the Postwar Period. Computer and electronic products, wholesale and retail trade, farms, and broadcasting and telecommunications were among the leading contributors to US productivity growth
during the Postwar Period. Many industries made negative contributions
to aggregate productivity growth. These included nonmarket services such
as health care, as well as resource industries affected by depletion, such as
oil and gas extraction and mining. Other negative contributions reflect the
growth of barriers to resource mobility in product and factor markets due,
in some cases, to more stringent government regulations.

Finally, we consider the growth of capital and labor inputs, as well as
growth in productivity, as sources of growth of the US economy. The contributions of college-educated and non-college-educated workers to US
economic growth are given by the relative shares of these workers in the
value of output, multiplied by the growth rates of their labor inputs. Workers with a college degree or higher level of education correspond closely
with "knowledge workers" who deal with information. Of course, not every
knowledge worker is college educated and not every college graduate is a
knowledge worker.

Figure 1.17 shows that contribution of college-educated workers predom-
inated in the growth of labor input during the Postwar Period, 1947–2014.
The contribution of non-college-educated workers was greater during the
Postwar Recovery, 1947–1973, but declined substantially during the Long
Slump of 1973–1995, and almost disappeared during the period 1995–2014

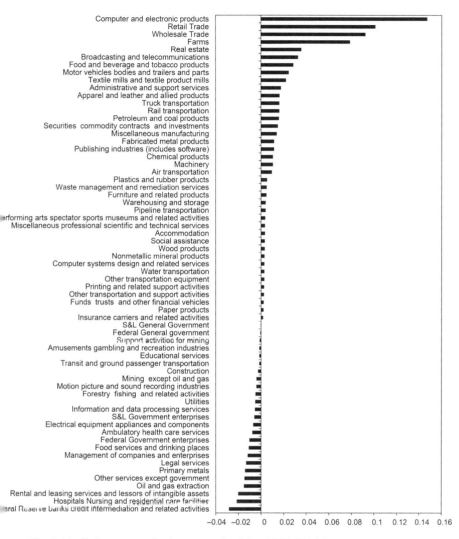

Fig. 1.16   Industry contributions to productivity, 1947–2014 (percentage per year)

of Growth and Recession. The contribution of college-educated workers was the dominant source of growth of labor input during the Long Slump and the period of Growth and Recession.

Capital input was the predominant source of US economic growth for the Postwar Period, 1947–2014, as we show in figure 1.17. Capital input was also predominant during the Postwar Recovery, the Long Slump, and the period of Growth and Recession. Considering the period of Growth and Recession in greater detail, figure 1.18 reveals that the contribution of capital input was about half of US economic growth during the Investment Boom

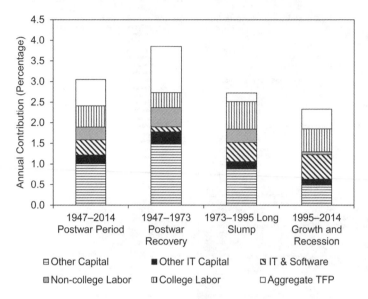

Fig. 1.17   Sources of US economic growth, 1947–2014

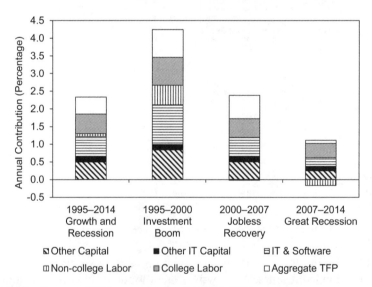

Fig. 1.18   Sources of US economic growth, 1995–2014

and increased in relative importance as the growth rate fell in the Jobless Recovery and again in the Great Recession.

Figure 1.17 also provides greater detail on important changes in the composition of the contribution of capital input. For the Postwar Period as a whole, the contribution of research and development to US economic growth was considerably less than the contribution of IT. However, the contributions of other forms of capital input predominated over both. While the contribution of research and development exceeded that of IT during the Postwar Recovery, the contribution of IT grew rapidly during the Long Slump and jumped to nearly half the contribution of capital input during the period of Growth and Recession. By contrast, the contribution of research and development shrank during both periods and became relatively insignificant. Figure 1.18 reveals that the contribution of capital input peaked during the Investment Boom, declined during the Jobless Recovery, and collapsed during the Great Recession, but the relative importance of IT remained the same throughout the period of Growth and Recession.

Figure 1.18 shows that all of the sources of economic growth contributed to the US growth resurgence after 1995, relative to the Long Slump represented in figure 1.17. Both IT and non-IT capital inputs contributed substantially to growth during the Jobless Recovery of 2000–2007, but the contribution of labor input dropped precipitously and the contribution of noncollege workers became slightly negative. The most remarkable feature of the Jobless Recovery was the sustained growth in productivity, indicating an ongoing surge of innovation.

Despite the slowdown of investment during the Great Recession, both IT and non-IT capital inputs continued to contribute substantially to US economic growth during the period 2007–2014. Productivity growth almost disappeared, reflecting a widening gap between actual and potential growth of output. The contribution of college-educated workers remained positive and substantial, while the contribution of noncollege workers became strongly negative. These trends represent increased rates of substitution of capital for labor and college-educated workers for noncollege workers.

We have now identified the sources of the growth of the US economy. The predominant source of US economic growth is the growth of capital and labor inputs. This characterizes the Postwar Period, 1947–2014, and the subperiods we have considered. Second, the growth of capital input is considerably more important than the growth of labor input as a source of US economic growth. Finally, investment in information technology equipment and software is the most important component of the growth of capital input as a source of growth of the US economy.

Productivity growth, while a much less important source of US economic growth than the growth of capital and labor inputs, is essential for sustaining economic growth in the long run. We have seen that productivity growth in the IT-producing industries has been the most important source of US productivity growth during the Postwar Period, 1947–2014. The contribution of the IT-producing industries can be traced to developments in technology that were successfully commercialized after the Postwar Recovery, 1947–1973.

1.4   Future US Economic Growth

Our final objective is to assess the prospects for revival of US economic growth. We present three alternative projections for US economic growth for the period 2014–2024: Base Case, Low Growth, and High Growth. This enables us to quantify the uncertainty in projections of the growth of capital quality and productivity growth. We present the three alternative projections in figures 1.19, 1.20, and 1.21. We compare these projections with historical data for the period 1990–2014.

Figure 1.19 includes three alternative projections of productivity growth for the period 2014–2024. For the Base Case, we set future productivity growth rates for IT-producing, IT-using, and non-IT industries equal to growth rates for the period of Growth and Recession, 1995–2014. The Low Growth projection is based on productivity growth rates for the period 1973–2014, including the Long Slump of 1973–1995. The High Growth projection incorporates productivity growth rates for the recent period, 2000–2014, including the Jobless Recovery of 2000–2007 and the Recession and Recovery of 2007–2014.

We use the following assumptions for all three projections: We set the capital share in value added and the share of reproducible capital in total capital stock equal to the averages for the Postwar Period, 1947–2014. We fix the shares of nominal GDP for IT-producing, IT-using, and non-IT sectors at the averages for the recent period, 2000–2014, to reflect changes in the relative importance of information technology. More details about the projections are provided in the appendix.

We define average labor productivity as output per hour worked. The growth rate of labor productivity is the sum of growth rates of labor quality,

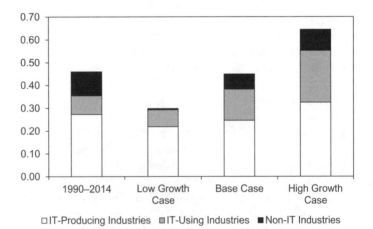

Fig. 1.19   Contribution of industry groups to aggregate productivity, 2014–2024

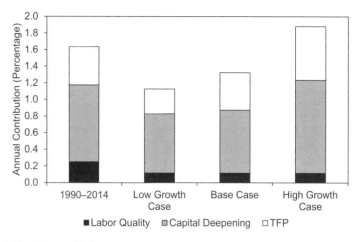

Fig. 1.20   Range of labor productivity projections, 2014–2024

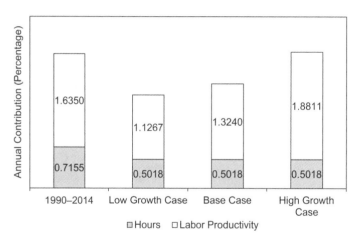

Fig. 1.21   Range of US potential output projections, 2014–2024

capital deepening, and total factor productivity, where capital deepening is defined as capital input per hour worked. We project growth rates of labor productivity and hours worked for the period 2014–2024, which sum to the growth rate of output for the US economy. Figure 1.20 gives the growth rates of labor productivity for the Base Case, Low Growth, and High Growth projections, while figure 1.21 presents the projected growth rates of output.

1.4.1   Base Case

Our projections of US economic growth incorporate trends in employment rates by gender, age, and education. For each gender-age-education category we assume that the employment rate remains equal to the rate in

2014, when the unemployment rate stabilized. We fix weekly hours for each gender-age-education group at the 2014 level, when the US economy reached full employment. Our projections of the growth rates of labor quality for 2014–2024 are considerably below the averages for the period 1990–2014, due to declines in the rates of growth of average educational attainment.

In the Base Case, we assume that the growth rates of capital quality and productivity for the next ten years will equal average growth rates for the period of Growth and Recession, 1995–2014. The Investment Boom of 1995–2000 combined rapid accumulation of IT capital and robust productivity growth. The Jobless Recovery of 2000–2007 had strong productivity growth, but slower growth of IT capital. The Recession and Recovery of 2007–2014 had weak productivity growth and much slower accumulation of IT capital.

The growth rate of capital quality during the period 1995–2014 that is used in the projection is slightly below the growth rate for the period 1990–2014. Capital deepening makes the biggest contribution to labor productivity growth, while the growth of productivity in the IT-producing sector will make the second-largest contribution during the period 2014–2024. We project that productivity growth in the IT-using sector during the period 2014–2024 will exceed its contribution during 1990–2014, reflecting more rapid productivity growth and the higher value share of this sector. Finally, total factor productivity (TFP) of the non-IT sector of the economy will contribute relatively little to labor productivity growth, even compared to the period 1990–2014.

Our Base Case projection of labor productivity growth over the next ten years, 2014–2024, is markedly lower than growth during the period 1990–2014. Our projection of labor-quality growth in the Base Case is also well below growth in 1990–2014. Total hours worked is projected to grow at 0.50 percent per year compared to 0.71 percent during 1990–2014, reflecting the future changes in the age structure and the assumption of fixed annual hours at 2014 levels for each age-gender-education group.

Combining our projected growth rates in hours worked and average labor productivity, we project the GDP growth rate at 1.83 percent per year over the next ten years. This is a substantial decline from the growth rate of 2.35 percent per year during the period 1990–2014. The slower growth in hours worked is reinforced by the slower growth of average labor productivity. We conclude by emphasizing that we do not model the determinants of employment, but rely on extrapolations of trends from the historical data.

1.4.2   Low Growth Case

Our first alternative assumption to the Base Case is that capital-quality and productivity growth over the next ten years will equal the averages over 1973–2014, a period that includes the Long Slump and the Recession and Recovery. The period of Recession and Recovery can be subdivided among the IT Boom, the Jobless Recovery, and the Recession and Recovery. By including the Long Slump and the Recession and Recovery periods, we dampen the growth rates in this low scenario. Taking averages over 1973–2014 yields a capital-quality growth rate that is nearly equal to the growth rate for the period 1990–2014.

We project that productivity growth in the IT-producing sector will be only slightly below the rate for 1990–2014. Using the 2000–2014 average share of the IT-producing sector in output, we obtain a substantial contribution of productivity growth from the IT-producing sector to growth of labor productivity. We project that the growth of productivity in the IT-using sector will be almost equal to the contribution for the period 1990–2014. Finally, we project that productivity growth from the non-IT sector will contribute very little to average labor productivity growth, even less than during the period 1990–2014.

In the Low Growth Case, our projected labor productivity growth for the next ten years is below the Base Case projection. Both the Base Case and the Low Growth projections are markedly below the growth of labor productivity during the period 1990–2014. The growth of hours worked in both scenarios is below the growth of hours for the period 1990–2014. Summing the growth rates in hours worked and labor productivity, the Low Growth Case projects output growth at 1.63 percent over the next ten years. This is a marked deceleration from the growth rate of 2.35 percent for the period 1990–2014.

1.4.3   High Growth Case

For the High Growth Case we assume that employment rates for each gender-age-education group are the same as in the Base Case for the ten-year period 2014–2024. Hours worked is also projected to grow at 0.50 percent over the next decade as in the Base Case, and the growth rate of labor quality will be substantially lower than during the period 1990–2014. We assume that growth rates of capital quality and productivity for the next ten years will equal their averages over the period 1995–2007. This includes the Investment Boom and the Jobless Growth periods, but excludes the Long Slump and the Great Recession as temporary slowdowns in economic growth. Taking averages over 1995–2007 yields a capital-quality growth rate significantly higher than the growth rate over the period 1990–2014.

In the High Growth Case, productivity growth in the IT-producing sector is more rapid than in the Base Case. This translates into a relatively high contribution of growth in total factor productivity to growth in average labor productivity. The growth of total factor productivity in the IT-using sector is also projected at a higher rate than in the Base Case. Finally, we project that productivity growth in the remainder of the economy will contribute more to labor productivity growth than in the Base Case.

Combining projections of growth in labor productivity and hours worked, the High Growth projection of GDP growth is 2.38 percent per year, only slightly above the growth rate of 2.35 percent during the period 1990–2014. Higher growth of productivity and capital quality are offset by lower growth of labor quality and slower capital deepening. It is important to recall that our projections of employment rates differ by demographic group, therefore the rapid growth in hours worked reflects the disparate impacts of the Great Recession on different types of workers.

1.4.4   Alternative Projections

Byrne, Oliner, and Sichel (2013) survey contributions to the debate over prospects for future US economic growth since the Great Recession. Cowen (2011) presents a pessimistic outlook in his book, *The Great Stagnation: How America Ate All the Low-Hanging Fruit, Got Sick, and Will (Eventually) Feel Better*. Cowen (2013) expresses a more sanguine view in his book, *Average Is Over: Powering America Beyond the Age of the Great Stagnation*. Robert Gordon (2016) analyzes headwinds facing the US economy in his book, *The Rise and Fall of American Economic Growth: The US Standard of Living since the Civil War*.

Byrne, Oliner, and Sichel (2013) provide detailed evidence on the recent behavior of IT prices. This is based on research at the Federal Reserve Board to provide deflators for the Index of Industrial Production. While the size of transistors has continued to shrink, performance of semiconductor devices has improved less rapidly, severing the close link that had characterized Moore's Law as a description of the development of semiconductor technology.[8] This view is supported by Pillai (2011) and by computer scientists Hennessey and Patterson (2012).[9]

Gordon's pessimism about the future development of technology in the IT-producing industries is forcefully rebutted by Brynjolfsson and McAfee (2014) in the *Second Machine Age: Work, Progress, and Prosperity in a Time of Brilliant Technologies*.[10] Baily, Manyika, and Gupta (2013) summarize an extensive series of studies of the prospects for technology in American industries, including the IT-producing industries, conducted by the McKinsey Global Institute and summarized by Manyika et al. (2011). These studies also present a more optimistic view of future technological developments.

Fernald (2016) presents a number of alternative projections and of US GDP growth and chooses a modal forecast of 1.6 percent per year as the most likely outcome. The Congressional Budget Office (CBO 2016) presents GDP projections for ten to thirty years. The thirty-year projection is 2.1 percent per year. The projections of Fernald and the CBO are compared with our three alternative projections—Low, Base Case, and High—in table 1.2.

8. Moore's Law is discussed by Jorgenson, Ho, and Stiroh (2005, chapter 1).
9. See Hennessey and Patterson (2012, figure 1.16, 46). An excellent journalistic account of the slowdown in the development of Intel microprocessors is presented by John Markoff in the *New York Times* for September 27, 2015.
10. Brynjolfsson and Gordon have debated the future of information technology on TED. See http://blog.ted.com/2013/02/26/debate-erik-brynjolfsson-and-robert-j-gordon-at-ted2013/.

Table 1.2   Comparison of growth projections (percent per year)

Source	Projection period	ALP	Hours	GDP	TFP	Capital deepening	Labor quality
CBO (2016)	2015–25	1.6	0.5	2.1	1.4 (NFB)		
Fernald (2016)	7–10 years	1.06	0.55	1.6			0.20
Jorgenson, Ho, and	Low case	1.13	0.50	1.63	0.30	0.71	0.21
Samuels (2016)	Base case	1.32	0.50	1.83	0.45	0.76	0.21
	High case	1.88	0.50	2.38	0.64	1.12	0.21

All three sets of projections are based on the analysis of sources of US economic growth.

The methodology employed by the CBO is inconsistent with the methodology used in the US National Income and Product Accounts and employed by Fernald, as well as by ourselves. The CBO does not include growth of labor quality in its analysis of the sources of growth. The CBO projections omit the slowdown in the growth of labor quality due to the leveling of average educational attainment for the US population that we have analyzed. Unfortunately, this has a major impact on the CBO's long-term projections of the federal government budget and, in particular, the CBO's projections of the government deficit, which determines whether the US budget is fiscally sustainable.

The CBO's Extended Baseline scenario, which corresponds to our Base Case projection, assumes a growth rate of total factor productivity of 1.3 percent per year. Under this assumption, the CBO projects that federal debt held by the public will reach 141 percent of the US GDP in 2046.[11] The CBO also presents an alternative projection, based on a growth rate of total factor productivity of 0.8 percent per year. For this projection, federal debt held by the public will reach of 173 percent of the GDP in 2046. In contrast, our Base Case estimate of total factor productivity growth is 0.46 percent per year, outside the range of estimates of productivity growth considered by the CBO. This would raise the Base Case estimate of federal debt held by the public in 2046 to 195 percent of the GDP. A refinement of this estimate would involve adding our estimate of the contribution of labor-quality growth omitted by the CBO of 0.12 percent per year to our Base Case estimate of total factor productivity growth. This would reduce the 2016 estimate of federal debt held by the public to 187 percent of the GDP in 2046.

1.5 Conclusions

Our industry-level data set for the Postwar Period shows that the growth of capital and labor inputs, recently through the growth of college-educated

11. Congressional Budget Office (2016, figure 7-3, 83).

workers and investments in both IT and non-IT capital, explains by far the largest proportion of US economic growth. International productivity comparisons reveal similar patterns for the world economy, its major regions, and leading industrialized, developing, and emerging economies.[12] Studies for more than forty countries have extended these comparisons to individual industries for the countries included in the World KLEMS Initiative. The results are reported in detail in Jorgenson, Fukao, and Timmer (2016).

Conflicting interpretations of the Great Recession can be evaluated from the perspective of our new data set. We do not share the technological pessimism of Cowen (2011) and Gordon (2016), especially for the IT-producing industries. Careful studies of the development of semiconductor and computer technology show that the accelerated pace of innovation that began in 1995 has reverted to lower, but still substantial, rates of innovation. Productivity growth in the IT-producing industries made a substantial positive contribution to aggregate productivity growth during the Great Recession.

Our findings also contribute to an understanding of the future potential for US economic growth. Our new projections are consistent with the perspective of Jorgenson, Ho, and Stiroh (2008), who showed that the peak growth rates of the Investment Boom of 1995–2000 were not sustainable. However, our projections are similar to those we presented earlier in Jorgenson, Ho, and Samuels (2016). While the low productivity growth of the Great Recession will be transitory, productivity growth is unlikely to return to the high growth rates of the Investment Boom and the Jobless Recovery.

Finally, we conclude that the new findings presented in this chapter have important implications for US economic policy. Maintaining the gradual recovery from the Great Recession will require a revival of investment in IT equipment and software, and non-IT capital as well. Enhancing opportunities for employment is also essential. While this is likely to be most successful for highly educated workers, raising participation rates for the less educated workers and the young will be needed for a revival of US economic growth.

Appendix

Projections

We adopt the methodology of Jorgenson, Ho, and Stiroh (2008) to utilize data for the sixty-five industries included in the US National Income and Product Accounts. The growth in aggregate value added (Y) is an index of the growth of capital (K) and labor (L) services and aggregate growth in productivity (A):

---

12. See Jorgenson and Vu (2017).

(1A.1) $\Delta \ln Y = \bar{v}_K \Delta \ln K + \bar{v}_L \Delta \ln L + \Delta \ln A$.

To distinguish between the growth of primary factors and changes in composition, we decompose aggregate capital input into the capital stock (Z) and capital quality (KQ), and labor input into hours (H) and labor quality (LQ). We also decompose the aggregate productivity growth into the contributions from the IT-producing industries, the IT-using industries, and the non-IT industries. The growth of aggregate output becomes

(1A.2) $\Delta \ln Y = \bar{v}_K \Delta \ln Z + \bar{v}_K \Delta \ln KQ + \bar{v}_L \Delta \ln H + \bar{v}_L \Delta \ln LQ,$

$\qquad\qquad + \bar{u}_{ITP} \Delta \ln A_{ITP} + \bar{u}_{ITU} \Delta \ln A_{ITU} + \bar{u}_{NIT} \Delta \ln A_{NIT}$

where the $\Delta \ln A_i$'s are productivity growth rates in the IT-producing, IT-using, and non-IT groups, and the u's are the appropriate weights. Labor productivity, defined as value added per hour worked, is expressed as

(1A.3) $\Delta \ln y = \Delta \ln Y - \Delta \ln H$.

We recognize the fact that a significant component of capital income goes to land rent. In our projections we assume that land input is fixed, and thus the growth of aggregate capital stock is

(1A.4) $\Delta \ln Z = \bar{\mu}_R \Delta \ln Z_R + (1 - \bar{\mu}_R) \Delta \ln LAND = \bar{\mu}_R \Delta \ln Z_R,$

where Z_R is the reproducible capital stock and $\bar{\mu}_R$ is the value share of reproducible capital in total capital stock.

We project growth using equation (1A.2), assuming that the growth of reproducible capital is equal to the growth of output, $\Delta \ln Y^P = \Delta \ln Z_R^P$, where the P superscript denotes projected variables. With this assumption, the projected growth rate of average labor productivity is given by

(1A.5) $\Delta \ln y^P = \dfrac{1}{1 - \bar{v}_K \bar{\mu}_R} \times \begin{bmatrix} \bar{v}_K \Delta \ln KQ - \bar{v}_K(1 - \bar{\mu}_R)\Delta \ln H + \bar{v}_l \Delta \ln LQ \\ + \bar{u}_{ITP} \Delta \ln A_{ITP} + \bar{u}_{ITU} \Delta \ln A_{ITU} + \bar{u}_{NIT} \Delta \ln A_{NIT} \end{bmatrix}$.

We emphasize that this is a long-run relationship that removes the transitional dynamics related to capital accumulation.

To employ equation (1A.5), we first project the growth in hours worked and labor quality. We obtain population projections by age, race, and gender from the US Census Bureau[13] and organize the data to match the classifications in our labor database (eight age groups, two genders). We read the 2010

---

13. The projections made by the US Census Bureau in 2012 are given on their website (http://www.census.gov/population/projections/data/national/2012.html). The resident population is projected to be 420 million in 2060. We make an adjustment to give the total population including Armed Forces overseas.

Census of Population to construct the educational attainment distribution by age, based on the 1 percent sample of individuals. We use the microdata in the Annual Social and Economic Supplement (ASEC) of the Current Population Survey to extrapolate the educational distribution for all years after 2010 and to interpolate between the 2000 and 2010 Censuses. This establishes the actual trends in educational attainment for the sample period.

Educational attainment derived from the 2010 Census shows little improvement for males compared to the 2000 Census with some age groups showing a smaller fraction with professional degrees. However, the proportion of females with BA degrees is higher in 2010 than 2000. Our next step is to project the educational distribution for each gender-age group. For this purpose we use the historical improvements in educational attainment by these groups shown in figure 1.6.

Educational attainment of workers at the end of our sample period is dominated by the effects of the Great Recession. Less educated workers experienced much higher unemployment rates than those with college degrees and had lower rates of participation. Second, improvement in the share of men with BA or MA or higher degrees between 2000 and 2010 is modest, with some age groups falling behind. The improvement in women's education is more pronounced, especially in the older age groups, but there are also certain age groups of women that regressed.

Given these observations, we assume continuing improvement for all ages. We allow a continuing rise in the share of people in each age group with BAs or MAs, based on the observed educational attainment in 2000 and 2010. The gain in the share with BAs and MAs among men during these ten years was very small, even negative for some age groups. The gain among women is greater, but not uniformly positive for all ages.

We establish a long-run target of maximum educational attainment for 2030 e^{max}_{saet} by assuming that there will be higher shares of people with BA degrees, MA degrees, professional degrees, or PhD degrees, with offsetting lower shares in the other categories (associate degree, some college, high school diploma, some high school). We impose a target education-age profile that is changing smoothly for two groups of men—those with BA degrees and professional degrees.

For men, we assume that the increase in the share of BAs by 2030 is similar to the change between 2000 and 2010 for those between twenty-four and forty-four years old. Given that the education-age profiles are somewhat erratic, this projection results in a somewhat uneven improvement by age. For the professional degree target for men, we assume that the future increase in the share is similar to the improvement between 2000 and 2010 for ages twenty-seven to thirty-seven. We apply similar rules for the associate degrees, BA, MA, and PhD categories. We then apply a reverse rule that lowers the share of those with elementary school, some high school without diploma, and high school diploma.

We apply a similar procedure for women. We impose a smooth increase for the share of women with MA degrees that covers both the 2000 and 2010 lines. We also assume higher shares for professional degrees and PhDs and offset this with shares of BAs and associate degrees that are very close to the 2010 values, and lower shares for high school diploma and lower categories. After establishing the e_{saet}^{max} target for 2030, we interpolate the 2014–2030 projected matrices linearly using the actual 2014 values and the target:

$$(1A.6) \qquad e_{saet}^{p} = \omega_t e_{saet}^{2012} + (1 - \omega_t)e_{saet}^{max} t = 2014, \ldots, 2030.$$

We apply this projected improvement to those age sixty and younger, and allow those age sixty-one and older to carry their educational attainment as they age:

$$(1A.7) \qquad e_{saet} = e_{saet}^{p} a = 0, \ldots, 60$$

$$e_{saet} = e_{s,a-1,e,t-1}^{p} a = 61, \ldots, 90+.$$

Given that those age a (> 60) in 2014 have higher educational attainment than those age $a - 1$ in 2014, this assumption generates a rising level of attainment in the population.

We assume that the educational attainment for men age thirty-nine or younger will be the same as the last year of the sample period; that is, a man who becomes twenty-two years old in 2024 will have the same chance of having a BA degree as a twenty-two-year-old man in 2014. For women, this cutoff age is set at thirty-three. For men older than thirty-nine years, and women older than thirty-three, we assume that they carry their education attainment with them as they age. For example, the educational distribution of fifty-year-olds in 2024 is the same as that of forty-year-olds in 2014, assuming that death rates are independent of educational attainment. Since a fifty-year-old in 2024 has a slightly higher attainment than a fifty-one-year-old in 2022, these assumptions result in a smooth improvement in educational attainment that is consistent with the observed profile in the 2010 Census.

After projecting the population matrix by gender, age, and education for each year, our next step is to project the hours-worked matrices by these characteristics. We use the weekly hours, weeks per year, and compensation matrices in 2014 described in Jorgenson, Ho, and Samuels (2016). We assume there are no further changes in the annual hours worked and relative wages for each age-gender-education cell. We calculate the effective labor input in the projection period by multiplying the 2014 hours per year by the projected population in each cell and weighting the hours per year by the 2014 compensation matrix. The ratio of labor input to hours worked is our labor-quality index.

The growth rate of capital input is a weighted average of the stocks of various assets weighted by their shares of capital income. The ratio of total capital input to the total stock is the capital-quality index that rises as the composition of the stock moves toward short-lived assets with high rental costs. The growth of capital quality during the period 1995–2000 was clearly unsustainable. For our Base Case projection we assume that capital quality grows at the average rate observed for 1995–2014. For the High Growth Case we use the rate for 1995–2007. Finally, we use the rate for 1990–2014 for the Low Growth Case.

References

Aaronson, Daniel, Loujia Hu, Arian Seifoddini, and Daniel Sullivan. 2014. "Declining Labor Force Participation and Its Implications for Unemployment and Employment Growth." *Federal Reserve Bank of Chicago Economic Perspectives* 38 (4): 11–138.

Aaronson, Stephanie, Tomaz Cajner, Bruce Fallick, Felix Galbis-Reig, Christopher Smith, and William Wascher. 2014. "Labor Force Participation: Recent Developments and Future Prospects." *Brookings Papers on Economic Activity* Fall: 197–255.

Baily, Martin, James Manyika, and Shalabh Gupta. 2013. "U.S. Productivity Growth: An Optimistic Perspective." *International Productivity Monitor* 25:3–12.

Brynjolfsson, Erik, and Andrew McAfee. 2014. *The Second Machine Age: Work, Progress, and Prosperity in a Time of Brilliant Technologies*. New York: W. W. Norton.

Byrne, David, Steven Oliner, and Dan Sichel. 2013. "Is the Information Technology Revolution Over?" *International Productivity Monitor* 25:20–36.

Congressional Budget Office (CBO). 2016. "The 2016 Long-Term Budget Outlook." Washington, DC, Congressional Budget Office, July.

Council of Economic Advisers. 2016. "The Long-Term Decline in Prime-Age Male Labor Force Participation." Washington, DC, Executive Office of the President.

Cowen, Tyler. 2011. *The Great Stagnation: How America Ate All the Low-Hanging Fruit, Got Sick, and Will (Eventually) Feel Better*. New York: Dutton.

———. 2013. *Average Is Over: Powering America beyond the Age of the Great Stagnation*. New York: Dutton.

Domar, Evsey. 1961. "On the Measurement of Technological Change." *Economic Journal* 71 (284): 709–29.

Fernald, John. 2014. "Productivity and Potential Output: Before, during, and after the Great Recession." Working paper, Federal Reserve Bank of San Francisco, June.

———. 2016. "Re-assessing Longer-Run U.S. Economic Growth: How Low?" Working Paper no. 2016-18, Federal Reserve Bank of San Francisco, August. https://www.frbsf.org/economic-research/files/wp2016-18.pdf.

Goldin, Claudia, and Lawrence F. Katz. 2008. *The Race between Education and Technology*. Cambridge, MA: Harvard University Press.

Gordon, Robert. 2016. *The Rise and Fall of American Economic Growth: The US Standard of Living Since the Civil War*. Princeton, NJ: Princeton University Press.

Harper, Michael, Brent Moulton, Steven Rosenthal, and David B. Wasshausen. 2009. "Integrated GDP-Productivity Accounts." *American Economic Review* 99 (2): 74–79.

Hennessey, John L., and David A. Patterson. 2012. *Computer Organization and Design*, 4th ed. Waltham, MA: Morgan Kaufmann.

Jorgenson, Dale W. 1963. "Capital Theory and Investment Behavior." *American Economic Review* 53 (2): 247–59.

———. 2009. *The Economics of Productivity*. Northampton, MA: Edward Elgar.

Jorgenson, Dale W., Kyoji Fukao, and Marcel P. Timmer, eds. 2016. *The World Economy: Growth or Stagnation?* Cambridge: Cambridge University Press.

Jorgenson, Dale W., Frank M. Gollop, and Barbara M. Fraumeni. 1987. *Productivity and U.S. Economic Growth*. Cambridge, MA: Harvard University Press.

Jorgenson, Dale W., Mun S. Ho, and Jon Samuels. 2012. "Information Technology and U.S. Productivity Growth." In *Industrial Productivity in Europe*, edited by Matilde Mas and Robert Stehrer, 34–65. Northampton, MA: Edward Elgar.

———. 2016. "U.S. Economic Growth—Retrospect and Prospect: Lessons from a Prototype Industry-Level Production Account for the U.S." In *The World Economy: Growth or Stagnation?*, edited by Dale W. Jorgenson, Kyoji Fukao, and Marcel P. Timmer, 34–69. Cambridge: Cambridge University Press.

Jorgenson, Dale W., Mun S. Ho, and Kevin J. Stiroh. 2005. *Information Technology and the American Growth Resurgence*. Cambridge, MA: MIT Press.

———. 2008. "A Retrospective Look at the U.S. Productivity Growth Resurgence." *Journal of Economic Perspectives* 22 (1): 3–24.

Jorgenson, Dale W., and Paul Schreyer. 2013. "Industry-Level Productivity Measurement and the 2008 System of National Accounts." *Review of Income and Wealth* 58 (4): 185–211.

Jorgenson, Dale W., and Khuong M. Vu. 2017. "The Outlook for Advanced Economies." *Journal of Policy Modeling* 39 (2): 660–72.

Kim, Donald D., Erich H. Strassner, and David B. Wasshausen. 2014. "Industry Economic Accounts: Results of the Comprehensive Revision Revised Statistics for 1997–2012." *Survey of Current Business* 94 (2): 1–18.

Kroft, Kory, Fabian Lange, Matthew J. Notowidigdo, and Lawrence F. Katz. 2016. "Long-Term Unemployment and the Great Recession: The Role of Composition, Duration Dependence, and Non-participation." *Journal of Labor Economics* 34 (S1): S7–54.

Krueger, Alan B., Judd Cramer, and David Cho. 2014. "Are the Long-Term Unemployed on the Margins of the Labor Market?" *Brookings Economic Papers* Spring: 229–300.

Kuznets, Simon. 1971. *Economic Growth of Nations: Total Output and Production Structure.* Cambridge, MA: Harvard University Press.

Lyndaker, Amanda S., Thomas Howells III, Erich H. Strassner, and David B. Wasshausen. 2016. "Integrated Input-Output and GDP by Industry Accounts, 1947–1996." *Survey of Current Business* 96 (2): 1–9.

Manyika, James, David Hunt, Scott Nyquest, Jaana Remes, Vikrram Malhotra, Lenny Mendonca, Byron August, and Samantha Test. 2011. *Growth and Renewal in the United States*. Washington, DC: McKinsey Global Institute.

Markoff, John. 2015. "Smaller, Faster, Cheaper, Over: The Future of Computer Chips." *New York Times*, Sept. 26. http://www.nytimes.com/2015/09/27/technology/smaller-faster-cheaper-over-the-future-of-computer-chips.html.

Mayerhauser, Nicole M., and Erich H. Strassner. 2010. "Preview of the Comprehensive Revision of the Annual Industry Accounts: Changes in Definitions, Classification, and Statistical Methods." *Survey of Current Business* 90 (3): 21–34.

McCulla, Stephanie H., Alyssa E. Holdren, and Shelly Smith. 2013. "Improved Estimates of the National Income and Product Accounts: Results of the 2013 Comprehensive Revision." *Survey of Current Business* 93 (9): 14–45.

Organisation for Economic Co-operation and Development. (OECD). 2001. *Statistical Working Party of the OECD Industry Committee.* Paris: Organisation for Economic Co-operation and Development.

Pillai, Unni. 2011. "Technological Progress in the Microprocessor Industry." *Survey of Current Business* 91 (2): 13–16.

Rosenthal, Steven, Matthew Russell, Jon D. Samuels, Erich H. Strassner, and Lisa Usher. 2016. "BEA/BLS Industry-Level Production Account for the U.S.: Integrated Sources of Growth, Intangible Capital, and the U.S. Economy." In *The World Economy: Growth or Stagnation?*, edited by Dale W. Jorgenson, Kyoji Fukao, and Marcel P. Timmer, 377–428. Cambridge: Cambridge University Press.

Schreyer, Paul. 2001. *OECD Manual: Measuring Productivity: Measurement of Aggregate and Industry-Level Productivity Growth.* Paris: Organisation for Economic Co-operation and Development.

———. 2009. *OECD Manual: Measuring Capital.* Paris: Organisation for Economic Co-operation and Development.

Solow, Robert M. 1957. "Technical Change and the Aggregate Production Function." *Review of Economics and Statistics* 39 (3): 312–20.

United Nations, Commission of the European Communities, International Monetary Fund, Organisation for Economic Co-operation and Development, and World Bank. 2009. *System of National Accounts 2008.* New York, United Nations. http://unstats.un.org/unsd/nationalaccount/sna2008.asp.

2

The Outlook for US Labor-Quality Growth

Canyon Bosler, Mary C. Daly, John G. Fernald, and Bart Hobijn

2.1 Introduction

Economists have long recognized the importance of human capital accumulation for economic growth. And since the seminal analysis of Jorgenson and Griliches (1967), which provided a straightforward measurement framework, indices of human capital, or labor quality, have become standard in growth-accounting studies for many countries. In this chapter, we assess alternative methods for estimating US labor quality and provide projections for the future. We also identify key uncertainties that will determine the actual path of US labor quality in the medium and longer run. In almost all scenarios we consider, labor quality adds less to growth over the next decade than it has historically—in some scenarios, much less.

We begin by reviewing commonly used methods for measuring labor quality. Since labor quality is not directly observable, measuring it requires researchers to find an observable proxy. Not surprisingly, the best proxy is wages, which should move closely with marginal products. For example, a neurosurgeon is likely to have a higher marginal product than a grocery clerk. This difference in marginal products is, in turn, arguably the main reason why the neurosurgeon is paid more.

Canyon Bosler is a graduate student in economics at the University of Michigan. Mary C. Daly is executive vice president and director of research at the Federal Reserve Bank of San Francisco. John G. Fernald is professor of economics at INSEAD. Bart Hobijn is professor of economics at Arizona State University.

We thank Doug Elmendorf, Chuck Hulten, Valerie Ramey, Todd Schoellman, and conference participants for helpful comments. The views expressed in this chapter are those of the authors and do not necessarily reflect the position of the Federal Reserve Bank of San Francisco or the Federal Reserve System. For acknowledgments, sources of research support, and disclosure of the authors' material financial relationships, if any, please see http://www.nber.org/chapters/c13694.ack.

The question is how best to impute the relative marginal products of workers based on different characteristics. We develop a novel statistical metric that evaluates the reliability of alternative approaches to imputing relative marginal products. Specifically, we examine the trade-off that each approach implicitly makes between (a) the share of the productivity-related variation in observed wages that is explained, and (b) the precision of the imputed estimates of relative marginal products of different workers.[1]

In our statistical assessment, the best-performing model is a parsimonious Mincer specification that includes experience, education, and, when accurate data are available, occupation. Experience and education are clearly related to productivity differentials across workers, and are empirically important for explaining the patterns of wages in the data. Other commonly used variables raise challenges. For example, both occupation and gender add explanatory power with little cost in terms of precision. But, historically, occupation has been challenging to forecast with any degree of accuracy so, for the purpose of projections, we exclude it. For gender, it is unclear to what degree gender-related wage differentials reflect marginal products, so we again prefer to exclude it. (In any case, including gender turns out to make little difference empirically to our estimates of labor quality.) Other variables (such as industry or race) add little to explanatory power while substantially reducing the precision of estimated marginal products.

We then use our preferred parsimonious Mincer specification to estimate labor-quality growth from 2002 to 2013 across three alternative data sets.[2] We find that labor quality grew about 0.5 percent per year—somewhat faster than its postwar average of about 0.4. Indeed, labor quality arguably explains a bit under one-third of labor productivity growth of 1.8 percent per year over the 2002–2013 period.[3] This finding is robust across data sources.

Strikingly, the growth and acceleration of labor quality since 2002 has a very different source than it did in the half century before that. In the twentieth century, the primary driver of labor-quality increases was rising educational attainment (Ho and Jorgenson 1999; Goldin and Katz 2009; Fernald and Jones 2014). In contrast, since 2002, the source of labor-quality growth has been a shift in the composition of employment away from lower-

---

1. For example, adding an additional variable might add explanatory power for wages, but at the cost of sharply reducing precision of imputed marginal products.
2. The time period is constrained by our desire to compare results across three publicly available data sources.
3. This contribution is calculated assuming that growth in output per hour rises one-for-one with growth in labor quality. That is, the growth in labor quality is *not* multiplied by labor's share, which would give the proximate growth-accounting contribution. The one-to-one mapping comes from standard economic models, where there is an indirect effect from endogenous growth in capital. The reason is that capital deepening in the models is typically in terms of "effective labor." Fernald and Jones (2014) discuss this accounting and estimate that increases in labor quality explained 0.4 percent per year of the 2.0 percent annualized growth in US GDP per hour between 1950 and 2007.

skilled and toward higher-skilled workers. This change owed to ongoing secular changes in the labor force as well as cyclical adjustments associated with the Great Recession.

Building on this analysis, we provide alternative scenarios for the evolution of labor-quality growth over the medium and longer run. Our work reinforces the view that labor-quality growth will add less to growth in productivity and output than it has historically. That said, the actual path of labor-quality growth is sensitive to uncertainties about trends in employment rates and, to a lesser extent, educational attainment. These differences will show up in productivity growth, but whether they matter for output growth depends on the degree to which they are offset by hours growth. This highlights a takeaway from our analysis, namely that labor-quality growth and hours growth are often negatively correlated. An important implication of this is that forecasts of overall labor-input growth, or quality-adjusted hours, are preferable to independent projections of labor quality and hours.

Section 2.2 reviews the growth-accounting definition of labor quality that we apply in this chapter. Section 2.3 then discusses the practical challenges involved in empirically applying our conceptual framework and assesses alternative approaches and data sets. Section 2.4 examines the evolution of labor quality since 2002, and compares approaches and data sets. Over this period, labor-quality growth was boosted by disproportionate declines in employment rates among low-skilled workers, especially during and after the Great Recession.

With a framework in place, section 2.5 turns to projections of labor-quality growth over the medium to long term. We forecast that labor-quality growth is likely to slow to somewhere in the range of 0.1 to 0.25 percentage points a year over the next ten years. Should employment composition return to its prerecession levels, medium-term labor-quality growth will fall below this baseline and could even turn negative. In the longer run, trends in education and employment rates are central. To generate labor-quality growth at close to its historical pace requires not just a continuing shift in the composition of employment from low-skilled toward high-skilled workers, but also a resumed upward trend in educational attainment. Although such a scenario is possible, we think it unlikely. In particular, although educational attainment has picked up since 2007, our preferred interpretation is that the rise represents a transitory reaction to a poor economy, not a new upward trend.

2.2   Definition of Labor-Quality Growth

Indices of labor quality are based on standard neoclassical production theory.[4] Consider a neoclassical value-added production function of the form

---

4. Ho and Jorgenson (1999) survey the history of labor-quality measurement and discuss several semantic and/or conceptual confusions.

(1) $$Y = F(A, K, H_1, \ldots, H_n).$$

Output, Y, is produced by combining the n types of labor inputs, $H_1, \ldots,$ H_n, with a capital input, K; A denotes the level of technological efficiency with which the inputs are combined.[5]

To quantify how changes in inputs affect output growth, we apply a first-order logarithmic Taylor approximation. Small letters denote the natural logarithms of the capitalized variables such that y is the log of output, Y. Applying the first-difference operator, Δ, we can write

(2) $$\Delta y = \ln Y_t - \ln Y_{t-1}.$$

This is simply the growth rate of output, as measured by the change in the logarithm of output. The Taylor approximation then reads

(3) $$\Delta y = \frac{\partial F}{\partial A} \frac{A}{Y} \Delta a + \frac{\partial F}{\partial K} \frac{K}{Y} \Delta k + \sum_{i=1}^{n} \frac{\partial F}{\partial H_i} \frac{H_i}{Y} \Delta h_i.$$

Output growth depends on technology growth plus the contribution of the various factors of production. The final term in this expression is the effect of changes in labor inputs on output growth, where growth in each type of labor is multiplied by its respective output elasticity.

The contributions of labor inputs can be further decomposed into the effect of growth in total hours (i.e., growth in $\sum_{i=1}^{n} H_i$) and changes in the composition of total hours. To do this we rewrite equation (3) as

(4) $$\Delta y = \frac{\partial F}{\partial A} \frac{A}{Y} \Delta a + \frac{\partial F}{\partial K} \frac{K}{Y} \Delta k$$

$$+ \left(\sum_{j=1}^{n} \frac{\partial F}{\partial H_j} \frac{H_j}{Y} \right) \left(\Delta h + \sum_{i=1}^{n} \frac{(\partial F / \partial H_i) H_i}{\sum_{j=1}^{n} (\partial F / \partial H_j) H_j} (\Delta h_i - \Delta h) \right).$$

Growth in total hours is Δh and the change in the composition of hours worked is

(5) $$\sum_{i=1}^{n} \frac{(\partial F / \partial H_i) H_i}{\sum_{j=1}^{n} (\partial F / \partial H_j) H_j} (\Delta h_i - \Delta h).$$

The change in the composition of hours worked in equation (5) amplifies or attenuates growth in total labor input relative to growth in total hours. This wedge between growth in labor input and growth in hours is commonly interpreted as labor-quality growth. Intuitively, if all types of labor inputs, H_i, grow at the same rate, then the composition of total hours does not change and labor-quality growth is zero. But if, instead, hours of relatively

5. Assuming a single capital input is for simplicity and does not affect the results that follow for labor input.

more productive workers (with high $[\partial F/\partial H_i]$) grow more quickly than hours of less productive workers, then labor-quality growth will be positive.

Empirically, the marginal products of labor, $(\partial F/\partial H_i)$, in equation (5) are not observed. Under standard neoclassical conditions, the $(\partial F/\partial H_i)$ are proportional to the nominal hourly wage earned by workers of type i, denoted W_i. We assume that the proportionality constant is equal across types of labor.[6] If this is the case then

(6)
$$\frac{(\partial F/\partial H_i)H_i}{\sum_{j=1}^{n}(\partial F/\partial H_j)H_j} = \frac{W_i H_i}{\sum_{j=1}^{n}W_j H_j},$$

which is the share of total compensation that gets paid to workers of type i.

Under these assumptions, labor-quality growth, denoted by g_{LQ}, is the compensation-share-weighted average deviation of labor input from total hours growth by type, that is,

(7)
$$g^{LQ} = \sum_{j=1}^{n}\frac{W_i H_i}{\sum_{j=1}^{n}W_j H_j}(\Delta h_i - \Delta h).$$

This is the measure of labor-quality growth that we analyze. It is the same as the one used in range of growth-accounting data sets for many countries.[7]

Note that growth in total labor input, or "quality-adjusted" hours, is simply the share-weighted growth in hours:

(8)
$$g^{LQ} + \Delta h = \sum_{j=1}^{n}\frac{W_i H_i}{\sum_{j=1}^{n}W_j H_j}\Delta h_i.$$

2.3   Measurement of Labor-Quality Growth

To implement equation (7) and obtain an empirical estimate of labor-quality growth requires three things:

1. **Definition of worker types:** decision regarding the specific types of workers, $i = 1, \ldots, n$, the labor-quality index will distinguish between.

2. **Estimate of wage by worker type:** estimate of average hourly earnings

6. In competitive markets, standard neoclassical assumptions imply that real (output-price-deflated) wages equal marginal products, so the assumption holds (with proportionality given by the output price). Imperfect competition in the output market allows firms to charge a markup of price over marginal cost, but the markup is constant across types of workers so the assumption again holds. It also holds if firms have some monopsony power in the labor market, as long as the wedge is constant across types of labor.

7. For the United States, examples include Jorgenson, Gollop, and Fraumeni (1987), Jorgenson, Ho, and Samuels (2014), Ho and Jorgenson (1999), Zoghi (2010), and the Bureau of Labor Statistics (2015a, 2015b). Notable examples for a wider set of countries include EUKLEMS (O'Mahony and Timmer 2009), the Conference Board's Total Economy Database (van Ark and Erumban 2015), and the Penn World Tables (Feenstra, Inklaar, and Timmer 2015).

for each worker type, W_i, used to construct the share of each worker type in total compensation.

3. **Measure of hours:** measure of hours worked by worker type, H_i, used to calculate the deviation of hours growth by worker type, Δh_i, from overall hours growth, Δh.

Item (3) is relatively straightforward. Measures of hours worked by individuals are available in many data sets. Once the worker types are defined, calculation of H_i simply involves aggregation of hours across individuals in each of the n groups.

Items (1) and (2) are less straightforward than (3), and we discuss the different options for dealing with them in this section. We are not the first to discuss the choice of worker types and wage measures in the context of the construction of labor-quality indices (e.g., see Zoghi 2010). Our contribution relative to that work is to introduce a framework that allows us to make tractable choices for (1) and (2) and "test" those choices against each other using standard statistical techniques.

In terms of data sets, we focus primarily on the American Community Survey (ACS). The ACS is a smaller, annual version of the decennial census and collects a relatively narrow range of demographic and socioeconomic data on a sample of about 1 percent of the US population (approximately three million individuals) each year.[8] We also consider two other data sets. The first is the Current Population Survey's Output Rotation Groups (CPS-ORG), which consists of the outgoing rotation groups from the Current Population Survey (CPS). This is the quarter of the CPS respondents that are asked about their earnings and income in any given month. This results in an annual sample of about 135,000 individuals. The second, the Current Population Survey's Annual Social and Economic Supplement (CPS-ASEC), is the Annual Social and Economic Supplement to the Current Population Survey, also known as the March supplement. It contains annual earnings and income data from the full March CPS sample (70,000 individuals).

Though based on different samples and sampling methods, each of the data sets allows for the construction of similar hourly wages, as well as the six variables of education, age, sex, race/ethnicity, industry, and occupation, that are our main focus. In all cases, we measure hours as usual hours worked per week, which is available in all three data sets.

2.3.1   Criteria for Choosing Worker Types and Wage Estimates

Indices of labor quality are built by dividing workers into groups based on their marginal products of labor, $(\partial F/\partial L_i)$. The decision about how many

---

8. The sample of the ACS has been expanded twice and has only been a 1 percent sample of the population since 2006. In 2000, its first year, the sample was just under 400,000 individuals and between 2001 and 2005 the sample was slightly over one million.

and which worker types, $i = 1, \ldots, n$, to use depends on (a) the degree to which the types distinguish between workers with different marginal products, and (b) the degree to which the different worker types capture the cross-individual variation in wages.

A simple way to quantitatively assess the degree to which these criteria are met for any particular grouping is a regression. To see this, consider j individuals and denote the log of their individual hourly wage by w_j. For each individual we also observe a vector $\mathbf{x}_j$ of individual-level characteristics based on their worker type, i. Under the assumption that relative wages reflect relative marginal products, the extent to which the characteristics in the vector, $\mathbf{x}_j$, capture cross-individual differences in marginal products can be measured as the fraction of individual-level log-wage variation that is explained by the variables in $\mathbf{x}_j$. This measure is equal to the R^2 of the following standard log-wage regression

$$(9) \qquad w_j = \mathbf{x}_j'\boldsymbol{\beta} + \varepsilon_j.$$

Here, $\mathbf{x}_j'\boldsymbol{\beta}$ is the part of the wage variation captured by the variables in $\mathbf{x}_j$.

Though simple, this specification is very general. It subsumes the case in which the elements of $\mathbf{x}_j$ are dummy variables that span the set of worker types. In this version, every type is a stratum made up of individuals with the characteristics as in Jorgenson, Gollop, and Fraumeni (1987).[9] It also includes the case where $\mathbf{x}_j$ contains polynomial terms of variables affecting workers' marginal product. In this case, equation (9) is a form of a Mincer (1974) regression. This is the model used by Aaronson and Sullivan (2001), among others.

Of course, in practice we do not know the true parameter vector $\boldsymbol{\beta}$ and the log-wage regression (9) is estimated using a sample of workers of finite size. This means that, at best, we can obtain an estimate $\hat{\boldsymbol{\beta}}$ of the parameter vector and that we thus infer the part of wages captured by our explanatory variables with error. To formalize this mathematically, we denote the standard deviation of the estimation error of the explained part as

$$(10) \qquad \sigma_j = \sqrt{E[(\mathbf{x}_j'(\hat{\boldsymbol{\beta}} - \boldsymbol{\beta}))^2]}.$$

Since it is important to have a reliable estimate, the smaller σ_j the better. However, for the construction of the labor-quality index, we are not interested in one particular worker, j, but instead in the reliability of the relative marginal product estimate, $\mathbf{x}_j'\hat{\boldsymbol{\beta}}$, across the whole sample. To gauge the reliability of the marginal product estimate across the sample, we consider the pth percentile of the standard errors, σ_j, across individuals. We denote this percentile by $\tilde{\sigma}_p$.

Based on this simple framework, we suggest two statistical criteria for

---

9. Most stratum-based studies use median rather than mean wages. Our results are not sensitive to this choice.

determining the types of workers to distinguish and the method to use when estimating wages.

1. **R^2 of log-wage regression**. This measures the share of cross-individual wage variation that is captured by our choice of worker types and specification of the log-wage equation.

2. **Percentile of standard error, $\tilde{\sigma}_p$, of marginal product estimates**. This captures how reliably we estimate the (relative) marginal product of labor across workers. Higher R^2's and lower $\tilde{\sigma}_p$'s are preferred.

Importantly, there is a direct trade-off between these two measures. In principle, we can obtain an $R^2 = 1$ in the estimated regression (9) by including as many linearly independent variables in x_j as we have observations, m. However, this would result in a regression with zero degrees of freedom and $\tilde{\sigma}_p \to \infty$. Alternatively, we can aim for a very low $\tilde{\sigma}_p$ at the expense of a R^2.

Using these tools we can directly compare different choices of (a) worker types and (b) wage estimates by worker type. We do so using scatterplots that plot the R^2 and $\tilde{\sigma}_p$ for each choice that we consider. Before we construct the scatterplots, we first describe the choices of worker types and wage-regression specifications we consider.

2.3.2   Choice of Worker Types and Wage-Regression Specifications

So far, we have discussed the choices of worker types, i, and the regression specification, that is, x_j, as two distinct decisions. In practice, however, they are one and the same. This is because for the variables that are commonly considered in log-wage regressions there are only a finite number of values. Consequently, for a given regression specification in terms of these variables there is only a finite number of permutations of x_j across individuals. In this context, a worker type, i, corresponds to a permutation of the covariates vector x_j.

With this in mind, two questions remain: (a) which variables should be included in the vector x_j, and (b) what functional form of these variables works best?

Choice of Variables in Wage Equation

The decision regarding which variables should be included in the regression is guided by the assumption, underlying the labor-quality growth derivation, that wage differentials between worker types reflect differences in relative marginal products of labor. This means that the variables we include in the wage equation should have two properties. First, they should explain a substantial part of the variation in wages across worker types. Second, the part of wage variation they explain should reflect only differences in marginal products.

Whether a variable has the first property is straightforward to verify statistically. The second property—that is, which variables capture marginal product differentials—is more controversial. This is because certain observable characteristics may be correlated with wedges between wages and marginal products.[10] Though such variables might improve the fit of the wage regression, (9), including them in our measure of labor quality would bias our results.

The most obvious variables to consider for inclusion in equation (9) are education and experience. Several decades of running Mincer regressions has demonstrated a robust correlation between education and potential experience (or age) and wages (Psacharopoulos and Patrinos 2004).[11] Although there is some controversy over the degree to which returns to education are derived from improved human capital as opposed to the signaling of unobservable worker characteristics, both perspectives tend to attribute educational wage differentials to differences in marginal products (Weiss 1995).[12] Overall, there is broad agreement that the correlation between wages and education or experience is driven by real productivity differentials.[13]

A substantial literature, summarized in Altonji and Blank (1999), has also pointed to a role for gender, race, and ethnicity in explaining wage differentials. Here we encounter substantial controversy as to whether, or to what degree, these wage differentials reflect differentials in productivity as opposed to discrimination. On the one hand, gender differentials may capture the fact that women are more likely to work part time or leave the labor force temporarily, which is not captured in the measures of experience available in standard data sets (Light and Ureta 1995). And ethnic differentials may proxy for unobserved language barriers that have a real impact on productivity (Hellerstein and Neumark 2008).[14] Yet, there is also a substantial literature documenting the existence of labor market discrimination, particularly on the basis of race and ethnicity, in both hiring and wages (Bertrand and Mullainathan 2003; Pager, Western, and Bonikowski 2009; Hellerstein, Neumark, and Troske 2002; Oaxaca and Ransom 1994).

10. See Boeri and van Ours (2013) for a textbook treatment of many possible sources of such wedges.

11. Some of the recent Mincer-regression literature has suggested that there are important differences in the education-experience return profiles between cohorts (Lemieux 2006; Heckman, Lochner, and Todd 2008). We allow for such cohort effects in that we estimate wage regressions on annual cross-sectional data. Thus, in our analysis, cohort and age effects are indistinguishable. This is appropriate for our application because we are only interested in making robust wage predictions and not in isolating specific returns.

12. Outside of developing countries there has been little empirical research that even asks the question of whether educational wage differentials might reflect something other than productivity, and the research in developing countries has generally concluded that the differentials are consistent with differences in productivity (Jones 2001; Hellerstein and Neumark 1995).

13. Broad as the agreement is, it is not entirely universal: incomplete labor contracts, labor market segmentation, or cultural factors could potentially drive a wedge between wage premiums associated with education and experience and differentials in marginal product (Blaug 1985).

14. Skrentny (2013) and Lang (2015) discuss the theoretical and empirical evidence on race and worker productivity.

Finally, there is also a body of literature suggesting that there are inter-industry wage differentials that persist even after controlling for education and experience (Dickens and Katz 1987; Krueger and Summers 1988).[15] Once again, such differentials could originate from genuine differences in productivity (e.g., the matching of a worker to a particular job may reflect differences in social skills; Deming [2015]) or from non-productivity-related features of an industry (such as profit sharing). Interestingly, although similar arguments could apply to occupational differences, there has been little research that considers whether there are persistent interoccupation wage differentials independent of educational and experience prerequisites. Though not the main purpose of our analysis, our estimates of equation (9) partially fill this void by including occupation in our analysis.

Thus, the observables we focus on are age, education, gender, race, industry, and occupation. We are aware that there are many other variables that could be interpreted as reflecting differences in marginal product of labor across workers. Examples include marital status, rural-urban location, or family structure. However, given the limited evidence that these variables are of first-order importance in explaining cross-individual variation in wages, we omit them from our analysis.

There is also a wide range of potentially influential unobservable characteristics (such as entrepreneurial talent [Silva 2007]; cognitive and noncognitive abilities [Heckman, Stixrud, and Urzua 2006]; and physical attractiveness [Hamermesh and Biddle 1994]).[16] Although it would be ideal to include measurements of, or proxies for, these characteristics in our analysis, that is not possible in the data sets available.

Choice of Functional Form

With the set of variables to include in x_j in hand, the last thing to consider is the specific functional form imposed on these variables. For example, is the traditional Mincer regression with a constant, linear years of education, and a quadratic polynomial in experience, the appropriate functional form or should dummies for high school graduation and college graduation be included to account for sheepskin effects (Hungerford and Solon 1987)? Are education and experience additively separable, or is there a nonlinear interaction between the two? These questions have been investigated quite carefully for the traditional Mincer regression variables of education and experience (Lemieux 2006), but less attention has been paid to the other variables.

Given this uncertainty around the appropriate functional form, one

15. Gibbons et al. (2005), however, suggest that sectoral wage differentials can be accounted for by allowing for sector-specific returns to skill.
16. These characteristics are unobservable in the sense that they are not measured as part of the standard data sets (ACS, CPS-ASEC, and CPS-ORG) that we use for our analysis.

approach is to allow for the maximum flexibility in the log-wage regression, (9). To do this, one would treat each possible combination of values of the included variables as a worker type. This boils down to running a fully nonparametric regression in which $\mathbf{x}_j$ is a vector with separate dummies for each worker type. The fitted log wage, $\mathbf{x}_j'\boldsymbol{\beta}$, for each worker type in that case is the average log wage for workers with that combination of values for the included variables. This approach, though flexible, results in a significant loss of degrees of freedom.

For example, if we only consider age and education, restrict the population under consideration to sixteen- to sixty-four-year-olds, and distinguish sixteen educational categories (as is the case with most standard US micro data sets), then this regression has 768 estimated parameters corresponding to the 768 possible permutations of age and education in the data. In practice, many of these worker types will contain very few observations in the data. For those worker types for which there is only one observation, the standard error of the estimated mean log wage is infinite, that is, $\sigma_i = \infty$.

Though such a nonparametric regression might result in a very good fit, the heterogeneity in marginal products of labor across worker types will be estimated with a high degree of uncertainty.

Stratum-based methodologies, which have been used extensively in prior growth-accounting exercises that account for labor quality (Gollop and Jorgenson 1983; Jorgenson, Gollop, and Fraumeni 1987; Ho and Jorgenson 1999; Jorgenson, Ho, and Samuels 2014), are a form of this type of dummy regression. Stratum-based studies define worker types by partitioning the population by observable characteristics, with the mean wage of each partition being interpreted as the wage for workers of that type.

In practice, in order not to run into the curse of dimensionality described above, stratum-based studies do not treat each value of a variable as distinct. Instead, they group different values of the variables together. For example, the sixteen educational categories are often collapsed into less than high school, high school, some college, and college categories. Using a less granular partition regains some degrees of freedom, but with a loss of some flexibility in the functional form. How granular a partition can be used largely depends on the sample size of the data set used.

In the context of the regression framework that we use here, this grouping of values imposes multidimensional step functions on the data. Thus, although the most granular partitions result in a nonparametric regression that will have an R^2 that is at least as high as any other regression specification, the partitions used in practice actually impose a restrictive functional form that does not necessarily fit the data better than alternative model specifications.

Concerns about the step functions imposed by partitioned dummy regressions have led some researchers to hew more closely to the Mincer regression literature (Aaronson and Sullivan 2001; Bureau of Labor Statistics 1993).

Table 2.1   Different levels of granularity of classification of variables

| | Number of | Groups per classification | | | |
Variable	classifications	(I)	(II)	(III)	(IV)
1. Gender	1	2	—	—	—
2. Age	2	9	13	—	—
3. Education	4	4	5	7	16
4. Race/ethnicity	4	2	3	5	8
5. Industry	2	12	50	—	—
6. Occupation	3	10	22	51	—

Notes: Total number of possible stratum specifications (including omission of one or more variables) is 1,799. Most granular definition includes 8,486,400 strata.

These specifications focus on education and experience as the fundamental drivers of human capital, marginal product, and wages.[17] These regressions generally include education (either as a polynomial in years of education or as a set of dummies indicating levels of educational attainment) and a polynomial in experience.

In addition to the baseline education and experience variables, these human capital specifications often include some interaction between gender and experience to account for women's higher rate of part-time work and temporary withdrawal from the labor force (either as an interaction between gender and experience or by estimating the regression on men and women separately). In some cases (Aaronson and Sullivan 2001; Bureau of Labor Statistics 1993, 2015a, 2015b) they also include control variables like part-time status, marital status, veteran's status, race, and rural location. These variables are not included to capture differences in marginal products across workers, but instead to reduce omitted variable bias in the education and experience coefficients.

Comparison of Specifications

Between the question of which variables to include and what functional form to impose, the task of selecting a preferred regression specification for a labor-quality measure is quite daunting. Even in the narrowed down set of variables we consider, age, education, gender, race, industry, and occupation, there are several options on how to group their values. For each of the six variables we use, table 2.1 lists how many different classifications we consider for our comparison of model specifications. In the last four columns of each row, the table lists how many groups are defined for each classification. For example, for age we consider two classifications: one that splits the individu-

17. As commonly done, we define experience as the difference between age and years of education (plus six).

als up into nine age groups and another into thirteen age groups. The number of permutations across the different classifications of variables is 192. This includes one classification for each of the variables. Once one allows for dropping variables, then the possible number of stratum specifications increases to 1,799. The most detailed one, which includes the most granular classification for all variables, consists of 8,486,400 worker types.[18]

As noted, we apply the statistical tools R^2 and $\tilde{\sigma}_p$ as two clear criteria on which we can base our model-specification decision. For our application we use the adjusted R^2, that is, $\bar{R}^2$, as it penalizes for overfitting the data. We consider the 80th percentile of the standard errors of the estimated relative marginal product of labor across workers; that is, we use $\tilde{\sigma}_{80}$ as our measure of the reliability of the imputed wages.[19]

We complete our analysis using three different data sets. Our results are qualitatively very similar across data sets. For the sake of brevity, we present results obtained using the ACS, since this is the data set with the largest sample size.[20]

Figure 2.1 illustrates the trade-off between the goodness of fit, $\bar{R}^2$, and the precision of the wage imputation, $\tilde{\sigma}_{80}$. Panel A shows the scatter plot in the $(\tilde{\sigma}_{80}, \bar{R}^2)$ space for all 1,799 stratum-based model specifications from table 2.1. This panel shows how increasing the $\hat{R}^2$ of the model specification comes at the cost of the precision with which the relative marginal products are imputed, that is, an increase in $\tilde{\sigma}_{80}$. Because a higher $\hat{R}^2$ and lower $\tilde{\sigma}_{80}$ are preferred, we are focusing on specifications that move us to the upper left in the plotted $(\tilde{\sigma}_{80}, \bar{R}^2)$ space.

Panel B shows the same 1,799 points as panel A with two sets of points highlighted. The crosses are the 192 stratum specifications that include all six variables we consider, with the difference being the level of granularity at which the variables are classified. These points are the ones where $\tilde{\sigma}_{80}$ is high, compared to $\bar{R}^2$, and thus correspond to specifications that overfit the data. At the other end of the cloud of points are the ones highlighted as circles. These are the specifications that do not include age and education. The gray points are specifications that include age and education, but not all four of the other variables. When we compare the circles with the gray points we find that, among the gray points, there are several specifications that have a substantially higher $\bar{R}^2$ and not much higher levels of $\tilde{\sigma}_{80}$.

We find that adding occupational dummies to the stratum definitions that

18. To put the amount of potential overfitting in perspective, this most granular definition of strata means that, on average, there are less than twenty workers per worker type in the United States, since civilian employment has never exceeded 150 million.

19. In principle, the choice of p for the percentile is arbitrary. However, qualitatively all results that we emphasize in this section hold for choices of $p > 75$. The reason we do not use the mean is that, in the case of the stratum-based methods, $\sigma_j = \infty$ for all worker types with one observation. This would also make the sample mean of the σ_j's go to ∞.

20. See appendix for results based on CPS-ORG and CPS-ASEC data.

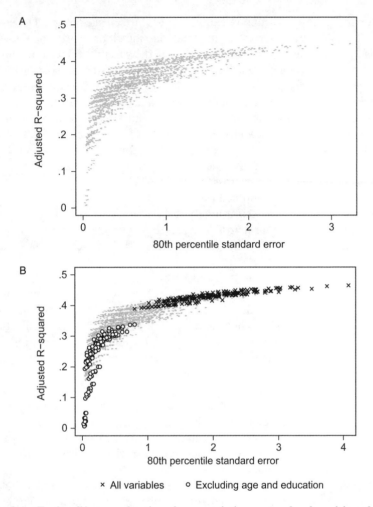

Fig. 2.1   Trade-off between fraction of wage variation captured and precision of imputed wages. *A*, all 1,799 specifications; *B*, all variables and specifications excluding age and education; *C*, age, education, and industry and/or occupation; *D*, age, education, and gender and/or race.

already condition on age and education yields the greatest improvement in fit and a relatively small decline in the precision of the imputed wages. This can be seen from panel C, which highlights the specifications that add only occupations as circles. As can be seen from the figure, adding occupation adds about 0.1 to the $\bar{R}^2$, but increases $\tilde{\sigma}_{80}$ only slightly. In contrast, adding industry alone, depicted by the empty squares, does not improve the fit as much as adding occupation, and results in lower precision with which the

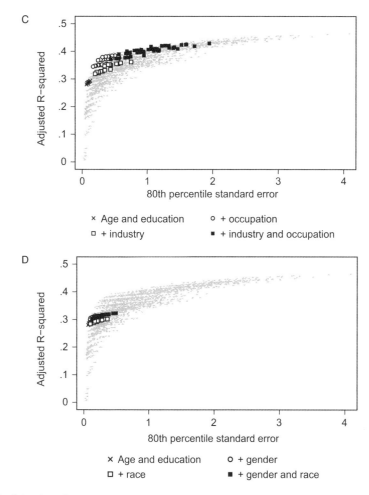

C

D

Fig. 2.1   (cont.)

wages are imputed.[21] Adding both industry and occupation results in values of $\tilde{\sigma}_{80}$ well above 0.5. This means that for more than 20 percent of the strata, log wages are imputed with a standard error of more than 0.5 (65 percent).

Panel D adds gender and race/ethnicity to the stratum definitions that include education and age. Race/ethnicity only slightly increases the fit at the cost of a substantial reduction in the precision of the marginal product imputation. Gender also increases the fit, but it is hard to know whether this

---

21. Of course, for some purposes, such as estimating industry-specific labor-quality indices, including industry dummies may still be necessary.

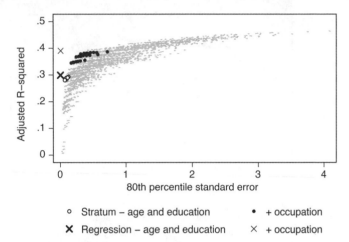

Fig. 2.2   Regression-based fit and precision compared to stratum-based specifications

reflects marginal product differentials or other factors. Since, in our analysis, the in- or exclusion of gender does not have a large effect on estimates of labor-quality growth we exclude gender from our specifications in the rest of this chapter.

In addition to the stratum-based model specifications, we also consider Mincer-type regressions. In particular, the *baseline* Mincer specification on which we settled includes a quadratic polynomial in experience and five education dummies.[22] Because our stratum-based analysis suggests that occupation is an important determinant of wages, we also consider a *baseline-plus-occupation* specification, which adds fifty-one occupation dummies.[23]

Figure 2.2 compares the regression-based fit and precision of imputed wages for the *baseline* and *baseline-plus-occupation* specifications with the stratum-based specifications. The lower cross in the figure shows the point for the baseline specification and the empty circles are the stratum-based points that only include age and education. Because the Mincer-type regression is more parsimonious than the semiparametric regressions, it results in more precisely imputed marginal product levels across workers, that is, it has a smaller $\tilde{\sigma}_{80}$. Moreover, the quartic polynomial in experience captures more of the variations in wages across workers than the piecewise linear specifications implied by the stratum-based methods. Consequently, the regression results in a higher $\bar{R}^2$. Thus, the flexibility of the semiparametric specifica-

22. A similar Mincer specification, with the addition of several control variables, was also used by Aaronson and Sullivan (2001).
23. We focus on this parsimonious baseline specification in the main text and illustrate that our main qualitative results are unaltered when additional covariates are included as controls in the appendix.

tion that Zoghi (2010) emphasizes when she proposes to use stratum-based medians as estimates of wages[24] is outperformed by the quartic polynomial in experience that we use here. As a result, the Mincer-regression-based way of imputing wages dominates the stratum-based methods in terms of both model-selection criteria.

This is not only true for the baseline regression specification, it is also true for the one that includes occupational dummies. In figure 2.2 the upper cross corresponds to the baseline-plus-occupation regression and the upper cloud of gray dots to the corresponding stratum-based regressions that include age, education, and occupation. Again, the Mincer-regression-based specification outperforms the stratum-based ones.

This evidence shows that our baseline and baseline-plus-occupation specifications perform well in terms of our two model-selection criteria.

2.3.3   Index Formula

Given the choice of the vector $\mathbf{x}_j$ and the period-by-period estimates of the parameter vector $\hat{\boldsymbol{\beta}}_t$, based on equation (9), the final choice to be made for the calculation of the labor-quality index is the index formula.

In line with the log-linear approximation of equation (4), the index formula that is used for most labor-quality index calculations is of the translog form and estimates labor-quality growth as the compensation-share weighted average of log changes in hours across worker types.[25] That is,

$$(11) \qquad \hat{g}_t^{LQ} = \sum_{i=1}^{n} \left(\frac{s_{i,t} + s_{i,t-1}}{2} \right) (\Delta h_i - \Delta h), \text{ where } \hat{W}_t(\mathbf{x}_i) = \exp(\mathbf{x}_i' \hat{\boldsymbol{\beta}}_t)^{26}$$

$$(12) \qquad \text{and } s_{i,t} = \frac{\hat{W}_t(\mathbf{x}_i) H_{i,t}}{\sum_{s=1}^{n} \hat{W}_t(\mathbf{x}_s) H_{s,t}}.$$

---

24. The regression framework we use here results in the conditional mean for a stratum to be the imputed wage. In unreported results, we redid our analysis with the conditional median as the wage estimate and obtained the same results compared to the Mincer specifications.

25. Compensation shares are averaged across the two periods between which growth rates are calculated.

26. Note that exponentiating the predicted logwage would not normally be sufficient to get a predicted wage in levels because

$$E[w_j] = E[\exp(\mathbf{x}_j \boldsymbol{\beta} + \varepsilon_j)] = E[\exp(\mathbf{x}_j \boldsymbol{\beta}) + (\varepsilon_j)] = \exp(\mathbf{x}_j \boldsymbol{\beta}) \cdot E[\exp(\varepsilon_j)]$$

and $E[\exp(\varepsilon_j)]$ is not 1. It is, however, a constant if the residuals are assumed to be independently and identically distributed. So if $\hat{W}_i = \exp(\mathbf{x}_j \boldsymbol{\beta})$ and $c = E[\exp(\varepsilon_j)]$, then plugging the predictions into the share of the wage bill calculation from equation (7) gives

$$\frac{W_i H_i}{\sum_{j=1}^{n} W_j H_j} \equiv \frac{c \hat{W}_i H_i}{\sum_{j=1}^{n} c \hat{W}_j H_j} = \frac{c \hat{W}_i H_i}{c \sum_{j=1}^{n} \hat{W}_j H_j} = \frac{\hat{W}_i H_i}{\sum_{j=1}^{n} \hat{W}_j H_j}$$

Therefore we need not make any adjustments to the predictions, nor do we need to impose an assumption on the distribution of the residuals beyond the standard assumption that they are IID.

This translog index formula has the desirable property that it is a so-called superlative index (Diewert 1978). That is, it is an exact index for a function (the translog) that provides a general second-order approximation of the production function. In other words, the labor-quality index does not rely simply on a first-order approximation (though we used such an approximation in our derivation in section 2.2 for expositional clarity).

For labor quality, implementation of the translog formula is complicated by the fact that in some cases the number of hours worked by a worker type, i, is zero. In that case, Δh_i cannot be calculated and such worker types are dropped from the calculations. Though dropping these worker types is a reasonable option because their compensation share is, presumably, small, one can also use another superlative price-index formula that does not suffer from this problem.

This is what we do in this chapter. In particular, we follow Aaronson and Sullivan (2001) and use a Fisher Ideal index formula of the form

$$(13) \quad \hat{g}_t^{LQ} = \left\{\frac{H_{t-1}}{H_t}\right\}\left\{\frac{\sum_i \hat{W}_t(\mathbf{x}_i)H_{i,t}}{\sum_i \hat{W}_t(\mathbf{x}_i)H_{i,t-1}}\right\}^{1/2}\left\{\frac{\sum_i \hat{W}_{t-1}(\mathbf{x}_i)H_{i,t}}{\sum_i \hat{W}_{t-1}(\mathbf{x}_i)H_{i,t-1}}\right\}^{1/2} - 1.$$

This formula allows us to include all worker types, i, in our calculations even if $H_{i,t} = 0$ or $H_{i,t-1} = 0$.[27]

2.4   Historical Labor-Quality Growth

Before we consider projections of labor-quality growth, we first examine its behavior over the past fifteen years. This is useful for two reasons. First, by comparing historical results for different specifications and data sets, we can assess how sensitive the labor-quality growth estimates are to the different choices discussed in section 2.3. Second, and most importantly, the concerns about plateauing educational attainment and the retirement of experienced older workers that many observers currently express were also raised as concerns early in the first decade of the twenty-first century. Our historical analysis shows that, contrary to these concerns, labor-quality growth barely slowed over the past fifteen years. This realization of labor-quality growth owes much to a reduction in the employment rates of less productive individuals, especially during and after the Great Recession. We will return to this point in the projection section.

2.4.1   Comparison across Methods and Data Sets

As we discussed in section 2.3, we construct our benchmark labor-quality index using ACS data based on our baseline Mincer specification. The index

---

27. For our benchmark specification, the problem of zeros does not occur, and the Translog and Fisher are virtually identical. It can make a little more difference in cases with extremely large numbers of cells, where there are more zeros.

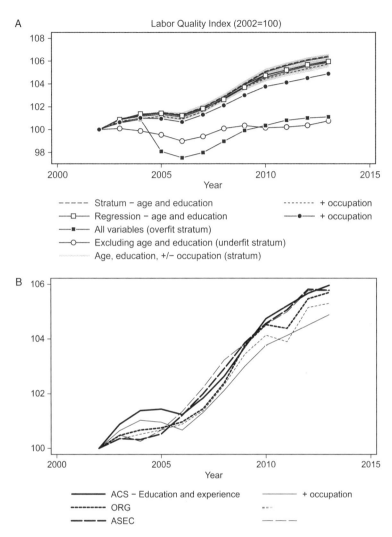

Fig. 2.3   **Comparison of results across specifications and data sets, 2002–2013.**
A, **different specifications using ACS data;** *B*, **ACS, CPS-ASEC, and CPS-ORG.**

for labor quality obtained from this specification is plotted as the line with squares, labeled "Regression—age and education," in figure 2.3, panel A.

From 2002 through 2013 the cumulative growth in the index was 5.96 percent, which is 0.53 annually. As the figure shows, labor-quality growth has been far from constant at this average during our sample period. Its standard deviation across years is 0.39. From 2002 to 2006 labor quality by this measure grew relatively slowly, about 0.37 percent per year. Subsequently, during the Great Recession, from 2008 to 2010, labor-quality growth logged in at 0.94 percent a year. Since then it has come down to 0.36 percent.

In section 2.3 we showed how our baseline specification outperformed many others in terms of goodness of fit of the log-wage regression, as well as the precision of imputed wages. In terms of labor-quality growth our baseline specification yields an estimate that is very close to those obtained using other specifications that include age and education. This can also be seen in figure 2.3, panel A. As the figure plots, the stratum- and regression-based methods give very similar estimates of the labor-quality index when both age and education are included in the vector x_i. Moreover, the index constructed does not change very much when we use the baseline-plus-occupation specification instead of the baseline specification.

Among the series plotted in figure 2.3, panel A, there are two clear outliers that exhibit much less cumulative labor-quality growth. The first is the stratum specification that includes all variables. Such a specification results in large errors in imputed wages, which reduces the correlation between hours growth and wages that drives labor-quality growth. As a result, the overfitted specification yields much less labor-quality growth than our baseline model. The other outlier series is the version that excludes age and education entirely (the underfit stratum). That series is flat, confirming that age and education are what drive the series.

Excluding the two outlier series, the cross-specification mean of average annual growth rates of labor quality is equal to the average annual labor-quality growth rate implied by our baseline index, namely 0.53 annually. The cross-specification standard deviation in these average annual rates is 0.03. Besides very similar mean growth rates, all these indices also show a very similar qualitative pattern over the sample period: slow growth from 2002 to 2006, an acceleration during the Great Recession, and a subsequent slowdown in 2011 and 2012.

The results in figure 2.3, panel A, are reminiscent of Zoghi (2010)[28] in that she suggests that estimated average annual labor-quality growth rates are fairly robust to the choice of model specification. This robustness of estimated average annual labor-quality growth rates also translates across data sets.

This can be seen from figure 2.3, panel B. It plots the baseline and baseline-plus-occupation results for the three data sets that we consider in this chapter, that is, for ACS, CPS-ASEC, and CPS-ORG. The six indices plotted look very similar.[29] In terms of their summary statistics, the mean average annual labor-quality growth rate across series in the figure is 0.49 percent with a standard deviation of 0.03.

Together, these results suggest that the pattern of labor-quality growth

---

28. See Zoghi (2010, table 12.2, 478).
29. The only exception is the ACS-based indices in 2005–2006. In this year, the sample size of the ACS was expanded from one to three million respondents, which appears to have resulted in a sample with a slightly lower level of labor quality than before.

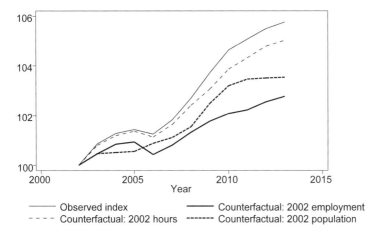

- - - - - - Observed index
- - - - Counterfactual: 2002 hours
———— Counterfactual: 2002 employment
------- Counterfactual: 2002 population

Fig. 2.4   Counterfactual indices for 2002 base-year hours, employment, and population
Note: Betas from 2002 wage regression.

from 2002 through 2013 we find using our baseline case is not the result of the particular specification or data set chosen. Indeed, we find this pattern for all reasonable model specifications and across all data sets. Overall, we conclude that from 2002 to 2013 labor quality has grown around 0.5 percent a year. This is about the same as the average of about 0.5 percent labor-quality growth between 1992 and 2002 (Bureau of Labor Statistics 2015a; Fernald 2015).

2.4.2   Counterfactuals to Identify the Sources of Growth

The fact that we find no substantial deceleration in labor-quality growth since 2002 is surprising, especially given the slow growth of educational attainment and the beginning of retirement among the oldest baby boomers during the period. Our analysis shows that as these adverse demographic and educational trends were pulling down labor-quality growth, a disproportionate decline in the employment-to-population (EPOP) ratio of lower-quality worker types was pushing it up. To illustrate this, we calculate three counterfactual historical indices, which are plotted in figure 2.4.

These counterfactuals take advantage of the fact that hours worked by workers of type i, H_i, are the product of (a) average hours worked per year by workers of this type, η_i, (b) the EPOP of these workers, E_i, and (c) the population of these workers, P_i. That is,

$$(14) \qquad H_i = \eta_i E_i P_i.$$

Using this expression, we can create different counterfactuals by holding one of the three factors, that is, η_i, E_i, and P_i, fixed at its 2002 level. We then allow the other two factors to change as observed in the data.

Figure 2.4 shows our baseline estimate, labeled "Observed index," as well as the three counterfactual indices. As can be seen from the figure, changes in average hours worked across worker types have had relatively little impact on labor-quality growth. In contrast, if the composition of the population had not changed since 2002, then labor-quality growth would have been about a third lower. This is because removing population changes eliminates the continued accumulation of experience of the baby boom generation from the calculations.

The most striking of the three counterfactuals, however, is the one for the EPOP ratio. From figure 2.4 it is clear that, if EPOP ratios by worker type had remained at their 2002 levels, labor-quality growth would have been half of what we observed over the past decade. Notably, the wedge between the observed index and the counterfactual with constant EPOP ratios increased most rapidly during the Great Recession. This wedge is consistent with the extensively documented composition effect of recessions on real wages. Many studies, including those by Bils (1985) and Solon, Barsky, and Parker (1994), find that the incidence of unemployment is more cyclical among low-wage workers.

In growth-accounting terms, this cyclical composition effect means that labor quality has a countercyclical component (Ferraro 2014). This is reflected in the strong negative correlation of around −0.9 between labor-quality growth and hours growth as measured by our baseline specification. This negative correlation is quite robust across specifications: figure 2.5 plots the correlations for all of the labor-quality specifications plotted in figure 2.3, panel A, except the overfit and underfit stratum specifications, and all of the correlations are strongly negative. An implication of this negative correlation is that it is important to jointly forecast labor quality and hours worked to get a robust estimate of labor input going forward.

As discussed, our labor-quality index captures the fact that EPOP ratios among lower-quality worker types are more cyclical. And our counterfactuals show that the disproportionate decline in employment rates among less skilled workers led to a recession-driven increase in labor-quality growth. Therefore, an important question for any medium-term forecast of labor-quality growth is to what extent these movements in EPOP ratios by worker types are transitory or permanent. Since a large part of the decline in these EPOP ratios reflects declines in labor force participation rates, this is largely a question of what fraction of recent movements in labor force participation is structural versus cyclical.

If labor force participation rebounds substantially, as the Congressional Budget Office (2015) projects, this will put downward pressure on labor-quality growth over our forecast horizon. However, if, as Aaronson et al. (2014) suggest, the bulk of the movements in participation rates across groups since 2007 have been structural, then our labor-quality index would be largely unaffected. In that case, there would be no downward pressure

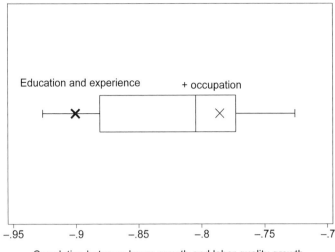

Correlation between hours growth and labor quality growth

Fig. 2.5   Correlation between labor-quality growth and hours growth for key indices
Notes: The plotted correlations are from the age and education and the age, education, and occupation specifications by both the stratum- and regression-based methods. That is, they are all of the specifications plotted in figure 2.2, except for the overfit and underfit stratum specifications. The bold X identifies our baseline specification and the thin X identifies our baseline-plus-occupation specification.

on labor-quality growth coming from changes in labor force participation by skill level.

This finding highlights an important lesson from our analysis. We should not be misled by the positive sound of *"increases* in labor quality" due to composition effects. Often, labor quality is discussed assuming a path of total hours. But an important factor driving labor-quality growth since early in the first decade of the twentieth century has been declines in hours (or a slowdown in hours growth) for lower-skilled workers. From equation (4) we know that what matters for output growth is the growth rate of the total labor input, which is hours growth plus labor-quality growth. Hence, if labor quality grows as a result of a selection effect among workers when total hours decline, then this is neither necessarily good news for growth of overall labor input nor for output growth.

2.5   Projecting Labor-Quality Growth

In this section we consider the outlook for labor-quality growth over the next ten years. We begin by reviewing the components of labor-quality growth projections. We then evaluate the performance of our baseline specification for 2002–2013, paying particular attention to the components that have contributed most to historical projection errors. Guided by these find-

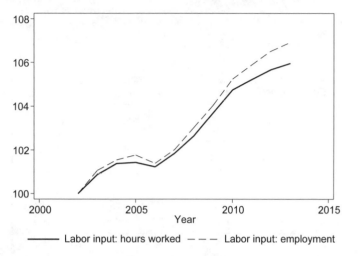

Fig. 2.6   Hours- versus employment-based historical labor-quality indices
Note: Employment-based indices ignore variation in average hours worked, $\eta_{i,t}$, across worker types.

ings we provide a range of alternative scenarios for future labor-quality growth in both the medium and longer run.

2.5.1   Components of Labor-Quality Growth Projections

As previously discussed, the index for labor quality, equation (13), is a highly nonlinear function of the parameter vector β_t and hours worked by worker type H_i. The fact that wages and hours are endogenous to one another further complicates the problem. In practice, producing an optimal forecast of labor-quality growth based on the joint distribution of future log-wage regression coefficients and future hours worked by worker type is not feasible.

In its place, researchers generally project labor-quality growth by projecting independently the log-wage parameter vector, β, and the hours worked by worker type, H_i, and substitute them into equation (13).[30] Given that time variation in the βs accounts for a very small portion of labor-quality growth over time, the convention is to hold log-wage parameters constant (Aaronson and Sullivan 2001; Jorgenson, Ho, and Samuels 2015). We follow this convention and set $\hat{\beta}_{t+h} = \hat{\beta}_{2013}$.

Turning to hours, recall that hours worked by worker type, $\hat{H}_{i,t+h}$, can be decomposed into the three factors as in equation (14), namely (a) average hours, $\hat{\eta}_{i,t+h}$, (b) the EPOP rate, $\hat{E}_{i,t+h}$, and (c) population, $\hat{P}_{i,t+h}$. Historically, accounting for heterogeneity in average hours worked by worker type does not make a material difference. This is highlighted in figure 2.6, which plots

---

30. This gives a joint projection of hours and labor quality, which is important given the negative correlation between hours and quality documented in section 2.4.2.

the observed baseline index against an employment-based index constructed under the assumption that all workers work the same number of hours, that is, $\eta_{i,t} = \eta_t$ for all i. The employment-based index shows average annual labor-quality growth of 0.61 percent, about a tenth of a percentage point higher than the 0.53 obtained from the hours-based index.[31] Given the modest difference, and the significant challenges associated with projecting heterogeneous hours worked, we set $\hat{\eta}_{i,t+h} = \eta_{t+h}$ for all worker types i. We use this 0.61 percent observed average annual growth of labor quality as our baseline for comparing the observed index with forecasts.[32]

2.5.2   Historical Projection Accuracy and Sources of Error

In this section we examine how our baseline projection specification would have performed for the 2002–2013 sample period. Specifically, we compare our projection to observed labor-quality growth and use an informal decomposition to evaluate the sources of forecast errors. Following Aaronson and Sullivan (2001), we build our projections using Census Bureau 2000 ("middle") National Population Projections by age, gender, and race.[33] To obtain population projections for all age and education combinations, we apply a multinomial logit model that estimates the probability distribution of our five educational levels based on age, cohort, gender, and race. We use these estimated probabilities to construct population projections by age and education, that is, to construct $\hat{P}_{i,t+h}$, for each year. Finally, to project the age- and education-specific EPOP ratios, $\hat{E}_{i,t+h}$, we estimate the probability that an individual is employed as a function of age, cohort, and education, using logit models that vary by gender and race.[34]

The results are shown in figure 2.7. The top line in panel A shows the observed employment-based index of labor quality, which grew at an average annual pace of 0.61 percent. The bottom line in panel A shows our projection of labor-quality growth as of 2002. The results are strikingly different; our projection expected average annual labor-quality growth to rise just 0.19 percent, well below the pace observed over the period. This large

---

31. This difference between the hours-worked-based and employment-based indices is even smaller in the CPS-ORG and CPS-ASEC data than in the ACS (see appendix).

32. Note that our baseline specification does not include occupation. Including occupation requires projecting population and EPOP ratios by age, education, and occupation. This turns out to result in very imprecise projections, since projections of employment by occupation, without considerations by age and education, already have large errors. To avoid introducing these errors into our projections, we limit ourselves to projections using our baseline specification.

33. Our projection method differs from Aaronson and Sullivan (2001) in the following ways: we distinguish five racial groups instead of four, define employment more narrowly to be consistent with our sample selection, and use ACS data.

34. Because the first ACS data were released in 2002, we cannot use ACS data for the estimation of the EPOP and educational attainment models. Instead, we estimate these models using 1992–1997 data from the CPS-ORG for this historical forecast. The full technical details of this projection are provided in the Projections of Educational Attainment and Employment subsection of the appendix.

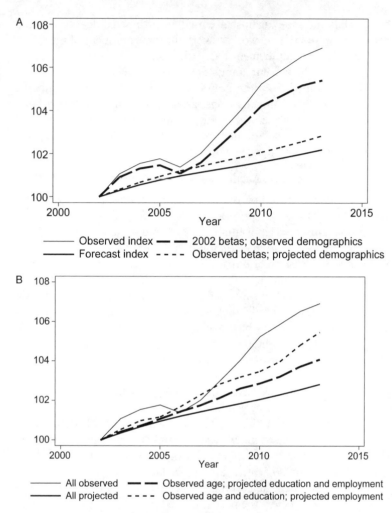

Fig. 2.7   **Decomposition of forecast errors from 2002 to 2013. *A*, projected hours distribution of x_i and projected β_t; *B*, projected EPOP ratios and education rates for observed βs.**

Notes: The EPOP and educational attainment models used to construct historical forecasts are based on 1992–1997 CPS-ORG data. *Forecast index* in panel A is based on extrapolated cohort effects holding βs constant at their 2002 values. *Observed betas* in panel A is the same as *all projected* in panel B. Real-time betas used for all four indices.

difference result is consistent with projections by Aaronson and Sullivan (2001), which used a slightly different model specification and CPS-ASEC data rather than ACS.

The remaining lines in figure 2.7, panel A, plot counterfactual indices that replace (a) projected demographics with observed demographics, and (b) projected log-wage regression parameters with observed parameters. The line labeled "2002 betas; observed demographics" is much closer to the actual index than the forecast index, suggesting that errors in the projected demographic variables account for a substantial portion of the forecast error over the period. In contrast, the line labeled "Observed betas; projected demographics" is very close to our baseline projection and far from the observed index. This suggests that time variation in the log-wage regression parameters accounts for a very small portion of the forecast errors. Panel A also shows that the bulk of the forecast errors accumulate during the Great Recession. In other words, deviations in demographics, $H_{i,t}$, from their projections, $\hat{H}_{i,t}$, in the Great Recession account for much of the forecast error.

Figure 2.7, panel B, takes a closer look at the specific demographic variables contributing to the large projection errors. The lines labeled "All observed" and "All projected" are the "Observed index" and "Observed betas; projected demographics" lines from panel A. The line labeled "Observed age and education; projected employment" reflects an alternative index based on observed components of the demographics, less the EPOP ratios, for which we use projections. The difference between this line and the "All observed" index isolates the effect of projection errors in EPOP ratios across worker types. As can be seen from the figure, these errors account for about one-third of the cumulative forecast error in labor-quality growth and are especially important after the onset of the Great Recession in 2008. The line "Observed age; projected education and employment" shows that projection errors in educational attainment also account for about one-third of the forecast error in labor-quality growth. The remaining error owes to misses in the census's population projections.[35] Notably, the projection errors for education and population accumulate relatively smoothly over our sample period.

2.5.3   Projections of Future Labor-Quality Growth

Going forward most commentators project labor-quality growth will be slower than its historical pace. This view stems from the fact that the exceptional increases in US educational attainment during the twentieth century seem unlikely to be repeated (Goldin and Katz 2009). However, as we will show, this oversimplifies the uncertainties surrounding the future path of labor-quality growth both in the medium and the longer run. To illustrate these uncertainties and how they relate to various components of labor-quality growth, we consider three potential future paths for educational attainment and employment-to-population rates and assess how these paths affect estimates of future labor-quality growth in the medium and longer run. These alternative paths, which are briefly described below and fully explained in the third section of the appendix, illustrate the mechanics of how different economic forces influence future labor quality. Given the lim-

---

35. Since this is not a formal decomposition, we are not accounting for the nonlinear contributions associated with interactions between the census demographics, distribution of education, and employment rates. These interactions, however, appear to be relatively minor compared to the first-order contributions of demographics, education, and employment.

ited role of the β (which capture relative returns to experience and education) in the accuracy of the historical projections, we hold at them fixed at their 2013 values.[36] To allow sufficient time for the economy to recover from the effects of the Great Recession, we define the medium term as 2015–2022. The longer run is 2022–2025.

For employment to population, we consider three alternative paths. The paths are meant to illustrate a range of potential outcomes.

1. **Cyclical rebound, or "revert"**: Age-education-specific EPOP rates return to 2007 values, between 2015 and 2022, and remain there. This scenario corresponds to the view that the changes in EPOP rates for specific age-education groups were cyclical.[37]

2. **Structural change, or "persist"**: Age-education-specific EPOP rates remain at 2013 levels. This scenario corresponds to the view that much of the decline in EPOP rates following the Great Recession is permanent.

3. **Extrapolated 2002–2007 structural trends in EPOPs**: The final path allows for heterogeneous paths across groups. Specifically, it extrapolates the declining EPOP rates of young people (with heterogeneity across education groups), the increasing EPOP rates of older people (particularly the more educated), and the widening gap between the EPOP rates of more and less educated prime-age people (Dennett and Modestino 2013; Burtless 2013; Aaronson et al. 2014).

The paths above illustrate how changes in various EPOP rates affect future US labor-quality growth.

We also consider three alternative paths for educational attainment. Again, these paths are meant to highlight a range of potential outcomes.

1. **Revert to precrisis levels**. During the Great Recession enrollment and graduation rates rose. This path assumes that the increase was a temporary cyclical effect and rates will return to their precrisis levels.

2. **Persist at 2013 educational plateau**. This alternative assumes that the uptick in educational attainment in recent years persists through future cohorts. Specifically, 2013 rates of educational attainment carry forward for each cohort over the next decade.

3. **Extrapolate 2007–2013 trends in education**. The final path assumes that the uptick in educational attainment over the past several years represents a resumed upward trend. Projections are based on age-specific time trends in educational attainment from logistic regressions.

36. An alternative approach to projections of education and employment would be to use a statistical model, following Aaronson and Sullivan (2001). Experiments with this methodology produced variable results that appear less reliable, especially in the more distant future, than the methods we employ here.
37. Both the cyclical (a) and structural (b) paths allow for a demographically (or educationally) driven structural decline in the aggregate employment-to-population ratio. However, they do not allow a structural decline (or increase) in age-education specific EPOPs.

Table 2.2   Labor-quality growth projections, 2015–2022

	EPOP ratio		
	Trend	Level	
Education	Extrapolated 2002–2007 trends in EPOP (I)	Revert to precrisis EPOPs (II)	Persist at 2013 EPOPs (III)
1. Extrapolate 2007–2013 trends in education	0.25/1.21	−0.20/0.86	0.15/0.51
2. Revert to precrisis educational attainment	0.19/1.20	−0.27/0.88	0.09/0.50
3. Persist at 2013 educational attainment	0.25/1.22	−0.21/0.88	0.14/0.51

Notes: Reported are average annual log-growth rates for 2015–2022.
Row scenarios: (1) Uptick in educational attainment since Great Recession reflects a permanent acceleration in educational attainment of those age thirty and younger. (2) Uptick since 2008 is fully cyclical and educational attainment reverts to its 2008 level. (3) Uptick in educational attainment since Great Recession reflects a step increase in educational attainment of those age thirty and younger.
Column scenarios: (I) Pre–Great Recession (2002–2007) trends in EPOP continue. (II) EPOP ratios in 2025 have reverted to their 2007 levels. (III) EPOP ratios will remain at their 2013 levels. More details about these scenarios can be found in section 2.5.3.

Table 2.2 shows projections for 2015–2022. All scenarios incorporate the Census Bureau's population projections by age group. In addition, the scenarios incorporate differing medium-run cyclical dynamics for employment rates and education. The columns of the table show the three alternative EPOP assumptions. The rows show the three educational attainment assumptions. For each cell, the first number shows growth in labor quality and the second shows growth in hours. (Note that, since we do not model average hours worked, hours grow at the same rate as employment growth.)[38]

A notable takeaway from table 2.2 is the potentially negative correlation in the medium run between growth in hours and growth in labor quality. The negative correlation appears in the two "level" EPOP scenarios. The "persist" (structural) scenario results in 0.35–0.36 percentage point faster labor-quality growth than the "revert" (cyclical) scenario. However, this is fully offset by 0.35–0.39 percentage point slower growth in hours. As a result, growth of total labor input grows at 0.59–0.67 percentage point per year in all of the scenarios in columns (2) and (3). This negative correlation highlights the importance of jointly modeling these two variables to obtain a forecast for quality-adjusted hours.

The near invariance of quality-adjusted-hours growth across the level scenarios seems surprising at first glance. Intuitively, an extra hour of work

38. This is equivalent to assuming that all workers work the same number of hours, a counterfactual assumption, but one that has been relatively innocuous historically (see figures 2.6 and 2A.4).

should add *something* to quality-adjusted hours—albeit more if it involves higher-skilled workers. The reason for the near invariance in table 2.2 is that low-skilled and high-skilled workers have seen an opposite pattern in EPOP ratios since 2007. Employment rates of lower-skilled workers have fallen, while rates for higher-skilled workers have risen. Thus, the "revert" scenario includes not only a rise in employment by lower-skilled workers, but also a decline by higher-skilled workers.

The first column of table 2.2, which extrapolates 2002–2007 EPOP trends, looks quite different from the others. In this case, we see markedly stronger growth in both labor quality and hours. For lower-skilled workers, there was little prerecession trend in EPOP rates. For this group of workers, this extrapolation-based scenario thus looks similar to the "revert" scenario, which boosts hours but holds labor quality down. But for higher-skilled and older workers, the prerecession trend was to increase employment rates. These workers tended to be below their estimated trend in 2013. Hence, in this scenario, these workers add both hours and skills to the labor force between 2015 and 2022. For both groups, hours increase quickly as employment rates rise. For labor quality, the extra hours of high-skilled workers dominate and labor quality rises more quickly.

Finally, looking down the columns, for none of the cases do the education scenarios matter much between 2015 and 2022. Extrapolating the rising educational trend from 2007 to 2013 (row 1) matters only a few basis points over this time period. The dominant force in the medium run is thus what happens to employment rates.

Turning to the longer run, table 2.3 shows projections for 2022–2025. These scenarios assume that all cyclical/transitional dynamics will have taken place by 2022. In the longer run, educational trends do matter. Looking down the three columns, the educational-extrapolation row implies almost two-tenths percentage point faster growth in labor quality than the "revert" or "persist" rows, with minimal difference in hours worked. Of course, this educational-extrapolation path assumes a considerable acceleration in educational attainment relative to what we have seen since World War II. Our reading of the data so far is that there is little indication that such an educational acceleration is actually happening. Rather, we view one of the plateau scenarios for educational attainment as more plausible—either the scenario where educational attainment for entering cohorts reverts to its 2007 levels, or where it persists at its 2013 levels. The CPS data suggest that some of the Great Recession-induced increase in educational attainment of younger cohorts may already be reversing.[39] The "revert" and "persist" rows of table 2.3 are very similar for both labor quality and hours. Relative to the revert or persist scenarios—which are very similar—we take the

---

39. Additional evidence for a reversal comes from census data on college enrollments relative to the population age sixteen to twenty-four. That enrollment rate peaked in 2011 and has since retreated somewhat.

Table 2.3   Labor-quality growth projections, 2022–2025

| | EPOP ratio | | |
| | Trend | Level | |
Education	Extrapolated 2002–2007 trends in EPOP (I)	Revert to precrisis EPOPs (II)	Persist at 2013 EPOPs (III)
1. Extrapolate 2007–2013 trends in education	0.47/0.60	0.28/0.37	0.27/0.41
2. Revert to precrisis educational attainment	0.30/0.59	0.09/0.42	0.09/0.42
3. Persist at 2013 educational attainment	0.32/0.60	0.12/0.41	0.12/0.42

Notes: Reported are average annual log-growth rates for 2022–2025.
Row scenarios: (1) Uptick in educational attainment since Great Recession reflects a permanent acceleration in educational attainment of those age thirty and younger. (2) Uptick since 2008 is fully cyclical and educational attainment reverts to its 2008 level. (3) Uptick in educational attainment since Great Recession reflects a step increase in educational attainment of those age thirty and younger.
Column scenarios: (I) Pre–Great Recession (2002–2007) trends in EPOP continue. (II) EPOP ratios in 2025 have reverted to their 2007 levels. (III) EPOP ratios will remain at their 2013 levels. More details about these scenarios can be found in the third section of the appendix.

predictions from the educational-extrapolation scenario as an upside risk for labor quality.

Finally, we consider the importance of employment rates for longer-run projections of labor quality. In the longer run, only *trends* in EPOP rates matter. Indeed, the two "level" columns look very similar to each other, showing that in the longer run it makes little difference whether we revert to precrisis EPOPs or remain at 2013 EPOPs.

2.5.4   Putting It All Together

The previous section highlighted the uncertainties around any forecast of labor-quality growth both in the medium and longer run. Here we provide a judgmental assessment of the most likely path for labor quality in the longer run. Looking at the bottom right two cells of table 2.3, where education plateaus and EPOPs remain level, we project labor-quality growth of about 0.1 percent per year and hours growth of a little above 0.4 percent per year. Quality-adjusted hours in these scenarios grows a little above 0.5 percent per year.

Although these "level" scenarios are a reasonable benchmark for the future, continuing shifts in EPOPs also seem plausible. Earlier, we found that these shifts were central to driving labor-quality growth from 2002 to 2013. This was also the case for the 2002–2007 period, before the employment effects of the Great Recession. Going forward, there is certainly the potential for technological advances to continue to generate job polarization, to displace low- and medium-skilled workers, and/or to entice high-skilled workers to increase their labor supply. If these trends were all to continue at their 2002–2007 pace, then it would lead to some longer-run boost in labor quality, though the effect on hours is ambiguous.

One particular unknown in this regard is whether older, more educated workers will continue to work longer than they have historically. For example, suppose we extrapolate EPOP trends only for those older than fifty-five years of age—a situation that would boost both labor quality and hours. With that limited extrapolation, we would see hours growth of about 0.55 percent and labor-quality growth of about 0.15 percent, implying quality-adjusted hours growth of about 0.70 percent per year. In the 2022–2025 period, these figures are not affected by whether other employment rates revert to precrisis levels or remain at 2013 levels.

Trends for individuals younger than age fifty-five are more nuanced and challenging to predict. In the extrapolation scenarios, educated prime-age workers tend to work more, while less educated prime-age workers tend to work less. We think it is unlikely that the trends continue at the earlier pace captured by the extrapolation column in table 2.3, but, qualitatively, the trends might continue in the same direction. That would suggest that it is plausible labor quality grows a little faster than 0.15 percent per year (the pace in the previous paragraph, where we extrapolate EPOP trends only for those older than age fifty-five). The effect on hours would be small, since the trends somewhat offset.

Thus, in the longer run, a projection of 0.10–0.25 percent growth in labor quality and perhaps 0.4 to 0.55 for hours is a plausible judgmental baseline.

2.6   Conclusion

Historically, rising labor quality was an important source of growth in US GDP per hour. Going forward, this source of growth is likely to slow markedly. Indeed, our preferred forecast is that, in the longer run (2022–2025), labor quality is likely to rise in the range of 0.10 to 0.25 percent per year. This implies that growth in quality-adjusted hours in the range of 0.5 to 0.8 percent per year is plausible, with a range of 0.7 to 0.8 percent per year seeming perhaps most likely. To see a faster pace of labor-quality growth, closer to its historical average pace, would require a renewed, and sustained, upward trend in educational attainment. In a typical macro model, the slow-down in labor-quality growth passes through one-for-one to slower growth in productivity and GDP.

In the twentieth century, the main driver of labor-quality growth was rising educational attainment (Fernald and Jones 2014; Ho and Jorgenson 1999). In contrast, in our empirical estimates and forecasts for the twenty-first century, we find a very different source of labor-quality growth: the diverging trends in employment rates for workers of different skills. Since 2002, employment rates for more educated, older individuals have risen, whereas employment rates for less educated, younger individuals have fallen.

These diverging trends explain why previous forecasts that labor quality would plateau (Aaronson and Sullivan 2001) went awry—from 2002 to 2013, labor quality turned out to grow at a pace even faster than it did in the second half of the twentieth century because of changing employment dynamics.

These forecast misses point to a broader lesson: it is essential to jointly examine growth in hours and growth in labor quality. Labor quality and hours are strongly negatively correlated in the short run, which implies that quality-adjusted hours are less variable than either quality or hours alone. Looking at hours or labor-quality growth independently can lead to inaccurate projections of potential output growth.

Going forward, movements in employment-to-population rates for different worker types continue to be central to how future labor quality will evolve. In the medium run (2015–2022), an important source of uncertainty is whether the diverging employment-rate movements seen since 2007 are cyclical or structural. If employment rates (based on age and education) revert to 2007 levels, then growth in labor quality is likely to be negative as lower-skilled workers return to employment. In this case, labor quality in the next few years will at least partially offset the strong growth since 2007. In contrast, if the changes since 2007 are structural, then growth in labor quality will be considerably stronger, albeit not at rates seen historically.

But, once again, these alternative paths illustrate the importance of jointly modeling labor quality and hours. Quality-adjusted labor input turns out to grow at remarkably similar rates in the scenarios where employment-to-population rates revert to 2007 values (cyclical), or remain at 2013 values (structural), leaving overall output growth unchanged.

Appendix
Data Details

ACS, CPS-ASEC, and CPS-ORG

To verify the robustness of our results, we calculate them for three commonly used US data sets that each allow for the construction of measures of labor-quality growth. The first is the American Community Survey (ACS), which is a smaller, annual version of the decennial census and collects a relatively narrow range of demographic and socioeconomic data on a sample of about 1 percent of the population (approximately three million individuals) each year.[40] The second, the CPS-ORG, consists of the outgoing rotation groups from the Current Population Survey (CPS). This is the quarter of

---

40. The sample of the ACS has been expanded twice and has only been a 1 percent sample of the population since 2006. In 2000, its first year, the sample was just under 400,000 individuals, and between 2001 and 2005 the sample was slightly over one million.

CPS respondents that are asked about their earnings and income in any given month. This results in an annual sample of about 135,000 individuals. The final data set, CPS-ASEC, is the Annual Social and Economic Supplement to the Current Population Survey, also known as the March supplement. It contains annual earnings and income data from the full March CPS sample (70,000 individuals). Though based on different samples and sampling methods, each of the data sets allows for the construction of similar hourly wages, as well as the six variables of education, age, sex, race/ethnicity, industry, and occupation, that are our main focus.

For each data set we construct the *sample* of workers to cover those in the civilian noninstitutional population ages sixteen and older that are employed in the private business sector (specifically, excluding anyone with self-employment or government employment earnings) and have both positive earnings and positive hours. The sample period is 2002–2013, because that is the period for which we have a consistent set of occupation and industry crosswalks and data from all three data sets.[41]

We define *wages* as hourly wages. Wages are constructed in slightly different ways in each of the data sets because of differences in reference period and questions asked. In the CPS-ORG, we use the hourly wage as constructed in the National Bureau of Economic Research's CPS Labor Extracts (Feenberg and Roth 2007). For the CPS-ASEC and ACS we define hourly wages as total annual earnings divided by the product of usual hours worked per week and weeks worked per year.[42] All wages are deflated into real 2005 dollars using the Consumer Price Index for All Urban Consumers, and wages exclude self-employment, self-owned business, and farm income.

Projections of Educational Attainment and Employment

The Census Bureau's 2000 National Population Projections provides projections of the age, gender, and race/ethnicity distribution of the population, but to forecast labor quality we need to further break these cells down by educational attainment and employment rates. To do so we follow a methodology similar to that used by Aaronson and Sullivan (2001)—our primary

41. In principle, the CPS-ASEC is available starting in 1962 onward and the CPS-ORG from 1979 onward if industry and occupation are omitted or approximate crosswalks are used. The ACS is available from 2000 onward without any need for adjustments.

42. In 2008 the ACS switched from collecting weeks worked as a continuous to a categorical value (thirteen weeks or less, fourteen to twenty-six weeks, twenty-seven to thirty-nine weeks, forty to forty-seven weeks, forty-eight or forty-nine weeks, and fifty to fifty-two weeks). Prior to 2008, the distribution of weeks worked within those ranges was remarkably stable over time, so we imputed a continuous value of weeks worked using the pre-2008 mean of people reporting weeks worked within a given range. We also tested using a more complex regression model on demographic characteristics to impute weeks worked, but found that it gave little more variation or precision in predicted weeks worked than using the pre-2008 mean. The same approach is used by the BLS for pre-1975 data, which has the same issue (Bureau of Labor Statistics 1993, 77).

adjustments are that we use five race/ethnicity categories instead of four and we define employment more narrowly as being employed exclusively in the private business sector to match the sample selection stated in the previous section. Given that the methodology is substantively unchanged, this section is largely a restatement of box 1 from Aaronson and Sullivan (2001, 65).

Let $p_{it}^j = P[y_{it} = j]$ for $j = 1, \ldots, 5$ by the probability that individual i in year t has educational attainment j, where the five levels of attainment are less than high school, high school graduate (including GEDs), some college (including associate's degree holders), college graduates (bachelor's), and postgraduates, and let $q_{it}^j = P[y_{it} \geq j \mid y_{it} \geq j - 1]$ for $j = 2, \ldots, 5$ be the probability of attaining education j given that the individual has completed the "prerequisite" education (e.g., for $j = 4$ this is the probability of an individual having completed college given that they have completed some college). We predict $\hat{q}_{it}^j$ using a logistic regression of the form

(2A.1) $$\log \frac{q_{it}^j}{1 - q_{it}^j} = \sum_a D_{it}^a \alpha_{ja} + \sum_b D_{it}^b \beta_{jb} + \mathbf{x}_{it} \gamma_j,$$

(2A.2) $$\text{and } \hat{q}_{ab}^j = \frac{\exp(\alpha_{ja} + \beta_{jb})}{1 + \exp(\alpha_{ja} + \beta_{jb})}$$

where D_{it}^a and D_{it}^b are dummies for being age a and born in year b, and $\mathbf{x}_{it}$ is a vector of control variables. From $\hat{q}_{ab}^j$ it is possible to calculate $\hat{p}_{ab}^j = \prod_{k=2}^j \hat{q}_{ab}^k (1 - \hat{q}_{ab}^{j+1})$, which can be interpreted as the predicted share of people born in year b with education j at age a or, since age, year, and birth year are perfectly collinear, the predicted share of people of age a with education j in year $b + a$. The models for education level j are estimated on the sample of people with at least $j - 1$ education and who are above an education-level-specific age threshold.[43] For the projections for the forecast error decomposition exercises in section 2.5.2 the models are estimated on the CPS-ORGs from 1992 through 1999, the same period Aaronson and Sullivan used for their forecasts.[44]

The idea behind these models is that educational attainment follows some sort of life-cycle pattern, with the probability of completing a certain level of education increasing rapidly for people younger than age thirty and then more gradually for those who are older. This life-cycle pattern is assumed to be the same for different cohorts, but cohorts born in different years are allowed to have uniformly higher or lower log odds of completing a given level of education. For high school, some college, and college levels of educa-

---

43. The thresholds are eighteen for high school, nineteen for some college, twenty-two for college, and twenty-six for postgraduate.

44. Ideally, this would have been estimated on ACS data to ensure consistency between these projection models and the log-wage regression. However, in order to distinguish age and cohort effects the projection model must be estimated on multiple years of data. Since there is no pre-2000 data for the ACS, this forces us to rely on another data set to construct the education and employment projections.

tion the model is estimated separately for each of ten gender-race-ethnicity combinations without any control variables (x_{it}). For postgraduates some of the gender-race-ethnicity samples become quite small, so the model is estimated separately for men and women with race/ethnicity dummies included as controls. The estimated model is then used to predict the fraction of individuals with each level of educational attainment based on the Census Bureau projections of the age, gender, and race/ethnicity distribution of the population.

The projection model is only able to estimate birth-year coefficients (β_{jb}) for birth years that are observed in the sample. However, some birth years that are too young to be observed in the sample will be old enough to be in the sample by later years of the projections—a child born in 2000 is too young to be in any of our current samples, but by 2025 they will be twenty-five years old and of critical importance to our forecasts. Therefore, we define these unobserved cohort coefficients by a linear extrapolation using the last fifteen birth-year coefficients (not including the most recent).[45] In effect, this approach extrapolates recent trends in educational attainment into the future.

This process yields projections of the population distribution of age and educational attainment, the key variables for our baseline Mincer specification. However, to construct our forecast of labor quality we must also project the EPOP rates for these worker types. Our EPOP projection model is identical to the educational attainment projection model, except educational attainment is added as a control variable. Rather than using the standard BLS definition of employment, we define employment as being employed exclusively in the private business sector—this makes our definition consistent with the sample selection used to construct our labor-quality measures.

Projection Scenarios for Educational Attainment and Employment

The Fisher Ideal index does not have the circularity property, so the labor-quality growth calculated from comparing a target year to a base year is not necessarily the same as the growth calculated from cumulated year-over-year changes. However, this is not true for the labor-quality growth projections because our assumption that the log-wage regression coefficients are constant over time means that the Fisher Ideal index collapses into the Laspeyres index, which does have the circularity property. This allows us to construct alternative projection scenarios based on assumptions about the education and employment distribution in a target year alone, without having to make assumptions about the path of educational attainment or EPOP between now and then. Therefore our projection scenarios discussed in sec-

---

45. The most recent coefficient is omitted because it is based on just one year of observations, making the sample size quite small.

tion 2.5 are based on the Census Bureau age projections for the years 2022 and 2025 and the education and employment assumptions described below.

Baseline labor quality in 2015 is calculated by applying the empirical 2013 education and employment distributions by age from the ACS to the Census Bureau population projections for 2015. That is, we calculate the share of twenty-five-year-olds that have a college degree, the share of twenty-five-year-olds with college degrees that are employed, and then combine that with the census projection of the number of twenty-five-year olds in 2015 to estimate the number of college-educated, twenty-five-year-old workers in 2015. This same baseline distribution is used in all nine labor-quality projections.[46]

Education Scenarios

All three education scenarios assume that the educational distribution for those older than age thirty will stay the same as they age. For example, the educational attainment of fifty-two-year-olds in 2025 is assumed to be the same as that of forty-year-olds in 2013 (the most recent year in our data). Although nontraditional educational attainment, differential mortality rates, and immigration make it unlikely that this assumption strictly holds, those forces are marginal enough that they are unlikely to cause substantial deviations. Where the scenarios differ is in their assumptions on the educational attainment of (a) people age thirty and younger in the projection year (the "young group"), and (b) the educational attainment of people younger than age thirty in 2013 that will be older than thirty in the projection year (e.g., thirty-one- to forty-two-year-olds in 2025, the "middle group"). The educational attainment of the *young group*, which was in middle school or below during the Great Recession and thus unlikely to have been driven by cyclical factors—their educational attainment can be thought of as representing a "normal" level. Unlike the young group, those in the middle group were making critical education decisions (such as whether to drop out of high school or college and whether to enroll in college or graduate school) during the Great Recession and its aftermath. Therefore, if "educational sheltering" has been a strong force during and after the Great Recession, as posited by Barrow and Davis (2012), Sherk (2013), and Johnson (2013), then their attainment may deviate from the norm.

Revert to precrisis educational plateau. The first education scenario assumes that the educational attainment of young people reverts to its precrisis levels. This reflects the possibility that the uptick in enrollment and graduation rates over the past several years is simply a temporary cyclical effect of the Great Recession. For the young group, this scenario assumes they will have the same distribution of educational attainment as people of

---

46. Note that the differences between the growth rates in the different scenarios is completely independent of the baseline, since we report log growth.

the same age in 2007.[47] For those in the middle age group, whose attainment may have been increased by "educational sheltering" effects, this scenario assumes that they will either have the educational attainment of someone that age in 2007 or their current educational attainment, whichever is higher. That is, they will have at least the educational attainment that would have been expected of them before the recession, and they may have a little more if the recession encouraged them to stay in school. Specifically, let $\hat{q}_a^j$ be the probability of someone with age a having at least education j in 2007, let $\tilde{q}_{a-12}^j$ be the probability of someone that will be age a in 2025 having at least education j in 2013, and let $q_a^j = \max(\hat{q}_a^j, \tilde{q}_{a-12}^j)$. Then for this scenario the share of people of age $a = 31, \ldots, 42$ with education j will be $p_a^j = q_a^j - q_a^{j+1}$. This is the same for 2022, except using $\tilde{q}_{a-9}^j$.

Persist at 2013 educational plateau. The second scenario assumes that the educational attainment of young people persists at its 2013 rate, reflecting the possibility that there was a step increase in educational attainment over the past several years, but that attainment has once again reached a plateau. This scenario assumes that people in the young group will have the same distribution of educational attainment as someone of the same age in 2013. For the middle group we have to account for the fact that the increase in educational attainment was gradual and had not fully propagated through for those older than age thirty, but people younger than thirty will often go on to further education, meaning that there is no clear baseline group. To get a baseline for this group we calculate the probability q^j in 2013 of completing at least education j for the five-year age group that are young enough to have experienced a sheltering effect, but old enough that we would expect them to have completed that level of education already.[48] For this scenario we define the expected educational attainment distribution of the middle group as $p^j = q^j - q^{j+1}$.

Extrapolate 2007–2013 trends in education. The final scenario assumes that the uptick in educational attainment over the past several years represents a resumed upward trend in education attainment rather than a temporary cyclical boost or a one-off step increase. Age-specific time trends in educational attainment are estimated from logistic regressions of the form

(2A.3) $$\log \frac{q_{it}^j}{1 - q_{it}^j} = \sum_a [\text{year} \cdot D_{it}^a \beta_a + D_{it}^a \gamma_a].$$

47. Recent research suggests that the housing boom depressed educational attainment by providing good job opportunities to low-skilled workers, in which case the educational attainment patterns from the boom years would be unusually low (Charles, Hurst, and Notowidigdo 2015). That would suggest that this may be a particularly pessimistic implementation of this "cyclical uptick" hypothesis. However, we believe this is still a useful scenario to consider as it provides a plausible worst-case scenario for education trends.

48. For high school we use nineteen to twenty-three, for some college we use twenty-three to twenty-seven, for college we use twenty-five to twenty-nine, and for postgraduate we use thirty to thirty-four. Less than high school is the residual category.

As in the second section, these logits are estimated on the population of people with education $j - 1$ or higher, and they are estimated on 2007–2013 data. Let q_a^{2013} be the probability that a person of age a had education j or higher in 2013. Then this scenario assumes that the probability of having at least education j at age a in 2025 is the probability of having education j in 2013 plus the age-specific time trend—that is, they have probability $q_a^j =$ invlogit[logit(q_a^{2013}) + 12 · β_a] of having at least education j at age a in 2025. As in the other cases, we then recover the share of people with education j at age a in 2025 as $p_a^j = q_a^j - q_a^{j+1}$. This is the same for 2022 except $q_a^j =$ invlogit[logit(q_a^{2013}) + 9 · β_a].

Employment Scenarios

The employment scenarios are much more straightforward to construct because there is little to no need to keep track of the stock of employment—the fact that 85 percent of twenty-nine-year-old college graduates were employed in 2013 does not impose particularly binding constraints on our assumptions about the EPOP rate of forty-one-year-old college graduates in 2025. Therefore, our two baseline employment scenarios simply assume that the EPOP rates for specific age-education groups in the projection year will be the same as in some other base year. For the *revert to precrisis EPOPs* scenario, we assume that the probability of a person of age a with education j being employed in the projection year is the same as it would have been in 2007.[49] This scenario corresponds to the view that the entire decline in EPOP rates for specific age-education groups is cyclical.[50] The second employment scenario is the inverse of this and assumes that the entire change in the EPOP rates of specific age-education groups is structural and will *persist at 2013 EPOPs*.[51]

Extrapolated 2002–2007 structural trends in EPOPs. The final scenario extrapolates certain precrisis trends in employment patterns out to the projection year. In particular, it extrapolates the declining EPOP rates of young people (with heterogeneity across education groups), the increasing EPOP rates of older people (particularly the more educated), and the widening gap between the EPOP rates of more and less educated working-age people

49. As with the first education scenario, this may be an extreme assumption on what the precrisis norm was—if the housing boom boosted EPOP rates to abnormal levels, then this scenario overstates the baseline EPOP rates. Similar to the education case, we believe this remains a useful scenario to consider as it illustrates a sort of best-case scenario for employment rates.

50. This still allows for a demographically (or educationally) driven structural decline in the aggregate employment-to-population ratio. What it does not allow for is a structural decline (or increase) in EPOP for specific age-education groups. For example, it does not allow for a structural decline in students working part time, or a structural increase in older people staying employed past the traditional retirement age.

51. This scenario may be too pessimistic in that the labor market has clearly continued to improve since 2013. Once more recent ACS data becomes available we will revise this scenario to reflect the most recent year of data available. However, this once again provides a sort of outlier case with unusually low EPOP rates.

(Dennett and Modestino 2013; Burtless 2013; Aaronson et al. 2014). Given that we have preselected the trends that are extrapolated, this scenario can be accused of cherry picking. We do not deny that vulnerability, and we do not intend this scenario to be understood as a probable outcome. Again, these scenarios are primarily intended to illustrate the mechanics of labor-quality growth and what factors are most critical to the setting expectations about future labor-quality growth in the United States, as well as to impose certain bounds on plausible forecasts of labor-quality growth.

To implement the third employment scenario we follow an approach similar to that in the education trends scenario above. To extract age and education-specific time trends in employment we run the following logistic regression on the sample of sixteen- to twenty-four- and fifty-five- to sixty-nine-year-olds over the 2002–2007 period

$$(2A.4) \qquad \log\frac{p_{it}}{1 - p_{it}} = \sum_a \sum_j [\text{year} \cdot D_{it}^a \cdot D_{it}^j \beta_{aj} + D_{it}^a \cdot D_{it}^j \gamma_{aj}],$$

and to extract education-specific time trends in employment among prime-age workers, we run the following logistic regression on twenty-five- to fifty-four-year-olds over the same period

$$(2A.5) \qquad \log\frac{p_{it}}{1 - p_{it}} = \sum_j [\text{year} \cdot D_{it}^j \beta_j + D_{it}^j \gamma_j],$$

where p_{it} is the probability of individual i being employed in year t, D_{it}^a is an indicator for being age a, and D_{it}^j is an indicator for having education j. Let p_{aj}^{2007} be the probability that a person of age a with education j was employed in 2007. Then this employment trends scenario assumes that the probability of a person of age a and education j being employed in 2025 is the probability of being employed in 2007 plus the relevant age- and education-specific time trend. For sixteen- to twenty-four- and fifty-five- to sixty-nine-year-olds this is $p_{aj} = \text{invlogit}[\text{logit}(p_{aj}^{2007}) + 18 \cdot \beta_{aj}]$, and for twenty-five- to fifty-four-year-olds it is $p_{aj} = \text{invlogit}[\text{logit}(p_{aj}^{2007}) + 18 \cdot \beta_j]$.[52] The 2022 projection is the same, except the βs are multiplied by 15 instead.

Robustness Checks

In this section of the appendix we present additional results that illustrate that our qualitative results are unchanged when we change some of the underlying assumptions, specifications, and across data sets.

Adding Control Variables to the Baseline Mincer Regression

Throughout the main text we limited ourselves to parsimonious baseline Mincer regression. However, prior implementations of such specifications

---

52. For people age seventy and older there is no time trend added in and their EPOP rate in the projection year is assumed to be the same as it was in 2007.

have included control variables to ensure that only productivity-induced wage differentials are reflected in the estimated wages (Aaronson and Sullivan 2001; Bureau of Labor Statistics 1993). Here we consider the robustness of our results to including standard control variables, such as part-time status, marital status, veteran status, race, and geographic location.

As discussed in subsection 2.3.2, it is critical that the variables included in the labor-quality specification (x_j) be (a) correlated with wages, and (b) that the correlation is driven by differentials in the marginal product of labor. A desirable property of a regression-based framework like equation (9) is that it allows for the inclusion of control variables, z_j, that may be correlated with both individual wages, w_j, and the variables meant to quantify marginal product differentials, x_j. The resulting generalized regression framework is

(2A.6) $w_j = x_j'\beta + z_j'\gamma + \varepsilon_j.$

Because we attribute only the part of wage variations explained by the variables in x_j to marginal product differentials, we impute the log marginal product of a worker as $x_j'\hat{\beta}$. The inclusion of these control variables does not alter our definitions of σ_j and $\tilde{\sigma}_p$. They continue to be based on x_j and $\hat{\beta}$.

What is less clear is the appropriate measure of fit when considering a regression with controls. Consider, for example, a set of controls z that predict wages ($\gamma \neq 0$), but for which the correlation between any element x of x and any element z of z is zero (corr(x, z) = 0). In this case the regression $\bar{R}^2$ will increase, making the specification appear more appealing than the version without z despite the fact that substantive components of the regression, x and $\hat{\beta}$, remain unchanged.[53] An alternative approach would be to consider the partial R-squared with respect to x, $\bar{r}_x^2$. However, then maximizing $\bar{r}_x^2$ is not necessarily desirable. For example, if the association between a control variable z and the core variables x has the same sign as the association between z and wages w, then the $\bar{r}_x^2$ will decline in the regression with z. But the $\bar{r}_x^2$ declined precisely because z had been a source of omitted variable bias and we are now controlling for that.

Ultimately, the selection of z operates on an orthogonal basis from the selection of x in a *properly controlled regression*. As discussed in subsection 2.3.2, the desirability of higher $\bar{R}^2$ is entirely conditional on the assumption that $\hat{W}_j \equiv \exp(x_j'\beta) = c \cdot W$—if any omitted variable bias is loaded onto β then this assumption is violated. In principle, this means that one should optimize z for each separate specification of x, at which point we can compare the $\bar{r}_x^2$ of the controlled regressions as is done in subsection 2.3.2.

Rather than undertaking this highly multidimensional and daunting task, we consider whether it is likely to be of first-order importance to any of our results. Specifically, we consider the impact of including two standard sets of controls in our baseline Mincer and baseline + occupation specifications. The first set of controls is a set of indicators for part-time

---

53. The standard errors will also slightly increase because of the loss of degrees of freedom.

employment, marriage, and race, which are the controls included in the specification used by Aaronson and Sullivan (2001).[54] The second set of controls is similar to the Bureau of Labor Statistics (1993), and includes indicators for part-time employment, veteran status, and which census division the individual lives in.[55]

Figure 2A.1, panel A, which is comparable to figure 2.2, plots the adjusted R-squared ($\bar{R}^2$) against the 80th percentile standard error of the predictions ($\tilde{\sigma}_{80}$). This shows that, as expected, the inclusion of the additional variables increases both $\bar{R}^2$ and $\tilde{\sigma}_{80}$. The Aaronson and Sullivan (2001) controls improve the fit slightly more and increase imprecision slightly less than the Bureau of Labor Statistics (1993) controls. However, as can be seen in figure 2A.1, panel B, there is almost no change in the partial R-squared with respect to $\mathbf{x}$, suggesting that either the control variables are not a significant source of omitted variable bias or that the biases they induce balance out, on average. This suggests that the impact of including these control variables on measured labor-quality growth is likely to be quite limited. This is confirmed in figure 2A.1, panel C, which plots the resulting labor-quality indices. The indices with the Bureau of Labor Statistics (1993) controls are virtually indistinguishable from their uncontrolled counterparts, while the Aaronson and Sullivan (2001) controls appear to exert a modest negative drag on labor quality, on the order of a couple hundredths of a percentage point per year. These results suggest to us that control variables are not of first-order importance in measuring or forecasting labor-quality growth.

Additional Results for CPS-ORG and CPS-ASEC

The majority of the results presented in the main text were produced using data from the ACS, but it is possible to conduct the same exercises using both the CPS-ORG and CPS-ASEC. In this section, we evaluate the robustness of key results from the main text in these alternative data sources. All of the qualitative results hold up, with some minor differences in magnitude.

Figure 2A.2, panels A and B, plot the adjusted R-squared ($\bar{R}^2$) against the 80th percentile standard error ($\tilde{\sigma}_{80}$) of the same specifications considered in section 2.3 and figure 2.2 for the CPS-ORG and CPS-ASEC, respectively. As we note in the main text, the large sample size of the ACS is relatively favorable to stratum-based specifications: with the CPS data sets, which are more than an order of magnitude smaller, the standard errors are an order of magnitude higher. In fact, the CPS-ASEC is small enough that for some of the more granular specifications, more than 20 percent of the observations are in single-observation cells with infinite standard errors,

54. We use five race/ethnicity indicators where they used four race indicators—we distinguish Hispanics from non-Hispanic whites, blacks, Asians, and other.
55. The Bureau of Labor Statistics (1993) specification also includes indicators for whether the individual is in a central city or balance of a standard metropolitan statistical area (SMSA)/core-based statistical area (CBSA) or in a rural area, which we omit.

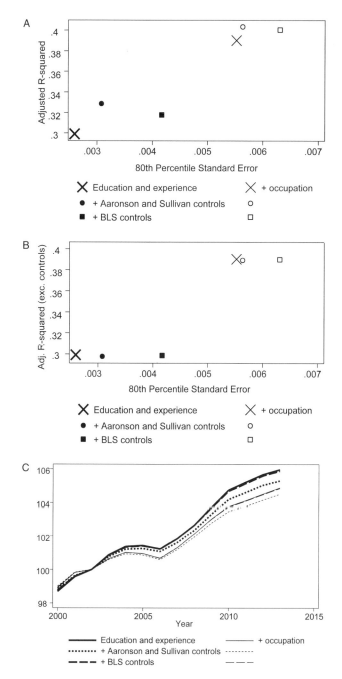

Fig. 2A.1 **Impact of including controls in Mincer specifications.** *A,* fit of both core and control variables; *B,* fit of the core variables only; *C,* labor-quality indices, with and without controls.

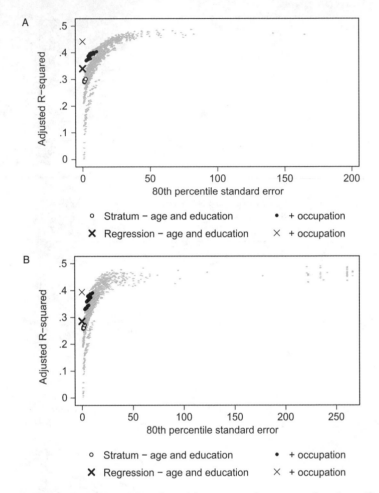

Fig. 2A.2   Regression-based fit and precision compared to stratum-based specifications (CPS data sets). *A*, **CPS-ORG;** *B*, **CPS-ASEC.**

leaving $\tilde{\sigma}_{80}$ undefined.[56] However, the trade-off between fit and precision is still clearly visible, the age and education or age, education, and occupation specifications strike a reasonable balance between fit and precision, and the baseline and baseline + occupation Mincer specifications dominate the stratum-based specifications. In short, the results are entirely consistent with our findings from the ACS.

Figure 2A.3, panels A and B, plot the 2002–2013 labor-quality indices presented in section 2.4 and figure 2.3 for the CPS-ORG and CPS-ASEC, respectively. Once again the results are quite similar to those found in the

56. We substitute the highest observed percentile standard error, which is the source of the vertical lines in the upper-right region of the figure.

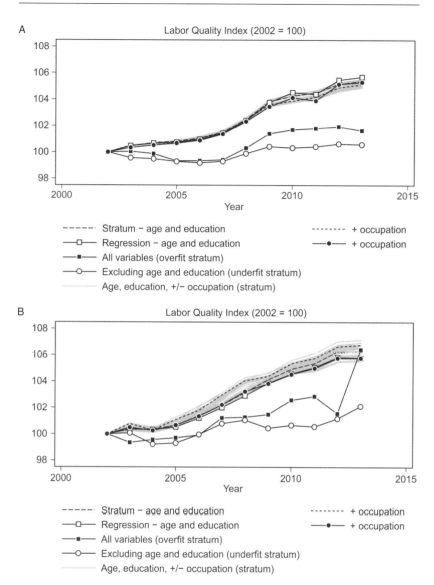

Fig. 2A.3   **Comparison of results across specifications, 2002–2013 (CPS data sets).** *A*, **CPS-ORG;** *B*, **CPS-ASEC.**

ACS. The overfit specification (which includes all six variables considered in section 2.3) and the underfit specification (which includes all variables except age and education) both show very little labor-quality growth over the first decade of the twenty-first century. In the case of the CPS-ASEC, the overfit and underfit specifications are quite noisy, with implausible jumps and changes in direction.

All of the age and education specifications (with or without occupation) and the baseline and baseline + occupation Mincer specifications, by contrast, are clustered together and quite similar to the ACS results in figure 2.3, panel A, although the CPS-ORG specifications show about 0.5 percent less cumulative labor-quality growth by 2013. The CPS-ORG results are also slightly more closely clustered than those for the other two data sets. This may be because hourly wages are measured directly in the CPS-ORG, whereas in the CPS-ASEC and ACS hourly wages are noisily derived from annual earnings divided by the product of usual weekly hours and weeks worked per year.

One notable difference is that ACS indices show an unexpected decline in labor quality between 2005 and 2006, while the CPS-based indices do not. This appears to be a data artifact induced by the tripling of the ACS sample size in 2006. A similar jump occurs when we calculate labor-quality growth between 2000 and 2001 in the ACS (not reported), and there also appears to be a slight tick in the 2012–2013 period for the ASEC, which saw a sample size change in 2013. Why changing sample size can induce these sharp adjustments in labor quality is somewhat unclear and bears more careful investigation.

Figure 2A.4, panels A and B, plot the 2002–2013 counterfactual labor-quality indices presented in section 2.4 and figure 2.4 for the CPS-ORG and CPS-ASEC, respectively. The results are qualitatively the same in the CPS data sets as in the ACS, with changes in average hours worked contributing relatively little to labor-quality growth, while changes in population demographics and demographic-specific EPOP rates both contributing significantly. However, there are two quantitative differences.

First, average hours appear to matter less in the CPS data sets. This is likely due to the fact that the ACS uses a categorical measure weeks worked after 2008, which induces additional noise in the measurement of average hours relative to the other two data sets. This is consistent with the fact that hours only make a significant difference after 2008. The relative unimportance of hours further strengthens our conviction that projecting average hours is not critical to a labor-quality forecast and that attempts to do so are likely to introduce as much forecast error as they address.

Second, whereas for the ACS the evolution of EPOP rates induced more labor-quality growth than changing demographics (compare the thick dashed and thick solid lines), in the two CPS data sets the contributions of employment and demographics are almost equal. Additionally, the contribution of EPOP rates, reflected in the thick dashed lines, is more obviously cyclical for the two CPS data sets—it is virtually flat before and after the Great Recession, with a substantial step increase during the Great Recession. The ACS, by contrast, shows significant labor-quality growth from EPOP rates even before the Great Recession, with the Great Recession simply accelerating the trend.

These observations have important implications for which of the sce-

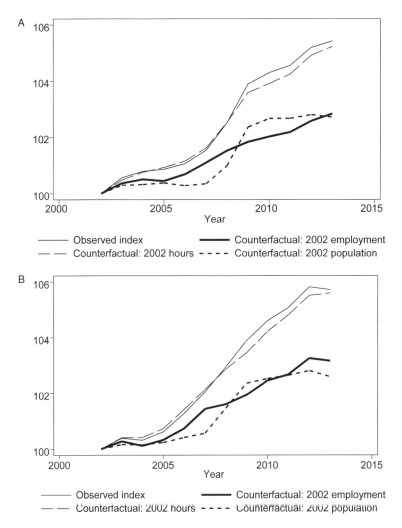

Fig. 2A.4   Counterfactual indices for 2002 base-year hours, employment, and population (CPS data sets). *A*, CPS-ORG; *B*, CPS-ASEC.
Note: Betas from 2002 wage regression.

narios presented in section 2.5 one finds most compelling. If one believes the CPS data sets more accurately reflect the role of the employment margin in driving labor-quality growth, then the two plateau scenarios appear most compelling: they suggest that the United States experienced an unusual upskilling of employment during the Great Recession that will either persist or unwind, while offering little evidence of a pre–Great Recession upskilling trend in employment. If, on the other hand, one believes that the ACS data more accurately reflects the contribution of employment composition to labor quality, then there appears to have been a significant pre–Great Recession structural trend, suggesting that the labor-quality growth from employment composition is unlikely to fully unwind and may even continue to drive a significant portion of labor-quality growth going forward.

References

Aaronson, Daniel, and Daniel Sullivan. 2001. "Growth in Worker Quality." *Economic Perspectives* 25 (4). Federal Reserve Bank of Chicago. https://www.chicagofed.org/publications/economic-perspectives/2001/4qepart5.

Aaronson, Stephanie, Tomaz Cajner, Bruce C. Fallick, Felix Galbis-Reig, Christopher Smith, and William L. Wascher. 2014. "Labor Force Participation: Recent Developments and Future Prospects." Finance and Economics Discussion Series no. 2014-64, Board of Governors of the Federal Reserve System.

Altonji, Joseph G., and Rebecca M. Blank. 1999. "Race and Gender in the Labor Market." In *Handbook of Labor Economics*, edited by O. Ashenfelter and D. Card, 3143–259. Amsterdam: Elsevier.

Barrow, Lisa, and Jonathan Davis. 2012. "The Upside of Down: Postsecondary Enrollment in the Great Recession." Research Report no. 36, Federal Reserve Bank of Chicago.

Bertrand, Marianne, and Sendhil Mullainathan. 2003. "Are Emily and Greg More Employable than Lakisha and Jamal? A Field Experiment on Labor Market Discrimination." NBER Working Paper no. 9873, Cambridge, MA.

Bils, Mark J. 1985. "Real Wages over the Business Cycle: Evidence from Panel Data." *Journal of Political Economy* 93 (4): 666–89.

Blaug, Mark. 1985. "Where Are We Now in the Economics of Education?" *Economics of Education Review* 4 (1): 17–28.

Boeri, Tito, and Jan van Ours. 2013. *The Economics of Imperfect Labor Markets*, 2nd ed., vol. 1 of Economics Books. Princeton, NJ: Princeton University Press.

Bureau of Labor Statistics. 1993. *Labor Composition and US Productivity Growth: 1948–90.* BLS Bulletin no. 2426, Washington, DC, US Department of Labor. https://www.bls.gov/mfp/labor_composition.pdf.

———. 2015a. "Changes in the Composition of Labor for BLS Multifactor Productivity." Washington, DC, US Department of Labor. http://www.bls.gov/mfp/mprlabor.pdf.

———. 2015b. "Multifactor Productivity." Washington, DC, US Department of Labor. http://www.bls.gov/mfp/.

Burtless, Gary. 2013. "Can Educational Attainment Explain the Rise in Labor Force Participation at Older Ages?" Technical Report no. 13-13, Center for Retirement Research at Boston College.

Charles, Kerwin Kofi, Erik Hurst, and Matthew J. Notowidigdo. 2015. "Housing Booms and Busts, Labor Market Opportunities, and College Attendance." NBER Working Paper no. 21587, Cambridge, MA.

Congressional Budget Office. 2015. "An Update to the Budget and Economic Outlook: 2015 to 2025." CBO Report no. 49892, Washington, DC. https://www.cbo.gov/publication/49892.

Deming, David J. 2015. "The Growing Importance of Social Skills in the Labor Market." NBER Working Paper no. 21473, Cambridge, MA.

Dennett, Julia, and Alicia Sasser Modestino. 2013. "Uncertain Futures? Youth Attachment to the Labor Market in the United States and New England." Research Report no. 13-3, Federal Reserve Bank of Boston.

Dickens, William, and Lawrence F. Katz. 1987. "Inter-Industry Wage Differences and Industry Characteristics." In *Unemployment and the Structure of Labor Markets*, edited by K. Lang and J. Leonard, 48–89. Hoboken, NJ: Blackwell.

Diewert, W. E. 1978. "Superlative Index Numbers and Consistency in Aggregation." *Econometrica* 46 (4): 883–900.

Feenberg, Daniel, and Jean Roth. 2007. "CPS Labor Extracts: 1979–2006." Technical report, Cambridge, MA, National Bureau of Economic Research. http://www.nber.org/data/morg.html.

Feenstra, Robert C., Robert Inklaar, and Marcel P. Timmer. 2015. "The Next Generation of the Penn World Table." *American Economic Review* 105 (10): 3150–182.

Fernald, John G. 2015. "Total Factor Productivity." Economic Research, Federal Reserve Bank of San Francisco. http://www.frbsf.org/economic-research/total-factor-productivity-tfp.

Fernald, John G., and Charles I. Jones. 2014. "The Future of US Economic Growth." *American Economic Review* 104 (5): 44–49.

Ferraro, Domenico. 2014. "The Asymmetric Cyclical Behavior of the U.S. Labor Market." 2014 Meeting Papers no. 1104, Society for Economic Dynamics.

Gibbons, Robert, Lawrence F. Katz, Thomas Lemieux, and Daniel Parent. 2005. "Comparative Advantage, Learning, and Sectoral Wage Determination." *Journal of Labor Economics* 23 (4): 681–724.

Goldin, Claudia Dale, and Lawrence F. Katz. 2009. *The Race between Education and Technology*. Cambridge, MA: Belknap Press.

Gollop, Frank, and Dale Jorgenson. 1983. "Sectoral Measures of Labor Cost for the United States, 1948–1978." In *The Measurement of Labor Cost*, edited by Jack E. Triplett, 185–236. Chicago: University of Chicago Press.

Hamermesh, Daniel S., and Jeff E. Biddle. 1994. "Beauty and the Labor Market." *American Economic Review* 84 (5): 1174–94.

Heckman, James J., Lance J. Lochner, and Petra E. Todd. 2008. "Earnings Functions and Rates of Return." *Journal of Human Capital* 2 (1): 1–31.

Heckman, James J., Jora Stixrud, and Sergio Urzua. 2006. "The Effects of Cognitive and Noncognitive Abilities on Labor Market Outcomes and Social Behavior." *Journal of Labor Economics* 24 (3): 411–82.

Hellerstein, Judith K., and David Neumark. 1995. "Are Earnings Profiles Steeper Than Productivity Profiles? Evidence from Israeli Firm-Level Data." *Journal of Human Resources* 30 (1): 89–112.

———. 2008. "Workplace Segregation in the United States: Race, Ethnicity, and Skill." *Review of Economics and Statistics* 90 (3): 459–77.

Hellerstein, Judith K., David Neumark, and Kenneth R. Troske. 2002. "Market Forces and Sex Discrimination." *Journal of Human Resources* 37 (2): 353–80.

Ho, Mun S., and Dale W. Jorgenson. 1999. "The Quality of the U.S. Work Force, 1948–95." Technical report, Harvard University.

Hungerford, Thomas, and Gary Solon. 1987. "Sheepskin Effects in the Returns to Education." *Review of Economics and Statistics* 69 (1): 175–77.

Johnson, Matthew T. 2013. "The Impact of Business Cycle Fluctuations on Graduate School Enrollment." *Economics of Education Review* 34:122–34.

Jones, Patricia. 2001. "Are Educated Workers Really More Productive?" *Journal of Development Economics* 64 (1): 57–79.

Jorgenson, Dale W., Frank M. Gollop, and Barbara M. Fraumeni. 1987. *Productivity and U.S. Economic Growth*. Cambridge, MA: Harvard University Press.

Jorgenson, Dale W., and Z. Griliches. 1967. "The Explanation of Productivity Change." *Review of Economic Studies* 34 (3): 249–83.

Jorgenson, Dale W., Mun S. Ho, and Jon D. Samuels. 2014. "What Will Revive U.S. Economic Growth? Lessons from a Prototype Industry-Level Production Account for the United States." In "Rapid Growth or Stagnation in the U.S. and World Economy?," *Journal of Policy Modeling* 36 (4): 674–91.

———. 2016. "Education, Participation, and the Revival of U.S. Economic Growth." NBER Working Paper no. 22453, Cambridge, MA.

Krueger, Alan B., and Lawrence H. Summers. 1988. "Efficiency Wages and the Inter-Industry Wage Structure." *Econometrica* 56 (2): 259–93.

Lang, Kevin. 2015. "Racial Realism: A Review Essay on John Skrentny's *After Civil Rights*." *Journal of Economic Literature* 53 (2): 351–59.

Lemieux, Thomas. 2006. "The 'Mincer Equation' Thirty Years after *Schooling, Experience, and Earnings*." In *Jacob Mincer: A Pioneer of Modern Labor Economics*, edited by Shoshana Grossbard, 127–45. New York: Springer US.

Light, Audrey, and Manuelita Ureta. 1995. "Early-Career Work Experience and Gender Wage Differentials." *Journal of Labor Economics* 13 (1): 121–54.

Mincer, Jacob A. 1974. *Schooling, Experience, and Earnings*. New York: National Bureau of Economic Research.

Oaxaca, Ronald, and Michael Ransom. 1994. "On Discrimination and the Decomposition of Wage Differentials." *Journal of Econometrics* 61 (1): 5–21.

O'Mahony, Mary, and Marcel P. Timmer. 2009. "Output, Input and Productivity Measures at the Industry Level: The EU KLEMS Database." *Economic Journal* 119 (538): F374–403.

Pager, Devah, Bruce Western, and Bart Bonikowski. 2009. "Discrimination in a Low-Wage Labor Market: A Field Experiment." *American Sociological Review* 74:777–99.

Psacharopoulos, George, and Harry Anthony Patrinos. 2004. "Returns to Investment in Education: A Further Update." *Education Economics* 12 (2): 111–34.

Sherk, James. 2013. "Not Looking for Work: Why Labor Force Participation has Fallen during the Recession." Technical Report no. 2722, Heritage Foundation.

Silva, Olmo. 2007. "The Jack-of-All-Trades Entrepreneur: Innate Talent or Acquired Skill?" *Economics Letters* 97 (2): 118–23.

Skrentny, John D. 2013. *After Civil Rights: Racial Realism in the New American Workplace*. Princeton, NJ: Princeton University Press.

Solon, Gary, Robert Barsky, and Jonathan A. Parker. 1994. "Measuring the Cyclicality of Real Wages: How Important is Composition Bias?" *Quarterly Journal of Economics* 109 (1): 1–25.

van Ark, Bart, and Abdul Erumban. 2015. "Productivity Brief 2015: Global Productivity Growth Stuck in the Slow Lane with No Signs of Recovery in Sight." Productivity Brief 2015, The Conference Board. https://www.conference-board.org/retrievefile.cfm?filename=the-conference-board-2015-productivity-brief.pdf&type=subsite.

Weiss, Andrew. 1995. "Human Capital vs. Signalling Explanations of Wages." *Journal of Economic Perspectives* 9 (4): 133–54.

Wyatt, Ian D. 2010. "Evaluating the 1996–2006 Employment Projections." *Monthly Labor Review* 133 (9): 33–69.

Zoghi, Cindy. 2010. "Measuring Labor Composition: A Comparison of Alternate Methodologies." In *Labor in the New Economy*, edited by Katharine G. Abraham, James R. Spletzer, and Michael Harper, 457–85. Chicago: University of Chicago Press.

Comment on Chapters 1 and 2   Douglas W. Elmendorf

I am pleased to have the opportunity to discuss these two terrific chapters. The chapters are wonderful examples of treating data with care and using smart empirical techniques, all in the service of addressing a crucial economic issue. It is preaching to the choir at the Conference on Research in Income and Wealth (CRIW), but still worth emphasizing, that this sort of research is incredibly valuable to both the economics profession and the broader world.

These authors are the perfect people to do this sort of analysis. Dale Jorgenson and Zvi Griliches wrote the seminal paper on human capital and economic growth in the late 1960s, and Dale has been a leader through his whole career in thinking hard about the data needed to do rigorous, quantitative analyses of economic growth and productivity, and then inducing those data to be collected by him and his coauthors and government statistical agencies around the world. Dale's coauthors today—Mun Ho and Jon Samuels—and the outstanding team of authors for the other chapter I will discuss—Canyon Bosler, Mary Daly, John Fernald, and Bart Hobijn—have made important contributions to our understanding of economic growth, and these chapters are another significant step forward.

I am grateful for the authors' work on labor quality and economic growth. I will not have much to say about the details of their empirical approaches. Instead, my aim is to provide some context about how the sorts of projections provided in these chapters matter for economic policy making.

As the director of the Congressional Budget Office (CBO) for six years ending this past March, I will focus on how CBO constructs and uses projections of output growth. The CBO's budget projections depend on its economic projections, and vice versa. The CBO formulates projections of potential output, and then projects that actual output will converge back toward potential output, usually within a few years. And CBO builds up its projections of potential output using projections of labor, capital, and productivity. Therefore, projections of faster or slower growth of labor quality have a direct impact on projected deficits and debt.

Currently, CBO projects that real gross domestic product (GDP) will increase by an average of 2.3 percent per year during the next ten years.

Douglas W. Elmendorf is dean of the Harvard Kennedy School and the Don K. Price Professor of Public Policy; he is also a research associate of the National Bureau of Economic Research.

These comments were delivered on October 16, 2015, and prepared for publication on October 10, 2016. The projections by the Congressional Budget Office cited in these comments were the most recently available when the comments were delivered; subsequent revisions to the projections have not altered the main points made here. For acknowledgments, sources of research support, and disclosure of the author's material financial relationships, if any, please see http://www.nber.org/chapters/c13696.ack.

That figure stems from CBO's estimates that GDP is currently a little below potential and that potential GDP will increase by an average of 2.1 percent per year. Looking further out, CBO projects that real GDP will increase by about 2.1 percent per year in the eleventh through twenty-fifth years of its long-term outlook.

Suppose that GDP increased one-half percentage point per year more slowly than CBO now projects. That would leave output after ten years 5 percent lower than projected, and after twenty-five years 12 percent lower than projected. Using the agency's published rules of thumb for assessing the impact on the budget of different economic outcomes over the next decade, that lower path for GDP would make the deficit ten years from now $345 billion larger than in the baseline projection and the cumulative deficit over the next ten years $1.5 trillion larger. Using the agency's alternative long-term projections based on different projections of key economic factors, that lower path for GDP would make federal debt twenty-five years from now 125 percent of GDP rather than the 107 percent in the basic projections. The effect is not even larger because slower GDP growth tends to lower health-care-spending growth, future Social Security benefits, and interest rates.

In fact, CBO has revised down its estimate for future GDP quite significantly in recent years. Since 2007, the agency has lowered its projection for potential output in 2017 by about 9 percent, which is equivalent to lowering average annual growth by nearly 1 percentage point. That downward revision widened the projected budget deficit in 2017, all else equal, by more than $500 billion. All else is not equal, because if the budget outlook had looked that much better three or four years ago, the policy actions that were taken probably would not have been taken. Still, it is clear that the budget outlook that drives so much debate depends very importantly on the agency's projections of output.

As I mentioned, CBO's projection of output over a decade varies one-for-one with its projection of potential output. That projection of potential output comes from a version of the growth accounting that these chapters do. However, in the agency's projections, changes in total factor productivity (TFP) include both changes in labor quality and changes in true TFP. Historically, improvements in labor quality have accounted for between a quarter and a third of growth in TFP as defined by CBO.

The CBO currently projects that growth of potential TFP over the coming decade will be close to the average growth of TFP over the past half century. In other words, CBO does not appear to be including any noticeable slowdown in the growth of labor quality. In CBO's long-term outlook, the agency projects a slight slowdown in TFP growth beyond the coming decade, attributing it to a slower rate of increase in educational attainment and other factors.

Of course, CBO is aware of the data suggesting a slowdown in the rate of improvement in educational attainment and thereby in labor quality.

My colleagues and I became concerned that we had not made a sufficient adjustment for a deceleration in labor quality in part because our approach did not address labor quality in a systematic way. Therefore, we launched an effort to model labor quality explicitly so we could break it out of TFP.

Now let me turn to the chapters. As the authors have explained, their work is a very careful application of growth accounting to understand how labor quality has evolved in the past and is likely to evolve in the future. Both chapters do a tremendous amount of detailed work with the data, and both chapters present alternative projections based on different assumptions so that readers can evaluate the robustness of the results. The central analytic issue is how well differences in wages across age-education groups and others capture differences in marginal products—that is, labor quality. I will come back to that issue in a minute.

What if wage differences do not reflect only differences in marginal products? Wage differences probably reflect differences in marginal products for the most part, but wage differences also reflect other factors, which may be important for correctly interpreting the results in these chapters.

For example, what if wages rise faster with age than marginal products do? I am paid more now than I was a decade ago, maybe not because I am more productive but because I am climbing a wage ladder. Under this view, the economy may have gained less from the increase in experience as baby boomers aged than it appears from this sort of analysis. Therefore, we will lose less as baby boomers retire, and we can be more optimistic about the future.

As another example, what if wages reflect marginal products better now than they did in the past? Social customs may have restrained wage dispersion a few decades ago more than they do today. Under this view, labor quality may have increased less over time than it appears from this sort of analysis. Therefore, the compositional shifts studied here have been less important, TFP growth has been more important, and we can be more optimistic about the future.

The studies present a wealth of interesting information, but the key findings are the following:

- *Growth of labor quality did not diminish during the past decade as had been expected.* One key reason is that employment losses during the Great Recession were concentrated among low-wage workers. That disproportionate job loss pushed up the average wage among people who remained employed.
- *Growth of labor quality will probably slow significantly in the coming decade.* The extent of the slowdown will depend on the extent to which low-wage workers return to the labor force and employment, with more returning workers implying a greater slowdown. From my perspective, the scenarios in which employment-population ratios or labor force participation rates return to their precrisis levels seem quite unlikely,

because so far employment-population ratios and labor force participation rates show only very partial bounce-backs. Instead, it seems much more likely that those ratios and rates will stay close to their current levels. In that case, the chapters suggest that we will see a slowdown in labor-quality growth of a few tenths of a percentage point per year.

Let me mention three other points. One is that a return of low-wage workers to employment would raise aggregate output and the income of these workers, even though it would depress growth in labor quality. I do not think there is any ambiguity about the effects on the economy and on these workers: they will only be paid if their marginal products are positive, and they will only come back if their wages exceed their opportunity costs, so their return would increase overall output and workers' income.

The second point is that policies to support advances in educational attainment would raise aggregate output and those workers' income. It concerns me a great deal that under the current caps on annual appropriations, federal investments—including in education—will soon fall to their lowest share of GDP in at least fifty years.

My third point is that policies to encourage greater labor force participation would raise aggregate output and could increase or decrease the well-being of those workers. Here is why. One can encourage more participation either by improving what one gets *in* the labor force or by diminishing what one gets *outside* the labor force. An expansion of the earned income tax credit is in the former category; it would raise aggregate output and increase the well-being of those workers. Repealing the health care subsidies under the Affordable Care Act is in the latter category; it would raise aggregate output and diminish the well-being of those workers. We need to think carefully about what sorts of policies to encourage labor force participation we want to pursue.

Let me conclude by thanking Dale, Mun, Jon, Canyon, Mary, John, and Bart again for their terrific work.

3

The Importance of Education and Skill Development for Economic Growth in the Information Era

Charles R. Hulten

3.1 Introduction

The rapid advances of information technology and globalization have led to major structural changes in the US economy. The extent of these changes is evident in the decline of manufacturing industry and the rise of selected service-producing sectors shown in figures 3.1 and 3.2. The share of manufacturing in private gross domestic product (GDP) has been cut in half over the last half century, from 30 percent in 1960 to less than 15 percent in 2015, and the share of private employment has fallen from around 34 percent to 10 percent. This decline was more than offset by increases in those service sectors that involve "expert" advice, information, or interventions—finance, business and professional, education, health, law, and information services: the share of value added rose from around 13 percent to 37 percent, while the share of employment rose from under 14 percent of total private employment to over 40 percent.[1] These shifting patterns reflect,

Charles R. Hulten is professor of economics emeritus at the University of Maryland and a research associate of the National Bureau of Economic Research.

I would like to thank Leonard Nakamura, Valerie Ramey, and Hal Varian for their comments on earlier drafts, as well as conference participants. Remaining errors and interpretations are my responsibility. For acknowledgments, sources of research support, and disclosure of the author's material financial relationships, if any, please see http://www.nber.org/chapters/c13937.ack.

1. The part of the service sector designated "expert" in figures 3.1 and 3.2 refers to those NAICS industries 51, 52, 54, 55, 56, 61, and 62 (the organization services include NAICS 54, 55, and 56). The statistics shown here are taken from the industry accounts of the Bureau of Economic Analysis. They are expressed as a share of the private economy because the focus of this chapter is on innovation, education, and growth accounting in the business sector. The ratio of private-to-total value added was 87 percent in 2015, and the corresponding ratio for full- and part-time employees was 86 percent, so the sectoral estimates are somewhat smaller when expressed as a ratio of the totals. The time series shown in figure 3.2 is pieced together from different parts of industry table 6.5 and is thus subject to some discrepancies.

116   Charles R. Hulten

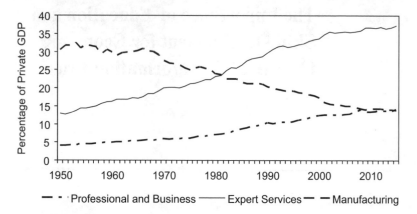

— ·Professional and Business ——— Expert Services — — Manufacturing

Fig. 3.1   US private GDP shares of manufacturing, expert services, and profes-sional and business services, 1950–2015
Source: Bureau of Economic Analysis, GDP-by-Industry, Industry Data, *Value Added by In-dustry as a Percentage of Gross Domestic Product.* The "expert" service sectors include the NAICS industries 51, 52, 54, 55, 61, and 62, and organizational service sectors 54, 55, and 56.

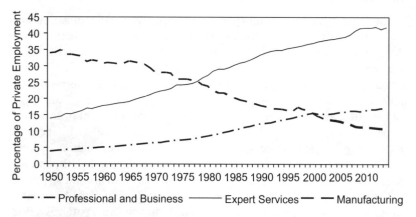

— ·— Professional and Business ——— Expert Services — — Manufacturing

Fig. 3.2   US private employment shares of manufacturing, expert services, and pro-fessional and business services, 1950–2015
Source: Bureau of Economic Analysis, Industry Economic Accounts, from various parts of table 6.5, *Full-Time Equivalent Employees by Industry.* See figure 3.1.

in part, the outsourcing of production to lower-wage countries, labor-saving technical change, and the evolution of demand for different products.[2] The trends in professional and business organizational services, also shown in figures 3.1 and 3.2, indicate a significant shift in employment within firms toward nonproduction activities, and reflect the growth of in-firm research

2. Haskel et al. (2012) and Autor, Dorn, and Hanson (2013).

and development, product design, and the emergence of sophisticated organizational management systems.

The change in the structure of employment and valued added occurred during a period that also saw a parallel increase in higher-order cognitive and noncognitive worker skills of the labor force, documented by Autor, Levy, and Murnane (2003) in their pathbreaking paper, as well as a significant increase in educational attainment. The fraction of the US population twenty-five years or older with at least a BA degree quadrupled (to 32 percent) over the period from 1960 to 2015; the fraction of those with at least a high school degree more than doubled (to almost 90 percent), according to data from the US Census Bureau's Current Population Survey (CPS). Evidence cited in this chapter suggests that the upward trends in educational attainment and the demand for more complex cognitive skills are connected to the structural changes in the economy evident in figures 3.1 and 3.2; those service sectors where the employment increase was most pronounced were also those where the high-skill, high-education professions are located. The observed structural shifts are thus consistent with the growth of the knowledge economy.

It is one thing to regard skill development and education as important for the functioning and growth of the economy, but how important are they compared to other factors that influence the growth of GDP? Surprisingly, estimates from the Bureau of Labor Statistics Multifactor Productivity Program (BLS 1983, and regular updates) suggest that educational attainment may not be as important for economic growth as the recent focus on education and skills implies. The BLS data indicate that changes in the composition of the labor force, largely due to education, accounted for only a small fraction (7 percent) of the growth in labor productivity in the US private business sector over the period 1995 to 2007 (the last year before the Great Recession). Robert Solow famously remarked in 1987 that "you can see the computer age everywhere but in the productivity statistics"; in the current context, one might say that we can see the revolution in educational attainment everywhere but in the productivity statistics.

Acemoglu and Autor (2012) have questioned how education can have played only a relatively small role in the growth of the economy, given the knowledge-intensive nature of the information revolution. Indeed, there is a large literature on the importance of education as a source of economic growth and on the importance of skill-biased technical change. However, most of this analysis does not stray far from a production function formulation of the problem and an emphasis on marginal productivities and factor substitutability.

The approach taken in this chapter builds on the contributions of Acemoglu and Autor (2011, 2012), who focus on the role of skills and education at the task and occupations levels of the production process, with the goal of linking the growth in complex nonroutine skills to skill-biased technical change. The activity-analysis model of this chapter also starts at the micro level of production, but focuses on the substitution possibilities among inputs; the goal is to show how limited substitution possibilities within the production techniques of an activity can lead to a much greater role for skill development and education than that implied by the neoclassical BLS approach, even though both use virtually the same growth-accounting methods. The basic idea is that the choice of technique determines the nature of the inputs required, and once a technique is adopted, substitution possibilities among the inputs are typically quite limited (accountants are not substitutes for neurosurgeons). The skills necessary for each type of activity come embodied in people, in part via their educational preparation, and access to people with the necessary skills and education becomes a critical factor enabling structural change and economic growth. Conversely, an inadequate supply of skilled workers with the requisite skills can serve as a drag on growth. Education provides a pool of general cognitive and occupational expertise, and in some cases, specific vocational skills from which firms can draw the workers they need. It is hard to imagine the economy of 2017 operating with a pool of workers in which less than half had a high school degree, as in 1960, and less than 10 percent had a college degree.

These points are developed in greater detail in the sections that follow. The Solow neoclassical growth-accounting model used by BLS is described in section 3.2, along with a critique of the theory underpinning its labor force composition adjustment in section 3.3. This is followed in section 3.4 by the activity-analysis framework proposed in this chapter. The fixed-proportion nature of the framework is described and illustrated using several examples. This "necessary input" model is contrasted with the aggregate production function approach, with special attention to its implication for skills and education. A sources-of-growth framework based on the activity-analysis model is derived and shown to be essentially equivalent to the neoclassical version of the growth-accounting model. This result allows the BLS growth-accounting estimates to be given a different interpretation, one that assigns a greater importance to labor skills and education than the conventional approach. The three sections that follow section 3.4 are empirical, and examine the evidence on the trends in labor and capital to see if they are consistent with the predictions of the activity-analysis framework. Section 3.5 traces the growing importance of higher educational attainment, higher-order cognitive and noncognitive skills, and professional occupations and employment over the last half century. Section 3.6 looks at the parallel development in the growth in information and communications technology equipment (ICT) and intangible knowledge capital like research and development (R&D). Sources-of-growth estimates expanded to include intangible capital are presented in section 3.7, and interpreted in light of the activity-analysis framework. A final section sums up.

3.2   The Neoclassical Growth-Accounting Model

Many factors affect the growth of GDP, including labor and its skills, but also capital formation and technical change. Any general assessment of the contribution of labor skills and education should therefore be framed in the context of all of the relevant factors. The main empirical framework that does this is the neoclassical growth-accounting model developed by Solow (1957) and greatly extended by Jorgenson and Griliches (1967), who laid the groundwork for the official productivity program at the BLS.

Neoclassical growth models share a common feature: they are rooted in the assumption of an aggregate production function relating aggregate outputs to the factor inputs of aggregate labor and capital, with a shift term that allows for changes in the productivity of the inputs: $Y_t = F(K_t, L_t, t)$. In describing the role of the shift term in the function, Solow states:

> The variable t for time appears in F to allow for technical change. It will be seen that I am using the phrase "technical change" as a short-hand expression for any kind of shift in the production function. Thus slowdowns, speed-ups, improvements in the education of the labor force, and all sorts of things will appear as "technical change." (1957, 312)

In its most succinct form, the aggregate formulation combines various types of capital into a single total K, and different types of labor into a single L. Once formed, they are treated as substitutes, implying that the same amount of output can be produced by different combinations of capital and labor.

The basic sources-of-growth model is derived from an aggregate production function, which is assumed to exhibit constant returns to scale in capital and labor, and Hicks-neutral productivity change as reflected in a shift term A_t. Under the further assumption that capital and labor are paid the value of their marginal products, the resulting $Y_t = A_t F(K_t, L_t)$ can be differentiated with respect to time to give the sources-of-growth equation

$$(1) \qquad \dot{y} = s_K \dot{k} + s_L \dot{\lambda} + \dot{a}.$$

Dots over variables indicate rates of growth, and time subscripts are dropped for ease of exposition. This formulation decomposes the growth rate of output into the growth rates of the inputs, weighted by their respective output elasticities (as proxied by income shares), and the growth in the productivity with which the inputs are used (total factor productivity, or TFP). The former is interpreted as a movement along the production function and the latter as a shift. Both processes are assumed to occur smoothly. All the elements of this equation except the last term can be measured using data on prices and quantities, or assumptions about parameters like capital depreciation. This allows the productivity variable to be measured as a residual.

There is no specific provision for the contributions of education or skills in the basic formulation. This issue was addressed by Jorgenson and Griliches (1967), who proposed a version of the production function that allowed for different types of labor, differentiated by worker characteristics like education, which have different wage rates and marginal products. The production function then becomes $Y = AF(K, L(H_1, \ldots, H_N))$, where the H_i's are the hours worked in each of the N categories, total hours are $H = \Sigma_i H_i$, and $L(\cdot)$ is a function that aggregates the N groups into an index of total labor input. The growth rate of L is the share-weighted contribution of each group's hours to total hours, where the s_{Hi} are each group's share of total *labor* income

(2)
$$\dot{\lambda} = \dot{h} + \sum_{i=1}^{N} s_{Hi}(\dot{h}_i - \dot{h}) = \dot{h} + q_{LC}.$$

The growth rate of labor input is thus the sum of the growth rate of total unweighted hours plus the labor-composition effect, q_{LC}. The associated growth equation is then

(3)
$$\dot{y} = s_K \dot{k} + s_L \dot{h} + s_L q_{LC} + \dot{a}.$$

The variable q_{LC} records the effect on output of a shift in worker hours among groups with different output elasticities (cum factor shares), and is positive when the composition of the labor force shifts toward higher productivity groups. In practice, multiple worker characteristics are included in the index.

It is this framework that produced the BLS estimates, cited in the introduction, that show q_{LC} accounted for only 7 percent of labor productivity growth in the private business sector over the period 1995 to 2007. The overall composition effect is dominated by the education effect, and the 7 percent estimate reflects the combined effect of the increase in the wage share of the educated (its weight in q_{LC}) and the growth rate of educational attainment as reflected in the H's. Estimates reported at the end of this chapter also show an acceleration in the q_{LC} effect in the 1970s, and a slowdown in the late 1990s averaging 7 percent for the period 1995–2007.

3.3   A Choice of Parables

The relatively small contribution of education in recent years seems inconsistent with the growth of the knowledge economy. Indeed, Hanushek and Woessmann (2015) begin their book on *The Knowledge Capital of Nations* with the statement that "knowledge is the key to economic growth" and go on to note the positive correlation between educational attainment and income per capita in a cross-sectional comparison of countries. Acemoglu and Autor (2012) have also expressed their reservations, as noted above.

Since it is hard to imagine the complex technologies and capital of the digital revolution being operated with a workforce equipped with only the most rudimentary cognitive skills and knowledge, it therefore seems appropriate to examine the sources-of-growth framework more closely to see what features of the model might lead to that result.

Solow himself recognized the simplification involved when he began his classic 1957 paper with "it takes something more than a 'willing suspension of disbelief' to talk seriously of the aggregate production function," and, in his 1987 Nobel Laureate Lecture, "I would be happy if you were to accept that [growth-accounting results] point to a qualitative truth and give perhaps some guide to orders of magnitude" (Solow 1988, xxii). Writing in defense of the aggregate approach, Samuelson (1962) argues that it is a parable whose purpose is insight building (more on this below).

Parables are neither inherently right nor wrong, just more or less useful for illustrating some underlying truth. The growth-accounting model has enjoyed great success for its insights into the general contours of economic growth. However, the aggregate model may be more successful in describing overall economic growth than in characterizing structural economic change and the implied role of education. The problem is that some of the assumptions underlying the neoclassical model require a particularly large suspension of disbelief. The first is the one-sector nature of the aggregate production function, $Y_t = F(K_t(\cdot), L_t(\cdot), t)$. The single product, Y_t, is a macroeconomic surrogate for the many products actually produced, and the surrogate aggregate production is a methodological parable for summarizing the complex processes that contribute to their production. This formulation is a useful, indeed, essential, part of the conceptual framework that underpins the circular flow of products and payments that characterize the aggregate economy. However, its usefulness is questionable for addressing issues concerning changes in the structure of the flows that make up the aggregate Y_t and the corresponding changes in the allocation of resources that are evident in figure 3.2.

A more general representation of the structure of production is needed in order to deal with these structural issues. A step in this direction can be made by formulating the production problem in terms of the production possibility frontier, $\phi[(Y_{1,t}, \ldots, Y_{m,t}); K_t(\cdot), L_t(\cdot), t]$. In this formulation, the collection of outputs at any point in time, $(Y_{1,t}, \ldots, Y_{m,t})$, is produced by aggregate capital, $K_t(\cdot)$, whose components are categories of capital identified by type and industry of use, and aggregate labor, $L_t(\cdot)$, whose components are categories of labor identified by their characteristics (including education) and industry of use. The technology shifter t is included to allow for overall increases in the efficiency with which labor and capital are used, although individual efficiency parameters $A_{i,t}$ might be used instead (or the factor augmentation equivalents). Underlying the production possibility frontier (PPF) are separate *industry* production functions for each sectoral output, $Y_{i,t} = F^i(K_{i,t}(\cdot), L_{i,t}(\cdot), t).$[3]

The multiproduct way of looking at the structure of production has an important implication for studying the importance of skills and education: a movement along the production possibility frontier not only changes the composition of output, it shifts the composition of the inputs required to produce the output. With these shifts come changes in the required composition of labor skills. This means that a change in the mix of skills may occur without technical change, as for example, when the movement along ϕ is caused by changes in the structure of consumer preferences or changes brought about by a shift in the pattern of global trade, or by non-unitary income elasticities. Indeed, aggregate output along the PPF may be unchanged.[4]

Then there is the question of technical change. This is represented in the conventional aggregate formulation as a shift in the production function holding inputs constant (or, a similar shift in the PPF). This convention implicitly views all technical change in terms of increases in the productivity of the input base, or "process innovation." This kind of innovation has made important contributions to economic growth during the course of the information revolution, but it is not the only kind of technical change, nor necessarily the most important. Innovation in new or improved products has also played a central role in the revolution.[5]

Product innovation changes the mix of outputs ($Y_{1,t}, \ldots, Y_{m,t}$) over time. Improved goods appear and ultimately displace their older counterparts, others drop out because of a lack of demand, while new goods enter the market. In the process, a new vector appears, ($Y_{1,t+1}, \ldots, Y_{m+k,t+1}$), with a product list expanded by k to allow for new items. The list of individual product functions is expanded accordingly, but with $Y_{i,t+1} = 0$ for displaced goods. The individual production functions for the new or improved goods may have a different set of skill requirements than those they displace. Evidence suggests that this was, indeed, the case during the information revolution, during which the growth in digital-economy goods has led to increases in the demand for more cognitively complex skills sets. However, it is important

---

3. The assumptions required to move from the individual sectoral production functions to an exact form of the aggregate production function are very restrictive (see Fisher [1969] for a detailed treatment and summary of this and other problems in the theory of aggregation).

4. The sources-of-growth equation (1) is, formally, a Divisia Index (Hulten 1973). A movement along the PPF frontier ϕ from one point to another involves line integration that does not change the value of the output index (the invariance property).

5. Data from the National Science Foundation's (2012) Business R&D and Innovation Survey (BRDIS) suggest that process-oriented business R&D is a small share of the total, accounting for only 15 percent of the $224 billion in domestic R&D paid for by companies (Wolfe 2012). The rest is for product development, though some of the new products are inputs to the production process (capital-embodied technical change, for example, or improved materials). The fraction of R&D devoted primarily to new consumer goods is not reported.

to note that, while the technology for producing the new goods became more complex and required more complex skills, the main impetus behind the increased demand for these skills was product innovation and not skill-biased process innovation.

Two further suspensions of disbelief are also needed. The first involves the assumption that the capital and labor are paid the values of their marginal products, thus allowing income shares to be used as a proxy for the underlying output elasticities in the sources-of-growth formulation. This is a very strong assumption, mainly defensible as a macroeconomic approximation. Prices may well deviate from marginal products due to monopolistic pricing, labor market rigidities, discrimination, and cyclical fluctuations in economic activity. Moreover, the marginal social return to education may exceed the marginal private return implied by market wages because of externalities of the type noted by Lucas (1988), a point elaborated in a subsequent section.

Second, the existence of separate aggregate labor and capital entities, $L(\cdot)$ and $K(\cdot)$, and of a unique q_{LC}, requires the assumption of weak separability in the aggregate production function. This, in turn, requires the marginal rate of substitution between one type of labor and another to be independent of the amount and composition of aggregate capital (Hulten 1973). This is a mathematical proposition, but in economic terms it means that if a worker in a lower education category acquires a higher degree in pursuit of a wage premium, output will increase without any change in capital or technology. This is problematic because those workers with higher educational attainment tend to end up in jobs or occupations with more complex technological requirements and capital. Simply educating more people will not, all else held equal, necessarily result in a significant increase in output, a point that will be elaborated in the activity-analysis model developed in the section that follows.

3.4   The Activity-Analysis Approach to Production

3.4.1   The Model

A close examination of the neoclassical model of production thus suggests that it may not capture the full effects of education buried in the underlying complexity of "reality." Indeed, one of the founders of the neoclassical aggregate approach, Paul Samuelson, has indeed "insisted" in his 1962 paper on "Parable and Realism in Capital Theory" that

> capital theory can be rigorously developed without using any Clark-like concept of aggregate "capital," instead relying upon a complete analysis of a great variety of heterogeneous physical capital goods and processes through time. Such an analysis leans heavily on the tools of modern linear and more general programming and might therefore be called neo-neo-classical. It takes the view that if we are to understand the trends in how

incomes are distributed among different kinds of labor and different kinds of property owners, both in the aggregate and in the detailed composition, then studies of changing technologies, human and natural resources availabilities, taste patterns, and all the other matters of microeconomics are likely to be very important. (193)

This is essentially the view taken in this chapter. But, he goes on to say

At the same time in various places I have subjected to detailed exposition certain simplified models involving only a few factors of production. Because of a Gresham's Law that operates in economics, one's easier expositions get more readers than one's harder. And it is partly for this reason that such simple models or parables do, I think, have considerable heuristic value in giving insights into the fundamentals of interest theory in all its complexities. (193)

The tension between the two perspectives over the appropriate level of analysis is central to the objections against the neoclassical production function and the concept of aggregate capital raised during the Cambridge Controversies of the 1950s and 1960s (Harcourt 1969).[6]

Given these questions and those that have been raised about the size of the neoclassical labor-composition effect, it seems reasonable to take a closer look at the foundations of the aggregate production framework, essentially disaggregating it to get at its "primitive" activity-analysis level. When approached at this foundational level, many of the issues raised in this chapter can be addressed, particularly those involving the way labor skills interact with capital to make educational attainment necessary for many activities. The "old-fashioned" activity-analysis model is well suited to this task.[7]

Where the neoclassical model offers a succinct and mathematically viable way of summarizing the supply side of the economy, activity analysis is neither succinct nor mathematically convenient. It does, however, provide a more detailed look into the underlying processes of growth and the shifting demands for the various skills and types of capital. It treats the firm and its various activities, not the aggregate production function, as the fundamental unit of analysis for studying those shifting demands.

6. Opposition to the aggregate production function and the neoclassical view of economic growth has a long history, and is by no means limited to the Cambridge Controversies. It is also present in the literatures on organizational theory, the importance of institutions in economic history, and in Schumpeterian analysis. Nelson and Winter (1982) provide an in-depth analysis of the evolutionary nature of the process of economic growth that focuses on the firm and its activities, and the skills and competence of its workers. The activity-analysis model sketched in this chapter is rooted in this view of the firm.

7. Activity analysis has had a long and honorable, though somewhat neglected, history. It was well positioned in the early 1950s to become the dominant supply-side paradigm for the economy. The 1951 Cowles Commission conference volume, *Activity Analysis of Production and Allocation*, edited by Tjalling C. Koopmans, contains papers authored by four future Nobel laureates in economics. Yet, it was neoclassical growth theory that prevailed over the next two decades.

An activity is defined in this chapter as an operational segment of a firm that has an identifiable output or outcome produced by a technique that specifies a certain mix of inputs. What gives the activity-analysis parable its distinctive feature is the assumption that the inputs are combined in a fixed proportion. This assumption implies that there is no substitution among inputs, so each input is necessary for the activity, and it thus contrasts strongly with the assumption of input substitutability in the neoclassical parable. A firm may operate several activities simultaneously, as, for example, both production and nonproduction (or overhead) activities, or the activities of multiple establishments within the firm, each producing a different product. In light of the model of Acemoglu and Autor (2011), it is worth noting that the way the labor input functions within an activity may involve a specific set of tasks requiring a specific set of skills.[8]

The following example illustrates the issues involved. A given amount of earth can be moved using different techniques: one technique uses a few skilled operators equipped with expensive bulldozers, the other uses many manual workers, each equipped with cheap shovels. In the aggregate representation of these different techniques, the "neoclassical" form of the technology for moving earth would be $Y = AF(K_H, K_L, H_S, H_U)$, or the factor augmentation equivalent (the capital subscripts denote "higher-technology bulldozers" and "lower-technology shovels," and "skilled" and "unskilled" for the labor subscripts). In order to speak of aggregate capital, K, and labor, L, this production function must have the previously noted separable form, which in this case is $Y = AF(K(K_H, K_L), L(H_S, H_U))$. The different types of labor are substitutable among each other within the labor aggregate $L(\cdot)$, as are the different types of capital within $K(\cdot)$, and the aggregates themselves are substitutable along an isoquant connecting K and L. The isoquant QQ shown in figure 3.3A allows for this substitution, which occurs as the movement along the isoquant from A to B as relative factor prices change from aa to bb. The broken L-shaped lines represent two activities that use different techniques for producing the same amount of output, Y, and illustrate a version of activity analysis in which the neoclassical isoquant is the envelope of the various activities.

As portrayed in figure 3.3A, activity analysis is seen to be conceptually consistent with the aggregate production function when capital is treated as

8. In their framework, a task is defined as a unit of work activity that produces an output. This use of the term "activity" in the context of job performance differs from the way an activity is conceived of in this chapter, which involves a fixed-proportions technology that may encompass many separate tasks and types of input. However, the task-based activity of Acemoglu and Autor and the production-based activity approach are mutually consistent and can operate simultaneously, although the former is used to motivate aggregate skill-biased technical change, with the implication that the BLS sources-of-growth estimates understate the role of complex-skill development, while the activity analysis in this chapter is used to motivate the "necessary input" framework that also implies that the role of complex-skill development is understated, though it operates through a different channel.

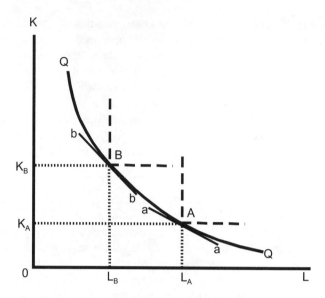

Fig. 3.3A   Activity-analysis model, two activities and malleable inputs

a homogeneous malleable entity that represents forgone consumption valued at investment cost. While this is a useful macroeconomic way of looking at capital and technology—Samuelson's surrogate production function—it glosses over the technical differences between shovels and bulldozers and the skill differences between the workers. It is therefore not a helpful framework for studying how the choice of technique affects the demand for skilled labor.

Figure 3.3B illustrates a less flexible version of activity analysis in which different types of capital work with the requisite types of labor and skills and cannot be substituted across activities without a corresponding change in labor.[9] This case implies that the separate inputs should not be combined using the $K(K_H, K_L)$ and $L(H_S, H_U)$ pairings of the aggregate production function approach, but instead by the functional pairings $a(K_L, H_U)$ and $b(K_H, H_S)$. This is represented in figure 3.3B by the broken L-shaped lines showing the two techniques for producing the same amount of output, Y. However, while both techniques produce the same kind (and amount) of output, the inputs on the axes refer to different types of capital *and* labor. One implication is that the factor price lines *aa* and *bb* refer to different input prices. Moreover, the strict complementarity of the techniques implies that the ratios of the marginal products of the different types of capital and labor are not well defined, and variations in the wage rental cannot affect the input

9. The two variants can be bridged under certain assumptions, as for example, Solow et al. (1966). However, Fisher (1969), who pays special attention to different types of labor input, shows that aggregation in the general case is problematic.

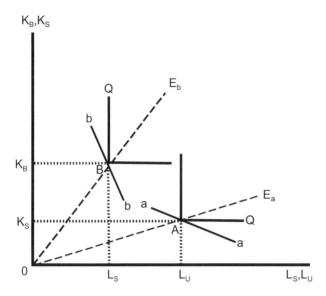

Fig. 3.3B   Activity-analysis model, two activities and nonmalleable inputs

ratio. Thus, if the wage rate increases, there can be no substitution of capital for labor within a technique.

An important implication for this chapter is that, since a shift in techniques from $a(K_L, H_U)$ to $b(K_H, H_S)$ cannot occur without a shift from unskilled to skilled workers and from less to more technologically sophisticated capital, a deficiency of skilled workers will slow or prevent the adoption of the $b(K_H, H_S)$ technology. It is also possible, in a more sophisticated rendering of the model, that a deficiency of workers with a particular skill set could induce innovation designed to compensate for the deficiency (the Habakkuk thesis), but the larger point is that in order for a firm to actually operate the activity $b(K_H, H_S)$, access to both K_H and H_S in the right proportions is necessary.

Figure 3.3C adds yet another complication. The activities in the first two figures represent different techniques for producing the same type of output. This is not a good assumption to apply to all activities in an era with a high rate of product innovation because switching from one quality, or model, of output to another often involves a switch in the way the goods are produced and in the inputs required. For example, in summarizing their study of new information technology (IT)–enhanced machinery, Bartel, Ichniowski, and Shaw (2007, 1721) make the following points:

> First, plants that adopt new IT-enhanced equipment also shift their business strategies by producing more customized valve products. Second, new IT investments improve the efficiency of all stages of the production process by reducing setup times, run times, and inspection times. The reductions in setup times are theoretically important because they

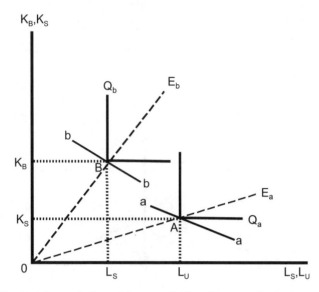

Fig. 3.3C   Activity-analysis model, two activities with nonmalleable inputs and different outputs

make it less costly to switch production from one product to another and support the change in business strategy to more customized production. Third, adoption of new IT-enhanced capital equipment coincides with increases in the skill requirements of machine operators, notably technical and problem-solving skills, and with the adoption of new human resource practices to support these skills.

The ability to customize output to suit the needs of the buyer represents an important change in product quality, and is linked, in this case, to increased skill requirements. The advent of the automatic teller machine, a labor-saving device from the standpoint of production, is another example of how the quality of a product was also improved, in this case by making money accessible at all times of day or night. These examples are illustrated in figure 3.3C by activity-specific output indexes.

3.4.2   Aggregation and Dynamics

The activities as portrayed in figures 3.3A, 3.3B, and 3.3C illustrate the logic of the activity-analysis model. From an operational standpoint, activities are generally combined to form a larger set that constitute the production plan of a firm. In formal terms, the technology of a firm j can be characterized at any point in time by the activity set $A_{j,t}$, whose elements are the totality of activities it operates $\{a_{i,j,t}(K_{i,j,t}, H_{i,j,t}, M_{i,j,t})\}$. An output or outcome is associated with each activity, although much of the output is delivered to other activities within the firm (e.g., overhead and different stages of production along an assembly line). The vector $M_{i,j,t}$ is added to allow for the presence of intermediate goods produced and used within the firm, but also the intermediate inputs acquired externally. The set $\{a_{i,j,t}(K_{i,j,t},$ $H_{i,j,t}, M_{i,j,t})\}$ is thus a disaggregated representation of the firm's technology, but it is not, strictly speaking, a neoclassical production function relating total output to aggregated inputs.

The firm is the organizational entity responsible for choosing the appropriate mix and level of activities for $A_{j,t}$ from a larger set of possible techniques. Selecting the right mix and level of activities is an essential organizational function of the firm, and once the selection has been made, the capital requirements of the firm $\{K_{i,j,t}\}$ and staffing needs $\{H_{i,j,t}\}$ are determined. Prescott and Visscher (1980) point to the acquisition and proper use of human capital as centrally important for the success of an organization, and Bloom and Van Reenen (2007) have pointed to the importance of good managers and management practices. The role of human agency can sometimes get lost in the formal mathematical presentation of the various models.

Firms can be grouped into industries for purposes of analysis, though again, there are aggregation issues. Indeed, many are similar to those encountered when aggregating the internally generated "output" of activities within firms, but with the additional complication posed by different ways of classifying industries (the company versus establishment problem). However, these difficulties are not germane to the main interests of this chapter, so we simply group firm-level activities into industry-level activities (however industry is defined), and then into an aggregate economy-wide activity set, A_t, whose elements include the totality of all activities, $\{a_{i,t}(K_{i,t}, H_{i,t}, M_{i,t})\}$. The significance of this formulation for the problem at hand is that, at any point in time, the total capital requirements $\{K_{i,t}\}$ and staffing needs $\{H_{i,t}\}$ of the *economy* are determined by the choice of activities at the firm level, the diversity of activities across firms in an industry, and the diversity of industries in the larger economy.

The mix of activities and skills can and does change over time, as witnessed by the structural changes in the economy evident in figures 3.1 and 3.2. This structural change is the visible result of the shifting composition of the aggregate activity set A_t occurring in response to the revolution in information and communication technology and the globalization of the world economy. New or improved products have made older goods obsolete, new processes and activities within firms have replaced older techniques, and new forms of product distribution have displaced older outlets. New firms and industries have appeared in this process of creative destruction, while older industries have declined and firms exited their industry or reinvented themselves. The changes occurring in A_t have also changed the demands for labor and capital. This has meant a larger demand for those higher-order skills, occupations, and education that have been made necessary by the information revolution. Again, one of the major implications of the activity-analysis framework, as it is set out above, is that *the observed structural changes could not have occurred without the parallel development of the appropriate skills*. In other words, the "necessary input" way of looking at structural change implies that skill development and the associated contribution of education is an organic part of the dynamic evolution of the changing economy.

Education also contributes to this evolution in another way. Much of the underlying innovation originates within firms through activities like R&D, product design, and strategic planning. Much of the innovation that drives the dynamics of firms and the economy comes in the form of product innovation. These activities are education intensive (Nelson and Phelps 1966), and some of the innovation may come in response to chronic deficits in some skill areas (e.g., process automation). And, even when innovation does not originate in the firm, it is implemented and sustained by the efforts of its management. The activities, and the people that operate them, endogenize the innovation process (as in Romer 1986, 1990), and, in turn, create a demand for the skills and occupations of the digital economy.

However, it is also important to stress (once more) that education by itself is not sufficient for creating more output growth. Moreover, it should also be noted that, while technical change and globalization have shifted the structure of activities toward those that require more complex skills, there are still activities that do not require higher levels of educational attainment (indeed, the large majority do not, as we will see in a subsequent section). The activity-analysis framework focuses on the necessity of the *appropriate* skills for the activity at hand, and this applies to the full range of activities in operation at any point in time, not just to those involving more complex labor skills.

3.4.3   Activities and the Measurement of GDP

An output is associated with each activity in a firm's activity set, $A_{j,t}$, even though some are shadow outputs delivered to other activities within the firm. The value of the output sold externally (intermediate and other) can be measured using market transaction prices and the resulting revenue divided between deliveries to final demand and deliveries to intermediate demand. This yields the accounting equation $P_{i,t}Q_{i,t} = P_{i,t}Q_{i,t}^D + \Sigma_j P_{i,t}Q_{i,j,t}^M$, where $Q_{i,j,t}^M$ is the delivery of the intermediate good from activity i to the other activities, and $Q_{i,t}^D$ is the external output delivered to final demand (for a one-product firm). The GDP is then defined as the summation across deliveries to final demand, giving $GDP_t = \Sigma_i P_{i,t}Q_{i,t}^D$.

On the input side, the cost of the inputs acquired externally—labor, capital, and intermediate inputs—can be summed to arrive at total cost, and this can be divided into the value added of labor and capital, on the one hand, and the cost of acquiring intermediate inputs on the other: $C_{i,t} = P_{i,t}^K K_{i,t} + P_{i,t}^L L_{i,t} + \Sigma_j P_{j,t}Q_{i,j,t}^M$. Gross domestic income (GDI) is then the sum of the value-added components, yielding: $GDI_t = \Sigma_i P_{i,t}^K K_{i,t} + \Sigma_i P_{i,t}^L L_{i,t}$. Because the production and use of intermediate inputs cancel out, the value of aggregate output equals the value of aggregate factor income in each year, or, GDP_t equals GDI_t.

Of what significance is this accounting result for the issues of importance to this chapter? It can be used to show that the growth-accounting results of BLS *do not depend on the existence of Solow's aggregate neoclassical production function*. The sources-of-growth decomposition in equation (1) can be derived directly from the accounting identities of the preceding paragraph that equates GDP and GDI, but only when each side of this equation is expressed in "real" inflation-corrected terms (that is, when nominal prices are replaced with a base-year price index). When this is done, $\text{GDP}_{0,t} = \Sigma_i P_{i,0}^D Q_{i,t}^D$ and $\text{GDI}_{0,t} = \Sigma_i P_{i,0}^K K_{i,t} + \Sigma_i P_{i,0}^L L_{i,t}$, where $\text{GDP}_{0,t}$ and $\text{GDI}_{0,t}$ are real GDP and real GDI in year t expressed in base-year prices. The annual final demand price indexes, $P_{i,t}^D$, and annual factor prices, $P_{i,t}^K$ and $P_{i,t}^L$, may have different time trends, and real $\text{GDP}_{0,t}$ does not in general equal real $\text{GDI}_{0,t}$, except in the base year. In other years, there is a wedge between the two that gives rise to a version of TFP. In its most general formulation, TFP is defined as the ratio of output per unit of total factor input, or equally, the ratio of real GDP to real GDI: $A_t = \text{GDP}_{0,t}/\text{GDI}_{0,t} = \Sigma_i P_{i,0} Q_{i,t}^D / [\Sigma_i P_{i,0}^K K_{i,t} + \Sigma_i P_{i,0}^L L_{i,t}]$. This, indeed, was the way growth accounting was formulated prior to Solow's 1957 paper (Hulten 2001).[10] The larger point is that the neoclassical production function approach is not necessary for the BLS-like growth-accounting results to be obtained, and it is not the only way the TFP results can be interpreted, particularly those relating to the role of skills and education.

3.5   Structural Changes in Education, Skills, and Occupations

The preceding sections are largely technical in nature. The three sections that follow are empirical, and make use of the existing literature to examine the evidence on the trends in labor and capital to see if they are consistent with the predictions of the activity-analysis framework. The third of these sections shows the results of a version of the sources-of-growth account expanded to include intangible capital, and interprets the role of skills and education in light of the "necessary input" activity-analysis model.

3.5.1   Educational Attainment

A look back over the last half century reveals major changes in the educational status of the US population and workforce. In 1960, only 40 percent

10. What Solow did in his 1957 paper was to provide an interpretation of the growth accounting ratio by assuming the existence of an aggregate production function, $Y = AF(K, L)$, in which case TFP = $A = Y/F(K, L)$. Solow's formulation of TFP is thus a special case of the more general formulation, one that summarizes and interprets the messy world of the full activity set, A_t, but also one that loses sight of the messy way activities are organized and the way different inputs and their characteristics actually relate to one another.

of the noninstitutionalized population age twenty-five or older had a high school degree or more, and only 8 percent had a college degree, according to 2015 CPS estimates (US Census Bureau 2015); by 1985, these figures rose to 74 percent and 19 percent; by 2013, almost 90 percent of this population had at least a high school degree, and more than 30 percent had at least a bachelor's degree. Similar numbers are reported in Valletta (chapter 9, this volume) on an employment basis. From 1980 to 2015, the portion of the employed with a high school degree or more went from 80 percent to 90 percent, those with a four-year college degree went from 16 percent to 25 percent, and those with a graduate degree went from 7 percent to 14 percent. In any case, there has been a significant and ongoing increase in educational attainment over the last three to five decades. Valletta also reports that the increase may have slowed in recent years.[11]

Many have noted that the growth in educational attainment coincides with a growth in the return to a college education (Acemoglu and Autor [2012] provide an excellent in-depth look at the data and survey of the associated literature). The estimates of Goldin and Katz (2010) show that the college wage premium relative to a high school degree increased from 40 percent in 1960 to almost 60 percent in 2005, and they attribute this growth to an imbalance in the demand for educated workers and the supply.[12] Valletta's estimates of wage premiums are, again, consistent with the Goldin-Katz results, and they also point to a very large premium for graduate degrees (particularly professional and doctorate degrees). Rising wage premiums are also consistent with an increase in the derived demand for more highly educated workers in conjunction with a lagged response in the supply of college-educated people. Limited substitution possibilities between skilled and underskilled workers in many of the emerging activities of the knowledge economy were a likely contributing factor.[13]

---

11. While the quantity of education, as measured by the growth in degrees, has increased significantly, it should be recognized that formal schooling is not identical to education or human capital accumulation (e.g., family and peer environment also matter). There is also an open question about the quality of education. The NAEP (2013) report card suggests that the literacy and numeracy skills of US 12th graders has been stagnant in recent years, and that a majority of students are stuck at skill levels that are rated below proficient, with one-quarter of students below "basic" in reading and one-third below "basic" in mathematics. Similar results were reported in the NAEP (2015) assessment. Indeed, the proportions have not changed significantly in recent decades. American students also lag those in many other countries, according to the OECD (2013) Programme of International Assessment of Adult Competencies (PIAAC). However, the same study also found that the United States stood out from other countries in its propensity to reward those with the highest skills (Broecke, Quintini, and Vandeweyer, chapter 7, this volume).

12. The Goldin-Katz college wage premium reflects an average across those with college degrees. This should not be confused with the marginal return to further education. Heckman, Humphries, and Veramendi (2016) find that factors like cognitive ability are a significant part of observed educational outcomes and argue that the marginal return may be well below the average.

13. The importance of educational externalities noted by Lucas (1988) is worth repeating here. Because of spillover externalities, the social return to education exceeds the already large private wage premium, and it is the total return that affects economic growth.

3.5.2   Task-Related Skills and Education

Structural changes in the distribution of task-related skills have received a great deal of attention in recent years, following the publication of Autor, Levy, and Murnane (2003). The authors distinguish between nonroutine and routine skills and manual versus analytical skills, and show that the non-routine analytical skills have grown in importance in the last five decades at the expense of the others. An updated version of these results from Autor and Price (2013) found that the gap between nonroutine cognitive and inter-personal skills and the other categories (routine and manual) increased from an index of 100 in 1960 to around 150 in 2010. In studying the college and graduate school wage premiums associated with these different skill categories, Valletta finds a growing premium for all skills, with the largest premiums for nonroutine cognitive skills. The premiums have increased over time, but also may have slowed in recent years.

There is an intuitive similarity between the patterns observed for higher education and higher-order skills, but the actual situation is more nuanced. Skill levels and education are not identical, a point often made in the literature.[14] Skills are appropriately defined as adeptness with respect to a specific task (complex or not), while education is a process though which information is transferred and capabilities developed. Moreover, it is widely recognized that education is only one of the channels through which skills are developed, and that other factors like family background and peer environment and idiosyncratic factors like health and cognitive ability, also matter.

Data from the recent BLS Occupational Requirements Survey (ORS) support this view. The ORS develops a metric specific vocational preparation (SVP) that measures the time spent in skill development, which is described as the time spent in preemployment training (formal education and certification and training programs), prior work experience in related jobs, and the time needed in the job itself to get to average performance (Gittleman, Monaco, and Nestoriak, chapter 5, this volume). When these three types of preparation are cross-classified with the actual time requirements, the authors report that postemployment training and prior work experience are the most important components of SVP, with formal education in third place. However, for those jobs requiring the highest levels of skills, formal education rises to second place behind prior work experience.

That study presents another important finding: those jobs requiring a BA degree or more account for less than 25 percent of all jobs (or less than 30 percent using the O*NET educational classification). It is interesting to note, in this regard, that only about 30 percent of the adult population has one of these degrees. Gittleman et al. also report that only 15 percent of

---

14. Cappelli (2015) argues that education is, at best, only a partial proxy for the full list of skill, ability, and knowledge requirements of most jobs. Education should not, therefore, be treated as equivalent to skills in discussions of skill development or deficits.

jobs were classified in the most complex category. This serves as a warning against an excessive focus on higher education and complex skills, as well as a reminder that a broad range of skills is needed for economic activity, and that those at the lower end are both economically and numerically significant.

However, while this evidence seems to downplay the importance of a college education, the ORS study also finds that higher educational attainment is positively correlated with the complexity of skills and choice of professional occupation. This comes from the part of the study that looks at three mental and cognitive dimensions of job requirements: "task complexity," "work control," and "regular contacts." The first is broken into categories ranging from very complex tasks to very simple; the second into categories ranging from very loose to very close control; the third ranges from structured and very structured regular contacts to very unstructured. One of the most interesting features of this analysis is the high correlation among the higher-skill segments of "task complexity," "work control," "regular contacts" dimensions, as well as the higher-skill components of educational attainment, SVP, and choice of occupation. The fit is not perfect, but a high degree of collinearity does suggest that certain regularities exist that characterize different jobs. Thus, while education is but one of several channels through which skills and expertise are developed, the collinearity suggests a link between higher education and higher-order skill sets. The ORS also reports data on the wage-skill gradient similar to those found in Autor and Handel (2013) and Goldin and Katz, and by Valletta. Those in jobs with the highest task complex skills, the loosest degree of work control, and the least structured interactions all earn significantly higher wages than those at the other end of these scales.[15]

3.5.3   Science, Technology, Engineering, and Mathematics

Developments in science and technology are at the heart of the information revolution and thus merit a close look. This is all the more important because science, technology, engineering, and mathematics (STEM) activities evoke highly educated workers in research labs and computer facilities working on complex problems. However, the study by Rothwell (2013) argues that there are actually two STEM economies. One is a "professional" STEM economy associated with higher education and high levels of compensation, which "plays a vital function in keeping American businesses on the cutting edge of technological development and deployment. Its work-

15. Much attention has been given to the importance of cognitive skills. However, recent research has also focused on the demand for noncognitive skills, which include characteristics like self-discipline, perseverance, attentiveness, dependability, orderliness, persistence in the pursuit of long-term goals, and the ability to get along with others. Deming (2017) shows that the labor market increasingly rewards social skills, and that jobs with high social-skill requirements have shown greater relative growth throughout the wage distribution since 1980. He also observes that the strongest employment and wage growth has occurred in jobs that require both high levels of hard cognitive skills and soft social skills. The importance of noncognitive skills is also noted in Lundberg (2013) and Heckman and Kautz (2012).

ers are generally compensated extremely well." The other STEM economy "draws from high schools, workshops, vocational schools, and community colleges," and its members are "less likely to be directly involved in invention, but they are critical to the implementation of new ideas, and advise researchers on the feasibility of design options, cost estimates, and other practical aspects of technological development." They "produce, install, and repair the products and production machines patented by professional researchers, allowing firms to reach their markets, reduce product defects, create process innovations, and enhance productivity."

Hanson and Slaughter (chapter 12, this volume) report that employment in the STEM professions has grown from around 3.5 percent of the total hours worked in the United States in 1993 to around 6 percent in 2013. In the broader view of STEM employment, Rothwell finds that 20 percent of all 2012 jobs required a "high level of knowledge in any one STEM field" based on his index of the STEM-skill content of various occupations (up from around 8 percent in 1900 and around 15 percent in 1950). He also finds that half of the STEM jobs are "available to workers without a four-year college degree."

The domestic supply of new professionals to the first STEM "economy" has expanded in recent years. National Center for Educational Statistics (NCES) data on STEM degrees completed show an expansion from 1990 to 2011 in BAs (39 percent for engineering, a doubling for science/math), in MAs (90 percent for engineering, 87 percent science/math), and in PhDs (76 percent for engineering, 60 percent science/math). This domestic growth in STEM skills has not, however, been sufficient to satisfy the demand for STEM workers. Hanson and Slaughter report that foreign-born workers currently account for one-half of the hours worked in STEM occupations among prime-age workers with an advanced degree, up from one-quarter in the 1990s and one-fifth in the 1980s. In other words, immigration is an important source of skills that supplements domestic efforts at skill development.

3.6   Structural Change in the Composition of Capital

The activity-analysis model of section 3.4 ties labor of various skills to the capital appropriate to those skills. The preceding section has documented the shift in the distribution of skills toward more complexity, as well as the occupations that embody them, and linked these shifts to the growth in educational attainment. This section documents a parallel shift on the capital side, consistent with the complementarity between capital and labor in the activity-analysis view of production.

The last forty years have seen a significant shift in the composition of investment in the US private business sector, away from tangible structures and equipment toward investments in intangible capital. There has also been a shift within tangible capital toward information and communications tech-

nology (ICT) equipment. Intangible capital is highly firm-specific and produced in-house, and includes such categories as computerized information,
innovative property like R&D, and economic competencies (the categories
proposed by Corrado, Hulten, and Sichel 2005, 2009). The first is mainly
software, and comprises 13 percent of the overall intangible investment rate
in 2010. Innovative property is a diverse group that includes not only the
conventional National Science Foundation (NSF) type of R&D, with its
orientation to science and technology, but also other important forms of
R&D such as investments in artistic originals (books, movies, and music),
development of new financial products, and architectural and engineering
designs. The largest category of intangible capital is economic competencies, divided into brand equity (advertising, marketing, customer support),
firm-specific human capital (worker training), and organizational structure,
a rather amorphous grouping that includes investments in management and
human resource systems, strategic planning, and management consulting.
Many of these intangibles are the source of a firm's intellectual property.

The rate of investment in these intangibles over the period 1977 to 2010
is shown in figure 3.4. The rate rose significantly over the period, starting at
just over 8 percent in 1977 and reaching just under 14 percent by the end of
the period. The growth in importance of this type of capital is in sharp contrast to the declining rate of tangible capital investment shown in the figure,
falling from the 11 percent to 13 percent range in the late 1970s to around 8
percent by the end of the period (9.6 percent in 2007, the last year before the
Great Recession). The overall trends reflect the decision by many companies
to move up the value chain to higher value-added overhead activities like

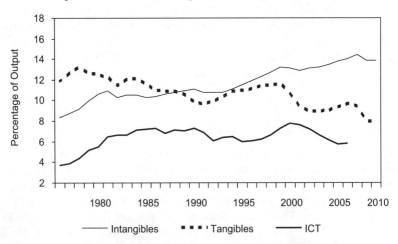

**Fig. 3.4   Investment in intangible capital, tangible capital, and ICT as a share of
private business GDP, 1977–2010**
Source: Data underlying Corrado and Hulten (2010, 2014).
Note: ICT refers to information and communications technology equipment.

R&D, product design, and marketing. It is interesting to note that the overall rate of investment, tangible and intangible combined, remained relatively constant over the period, heightening the importance of structural change for understanding dynamic changes in the economy, and not just the growth of the economy.

When the rate of investment of ICT capital is broken out of total tangible capital in figure 3.4 and shown separately, the ICT investment share is seen to have doubled between the mid-1970s and mid-1980s, then remained relatively constant, and then surged again in the late 1990s before falling back to its post-1980 trend (while the intangible rate continued to increase, though at a much slower pace). However, these patterns do not tell the whole story. While the investment rate of the non-ICT tangible category (not shown) has declined in relative importance in recent years, this category of capital is far from technologically stagnant. The digital revolution has found its way into such non-ICT tangible capital goods as autos and trucks, medical equipment, and machine tools (recall the 2007 paper by Bartel, Ichniowski, and Shaw), as well into some structures. The extent to which technology is embodied in capital is hard to determine, but my own rather dated estimate found a large embodiment effect for the period 1947–1983: the unadjusted annual growth rate of equipment, as estimated by the BLS, was 4.4 percent, while the quality-adjusted rate calculated in the paper was 7.3 percent (Hulten 1992). The BEA does make a quality adjustment to some types of equipment, with those for computing equipment and software being notably large.

The time path of the intangible investment rate is shown again in figure 3.5,

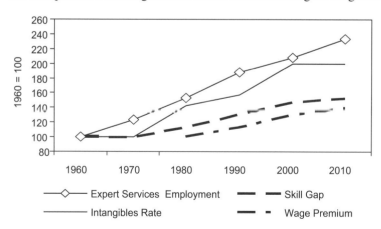

Fig. 3.5   Expert services employment, nonroutine skill gap, intangible investment, and college wage premiums during the expansion of the knowledge economy

Sources: Expert service industries employment: Bureau of Economic Analysis, Industry Economic Accounts, table 6.5, *Full-Time Equivalent Employees by Industry* (see figure 3.2). Skill gap (ratio of nonroutine cognitive and interpersonal indexes to the other indexes): Autor and Price (2013). Intangible investment rate (see figure 3.4). Wage premium based on Valletta (chapter 9, this volume) (average of college-only and graduate premiums, 1980 = 100).

with the 1960 value indexed to 100 in order to facilitate comparison with education and skill indicators. The four variables included in this figure—the rate of intangible investment, "expert" industry employment, the college wage premium, and the Autor-Price gap between nonroutine cognitive and noncognitive skills, each indexed to an initial 100—all show upward trends. The visible association of these trends over the course of the information revolution is far from dispositive, but it does not require much of Solow's suspension of disbelief to recognize in the aggregate data the reality that is readily apparent at the level of the research lab, corporate headquarters, or the plant floor.

3.7   Growth Accounting and Activity Analysis

3.7.1   The Sources-of-Growth Model with Intangible Capital

What does the importance of intangible capital, skills, and education in the activity-analysis parable imply for the sources of growth? While neoclassical and activity-analysis models operate through different mechanisms, the sources-of-growth estimates associated with the former are consistent with those of the latter, as discussed in section 3.4. The conventional BLS sources-of-growth estimates can thus be interpreted in light of either model. When this is done, the activity-analysis reinterpretation assigns a much greater role to education.

The sources-of-growth estimates of this chapter are shown in table 3.1. Unlike the conventional BLS growth accounts, the estimates of this table include the list of intangibles studied by Corrado, Hulten and Sichel (2009).[16] The expanded growth rate of output per hour in the US private business sector over the period 1948 to 2007 (the last year before the financial crisis) is decomposed into the contributions of tangible and intangible capital per labor hour, labor composition, and TFP growth.[17] The top panel shows the

16. The estimates shown in table 3.1 are based on Corrado and Hulten (2010) and updates. When the list of inputs is expanded to include the stock of intangible capital, the concept of output must be expanded to include the corresponding output of intangible investment.

17. When interpreting the capital-labor ratios in table 3.1 in terms of activity analysis, it is important to recognize that the table involves the ratio of different types of capital to *total* labor input; in the case of intangible capital, R, this is R/L. This is not the ratio relevant for the activity-analysis interpretation, which is, instead, the ratio of intangible capital to the labor actually used with intangible capital, R/L_r. The former is related to the latter by the equation $R/L = (R/L_r)(L_r/L)$. In pure activity analysis, R/L_r is given by the technology, and any growth in the ratio is zero. Growth in R/L, as seen in table 3.1, must therefore reflect a change in the employment ratio, L_r/L. The data on employment patterns in figure 3.2 show significant growth in the relative shares of both expert service and overhead organizational services, suggesting that this indeed may have happened. These types of jobs are precisely those most likely to be used with intangible capital, so it is not implausible that much of the observed change in R/L was largely due to an increase in L_r/L. However, this is only a surmise, since there is no tight match between different types of intangible capital (which are quite heterogeneous) and the requisite types of labor skills (also heterogeneous). Moreover, R/L_r itself may well have

Table 3.1   Sources of growth in US private business sector (average of annual growth rates)

	1948–2007	1948–1973	1973–1995	1995–2007
1. Output per hour	2.41	2.99	1.56	2.76
Percentage point contribution to output per hour of:				
2. Tangible capital	0.65	0.76	0.52	0.64
a. ICT equipment	0.23	0.11	0.28	0.36
b. Non-ICT tangible capital	0.42	0.65	0.24	0.27
3. Intangible capital	0.42	0.30	0.39	0.74
a. Computerized information	0.06	0.01	0.07	0.15
b. Innovative property	0.19	0.15	0.16	0.32
(1) R&D (NSF/BEA)	0.10	0.08	0.07	0.17
(2) Other (incl. non-NSF R&D)	0.09	0.07	0.09	0.15
c. Economic competencies	0.17	0.14	0.15	0.27
4. Labor composition	0.20	0.15	0.26	0.20
5. TFP	1.14	1.78	0.39	1.20
Percent of total contribution to output per hour of:				
2. Tangible capital (%)	27	25	33	23
a. ICT equipment (%)	10	4	18	13
b. Non-ICT tangible capital (%)	17	16	13	10
3. Intangible capital (%)	17	10	25	27
a. Computerized information (%)	2	0	4	5
b. Innovative property (%)	8	5	10	12
(1) R&D (NSF/BEA) (%)	4	3	4	6
(2) Other (%)	4	2	6	5
c. Economic competencies	7	5	10	10
4. Labor composition (%)	8	5	17	7
5. TFP (%)	47	60	25	43

Source: Corrado and Hulten (2010).
Notes: ICT refers to information and communications technology equipment, BEA to the Bureau of Economic Analysis, NSF to the National Science Foundation, and TFP to total factor productivity. Details may not add up due to rounding error.

percent contribution of the first four to the growth in output per hour, measured as the growth rate of each multiplied by its income share, with TFP measured as a residual. It is apparent that the sources of growth changed appreciably over the course of the whole period. The contribution of intangible capital increased almost threefold (10 percent to 27 percent) from the first subperiod, 1948–1973, to the last, 1995–2007. The ICT capital experienced a similar proportionate increase (4 percent to 13 percent) and the combined contribution was 40 percent in the last period. The contribution of TFP fell from 60 percent to 43 percent. Labor composition enjoyed a

increased during the ICT Revolution as superior types of intangible and ICT capital entered production and enabled new activities or, alternatively, as the mix of activities shifted to those with a greater degree of capital intensity.

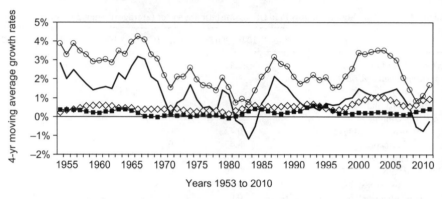

Fig. 3.6   Contribution to labor productivity growth from TFP, intangible capital, and labor composition
Source: Data underlying Corrado and Hulten (2010, 2014).
Note: Labor productivity is output per hour, and labor composition is the labor-composition term.

"boom" in the middle period that saw its contribution increase threefold to 17 percent, but this fell back to 7 percent during the last period.[18]

Figure 3.6 presents these trends in an annual time-series format. The annual growth rate of output per hour, shown at the top of the figure, follows a generally declining, but volatile, path. The same is true of the TFP growth path below it, with the volatility of the former reflected in the latter (no surprise, since TFP is measured as a residual). The growing importance of knowledge capital deepening via intangibles is evident, increasing to the point where its contribution to growth rivals that of the declining TFP trend. The relatively small contribution of labor quality is also shown, indicating an upward surge in the 1980s before falling back during the 1990s.

The neoclassical interpretation of table 3.1 and figure 3.6 suggests an important role for capital deepening via the substitution of capital for labor, and a relatively small role for labor-composition change. The activity-analysis interpretation suggests a different view of the matter, one that interprets the same patterns in terms of the structural change in the composition of activities. In this alternative view, the large contributions of intangible and ICT capital per worker hour evident in this table (and in figure 3.6) were enabled by the growth in educational attainment, skills, and professional occupations.[19] Indeed, the latter were necessary for the growth of the former,

18. Given the prominence of R&D spending in discussions of innovation, it is interesting to note the relatively small (6 percent) role played by scientific "NSF" R&D from 1995 to 2007.
19. Beaudry, Green, and Sand (2016) appeal to the link between knowledge capital and college-educated labor as an explanation for a slowdown in the demand for higher-order skills and higher education after 2000, which they term the "Great Reversal in the Demand for Skill and Cognitive Tasks." They attribute the "reversal" to the slowing growth in ICT

implying that education's role in the growth process was almost certainly much larger than the rather small amount assigned to it by the aggregate approach of neoclassical growth accounting. The contribution of education may be even greater still, since some of its effects may be suppressed in the residual measure of TFP. There are at least three channels through which this can happen. One of the most important for this chapter is the spillover externalities associated with an educated workforce identified in Lucas (1988). In his model of economic development, educated workers interact in ways not captured by private wage premiums, leading to a social return to education that exceeds the private return. The increase in GDP associated with the excess return is not captured by the measured contribution of labor growth or the labor-composition term, and is thus suppressed into the TFP residual (which is thereby overstated).

Much the same can be said of R&D spillovers (Romer 1986, 1990). By its nature, knowledge is nonrival and subject to diffusion, and the social rate of return may therefore exceed the private rate of return to the original innovator. Hall, Mairesse, and Mohnen (2010) review the literature on the relative private and social returns to R&D investment and conclude that the latter is "almost always estimated to be substantially greater than the private returns" (1073). This, too, is suppressed into the TFP residual. Finally, Acemoglu and Autor (2011, 2012) show that task-oriented skill-biased technical change may be suppressed into the TFP residual. Where the conventional Solow model assumes that technical change has the Hicks-neutral form and is thus without a factor bias, they show that when there is a bias that favors skilled workers and occupations, education's observed contribution to growth may be understated and measured TFP overstated.

3.7.2   The Sources of Growth: Firm Dynamics

The statistics of table 3.1 portray growth as a rather "bloodless" and formulaic process in which inputs and technology are mathematically transformed into output. The actual process of growth is anything but "bloodless," involving, as it does, the birth and death of firms and the struggle for survival and success of incumbent firms. Since this chapter has emphasized the importance of structural changes in the microactivities that underpin the aggregate flows of inputs and output, and emphasized the importance of human agency in organizing and staffing these activities, a closer look at the firm dynamics that underpin the evolution of these activities is warranted.

---

equipment and software (which are treated as a general purpose organization technology within the firm). They use a neoclassical optimization approach in their modeling of the link, and a more limited concept of intangible capital. The focus of this chapter is on the contribution of education and skills to economic growth and productivity, using a much broader conception of knowledge capital (all intangible capital and ICT equipment) and stocks as well as flows. The data underlying figure 3.6 of this chapter indicate that the contributions of ICT equipment and software did decline after 2000, but also there was not much of a decline in the contribution of the rest of nonsoftware intangible capital (although there was a large amount of cyclical variability).

The industries in the private economy are typically composed of both large and small firms, as well as older and newer ones. Research has shown that all firms are not equal when it comes to growth, and that those that are relatively young and rapidly growing are responsible for a disproportionate amount of net job creation (Haltiwanger, Jarmin, and Miranda 2013; Strangler 2010; and Sadeghi, Spletzer, and Talan 2012). Strangler finds that, in a typical year, fast-growing young firms ("gazelles") made up less than 1 percent of all companies, but generated about 10 percent of all new jobs. Sadeghi, Spletzer, and Talan report that the 0.5 percent of all companies classified as "high-growth firms" between 2008 and 2011 were responsible for a third of all gross job creation among firms whose employment increased over the period. Moreover, smaller firms are also an important source of R&D spending. According to NSF data, small companies with fewer than 500 employees in 2009 had an average R&D investment *rate* that was three times that of the largest firms and employed a third of R&D workers, despite their much smaller sales and overall employment.

Hathaway and Litan (2014) highlight the importance of firm births and deaths. They note that one new business is born approximately every minute, and that another business fails every eighty seconds. They go on to show that jobs are both created and destroyed in the process, with net job creation of 600,000 jobs in 2012. This "churn," as they call it, suggests a Schumpeterian view of firm dynamics in which growth is neither smooth nor formulaic. It is a process in which good decisions and good luck tend to be rewarded and inadequate or obsolete business models punished. By implication, human agency and competence in the formulation and execution of business models, and in making the investments needed to enhance a firm's capabilities and products, are critical in order for new entrants to become gazelles and for incumbents to prosper.

The churning of firms through entry and exit has implications for economic growth. It is an important mechanism through which new products and processes enter the economy, and through which new markets are developed. Intangible capital and higher-order skills, cognitive and noncognitive, play a major role in this process. The most important asset of a successful new enterprise is the capability (though not necessarily higher education) of those who start and guide its development, who manage its operation, and who foster technological and organizational innovation. The study by Kerr and Kerr (2017) shows that these key ingredients are sometime "imported," as witnessed by the finding that around a quarter of all entrepreneurs in 2008 were immigrants, up from some 17 percent in 1995.[20] They also report that

---

20. The notion of "entrepreneur" used here is defined as someone who is among the top three initial earners in the new business. Kerr and Kerr also report that their findings are roughly comparable to those in the large literature they review, though a few report appreciably lower percentages.

38 percent of new firms had at least one immigrant entrepreneur, and that the share of employees in new firms who were immigrants was 26 percent.

3.8 Summary and Conclusions

The neoclassical model and the activity-analysis model of this chapter offer different windows on the role of education in the process of economic growth, two ways of looking into the same complex processes involved. The activity-analysis perspective provides insights into the role of skill development and education in the functioning of the economy, a perspective that is important because workers with different skills and levels of education are not freestanding ingredients in a recipe for making aggregate output. They are the necessary ingredients of the specific recipe for which they are needed, in conjunction with the capital and other inputs required in order to operate the activity at a given scale. A deficit in either the requisite skills or the associated capital limits the operation or growth of those activities. To repeat, it is hard to imagine today's emerging knowledge economy operating with a workforce in which less than half the workers had a high school degree, and less than 10 percent had a college degree.

What the future actually holds for continued economic growth and employment is a matter of great conjecture. Powerful technological and global forces continue to shape the world of work, and one can only guess where they will lead in the "race against the machine" of Brynjolfsson and McAfee (2014). Looking backward at the data, the importance of the high-skill-occupation/education nexus for past economic growth seems well established. Looking ahead, it may well be that robots will ultimately make most human work skills obsolete. It may be that education will increasingly be seen as preparation for a productive life of leisure. But for now, it seems reasonable to conclude that a strong educational system—one that provides a full range of skill development—remains an essential part of America's economic prosperity. As Levy and Murnane (2013, 5) put it: "For the foreseeable future, the challenge of 'cybernation' is not mass unemployment but the need to educate many more young people for the jobs computers cannot do."

References

Acemoglu, D., and D. Autor. 2011. "Skills, Tasks and Technologies: Implications for Employment and Earnings." *Handbook of Labor Economics* 4:1043–171.
———. 2012. "What Does Human Capital Do? A Review of Goldin and Katz's The Race between Education and Technology." *Journal of Economic Literature* 50 (2): 426–63.

Autor, D. H., D. Dorn, and G. H. Hanson. 2013. "The China Syndrome: Local Labor Market Effects of Import Competition in the United States." *American Economic Review* 103 (6): 2121–68.

Autor, D. H., and M. J. Handel. 2013. "Putting Tasks to the Test: Human Capital, Job Tasks and Wages." *Journal of Labor Economics* 31 (2, pt. 2): S59–96.

Autor, D. H., F. Levy, and R. J. Murnane. 2003. "The Skill Content of Recent Technological Change: An Empirical Investigation." *Quarterly Journal of Economics* 118 (4): 1279–333.

Autor, D. H., and B. Price. 2013. "The Changing Task Composition of the US Labor Market: An Update of Autor, Levy, and Murnane (2003)." MIT working paper, Massachusetts Institute of Technology, June.

Bartel, A. P., C. Ichniowski, and K. L. Shaw. 2007. "How Does Information Technology Affect Productivity? Plant-Level Comparisons of Product Innovation, Process Improvement, and Worker Skills." *Quarterly Journal of Economics* 122 (4): 1721–58.

Beaudry, P., D. A. Green, and B. Sand. 2016. "The Great Reversal in the Demand for Skill and Cognitive Tasks." *Journal of Labor Economics* 34 (1, pt. 2): S199–247.

Bloom, N., and J. Van Reenen. 2007. "Measuring and Explaining Management Practices across Firms and Nations." *Quarterly Journal of Economics* 122 (4): 1351–408.

Brynjolfsson, E., and A. McAfee. 2014. *Race against the Machine: How the Digital Revolution Is Accelerating Innovation, Driving Productivity, and Irreversibly Transforming Employment and the Economy.* Digital Frontier Press, Jan. 29.

Bureau of Labor Statistics. 1983. *Trends in Multifactor Productivity, 1948–81,* Bulletin 2178. Washington, DC: US Government Printing Office. Sept. 1983, and updates.

Cappelli, P. 2015. "Skill Gaps, Skill Shortages, and Skill Mismatches." *ILR Review* 68 (2): 251–90.

Corrado, C. A., and C. R. Hulten. 2010. "How Do You Measure a Technological Revolution?" *American Economic Review* 100 (2): 99–104.

———. 2014. "Innovation Accounting." In *Measuring Economic Sustainability and Progress,* edited by D. W. Jorgenson, J. S. Landefeld, and P. Schreyer, 595–628. Chicago: University of Chicago Press.

Corrado, C., C. Hulten, and D. Sichel. 2005. "Measuring Capital and Technology: An Expanded Framework." In *Measuring Capital in the New Economy,* vol. 65, edited by C. Corrado, J. Haltiwanger, and D. Sichel, 11–41. Chicago: University of Chicago Press.

———. 2009. "Intangible Capital and Economic Growth." *Review of Income and Wealth* 55 (3): 661–85.

Cowles Commission. 1951. *Activity Analysis of Production and Allocation.* Proceedings of a Conference, edited by Tjalling C. Koopmans, in cooperation with Armen Alchian, George B. Danzig, Nicholas Georgescu-Roegen, Paul A. Samuelson, and Alert W. Tucker, The Cowles Commission. New York: John Wiley & Sons.

Deming, D. J. 2017. "The Growing Importance of Social Skills in the Labor Market." *Quarterly Journal of Economics* 132 (4): 1593–640.

Fisher, F. M. 1969. "The Existence of Aggregate Production Functions." *Econometrica* 37 (4): 553–77.

Goldin, C., and L. F. Katz. 2010. *The Race between Education and Technology.* Cambridge, MA: Belknap Press.

Hall, Bronwyn, Jacques Mairesse, and Pierre Mohnen. 2010. "Measuring the Return to R&D." In *Handbook of the Economics of Innovation,* edited by B. H. Hall and N. Rosenberg, 1033–82. Amsterdam: Elsevier-North Holland.

Haltiwanger, J. C., R. S. Jarmin, and J. Miranda. 2013. "Who Creates Jobs? Small versus Large versus Young." *Review of Economics and Statistics* 95 (2): 347–61.

Hanushek, E. A., and L. Woessmann. 2015. *The Knowledge Capital of Nations: Education and the Economics of Growth.* Cambridge, MA: MIT Press.

Harcourt, G. C. 1969. "Some Cambridge Controversies in the Theory of Capital." *Journal of Economic Literature* 7 (2): 369–405.

Haskel, J., R. Z. Lawrence, E. E. Leamer, and M. J. Slaughter. 2012. "Globalization and U.S. Wages: Modifying Classic Theory to Explain Recent Facts." *Journal of Economic Perspectives* 26 (2): 119–40.

Hathaway, I., and R. E. Litan. 2014. "Declining Business Dynamism in the United States: A Look at States and Metros." Economic Studies Report, Brookings Institution, May. https://www.brookings.edu/research/declining-business-dynamism-in-the-united-states-a-look-at-states-and-metros/.

Heckman, J. J., J. E. Humphries, and G. Veramendi. 2016. "Returns to Education: The Causal Effects of Education on Earnings, Health and Smoking." NBER Working Paper no. 22291, Cambridge, MA. http://www.nber.org/papers/w22291.

Heckman, J. J., and T. Kautz. 2012. "Hard Evidence on Soft Skills." *Labour Economics* 19 (4): 451–64.

Hulten, C. R. 1973. "Divisia Index Numbers." *Econometrica* 41 (6): 1017–25.

———. 1992. "Growth Accounting When Technical Change Is Embodied in Capital." *American Economic Review* 82 (4): 964–80.

———. 2001. "Total Factor Productivity: A Short Biography." In *New Developments in Productivity Analysis*, edited by Charles R. Hulten, Edwin R. Dean, and Michael J. Harper, 1–47. Chicago: University of Chicago Press.

Jorgenson, D. W., and Z. Griliches. 1967. "The Explanation of Productivity Change." *Review of Economic Studies* 34:349–83.

Kerr, S. P., and W. R. Kerr. 2017. "Immigrant Entrepreneurship." In *Measuring Entrepreneurial Business: Current Knowledge and Challenges*, edited by John Haltiwanger, Erik Hurst, Javier Miranda, and Antoinette Schoar, 187–249. Chicago: University of Chicago Press.

Levy, F., and R. Murnane. 2013. "Dancing with Robots: Human Skills for Computerized Work." *Third Way*, June 1. https://www.thirdway.org/report/dancing-with-robots-human-skills-for-computerized-work.

Lucas, R. E. Jr. 1988. "On the Mechanics of Economic Development." *Journal of Monetary Economics* 22:3–42.

Lundberg, S. 2013. "The College Type: Personality and Educational Inequality." *Journal of Labor Economics* 31 (3): 421–41.

National Assessment of Educational Progress (NAEP). 2013. The Nation's Report Card, *Are the Nation's Twelfth-Graders Making Progress in Mathematics and Reading?* https://www.nationsreportcard.gov/reading_math_g12_2013/#/student-progress.

———. 2015. The Nation's Report Card, *Math and Reading Scores at Grade 12*. https://www.nationsreportcard.gov/reading_math_g12_2015/.

National Science Foundation. 2012. *National Patterns of R&D Resources: 2009 Data Update*, National Center for Science and Engineering Statistics, Detailed Statistical Tables, NSF 12-321. June.

Nelson, R. R., and E. S. Phelps. 1966. "Investment in Humans, Technological Diffusion, and Economic Growth." *American Economic Review* 56 (1/2): 69–75.

Nelson, Richard R., and Sidney G. Winter. 1982. *An Evolutionary Theory of Economic Change.* Cambridge, MA: Belknap Press.

OECD. 2013. *OECD Skills Outlook 2013: First Results from the Survey of Adult Skills.* OECD Publishing. http://dx.doi.org/10.1787/9789264204256-en.

Prescott, E. C., and M. Visscher. 1980. "Organizational Capital." *Journal of Political Economy* 88 (3): 446–61.

Romer, P. M. 1986. "Increasing Returns and Long-Run Growth." *Journal of Political Economy* 94 (5): 1002–37.

———. 1990. "Endogenous Technological Change." *Journal of Political Economy* 98 (5, pt. 2): S71–102.

Rothwell, J. 2013. "The Hidden STEM Economy." Metropolitan Policy Program, Brookings Institution, June. https://www.brookings.edu/research/the-hidden -stem-economy/.

Sadeghi, A., J. R. Spletzer, and D. M. Talan. 2012. "High Growth Firms." Unpublished manuscript, US Bureau of Labor Statistics, Mar. 30.

Samuelson, P. A. 1962. "Parable and Realism in Capital Theory: The Surrogate Production Function." *Review of Economic Studies* 29 (3): 193–206.

Solow, R. M. 1957. "Technical Change and the Aggregate Production Function." *Review of Economics and Statistics* 39:312–20.

———. 1987. "We'd Better Watch Out." *New York Times Book Review*, July 12.

———. 1988. *Growth Theory: An Exposition*. New York: Oxford University Press.

Solow, R. M., J. Tobin, C. C. von Weizsacker, and M. Yaari. 1966. "Neoclassical Growth with Fixed Factor Proportions." *Review of Economic Studies* 33 (2): 79–115.

Strangler, D. 2010. "High-Growth Firms and the Future of the American Economy." Ewing Marion Kauffman Foundation, March. https://www.kauffman.org/what -we-do/research/firm-formation-and-growth-series/highgrowth-firms-and-the -future-of-the-american-economy.

US Census Bureau. 2015. *CPS Historical Time Series Tables*, Table A.1. "Years of School Completed by People 25 Years and over, by Age and Sex: Selected Years 1940 to 2015." https://www.census.gov/data/tables/time-series/demo/educational -attainment/cps-historical-time-series.html.

Wolfe, R. M. 2012. "Business R&D Performed in the United States Cost $291 Billion in 2008 and $282 Billion in 2009." *InfoBrief*, National Science Foundation, Social, Behavioral and Economic Sciences, NSF 12-309, March. https://www.nsf .gov/statistics/infbrief/nsf12309/.

II

Jobs and Skills Requirements

4

Underemployment in the Early Careers of College Graduates following the Great Recession

Jaison R. Abel and Richard Deitz

"Welcome to the Well-Educated-Barista Economy"
—Galston, *Wall Street Journal*

4.1 Introduction

The image of a young newly minted college graduate working behind the counter of a hip coffee shop has become a hallmark of the plight of college graduates following the Great Recession. Indeed, although economic conditions steadily improved through the recovery, significant slack remained in the labor market, and many recent graduates were not finding jobs commensurate with their education. The underemployment rate for recent college graduates—that is, the share working in jobs that typically do not require a college degree—continued to climb for several years following the Great Recession, topping out at nearly 50 percent, a level not seen since the early 1990s.

While underemployment among recent college graduates has attracted wide attention in the media and among policymakers, very little is actually known about the nature of college underemployment or what seems to make some college graduates more prone to being underemployed than others.[1] In this chapter, we examine the plight of college graduates in the aftermath of the Great Recession. We examine in detail the types of jobs underemployed

Jaison R. Abel is an assistant vice president and head of the Regional Analysis Function at the Federal Reserve Bank of New York. Richard Deitz serves as assistant vice president and senior economist for the Federal Reserve Bank of New York.

The views and opinions expressed here are solely those of the authors and do not necessarily reflect those of the Federal Reserve Bank of New York or the Federal Reserve System. For acknowledgments, sources of research support, and disclosure of the authors' material financial relationships, if any, please see http://www.nber.org/chapters/c13697.ack.

1. For example, a 2012 *Associated Press* article with the headline "Half of New Grads are Jobless or Underemployed" reignited an intense debate about the value of a college degree. Headlines such as "College Grads May Be Stuck in Low-Skill Jobs" (Casselman 2013) and "Welcome to the Well-Educated-Barista Economy" (Galston 2014) became commonplace after the Great Recession.

college graduates hold, and explore some of the factors associated with a greater likelihood of being underemployed.

We conclude that while there is *some* truth behind the popular image of the college-educated barista, this picture is not an accurate portrayal of the typical underemployed recent college graduate. Contrary to popular perception, we show that only a small fraction of recent graduates worked in a low-skilled service job following the Great Recession. Instead, we find that underemployed recent graduates held a wide range of jobs, and while most are clearly not equivalent to jobs that require a college degree, some are fairly skilled and well paid. In addition, we find that underemployed college graduates were more likely to be working in these higher-paying noncollege jobs than similarly aged young workers without a college degree. Still, we find that roughly 9 percent of recent graduates—or about one-fifth of the underemployed—start their careers working in a low-skilled service job.

We then explore the characteristics of underemployed recent college graduates, and examine correlates associated with being underemployed or working in a low-skilled service job. We find that men are more likely to be underemployed than women, though a larger share of underemployed men work in the highest-paying noncollege jobs. Further, we show that underemployment is far more likely for recent graduates with some college majors compared to others. For example, those with majors in liberal arts or general business are two to three times more likely to be underemployed than those with engineering or nursing majors. The patterns we uncover suggest that those recent graduates who major in more quantitatively oriented and occupation-specific fields tend to have much lower underemployment than those with majors that are more general. Finally, our analysis suggests that underemployment is a temporary phase for a good number of recent graduates, particularly among those who start their careers working in a low-skilled service job, as many transition to better jobs after spending a few years in the labor market.

Though underemployment appears to have become increasingly prevalent in the labor market, particularly among college graduates, only a small body of research on the subject currently exists. Much of this research focuses on underemployment among reemployed workers following layoffs, or those who work in part-time or temporary positions (see, e.g., Feldman 1996; McKee-Ryan and Harvey 2011). In addition, much of the existing underemployment literature emphasizes the emotional and psychological effects of underemployment, rather than its economic consequences. An early exception is Feldman and Turnley (1995), who study underemployment among a small sample of recent college graduates with business degrees, and more recently, Abel, Deitz, and Su (2014) provide some historical context by examining underemployment among recent college graduates over the past few decades. Our work builds on this small body of research by providing a more detailed analysis of the types of jobs held by underemployed graduates in the early stages of their careers, and by identifying the factors that make some graduates more prone to underemployment than others.

One strand of the literature that is closely related to underemployment examines overeducation in the labor market (see, e.g., Hersch 1991; Chevalier 2003; Chevalier and Lindley 2009; Green and Zhu 2010). However, unlike our work, this research typically relies on self-reported measures of whether there is a match between a worker's education and job to assess the extent and economic effects of overeducation.

Our work is also related to a small but growing literature documenting the economic consequences of graduating from college during recessions (see, e.g., Kahn 2010; Oreopoulos, von Wachter, and Heisz 2012; Altonji, Kahn, and Speer 2016). This research indicates that adverse labor market conditions in the early careers of college graduates can have significant long-term effects on earnings, and shows that these negative effects differ greatly by college major and ability. These studies generally do not directly examine the types of jobs graduates obtain in the early stages of their careers. However, differences in the quality of the initial placement of graduates with more challenging college majors or higher ability is believed to be an important contributor to differences in longer-term employment outcomes. Our work provides some support for this explanation by documenting that recent graduates with college majors that provide technical training and quantitative skills are far less likely to be underemployed in the early stages of their careers than those with majors that tend to be less quantitative in nature.

Indeed, the role of college major in finding a good job has become of considerable interest in recent years given the weak labor market following the Great Recession. Recent research has documented significant heterogeneity in the labor market outcomes of college graduates with different majors (see, e.g., Altonji, Blom, and Meghir 2012; Altonji, Kahn, and Speer 2014, 2016), and information on labor market outcomes by major has been shown to influence the choices students make while in college (see, e.g., Betts 1996; Zafar 2013; Wiswall and Zafar 2015a, 2015b). Our work adds to this body of research by providing new information about how one's college major is associated with an understudied labor market outcome—the likelihood of being underemployed upon graduation. Further, we are able to examine labor market outcomes for a more detailed set of college majors than has previously been studied.

4.2   The Labor Market for College Graduates following the Great Recession

The Great Recession was the deepest downturn experienced in the United States in the postwar era, and its effects on the labor market were swift and severe. Though labor market conditions started to improve in early 2010, the recovery that followed was slow and uneven, resulting in a large amount of slack that persisted for an extended period of time (see, e.g., Elsby, Hobijn, and Şahin 2010; Elsby et al. 2011; Şahin et al. 2014). Those unlucky college

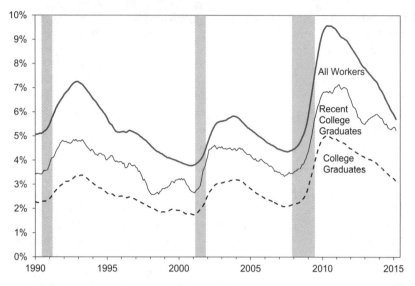

Fig. 4.1   Unemployment among college graduates

Source: US Census Bureau and US Bureau of Labor Statistics, Current Population Survey.
Notes: Rates are calculated as a twelve-month moving average. Recent college graduates are those age twenty-two to twenty-seven with a bachelor's degree or higher, while college graduates are those age twenty-two to sixty-five with a bachelor's degree or higher. All workers are those age sixteen to sixty-five regardless of education. All figures exclude those in the military or currently enrolled in school. Shaded area indicates period designated recession by the NBER.

graduates who started their careers in the aftermath of the Great Recession struggled to find jobs, let alone jobs that utilized their degrees. Much of this difficulty can be traced to relatively weak labor demand for college graduates during the recovery.

4.2.1   Unemployment among College Graduates

Though college graduates generally weathered the economic storm better than those without a degree, they were not immune from its effects. As figure 4.1 shows, unemployment rose sharply during the Great Recession and continued to climb in the early stages of the recovery to levels not seen in decades. Figure 4.1 also shows the unemployment rate for recent college graduates. For the purposes of our analysis, we define recent college graduates as those with at least a bachelor's degree who are twenty-two to twenty-seven years old. We select this group to capture college graduates within their first five years after graduation who are at the beginning of their careers.[2]

2. The typical age at which people earn a bachelor's degree in the United States is twenty-two. While some graduates receive their degree at ages beyond their early twenties, data limitations do not allow us to identify these older graduates. We exclude those in the military and individuals enrolled in school, whether full time or part time, to avoid confusion about whether someone's employment status is influenced by whether they are attending school.

Unemployment among recent college graduates, who are often more susceptible to cyclical changes in the labor market than college graduates as a whole, doubled from about 3.5 percent before the recession to a peak of more than 7 percent in 2011. However, unemployment among recent college graduates began to fall in late 2011, and generally continued to trend down thereafter. Even with this progress, unemployment among recent college graduates fell less steeply than for college graduates as a whole, underlying the more negative effects of labor market conditions for recent graduates compared to their more seasoned counterparts.

4.2.2   Underemployment among College Graduates

While the unemployment rate has declined, such a statistic reveals only part of the story about the plight of recent college graduates following the Great Recession. Indeed, the weak labor market prompted widespread concern that recent graduates were underemployed—that is, working in jobs that typically do not require a college degree (see, e.g., Fogg and Harrington 2011; Yen 2012; Vedder, Denhart, and Robe 2013).

We measure the underemployment rate as the share of employed college graduates working in jobs that do not require a college degree. To distinguish between college jobs and noncollege jobs, we rely on the Department of Labor's O*NET database.[3] The O⁺NET contains occupation-level data for hundreds of occupations collected via interviews of incumbent workers and input from professional occupational analysts on a wide array of job-related requirements. We use the following question from the O*NET Education and Training Questionnaire to determine whether an occupation requires a college degree: "If someone were being hired to perform this job, indicate the level of education that would be *required?*" (emphasis added). Respondents then select from twelve detailed education levels, ranging from less than a high school diploma to postdoctoral training. We consider a college education to be a requirement for a given occupation if more than 50 percent of the respondents working in that occupation indicated that at least a bachelor's degree was necessary to perform the job.[4]

We show the underemployment rate in figure 4.2 for both recent college graduates and college graduates as a whole. The underemployment rate for recent college graduates consistently holds well above the rate for all college graduates, which has hovered at around one-third for at least the past twenty-five years, reflecting the challenges faced by newly minted graduates as they enter the labor market. Focusing on the period following the Great

---

3. We use O*NET Version 18.1 for our analysis (see http://www.onetcenter.org/ for more information). The O*NET database is discussed in detail by Peterson et al. (2001).

4. We selected this threshold because it indicates that the majority of respondents believe that at least a bachelor's degree is required to perform a given job. In practice, however, few occupations are clustered around the 50 percent threshold. For most occupations, respondents either overwhelmingly believe that a bachelor's degree is required for the job or not.

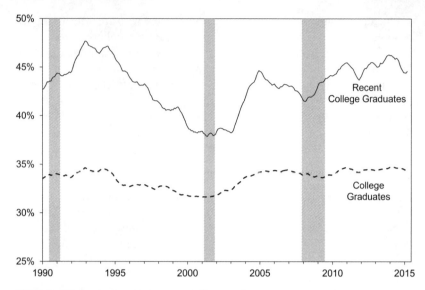

Fig. 4.2   Underemployment among college graduates

Sources: US Census Bureau and US Bureau of Labor Statistics, Current Population Survey; US Department of Labor, O*NET.

Notes: Rates are calculated as a twelve-month moving average. Recent college graduates are those age twenty-two to twenty-seven with a bachelor's degree or higher, while college graduates are those age twenty-two to sixty-five with a bachelor's degree or higher. All figures exclude those in the military or currently enrolled in school. Shaded area indicates period designated recession by the NBER.

Recession, apart from a brief dip in early 2011, the underemployment rate for recent college graduates continued to climb well into 2014, rising to more than 46 percent, a level not seen since the early 1990s. This divergence between falling unemployment and rising underemployment among recent college graduates between mid-2011 and mid-2014 suggests that more graduates were finding jobs during this time, just not necessarily good ones.

Of note, underemployment is not a new phenomenon facing young graduates in recent years. Indeed, underemployment among recent college graduates was on an upward trend for several years *before* the Great Recession. While there appears to be a cyclical component to underemployment among recent college graduates, the broader V-shaped pattern in the underemployment rate over the past twenty-five years is also consistent with recent research by Beaudry, Green, and Sand (2014, 2016) arguing that there has been a reversal in the demand for cognitive skills since 2000. According to this research, businesses ramped up their hiring of college-educated workers in an effort to adapt to the technological changes occurring during the 1990s. However, as the information technology revolution reached maturity, demand for cognitive skill fell accordingly. As a result, during the first decade of the twenty-first century, many college graduates were forced to move down the job ladder to take jobs typically performed by lower-skilled

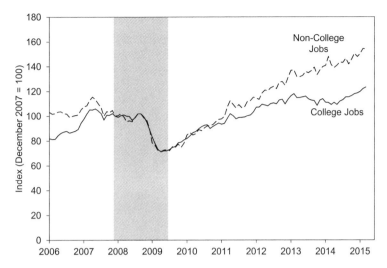

Fig. 4.3   The demand for college graduates through the Great Recession
Source: The Conference Board, Help Wanted OnLine; US Department of Labor, O*NET.
Note: Shaded area indicates period designated recession by the NBER.

workers. From this perspective, the relatively low underemployment rates among recent college graduates at the peak of the technology boom around 2000 may in fact be an outlier, while the rise in underemployment since then represents a return to more typical conditions.

4.2.3   The Demand for College Graduates after the Great Recession

To gain a better understanding of what is behind recent patterns in both unemployment and underemployment among college graduates, we measure the availability of college jobs and noncollege jobs around the Great Recession. We use data on online job postings from The Conference Board's Help Wanted OnLine (HWOL) database, which provides information on the full universe of online job postings during this period and serves as a comprehensive measure of labor demand.[5] We use monthly data measuring total advertised job postings. Importantly, for our purposes, the HWOL database assigns a detailed occupation code to each advertised posting. We use these occupation codes to distinguish between college jobs and noncollege jobs using the O*NET classification defined previously.

The trend in job postings for both types of jobs is shown in figure 4.3. Although postings for college jobs and noncollege jobs rebounded at

5. Advertised job vacancies are collected from more than 16,000 online job boards, including corporate job boards, and efforts are made to remove duplicate postings. (See https://www.conference-board.org/data/helpwantedonline.cfm for more information on the HWOL database.) Because the earliest available HWOL data start in 2005, we are not able to examine the extent to which the demand for college graduates started to decline around 2000, as suggested by Beaudry, Green, and Sand (2014, 2016).

roughly the same pace immediately following the Great Recession, by 2011 the demand for college graduates began to fall behind. In fact, postings for college jobs leveled off around 2013, and even declined slightly through mid-2014, while postings for noncollege jobs continued to rise at a fairly steady clip throughout the recovery.

The steady growth of noncollege jobs, coupled with the relatively soft demand for college graduates during this three-year period, appears to have forced many recent college graduates to take jobs not commensurate with their education. With the demand for college graduates rising again beginning in mid-2014, underemployment also started to come down. However, even with this modest improvement, 44.6 percent of college graduates—nearly one in two—found themselves underemployed in the early stages of their careers following the Great Recession. However, these data reveal little about the types of jobs these underemployed workers were performing.

4.3   Are All Underemployed College Graduates Working as Baristas?

To provide a deeper understanding of the types of jobs held by underemployed recent college graduates in the years following the Great Recession, we turn to the American Community Survey (ACS), a nationally representative 1 percent sample of the population conducted on an annual basis (Ruggles et al. 2015). These data include a variety of detailed economic and demographic information for individuals, including a person's occupation, wage, and education. We pool annual data for the years 2009 to 2013, leaving us with a roughly 5 percent random sample of the US population.

Our sample of recent college graduates contains nearly 180,000 observations representing more than 20 million individuals during the 2009 to 2013 period. For comparison purposes, we also construct a parallel sample of young workers age twenty-two to twenty-seven without a college degree. This sample contains roughly 346,000 observations representing about 44 million individuals over this same period. Because men and women may choose different career paths or have different experiences in the labor market, we perform all of our analyses overall and separately by gender.

4.3.1   Types of Jobs Held by Underemployed College Graduates

What types of jobs are underemployed recent graduates performing, and how common is it for such workers to be stuck in a low-paying job, such as a coffee house barista? To address these questions, we create ten underemployed occupation categories from the hundreds of detailed occupation codes identified in the data. In forming these occupation categories, we attempted to create groups with a reasonably comparable set of knowledge and skill requirements based on the nature of the work performed. In some cases, we also used average wages earned in these detailed occupations to assign them to these categories. Table 4.1 displays these groupings together

Table 4.1   Occupation categories of underemployed college graduates

Occupation category	Average wage, full-time workers ($)	Average monthly job postings	Percent growth in postings
Information processing and business support	59,059	188,000	63
Managers and supervisors	55,415	359,200	122
Public safety	52,567	31,300	76
Sales	52,474	293,700	66
Arts and entertainment	48,765	29,000	9
Skilled trades	47,268	158,000	162
Office and administrative support	37,207	351,000	57
Health care technicians and assistants	36,223	220,500	34
Physical laborers	33,006	275,200	285
Low-skilled service	23,584	271,100	133

Source: US Census Bureau, American Community Survey, 2009–2013; The Conference Board, Help Wanted OnLine; US Department of Labor, O*NET.
Notes: Average wages are calculated for all workers age twenty-two to sixty-five who usually work at least thirty-five hours per week for forty or more weeks per year. Average monthly job postings are calculated for the years 2009 to 2013. Percent growth in postings is calculated from mid-2009, the end of the Great Recession, through mid-2014.

with the average wage paid to all workers in each group, not just recent college graduates.[6]

These occupation categories fall into six tiers based on how well jobs in each group tend to pay. The first tier contains two groups of relatively high-paying jobs, where workers on average earn more than $55,000 per year. The highest-paying occupation category, Information Processing and Business Support, tends to emphasize cognitive skills, and workers in these jobs typically work with technology, use or produce information in their jobs, and often play a supporting role to others within their line of business. Examples of the kinds of jobs included in this category are human resource workers, computer support specialists, web developers, computer network architects, and paralegals. The next highest-paying category is Managers and Supervisors, which includes workers who have direct oversight of other employees within their organization, and are often responsible for managing part of a business. Some decision-making is typically required in these types of jobs, but such decisions are often fairly limited in scope. Examples of jobs that fall within this category include first-line supervisors of various types of workers (e.g., retail sales, administrative support, and production) and food service managers.

6. We focus on the average wages of all workers in these occupation categories to give a general sense about the relative differences in skill levels across the categories we create. While recent college graduates tend to earn less than these figures, largely because such workers are in the early stages of their careers, the pattern for recent graduates is similar to that for all workers.

The second tier of underemployed occupation categories tend to pay between $50,000 and $55,000 per year, and includes Public Safety and Sales jobs. Jobs in the Public Safety category emphasize a combination of physical and cognitive skills, and workers in these types of jobs tend to protect and serve the public. Examples of the kinds jobs included in this category are police officers, detectives, security guards, and firefighters. Jobs in the Sales category tend to require strong interpersonal skills and the ability to interact with customers. Workers in these jobs are responsible for selling a wide array of goods and services, ranging from physical products found on the shelves of retail stores to insurance policies and real estate. Examples of the kinds of jobs included in this category are sales representatives, insurance agents, real estate brokers, as well as retail salespersons.[7]

The third tier of underemployed occupations pays, on average, around $48,000, and includes Arts and Entertainment and Skilled Trades. Workers in these jobs are often highly skilled, but these are not the types of skills typically developed by earning a college degree. Examples of the types of jobs captured in this tier include professional athletes, musicians, actors, and dancers, as well as electricians, machine repairers, plumbers, and welders.

The fourth tier has average annual earnings ranging between $35,000 and $40,000. This tier includes two groups. First, Office and Administrative Support, which tends to emphasize clerical knowledge, oral and written communication skills, and basic proficiency with computers. While some cognitive skills are required, the demands are typically below what is required of workers in Information Processing and Business Support jobs. Examples of jobs in this category include secretaries, customer service representatives, and office clerks. Second, this tier includes Health Care Technicians and Assistants. Workers in these jobs provide care for others, but typically in a role that supports a health care practitioner. Many of these jobs require an associate's degree or some other type of training certificate. Examples of the jobs in this category are medical assistants, nursing aides, diagnostic technicians, and dental hygienists.

The fifth tier consists of Physical Laborers. Jobs in this category tend to emphasize the physical dimension of a worker's skill set, such as strength, agility, and dexterity. Examples of jobs in this category include construction laborers, truck drivers, roofers, and highway maintenance workers.

Finally, the lowest-paying tier consists of Low-Skilled Service jobs, which tend to pay around minimum wage.[8] These are the types of jobs that, rightly or wrongly, have become the poster child for underemployed young college

---

7. While retail sales jobs might be viewed as similar to low-skilled service jobs, retail sales jobs tend to require more skill, particularly in the areas of communication and persuasion, and pay significantly higher wages, even for young college graduates.
8. Autor and Dorn (2013) demonstrate that growth in these types of jobs has been strong in recent decades, which has contributed to the polarization of the US workforce.

Table 4.2   Share of underemployed recent college graduates by occupation category

Occupation category	Share of underemployed recent college graduates	Share of young workers without a college degree
Information processing and business support	11.4	2.0
Managers and supervisors	13.1	7.8
Public safety	3.7	2.8
Sales	11.7	5.1
Arts and entertainment	3.0	0.7
Skilled trades	2.7	8.2
Office and administrative support	25.2	15.0
Health care technicians and assistants	4.7	6.6
Physical laborers	5.4	24.1
Low-skilled service	19.3	27.6

Source: US Census Bureau, American Community Survey, 2009–2013.
Notes: Recent college graduates are those age twenty-two to twenty-seven with a bachelor's degree or higher, while young workers are those age twenty-two to twenty-seven without a bachelor's degree. All figures exclude those in the military or currently enrolled in school.

graduates in recent years. Examples of the kinds of jobs found in this category are waiters and waitresses, cashiers, bartenders, cooks, and, yes, baristas.

While demand in the noncollege segment of the labor market doubled in the years following the Great Recession, this growth was not merely in low-paying jobs. We turn back to the HWOL database to provide estimates of the number and growth of monthly job postings for each of the occupation categories identified above between 2009 and 2013, also shown in table 4.1. The Managers and Supervisors category had the largest number of job postings after the Great Recession, followed closely by Office and Administrative Support. The two lowest-paying categories, Physical Laborers and Low-Skilled Service, saw large increases in demand, as did Skilled Trades and Managers and Supervisors. These figures suggest that while many low-skilled service jobs were available during this time, there were plenty of opportunities in jobs that tended to pay higher wages. Next, we examine which jobs both underemployed college graduates and those without college degrees took.

4.3.2   What Jobs Did Underemployed Graduates Take?

Table 4.2 shows the share of underemployed recent college graduates across the ten occupation categories in the years following the Great Recession. Contrary to popular perception, most underemployed recent college graduates were not working in low-skilled service jobs. Indeed, nearly half were working in relatively high-paying jobs, with more than 10 percent each working in the Information Processing and Business Support, Managers and Supervisors, and Sales categories. At 25 percent, the largest share of underemployed workers were employed in the Office and Administrative

Support category. While these jobs may not be as desirable as the typical college job, which pays around $78,500 annually, they are significantly better than low-skilled service jobs. That said, about one-fifth of underemployed recent college graduates—roughly 9 percent of all recent graduates—were working in a low-skilled service job.[9]

Comparing the distribution of underemployed college graduates to young workers of the same age without a college degree yields some important insights about the value of a college degree for underemployed workers. Those with a college degree were much more likely to be working in higher-paying jobs than those without. This pattern is particularly evident in the highest-paying occupation categories that tend to emphasize cognitive skills and decision-making, such as the Information Processing and Business Support and Managers and Supervisors categories. While around 40 percent of recent college graduates were employed in the two highest-paid tiers of noncollege occupations, only 18 percent of young workers without degrees held these types of jobs. By contrast, among those working in these occupation categories, more than half of young workers without a college degree were working in the low-paying Physical Laborers and Low-Skilled Service occupation categories, double the share for recent college graduates. Moreover, though not shown in the table, we also find that underemployed recent college graduates tend to earn more than similarly aged young workers without a college degree *within* each occupation category.

While the same general patterns hold between the genders, there are some notable differences, as shown in table 4.3. Underemployed men are more likely to be working in the highest-paying occupation categories, including Information Processing and Business Support and Managers and Supervisors. The male-female ratio is also particularly large for jobs in the Public Safety and Skilled Trades categories, both of which tend to emphasize physical skills. By contrast, underemployed women are much more likely to be working in Office and Administrative Support jobs, and, to a lesser extent, the Health Care Technicians and Assistants category. In terms of the lower-paying categories, underemployed men are more likely than women to be working in jobs in the Physical Laborers category, while underemployed women are more likely to be working in jobs in the Low-Skilled Service category.

4.4 Which Graduates Are More Prone to Underemployment?

We next turn to the question of which recent college graduates are more likely to be underemployed. We use probit regressions to reveal which char-

---

9. As an alternative to the Low-Skilled Service category, we also measured the share of all underemployed workers earning around the minimum wage. We estimate this share to be roughly 20 to 25 percent, comparable to the share working in a low-skilled service job.

Table 4.3   Share of underemployed recent college graduates by occupation category and gender

Occupation category	Share of underemployed recent college graduates		Share of young workers without a college degree	
	Male	Female	Male	Female
Information processing and business support	12.1	10.7	2.1	2.0
Managers and supervisors	15.1	11.4	7.7	8.1
Public safety	5.9	1.9	3.9	1.3
Sales	12.6	11.0	4.6	5.9
Arts and entertainment	3.9	2.2	0.7	0.6
Skilled trades	5.0	0.8	13.1	1.0
Office and administrative support	17.8	31.3	9.3	23.4
Health care technicians and assistants	2.4	6.5	1.9	13.6
Physical laborers	9.2	2.3	35.5	7.5
Low-skilled service	16.1	22.0	21.3	36.7

Source: US Census Bureau, American Community Survey, 2009–2013.

Notes: Recent college graduates are those age twenty-two to twenty-seven with a bachelor's degree or higher, while young workers are those age twenty-two to twenty-seven without a bachelor's degree. All figures exclude those in the military or currently enrolled in school.

acteristics of recent college graduates are associated with a higher probability of being underemployed, with a particular focus on college major. Because men and women may choose different career paths or have different experiences in the labor market, we estimate our regression models using aggregate data and separately by gender. We wish to emphasize that our models are not meant to imply causation, but rather to uncover some of the correlates to the likelihood of being underemployed based on the characteristics of workers we are able to identify in the data we employ.

4.4.1   Estimation Approach

Because our measures of underemployment are binary variables, we use probit models to estimate the likelihood of underemployment among recent college graduates. Specifically, letting $UNDER_i$ represent the underemployment of individual i located in state j during year t, the probability that an individual is working in a job that does not require a college degree can be expressed as:

(1) $$\text{Prob}(UNDER_i = 1) = \Phi(\beta X_i + \delta M_i + \phi_j + \phi_t)$$

where X_i is a vector of individual-level worker characteristics, M_i is a vector of dummy variables denoting an individual's college major, ϕ_j is a state-level spatial fixed effect, ϕ_t is an annual time fixed effect, and β and δ are parameters to be estimated; $\Phi(\cdot)$ is a normal cumulative distribution function, and the estimated parameters are chosen to maximize the sum of the log likelihoods over all observations. We estimate our models using two different measures for $UNDER_i$, one that broadly includes graduates working in any noncollege job, and a second more narrowly defined measure of underemployment for those working in the Low-Skilled Service category.

Of particular interest for our purposes, the ACS began to include information on an individual's undergraduate degree major starting in 2009. Specifically, the ACS provides information for more than 170 detailed degree major categories. Since many of these detailed majors contain relatively few observations, we collapse this list into seventy-three majors to preserve large enough sample sizes to obtain meaningful results.

To explore how differences in worker characteristics, X_i, are related to the likelihood of underemployment, our probit models include a wide range of individual-level characteristics such as gender, age, marital status, the presence of children, race and ethnicity, and disability status.[10] In addition, when collecting information about college major, the ACS allows individuals to list up to two majors. We consider those individuals who listed two majors as having graduated with a double major, which we control for, and count the first listed as that person's college major. As another control, we are also able to identify recent college graduates who have earned a graduate degree.[11]

Table 4.4 provides descriptive statistics for the worker characteristics included in our study for three groups: all recent college graduates, those who are underemployed, and those working in a low-skilled service job. Interestingly, there are more underemployed women (55 percent) than men (45 percent). This differential partly reflects the fact that there are now more women college graduates than men in the overall population, though men seem to be slightly overrepresented among the underemployed. By contrast, men are underrepresented among low-skilled service workers. About 20 percent of the underemployed are married, 8 percent have children, 12 percent graduated with a double major, and 6 percent earned a graduate degree. Proportionally fewer recent college graduates working in a low-skilled service job were married, had children, graduated with a double major, or earned a graduate degree.

To account for differences in local economic conditions across time and space, which may influence the likelihood of being underemployed, we include state-level spatial fixed effects, ϕ_j, and annual time fixed effects, ϕ_t, in our models.[12] In all of our analysis, we report robust standard errors

10. To allow for nonlinear effects from gaining experience in the labor market, we follow the convention in wage studies and include both age and age-squared in our models.
11. The ACS indicates whether an individual holds a master's degree, professional degree, or doctoral degree, but does not provide information about the type of graduate degree (e.g., MA, MBA, JD, MD) or course of study while in graduate school.
12. For example, Mian and Sufi (2010, 2011) show that the most pronounced effects of the Great Recession were concentrated in the "Sand States," and that the pace of recovery generally

Table 4.4   **Characteristics of recent college graduates**

	All recent grads		Underemployed		Low-skilled service	
Variable	Mean	Std. dev.	Mean	Std. dev.	Mean	Std. dev.
Employment status						
Underemployed	0.446	0.497	1.000	0.000	1.000	0.000
Low-skilled service	0.086	0.281	0.193	0.395	1.000	0.000
Age and gender						
Age	25.1	1.5	24.9	1.5	24.6	1.6
Male	0.436	0.496	0.450	0.497	0.374	0.484
Family background						
Married	0.234	0.423	0.199	0.399	0.157	0.364
Children	0.082	0.274	0.078	0.269	0.068	0.251
Race and ethnicity						
White	0.800	0.400	0.795	0.403	0.797	0.402
Black	0.070	0.255	0.085	0.279	0.076	0.265
American Indian	0.003	0.052	0.003	0.057	0.003	0.053
Asian	0.083	0.275	0.064	0.244	0.061	0.239
Other race	0.045	0.207	0.053	0.223	0.063	0.243
Hispanic	0.079	0.270	0.092	0.289	0.108	0.310
Disability status						
Disabled	0.014	0.117	0.016	0.126	0.017	0.130
Education						
Double major	0.121	0.326	0.117	0.321	0.107	0.309
Graduate degree	0.148	0.355	0.064	0.244	0.057	0.232
N	20,233,500		9,031,408		1,744,695	

Source: US Census Bureau, American Community Survey, 2009–2013.

Notes: Recent college graduates are those age twenty-two to twenty-seven with a bachelor's degree or higher. All figures exclude those in the military or currently enrolled in school.

clustered at the state level, which tends to increase standard errors but does not affect the point estimates themselves.

Despite our efforts to control for differences in local economic performance and a wide range of individual worker characteristics, care must be taken when interpreting our findings. Most significantly, in part, students sort into their chosen field of study based on their ability to complete the required coursework (see, e.g., Arcidiacono 2004; Zafar 2011, 2013). Thus, not all majors are feasible for every college student, and graduates with different majors likely differ in other important ways that we are unable to measure, such as intelligence, perseverance, or motivation. Indeed, recent

---

differed across states. Further, Abel and Deitz (2015) show that local labor market conditions can influence the likelihood and quality of the match between an individual's education and job. We also estimated a model using spatial fixed effects at the local labor market area, which we defined as metropolitan areas and the rural portion of each state. Results were nearly identical to those reported in the paper, but small sample sizes within many local labor markets prevented us from estimating models using underemployed graduates working in low-skilled service jobs.

research has shown that graduating with a math or science major is more difficult than other fields of study (Stinebrickner and Stinebrickner 2014). In addition, our results represent average outcomes for graduates within each of the seventy-three college majors we analyze. Thus, by definition, some individuals within each major will have better or worse outcomes than our results suggest. Nonetheless, examining the typical experience within each major can provide useful insights into the correlates of the likelihood of underemployment.

4.4.2   Estimation Results

Because of the difficulties associated with interpreting raw coefficient estimates obtained via probit analysis, we instead present the corresponding average marginal effects and predicted probabilities obtained from our analysis. As such, our estimates can be interpreted as the average percentage point change in the probability of either being underemployed or working in a low-skilled service job. We first describe how the probability of being underemployed is correlated with the worker characteristics we are able to identify, and then turn to the role of college major.

Worker Characteristics

Table 4.5 presents the average marginal effects associated with the worker characteristics included in our analysis. Columns (1)–(3) show results using underemployment in general as the dependent variable, while columns (4)–(6) show results using Low-Skilled Service jobs only. Our results show that the likelihood of college underemployment differs significantly across a wide range of worker characteristics.

Regarding gender differences, our analysis indicates that male graduates are 1.2 percentage points more likely to be underemployed in the early stages of their careers than their female counterparts. Specifically, men have a predicted probability of 45.3 percent compared to 44.1 percent for women—a gap that represents about a 3 percent difference between these groups. This difference may stem in part from the recent success women have enjoyed relative to men while in college, but it could also reflect the fact that underemployed men tend to be more represented in the higher-paying noncollege occupation categories, and, therefore may have less incentive to seek a college job.[13] Indeed, women graduates are 1.1 percentage points (9.1 percent compared to 8.0 percent) more likely to be working in a low-skilled service job than men—a difference of more than 12 percent. For both men and

---

13. Goldin, Katz, and Kuziemko (2006) show that women are now much more likely to enroll in and complete college than men, reversing the college gender gap. Fortin, Oreopoulos, and Phipps (2015) demonstrate that the relatively strong academic performance of women compared to men in recent decades stems, in large part, from being better prepared for and focused on college.

Table 4.5   Average marginal effects from underemployment and low-skilled service probit models

	Underemployed			Working in low-skilled service jobs		
	Overall (1)	Male (2)	Female (3)	Overall (4)	Male (5)	Female (6)
Male	0.012***	—	—	−0.011***	—	—
	(0.003)			(0.003)		
Age	−0.015***	−0.013***	−0.016***	−0.011***	−0.009***	−0.012***
	(0.001)	(0.001)	(0.001)	(0.000)	(0.001)	(0.001)
Married	−0.040***	−0.044***	−0.039***	−0.026***	−0.033***	−0.022***
	(0.004)	(0.007)	(0.005)	(0.003)	(0.003)	(0.004)
Children	0.029***	0.044***	0.025***	0.007**	0.004	0.010**
	(0.007)	(0.011)	(0.008)	(0.003)	(0.007)	(0.005)
Black	0.075***	0.081***	0.070***	0.007	0.011	0.003
	(0.008)	(0.010)	(0.009)	(0.005)	(0.007)	(0.006)
American Indian	0.074***	0.058	0.082**	0.003	−0.036***	0.028
	(0.025)	(0.037)	(0.033)	(0.015)	(0.008)	(0.024)
Asian	−0.021***	−0.035***	−0.014	−0.002	−0.003	0.004
	(0.006)	(0.009)	(0.009)	(0.008)	(0.008)	(0.008)
Other race	0.039***	0.045***	0.034***	0.018***	0.018***	0.019***
	(0.011)	(0.017)	(0.010)	(0.005)	(0.007)	(0.007)

(continued)

Table 4.5　(continued)

	Underemployed			Working in low-skilled service jobs		
	Overall (1)	Male (2)	Female (3)	Overall (4)	Male (5)	Female (6)
Hispanic	0.045***	0.074***	0.023***	0.026***	0.034***	0.019***
	(0.009)	(0.012)	(0.009)	(0.005)	(0.007)	(0.004)
Disabled	0.042***	0.030*	0.054***	0.014*	0.004	0.022*
	(0.012)	(0.017)	(0.019)	(0.008)	(0.006)	(0.013)
Double major	−0.046***	−0.051***	−0.042***	−0.016***	−0.013***	−0.019***
	(0.005)	(0.009)	(0.004)	(0.002)	(0.003)	(0.003)
Graduate degree	−0.252***	−0.229***	−0.263***	−0.054***	−0.040***	−0.063***
	(0.006)	(0.007)	(0.006)	(0.002)	(0.003)	(0.003)
Log pseudo likelihood	−12,227,478***	−5,401,846***	−6,792,684***	−5,503,035***	−2,115,863***	−3,357,967***
Pseudo R-squared	0.121	0.112	0.131	0.074	0.090	0.068
Weighted N	20,233,500	8,818,586	11,414,914	20,233,500	8,818,586	11,414,914

Source: US Census Bureau, American Community Survey, 2009–2013.

Notes: Robust standard errors, clustered at the state level, are reported in parentheses. Models also include the following controls (coefficients not reported for brevity): individual's college major (seventy-three degree fields), state, and year. Marginal effects for dummy variables represent discrete change from 0 to 1.

***Significant at the 1 percent level.

**Significant at the 5 percent level.

*Significant at the 10 percent level.

women, the likelihood of being underemployed or working in a low-skilled service job declines sharply as workers age from twenty-two to twenty-seven.

In terms of family considerations, graduates who are married are less likely to be underemployed (41.5 percent compared to 45.6 percent) or working in a low-skilled service job (6.6 percent compared to 9.2 percent), and this is particularly true among married men. In addition, those graduates with children are more likely to be underemployed (47.4 percent compared to 44.4 percent). Women with children, in particular, are more likely to be working in a low-skilled service job. One potential explanation for these findings is that those who are married or without children have a greater ability to search for better jobs because they have more resources available, or face fewer constraints, and that these factors reduce the likelihood of being underemployed. However, more research is needed to disentangle the potentially complex relationships between gender, family, and the likelihood of underemployment.

Underemployment following the Great Recession also varied significantly across racial and ethnic groups. Compared to white graduates, who have a 44.1 percent likelihood of being underemployed, black and American Indian graduates are 17 percent more likely to be working in a noncollege job, while Asian graduates are 5 percent less likely. Our estimates also indicate nonwhite graduates are more likely to be working in low-skilled service jobs, though these differences are generally not statistically significant. Moreover, those of Hispanic origin are 10 percent more likely to be underemployed and 31 percent more likely to be working in a low-skilled service job than non-Hispanics. Looking across genders, the magnitudes of our estimates pertaining to race and ethnicity tend to be larger for men than women. These findings are broadly consistent with other research showing that minorities, particularly black and Hispanic men, tend to suffer the most during recessions (see, e.g., Elsby, Hobijn, and Şahin 2010; Elsby et al. 2011; Hoynes, Miller, and Schaller 2012; Nunley et al. 2015).

Graduates with a disability are 4.2 percentage points—or 10 percent—more likely to be underemployed than those who are not, and are 1.4 percentage points—or 16 percent—more likely to be working in a low-skilled service job. In both cases, the estimated effects are larger for women than for men.

Graduating with a double major or earning a graduate degree are both associated with a lower likelihood of being underemployed or working in a low-skilled service job. Graduates with a double major are 4.6 percentage points less likely to be underemployed than those with a single major, and are 1.6 percentage points less likely to be working in a low-skilled service job. Those with a graduate degree are 25.2 percentage points less likely to be underemployed than those without, and are 5.2 percentage points less likely to be working in a low-skilled service job. These results are expected as those with two majors or a graduate degree tend to have built more skills, and especially for those with a graduate degree, have developed occupation-specific skills and training that may allow them better access to employment opportunities. The reduced likelihood of college underemployment for those with a double major or graduate degree is similar for both men and women.

College Major

The role of college major in finding a good job has become of considerable interest in recent years given the weak labor market following the Great Recession. While not all students are willing and able to complete a degree in any major, some choice is involved, making information about the success of those with certain majors relative to others of value to students and parents. In tables 4.6 and 4.7, we present the predicted probabilities of being underemployed or working in a low-skilled service job, respectively, by college major, holding constant the other variables in our model. Given the large amount of information contained in these tables and the fact that the patterns do not appear to differ widely by gender, we also plot the overall predicted probabilities for selected college majors in figures 4.4 and 4.5. Though there are differences in the rankings of college majors for each measure of underemployment, five broad themes emerge.[14]

First, it is clear that college major is a significant correlate with the probability of being underemployed in the early careers of college graduates. While, on average, 44.6 percent of recent graduates work in a noncollege job, underemployment rates range from 70 percent for graduates with a criminal justice major to 9.5 percent for those with a nursing degree. Similarly, while on average, only 8.6 percent of recent college graduates work in a low-skilled service job, this figure ranges from 23.4 percent for those majoring in leisure and hospitality to 1.7 percent for graduates with a civil engineering major.

Second, graduates with college majors that provide technical training and quantitative skills are far less likely to be underemployed than those with majors that tend to be less quantitative in nature. Indeed, for both measures of college underemployment, graduates with majors in the science, technology, engineering, and mathematics (STEM) fields tend to have some of the lowest predicted probabilities of working in a noncollege job. In particular, graduates with any type of engineering major generally fared well in the labor market following the Great Recession. Outside of the traditional STEM majors, those with majors that are quantitatively oriented, such as accounting, business analytics, economics, and finance, also tend to have relatively low underemployment rates. By contrast, those with majors in less quantitative subjects such as English language, sociology, communications, art history, or anthropology tend to have relatively high rates of underemployment.

14. The Spearman rank correlation of the predicted probabilities of being underemployed and working in a low-skilled service job by college major is 0.57.

Table 4.6 Probability of underemployment among recent college graduates by major

Major	Overall	SE	Male	SE	Female	SE
Criminal justice	0.700	(0.011)	0.752	(0.017)	0.646	(0.013)
Performing arts	0.663	(0.013)	0.654	(0.025)	0.669	(0.012)
Leisure and hospitality	0.640	(0.019)	0.669	(0.026)	0.613	(0.016)
Anthropology	0.624	(0.019)	0.617	(0.026)	0.624	(0.024)
Art history	0.621	(0.021)	0.736	(0.047)	0.592	(0.023)
Public policy and law	0.618	(0.029)	0.547	(0.052)	0.674	(0.030)
Business management	0.601	(0.006)	0.592	(0.011)	0.613	(0.007)
Fine arts	0.591	(0.009)	0.604	(0.012)	0.580	(0.012)
History	0.575	(0.011)	0.581	(0.013)	0.573	(0.016)
Animal and plant sciences	0.572	(0.019)	0.548	(0.031)	0.587	(0.024)
Miscellaneous technologies	0.554	(0.020)	0.553	(0.023)	0.579	(0.027)
Communications	0.554	(0.007)	0.595	(0.012)	0.529	(0.009)
Liberal arts	0.553	(0.022)	0.611	(0.018)	0.519	(0.030)
General business	0.551	(0.013)	0.550	(0.014)	0.558	(0.014)
Political science	0.548	(0.011)	0.538	(0.013)	0.562	(0.012)
Marketing	0.545	(0.007)	0.543	(0.012)	0.544	(0.010)
Sociology	0.541	(0.017)	0.573	(0.030)	0.524	(0.016)
Mass media	0.539	(0.013)	0.563	(0.022)	0.522	(0.019)
Foreign language	0.538	(0.013)	0.561	(0.027)	0.525	(0.017)
Philosophy	0.537	(0.018)	0.563	(0.016)	0.507	(0.026)
English language	0.534	(0.009)	0.571	(0.019)	0.513	(0.013)
Agriculture	0.533	(0.030)	0.550	(0.032)	0.515	(0.042)
Advertising and public relations	0.511	(0.011)	0.547	(0.042)	0.493	(0.010)
Medical technicians	0.507	(0.027)	0.470	(0.055)	0.512	(0.030)
Environmental studies	0.504	(0.021)	0.553	(0.020)	0.446	(0.032)
Psychology	0.503	(0.009)	0.537	(0.013)	0.488	(0.010)
International affairs	0.502	(0.024)	0.511	(0.033)	0.495	(0.026)
Interdisciplinary studies	0.501	(0.018)	0.498	(0.021)	0.502	(0.024)
Theology and religion	0.500	(0.019)	0.495	(0.025)	0.510	(0.031)
Ethnic studies	0.498	(0.014)	0.486	(0.029)	0.497	(0.017)
General social sciences	0.492	(0.035)	0.524	(0.068)	0.463	(0.032)
Health services	0.488	(0.013)	0.537	(0.029)	0.475	(0.014)
Miscellaneous biological sciences	0.478	(0.013)	0.482	(0.026)	0.473	(0.018)
Geography	0.469	(0.030)	0.482	(0.045)	0.453	(0.036)
Biology	0.448	(0.009)	0.448	(0.011)	0.446	(0.011)
Earth sciences	0.446	(0.034)	0.438	(0.039)	0.463	(0.063)
Engineering technologies	0.445	(0.020)	0.444	(0.022)	0.492	(0.049)
Nutrition sciences	0.442	(0.025)	0.546	(0.068)	0.421	(0.025)
Information systems and management	0.441	(0.016)	0.440	(0.019)	0.474	(0.031)
Family and consumer sciences	0.440	(0.017)	0.453	(0.063)	0.431	(0.016)
Miscellaneous physical sciences	0.428	(0.042)	0.398	(0.047)	0.467	(0.056)
Journalism	0.425	(0.012)	0.452	(0.020)	0.406	(0.015)
Commercial art and graphic design	0.419	(0.011)	0.403	(0.017)	0.419	(0.014)
Economics	0.413	(0.021)	0.425	(0.021)	0.408	(0.027)
Biochemistry	0.402	(0.022)	0.373	(0.044)	0.428	(0.026)
Treatment therapy	0.394	(0.015)	0.483	(0.031)	0.358	(0.017)
Architecture	0.392	(0.017)	0.424	(0.021)	0.351	(0.021)
Business analytics	0.376	(0.015)	0.382	(0.019)	0.382	(0.024)
Chemistry	0.371	(0.016)	0.406	(0.021)	0.339	(0.026)

(*continued*)

Table 4.6   (continued)

Major	Overall	SE	Male	SE	Female	SE
Finance	0.370	(0.015)	0.368	(0.015)	0.388	(0.018)
Social services	0.357	(0.016)	0.424	(0.050)	0.347	(0.016)
Mathematics	0.330	(0.015)	0.350	(0.021)	0.311	(0.020)
Pharmacy	0.322	(0.037)	0.312	(0.045)	0.325	(0.039)
Physics	0.318	(0.025)	0.356	(0.032)	0.238	(0.034)
Miscellaneous engineering	0.287	(0.016)	0.292	(0.019)	0.294	(0.026)
Secondary education	0.280	(0.014)	0.311	(0.017)	0.260	(0.017)
Construction services	0.275	(0.028)	0.289	(0.027)	0.233	(0.081)
General engineering	0.263	(0.020)	0.267	(0.023)	0.277	(0.035)
Accounting	0.263	(0.009)	0.259	(0.014)	0.267	(0.010)
Computer science	0.262	(0.017)	0.260	(0.015)	0.316	(0.029)
General education	0.245	(0.013)	0.290	(0.024)	0.231	(0.015)
Industrial engineering	0.230	(0.023)	0.236	(0.032)	0.224	(0.038)
Early childhood education	0.227	(0.018)	0.341	(0.083)	0.218	(0.019)
Miscellaneous education	0.223	(0.015)	0.249	(0.035)	0.209	(0.015)
Aerospace engineering	0.218	(0.028)	0.245	(0.036)	0.110	(0.044)
Elementary education	0.215	(0.013)	0.262	(0.024)	0.207	(0.013)
Electrical engineering	0.205	(0.012)	0.209	(0.011)	0.211	(0.028)
Mechanical engineering	0.203	(0.014)	0.211	(0.017)	0.176	(0.025)
Chemical engineering	0.189	(0.021)	0.205	(0.025)	0.165	(0.028)
Civil engineering	0.187	(0.014)	0.188	(0.017)	0.191	(0.021)
Computer engineering	0.180	(0.018)	0.179	(0.019)	0.236	(0.044)
Special education	0.153	(0.020)	0.173	(0.066)	0.147	(0.020)
Nursing	0.095	(0.012)	0.159	(0.026)	0.087	(0.010)

Source: US Census Bureau, American Community Survey, 2009–2013.

Table 4.7   Probability of working in a low-skilled service job among recent college graduates by major

Major	Overall	SE	Male	SE	Female	SE
Leisure and hospitality	0.234	(0.010)	0.240	(0.019)	0.227	(0.011)
Performing arts	0.206	(0.017)	0.181	(0.037)	0.224	(0.013)
Fine arts	0.165	(0.009)	0.143	(0.012)	0.178	(0.009)
Anthropology	0.155	(0.011)	0.161	(0.020)	0.155	(0.015)
Nutrition sciences	0.152	(0.019)	0.310	(0.060)	0.135	(0.020)
Family and consumer sciences	0.152	(0.009)	0.128	(0.039)	0.158	(0.009)
Liberal arts	0.135	(0.009)	0.155	(0.017)	0.125	(0.011)
Animal and plant sciences	0.134	(0.012)	0.135	(0.021)	0.132	(0.014)
History	0.129	(0.007)	0.116	(0.008)	0.143	(0.012)
Philosophy	0.126	(0.016)	0.129	(0.017)	0.118	(0.019)
Early childhood education	0.125	(0.013)	0.068	(0.049)	0.129	(0.012)
Foreign language	0.123	(0.011)	0.124	(0.030)	0.126	(0.012)
General social sciences	0.122	(0.015)	0.093	(0.019)	0.145	(0.027)
Theology and religion	0.121	(0.015)	0.112	(0.019)	0.137	(0.020)
Earth sciences	0.119	(0.029)	0.099	(0.028)	0.145	(0.059)
English language	0.119	(0.006)	0.128	(0.011)	0.117	(0.007)
Psychology	0.118	(0.005)	0.108	(0.007)	0.124	(0.006)
Environmental studies	0.114	(0.012)	0.105	(0.016)	0.124	(0.019)
Social services	0.109	(0.010)	0.130	(0.039)	0.111	(0.009)
Sociology	0.108	(0.006)	0.111	(0.012)	0.109	(0.008)
Art history	0.106	(0.015)	0.227	(0.055)	0.090	(0.013)
Miscellaneous biological sciences	0.106	(0.009)	0.085	(0.010)	0.121	(0.012)

Table 4.7 (continued)

Major	Overall	SE	Male	SE	Female	SE
Treatment therapy	0.105	(0.010)	0.170	(0.027)	0.080	(0.009)
Ethnic studies	0.102	(0.012)	0.093	(0.017)	0.109	(0.016)
Elementary education	0.100	(0.008)	0.086	(0.015)	0.103	(0.008)
Interdisciplinary studies	0.099	(0.007)	0.070	(0.010)	0.118	(0.011)
Secondary education	0.095	(0.007)	0.090	(0.011)	0.099	(0.009)
Special education	0.093	(0.017)	0.090	(0.038)	0.096	(0.020)
Communications	0.092	(0.004)	0.089	(0.006)	0.096	(0.006)
Mass media	0.092	(0.011)	0.104	(0.017)	0.080	(0.014)
General education	0.091	(0.007)	0.076	(0.014)	0.098	(0.009)
Miscellaneous physical sciences	0.091	(0.018)	0.076	(0.027)	0.106	(0.035)
Biology	0.088	(0.004)	0.085	(0.007)	0.091	(0.007)
Health services	0.087	(0.006)	0.087	(0.009)	0.091	(0.007)
Criminal justice	0.085	(0.004)	0.068	(0.006)	0.105	(0.007)
Geography	0.084	(0.015)	0.086	(0.018)	0.080	(0.020)
Political science	0.083	(0.007)	0.089	(0.010)	0.074	(0.008)
Business management	0.082	(0.005)	0.076	(0.005)	0.088	(0.006)
Advertising and public relations	0.078	(0.007)	0.065	(0.014)	0.084	(0.008)
Commercial art and graphic design	0.077	(0.005)	0.062	(0.008)	0.085	(0.007)
Journalism	0.077	(0.006)	0.075	(0.011)	0.079	(0.008)
General business	0.077	(0.005)	0.070	(0.006)	0.082	(0.008)
Pharmacy	0.073	(0.017)	0.073	(0.023)	0.073	(0.027)
Architecture	0.072	(0.008)	0.074	(0.014)	0.066	(0.013)
Miscellaneous education	0.070	(0.010)	0.049	(0.021)	0.080	(0.011)
International affairs	0.070	(0.008)	0.081	(0.014)	0.063	(0.008)
Biochemistry	0.068	(0.011)	0.052	(0.022)	0.083	(0.016)
Agriculture	0.068	(0.010)	0.065	(0.014)	0.073	(0.017)
Mathematics	0.062	(0.009)	0.056	(0.010)	0.066	(0.013)
Marketing	0.061	(0.004)	0.061	(0.007)	0.061	(0.005)
Public policy and law	0.060	(0.011)	0.025	(0.010)	0.089	(0.018)
Chemistry	0.056	(0.009)	0.054	(0.012)	0.059	(0.016)
Miscellaneous technologies	0.054	(0.009)	0.043	(0.007)	0.074	(0.023)
Physics	0.049	(0.016)	0.059	(0.021)	0.016	(0.009)
Economics	0.046	(0.006)	0.043	(0.005)	0.046	(0.008)
Information systems and management	0.045	(0.007)	0.036	(0.009)	0.068	(0.014)
Engineering technologies	0.041	(0.007)	0.031	(0.007)	0.083	(0.028)
Accounting	0.038	(0.003)	0.033	(0.004)	0.043	(0.004)
General engineering	0.036	(0.006)	0.030	(0.006)	0.056	(0.019)
Finance	0.036	(0.003)	0.036	(0.004)	0.033	(0.004)
Chemical engineering	0.034	(0.010)	0.037	(0.013)	0.024	(0.014)
Medical technicians	0.034	(0.009)	0.032	(0.021)	0.035	(0.010)
Electrical engineering	0.029	(0.008)	0.024	(0.008)	0.044	(0.013)
Computer science	0.027	(0.004)	0.018	(0.004)	0.065	(0.015)
Computer engineering	0.027	(0.006)	0.023	(0.007)	0.041	(0.023)
Business analytics	0.025	(0.005)	0.019	(0.005)	0.038	(0.012)
Construction services	0.025	(0.007)	0.019	(0.005)	0.080	(0.053)
Nursing	0.025	(0.004)	0.054	(0.011)	0.022	(0.004)
Industrial engineering	0.024	(0.009)	0.019	(0.011)	0.033	(0.016)
Miscellaneous engineering	0.024	(0.005)	0.019	(0.006)	0.033	(0.008)
Aerospace engineering	0.021	(0.009)	0.021	(0.010)	0.010	(0.009)
Mechanical engineering	0.019	(0.004)	0.019	(0.004)	0.016	(0.006)
Civil engineering	0.017	(0.004)	0.016	(0.004)	0.015	(0.008)

Source: US Census Bureau, American Community Survey, 2009–2013.

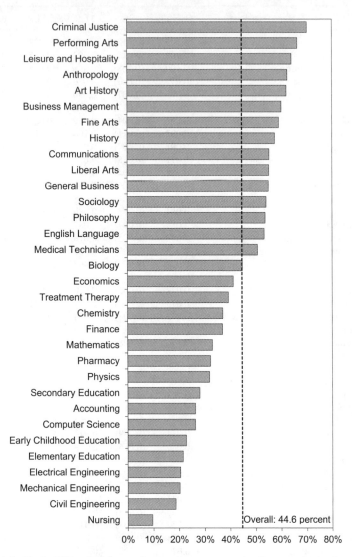

Fig. 4.4   Probability of underemployment among recent college graduates for selected majors

Source: US Census Bureau, American Community Survey, 2009–2013.

Third, graduates with college majors that provide occupation-specific training tend to be less likely to be underemployed than those with majors providing a more general education. For example, occupation-specific majors like education, engineering, and health-related fields, tended to have much lower rates of underemployment than those with majors in more general fields such as liberal arts, philosophy, or history. This pattern also emerges

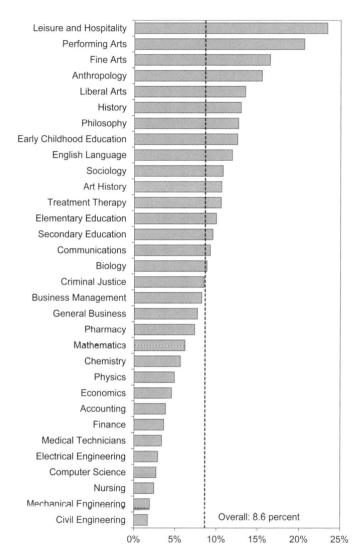

Fig. 4.5   Probability of working in a low-skilled service job among recent college graduates for selected majors
Source: US Census Bureau, American Community Survey, 2009–2013.

when examining the outcomes of graduates within a specific academic discipline that may offer both occupation-specific majors and majors that are more general. The business field provides a case in point: those with a more targeted major, such as accounting or finance, tend to have lower underemployment rates than those with majors that are less directly connected to specific jobs, such as business management or general business.

Fourth, however, there are some college majors that offer occupation-specific training that tends to be geared toward jobs that do not typically require a bachelor's degree, and graduates with these majors are more likely to be underemployed. For example, those who major in criminal justice may be expecting to take jobs in Public Safety (such as a police officer or detective) and those with a fine arts or performing arts major may be expecting to take jobs in Arts and Entertainment (such as a photographer or dancer). In addition, those with a leisure and hospitality major may be trained for a number of jobs that do not require a college degree, such as a restaurant manager or health and wellness instructor. Further, while those with health-care-related degrees generally tend to have relatively low underemployment, those with a medical technicians major, which likely prepares students to take jobs in the Health Care Technicians and Assistants category, have relatively high underemployment.

Finally, graduates with college majors geared toward growing parts of the economy are generally less likely to be underemployed. Indeed, the health and education sectors in particular continued to grow through both the downturn and recovery alike, creating job opportunities for people with skills oriented toward these types of jobs. As such, the likelihood of underemployment was fairly low for those with health-care-related majors, such as nursing, pharmacy, and treatment therapy. Similarly, those with an education-related major tend to experience below average underemployment in general, though such graduates tend to have higher rates of working in low-skilled service jobs, particularly those who major in elementary or early childhood education.

4.5   Transitioning to Better Jobs

A key finding from our empirical analysis is that, to some degree, under-employment is a temporary phase for many recent graduates as they transition from school to the labor market. This pattern is particularly evident for those who start their careers working in a low-skilled service job. Indeed, such adjustment is not merely a new phenomenon resulting from the Great Recession—research has shown that underemployment typically falls as new graduates spend time in the labor market, and that this pattern has been occurring for decades (Abel, Deitz, and Su 2014).

To illustrate this point, in figure 4.6 we use estimates from our probit analysis to plot the likelihood of being underemployed (panel A) and working in a low-skilled service job (panel B) by age, overall and separately by gender. In both cases, we identify a strong downward trend in the likelihood of working in a noncollege job as graduates gain more experience in the labor market. At age twenty-two, when fresh out of college, the likelihood of being underemployed is nearly 50 percent, but this figure falls to around 42 percent by age twenty-seven—a 15 percent decline. Not only are women generally less likely to be underemployed than men at any age, the decline in underem-

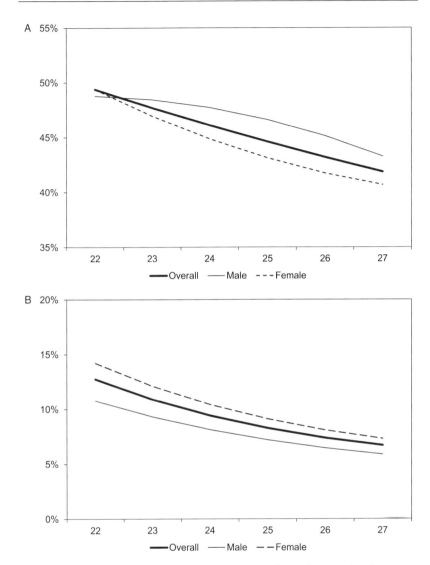

Fig. 4.6   **Employment outcomes of recent college graduates by age.** *A*, **under-employed;** *B*, **low-skilled service.**
Source: US Census Bureau, American Community Survey, 2009–2013.

ployment is also more pronounced for women than for men. The transition out of low-skilled service jobs is even more striking. At age twenty-two, the predicted probability of working in such a job is about 13 percent, but this figure falls to 6.7 percent by age twenty-seven—a nearly 50 percent decline. The likelihood of working in a low-skilled service job declines at a similar pace for men and women.

To examine more of the details of this transition, in table 4.8 we compare

Table 4.8   Share of younger and older recent college graduates by
            occupation category

	Share of underemployed recent college graduates, all	
Occupation category	Younger	Older
Information processing and business support	5.5	5.3
Managers and supervisors	5.1	5.8
Public safety	1.4	1.6
Sales	6.5	4.4
Arts and entertainment	1.4	1.5
Skilled trades	1.1	1.6
Office and administrative support	12.7	10.1
Health care technicians and assistants	2.4	2.1
Physical laborers	2.9	2.2
Low-skilled service	12.6	6.6
College jobs	48.4	59.0

Source: US Census Bureau, American Community Survey, 2009 and 2013.

Notes: Younger recent college graduates are those age twenty-two to twenty-three with a bachelor's degree or higher in 2009, while older recent college graduates are those age twenty-six to twenty-seven with a bachelor's degree or higher in 2013. All figures exclude those in the military or currently enrolled in school.

the jobs held by Younger Recent Graduates (age twenty-two to twenty-three in 2009) to Older Recent Graduates of the same cohort (age twenty-six to twenty-seven in 2013). Consistent with our analysis above, a larger share of graduates worked in college jobs in their midtwenties (59 percent) compared to their early twenties (48 percent). In addition, the composition of jobs held by recent graduates changed within the underemployed occupation categories as these workers aged. The share employed in the lowest-paying Low-Skilled Service group drops by half, suggesting that these jobs are temporary for a good number of recent graduates: by the age of twenty-six or twenty-seven, only 6.6 percent are still working in these types of jobs. The other two groups with the most significant declines include Office and Administrative Support and Sales. Though we cannot identify which jobs graduates tend to move into since our data are cross-sectional in nature—that is, workers may be shifting into other noncollege jobs or into college jobs—these figures suggest that many underemployed graduates, particularly those who start in a low-skilled service job, are able to transition to better jobs as they gain more experience in the labor market.

Table 4.9 presents this same information by gender. In general, these patterns continue to hold when looking at men and women separately. However, while a larger share of women transition out of underemployment to college jobs by their late twenties than men, we find that the share of underemployed graduates working in the high-paying Managers and Supervisors occupa-

Table 4.9   Share of younger and older recent college graduates by occupation category and gender

Occupation category	Share of underemployed recent college graduates, male		Share of underemployed recent college graduates, female	
	Younger	Older	Younger	Older
Information processing and business support	6.3	5.4	5.0	5.1
Managers and supervisors	5.0	6.4	5.2	5.3
Public safety	2.7	2.5	0.6	0.8
Sales	7.1	5.4	6.1	3.5
Arts and entertainment	2.3	2.1	0.8	0.9
Skilled trades	2.2	3.1	0.4	0.4
Office and administrative support	9.8	6.9	14.6	12.6
Health care technicians and assistants	1.4	1.3	3.0	2.8
Physical laborers	5.7	3.8	1.2	0.8
Low-skilled service	11.0	5.7	13.6	7.4
College jobs	46.4	57.4	49.7	60.4

Source: US Census Bureau, American Community Survey, 2009 and 2013.
Notes: Younger recent college graduates are those age twenty-two to twenty-three with a bachelor's degree or higher in 2009, while older recent college graduates are those age twenty-six to twenty-seven with a bachelor's degree or higher in 2013. All figures exclude those in the military or currently enrolled in school.

tion category tends to increase more for men than for women. This share rose about one and a half percentage points for men, but edged up only slightly for women. There was also a slight increase in the share of women working in the highest-paid category of Information Processing and Business Support, while men experienced almost a full percentage point decline.

Nonetheless, while underemployment appears to be a temporary phase for many recent graduates who are able to transition to better jobs, a large share of college graduates remain underemployed long after the initial transition into the labor market, and this was particularly true following the Great Recession. Indeed, even in the best of economic times, about one-third of all college graduates work in a noncollege job. This figure is fairly stable and does not appear to be particularly responsive to the business cycle. This suggests that at least *some* college graduates may simply prefer to work in such jobs, either because they like the nature of the work involved, or because of geographic or family considerations such as taking a lower-skilled job due to a dual labor market search, or while raising children.

4.6   Conclusions

With the Great Recession and weak labor market that followed in its wake, the prevalence of underemployment among recent college graduates reached highs not seen since the early 1990s. However, contrary to popular perception, our work reveals that most of these newly underemployed workers were not forced into low-skilled service jobs. In fact, many of the jobs such graduates took, while clearly not equivalent to jobs that require a college degree, appeared to be more oriented toward knowledge and skill when compared to the distribution of jobs held by young workers without a college degree. Indeed, our analysis also suggests that underemployment is a temporary phase for many young graduates when they enter the labor market, as it often takes time for newly minted graduates to find jobs suited to their education.

We also find that some college graduates have had much better luck finding a college-level job than others. In particular, the likelihood of being underemployed is relatively low for those with quantitatively oriented and occupation-specific majors, and much higher for those with degrees in more general fields. Those with STEM and health-care-related majors have done particularly well in recent years.

These findings raise some interesting questions about the relative supply and demand for specific skill sets obtained in college, and about the value of some majors relative to others in today's economy. While we do not present our findings in the context of a formal supply and demand model, our work does suggest that certain skills have a higher demand relative to supply than others—such as those majors related to the STEM fields and health care. Our findings also raise the specter that degrees in some majors, particularly those that are broad based such as liberal arts and general business, may be less sought after than others. Further, graduates with some majors seem to more easily fall into jobs that typically do not require their degrees, such as leisure and hospitality and criminal justice.

Why are graduates with certain majors faring so poorly upon graduation? Is high underemployment for those with these particular majors a consequence of the quality of the students who choose these majors, the quality of the programs and the skills that are developed (or not developed), or is it that the skills that these majors provide are not as valuable as others? More research is required to address these challenging questions.

More generally, today's high level of underemployment is concerning, and raises a number of questions about why it has continued to rise for more than a decade despite ongoing improvement in the labor market. No doubt, the depth of the Great Recession and the relatively lackluster demand for college graduates through the recovery has been a contributing factor. However, there are lingering questions about whether this soft demand is a long-term phenomenon, as opposed to cyclical in nature. Indeed, recent research suggests that structural changes in the economy may have reduced the demand for college graduates starting as early as 2000 (Beaudry, Green, and Sand 2014, 2016). On the supply side, there are questions about whether the quality of students graduating from college has deteriorated in recent years, with some research suggesting that many students gain little knowledge or skill from a college education (Arum and Roksa 2011, 2014). Our work suggests that these questions are complex, particularly since college graduates with certain skill sets seem to be doing much better in the labor market than others. Further research into these questions would be particularly valuable.

While this work provides more detailed information about the nature of underemployment than has previously been available, it does have its limitations. The most significant limitation is that we cannot fully account for potential unobserved heterogeneity across individuals, such as our inability to control for college grades or the quality of the educational institution attended. In particular, attendance at for-profit colleges increased dramatically during the Great Recession, which may have altered the composition of students graduating during the period we study. Further, we do not have information about innate ability, and so we do not know the value that a college degree is adding relative to one's baseline skill, or how ability factors into which college major people choose. Any of these factors could be contributing to the patterns we observe. In addition, it would be desirable to follow the same individuals over time to capture measures of ability and to track career progression. However, we are not able to do so with the data sets we employ, so we leave these issues for future research. Nonetheless, we believe this work takes an important step forward by providing a more complete picture of underemployment in the early careers of college graduates following the Great Recession.

References

Abel, Jaison R., and Richard Deitz. 2015. "Agglomeration and Job Matching among College Graduates." *Regional Science and Urban Economics* 51:14–24.

Abel, Jaison R., Richard Deitz, and Yaqin Su. 2014. "Are Recent College Graduates Finding Good Jobs?" Federal Reserve Bank of New York, *Current Issues In Economics and Finance* 20 (1): 1–8.

Altonji, Joseph G., Erica Blom, and Costas Meghir. 2012. "Heterogeneity in Human Capital Investments: High School Curriculum, College Major, and Careers." *Annual Review of Economics* 4:185–223.

Altonji, Joseph G., Lisa B. Kahn, and Jamin D. Speer. 2014. "Trends in Earnings Differentials across College Majors and the Changing Task Composition of Jobs." *American Economic Review* 104 (5): 387–93.

———. 2016. "Cashier or Consultant? Entry Labor Market Conditions, Field of Study, and Career Success." *Journal of Labor Economics* 34 (S1, pt. 2): S361–401.

Arcidiacono, Peter. 2004. "Ability Sorting and the Returns to College Major." *Journal of Econometrics* 121 (1–2): 343–75.

Arum, Richard, and Josipa Roksa. 2011. *Academically Adrift: Limited Learning on College Campuses.* Chicago: University of Chicago Press.

———. 2014. *Aspiring Adults Adrift: Tentative Transitions of College Graduates.* Chicago: University of Chicago Press.

Autor, David H., and David Dorn. 2013. "The Growth of Low-Skill Service Jobs and the Polarization of the US Labor Market." *American Economic Review* 103 (5): 1553–97.

Beaudry, Paul, David A. Green, and Benjamin M. Sand. 2014. "The Declining Fortunes of the Young Since 2000." *American Economic Review* 104 (5): 381–86.

———. 2016. "The Great Reversal in the Demand for Skill and Cognitive Tasks." *Journal of Labor Economics* 34 (S1, pt. 2): S199–247.

Betts, Julian R. 1996. "What Do Students Know about Wages?" *Journal of Human Resources* 31 (1): 27–56.

Casselman, Ben. 2013. "College Grads May Be Stuck in Low-Skill Jobs." *Wall Street Journal*, Mar. 26.

Chevalier, Arnaud. 2003. "Measuring Over-Education." *Economica* 70 (279): 509–31.

Chevalier, Arnaud, and Joanne Lindley. 2009. "Overeducation and the Skills of UK Graduates." *Journal of the Royal Statistical Society* 172 (2): 307–37.

Elsby, Michael W., Bart Hobijn, and Ayşegül Şahin. 2010. "The Labor Market in the Great Recession." *Brookings Papers on Economic Activity* Spring: 1–48.

Elsby, Michael W., Bart Hobijn, Ayşegül Şahin, and Robert G. Valletta. 2011. "The Labor Market in the Great Recession—An Update to September 2011." *Brookings Papers on Economic Activity* Fall: 353–84.

Feldman, Daniel C. 1996. "The Nature, Antecedents and Consequences of Underemployment." *Journal of Management* 22 (3): 385–407.

Feldman, Daniel C., and William H. Turnley. 1995. "Underemployment among Recent Business College Graduates." *Journal of Organizational Behavior* 16 (S1): 691–706.

Fogg, Neeta P., and Paul E. Harrington. 2011. "Rising Mal-Employment and the Great Recession: The Growing Disconnection between Recent College Graduates and the College Labor Market." *Continuing Higher Education Review* 75:51–65.

Fortin, Nicole M., Philip Oreopoulos, and Shelly Phipps. 2015. "Leaving Boys Behind: Gender Disparities in High Academic Achievement." *Journal of Human Resources* 50 (3): 549–79.

Galston, William A. 2014. "Welcome to the Well-Educated-Barista Economy." *Wall Street Journal*, Apr. 29. https://www.wsj.com/articles/william-a-galston-welcome-to-the-well-educated-barista-economy-1398813598.

Goldin, Claudia, Lawrence F. Katz, and Ilyana Kuziemko. 2006. "The Homecoming of American College Women: The Reversal of the College Gender Gap." *Journal of Economic Perspectives* 20 (4): 133–56.

Green, Francis, and Yu Zhu. 2010. "Overqualification, Job Dissatisfaction, and Increasing Dispersion in the Returns to Graduate Education." *Oxford Economic Papers* 62 (4): 740–63.

Hersch, Joni. 1991. "Education Match and Job Match." *Review of Economics and Statistics* 73 (1): 140–44.

Hoynes, Hilary, Douglas L. Miller, and Jessamyn Schaller. 2012. "Who Suffers during Recessions?" *Journal of Economic Perspectives* 26 (3): 27–48.

Kahn, Lisa B. 2010. "The Long-Term Labor Market Consequences of Graduating from College in a Bad Economy." *Labour Economics* 17 (2): 303–16.

McKee-Ryan, Frances M., and Jaron Harvey. 2011. "'I Have a Job, But . . .': A Review of Underemployment." *Journal of Management* 37 (4): 962–96.

Mian, Atif, and Amir Sufi. 2010. "Household Leverage and the Recession of 2007 to 2009." *IMF Economic Review* 58:74–117.

———. 2011. "Consumers and the Economy, Part II: Household Debt and the Weak U.S. Recovery." Federal Reserve Bank of San Francisco *Economic Letter* no. 2011-

02, 1–5, January. https://www.frbsf.org/economic-research/publications/economic -letter/2011/january/consumers-economy-household-debt-weak-us-recovery/.

Nunley, John M., Adam Pugh, Nicholas Romero, and R. Alan Seals. 2015. "Racial Discrimination in the Labor Market for Recent College Graduates: Evidence from a Field Experiment." *BE Journal of Economic Analysis & Policy* 15 (3): 1093–125.

Oreopoulos, Philip, Till von Wachter, and Andrew Heisz. 2012. "The Short- and Long-Term Career Effects of Graduating in a Recession." *American Economic Journal: Applied Economics* 4 (1): 1–29.

Peterson, Norman, Michael Mumford, Walter Borman, Richard Jeanneret, Edwin Fleishman, Kerry Levin, Michael Campion, et al. 2001. "Understanding Work Using the Occupational Information Network (O*NET): Implications for Practice and Research." *Personnel Psychology* 54 (2): 451–92.

Ruggles, Steven, Katie Genadek, Ronald Goeken, Josiah Grover, and Matthew Sobek. 2015. *Integrated Public Use Microdata Series: Version 6.0* [Machine-readable database]. Minneapolis: University of Minnesota.

Şahin, Ayşegül, Joseph Song, Giorgio Topa, and Giovanni L. Violante. 2014. "Mismatch Unemployment." *American Economic Review* 104 (11): 3529–64.

Stinebrickner, Ralph, and Todd R. Stinebrickner. 2014. "A Major in Science? Initial Beliefs and Final Outcomes for College Major and Dropout." *Review of Economic Studies* 81 (1): 426–72.

Vedder, Richard, Christopher Denhart, and Jonathan Robe. 2013. "Why Are Recent College Graduates Underemployed?" Center for College Affordability and Productivity Policy Paper, January.

Wiswall, Matthew, and Basit Zafar. 2015a. "Determinants of College Major Choice: Identification Using and Information Experiment." *Review of Economic Studies* 82 (2): 791–824.

———. 2015b. "How Do College Students Respond to Public Information about Earnings?" *Journal of Human Capital* 9 (2): 117–69.

Yen, Hope. 2012. "Half of New Grads are Jobless or Underemployed." *Associated Press*, Apr. 24.

Zafar, Basit. 2011. "How Do College Students Form Expectations?" *Journal of Labor Economics* 29 (2): 301–48.

———. 2013. "College Major Choice and the Gender Gap." *Journal of Human Resources* 48 (3): 545–95.

5

The Requirements of Jobs
Evidence from a Nationally
Representative Survey

Maury Gittleman, Kristen Monaco,
and Nicole Nestoriak

5.1 Introduction

Does the US workforce have the skills needed to be internationally competitive in the twenty-first century? Which jobs are vulnerable to loss as a result of the introduction of new technology, competition from trading partners, or offshoring (Autor 2015; Blinder 2009; Jensen and Kletzer 2010; Oldenski 2014)? Why have the differentials between the earnings of those with a college education and those without widened since 1979 (Bound and Johnson 1992; Katz and Murphy 1992)? What types of skills have a high and/or rising return in the labor market and what skills do not, and which skills are complementary with each other (Murnane, Willett, and Levy 1995; Borghans, ter Weel, and Weinberg 2014; Weinberger 2014; Deming 2015)? More generally, how are worker skills, job tasks, technological change, and international trade interacting to affect the earnings distribution and the employment structure (Acemoglu and Autor 2011; Firpo, Fortin, and Lemieux 2011)? To address these questions, it is useful and, in some cases, essential to have a solid understanding of the skills demanded of the workforce, as well as the tasks that must be performed.[1]

Maury Gittleman is a research economist at the Bureau of Labor Statistics. Kristen Monaco is Associate Commissioner for Compensation and Working Conditions at the Bureau of Labor Statistics. Nicole Nestoriak is a senior research economist at the Bureau of Labor Statistics.

The views expressed here are those of the authors and do not necessarily reflect the views or policies of the Bureau of Labor Statistics or any other agency of the US Department of Labor. The authors thank Bradley Rhein and Kristin Smyth for technical assistance. For acknowledgments, sources of research support, and disclosure of the authors' material financial relationships, if any, please see http://www.nber.org/chapters/c13699.ack.

1. Acemoglu and Autor (2011) distinguish between skills and tasks as follows. They define a task "as a unit of work activity to produce output." On the other hand, skill is considered to be a "worker's endowment of capabilities for performing various tasks."

While there are several data sets that researchers draw upon in studies of these kinds of questions—including the *Dictionary of Occupational Titles* (*DOT*), the Occupational Information Network (O*NET), and the OECD's Survey of Adult Skills (PIAAC)—the Bureau of Labor Statistics (BLS) is currently conducting the Occupational Requirements Survey (ORS), which promises to provide new information at the detailed occupation level. The ORS, developed in collaboration with the Social Security Administration (SSA), collects elements in four categories—educational requirements, mental and cognitive demands, physical demands, and environmental working conditions. While, as will be discussed in greater detail below, the primary reason for the initiation of the ORS is for potential use by SSA as a data source in disability adjudication, the data will be useful for numerous stakeholders due to the type of information collected and the level of detailed estimates that will be available as the first years of collection are completed.

In fiscal year (FY) 2015, BLS completed data collection for the ORS preproduction test. The preproduction test might better be described as a dress rehearsal as the sample design, collection procedures, data capture systems, and review were structured to be as close as possible to those that will be used in full-scale production, when there will be a larger sample size and the estimates will be intended for evaluation for use in the disability adjudication process. The preproduction sample, which is the source of the estimates presented in this chapter, is nationally representative when appropriate sample weights are used.[2]

This chapter is organized as follows: section 5.2 provides context for ORS by briefly describing the disability adjudication process, the data needs of this process, and how ORS is structured to meet those needs. Section 5.3 presents some initial estimates of occupational requirements, including educational, mental and cognitive, and physical demands. Section 5.4 exploits the linkage between ORS and BLS's National Compensation Survey to provide an exploratory analysis of the relationship between ORS elements and wages. Section 5.5 examines the relationship between job requirements and safety outcomes, while section 5.6 concludes and outlines additional potential uses for ORS data.

5.2   The Occupational Requirements Survey

5.2.1   *Dictionary of Occupational Titles* and Disability Determination

A brief history of the *Dictionary of Occupational Titles* (*DOT*) and disability determination by the Social Security Administration (SSA), which is recounted in Handel (2015a), will help to place the ORS data-collection efforts in context. Beginning in 1939, the Department of Labor (DOL) published the first edition of the *DOT*, which was designed as a tool to facilitate

---

2. The preproduction data will not be used in SSA's disability adjudication process.

matching job seekers to vacancies during the Great Depression. The second, third, and fourth editions of the *DOT* appeared in 1949, 1965, and 1977, respectively, with a partial update, called a "revised fourth edition," published in 1991. While the *DOT* retained its original purpose, beginning with the third edition, the SSA contracted with DOL to publish a supplement known as the *Selected Characteristics of Occupations* (*SCO*), to be used in disability determination. The *SCO* added information on specific vocational preparation (SVP)—the amount of time required for a worker to learn the techniques needed for average performance in a given job—along with elements on physical demands and environmental conditions. The *DOT* is still used in disability determination, though given that it was last updated in 1991, SSA has long wanted to find more current information.

For DOL's purposes, the *DOT* has been replaced by the Occupational Information Network, known as O*NET. As a bridge, early versions of O*NET reviewed raw data collected for the *DOT* in previous decades and recoded them in terms of the new O*NET variables. O*NET began collecting new data from surveys of job incumbents in 2001, replacing the recoded *DOT* data on a rolling basis until June 2008, when the first complete version of O*NET based on new data became available. In contrast to the *DOT*, where jobs were rated by job analysts, O*NET is largely based on responses by incumbents, although job analysts do complete certain sections of it (see Handel [2015b] for further details). O*NET, however, has not been usable from SSA's standpoint because it does not contain the full set of detailed job requirements needed to adjudicate disability claims under current Social Security regulations and policy.

For the purposes of Social Security Administration disability adjudication, the law defines disability as the inability to do any substantial gainful activity by reason of any medically determinable physical or mental impairment that can be expected to result in death or has lasted or can be expected to last for a continuous period of not less than twelve months. The SSA uses a five-step sequential process to determine disability. By the end of the third step,[3] the claimant who has met current earnings and medical hurdles has his/her residual functional capacity to perform work-related activities classified according to the five exertional levels of work: sedentary, light, medium, heavy, and very heavy. The final two steps require occupational information to compare the functional capacities of an individual to those required by available jobs:

- Step 4. *Previous work test.* Can the applicant do the work he or she had done in the past? If the individual's residual functional capacity equals the previous work performed, the claim is denied on the basis that the individual can return to his/her former work. If the claimant's residual

---

3. Step 1. Is the claimant engaging in substantial gainful activity? Step 2. Does the claimant have a severe impairment? Step 3. Does the impairment(s) meet or equal SSA's medical listings?

functional capacity is less than the demands of his or her previous work, the application moves to Step 5.

- Step 5. *Any work test.* Does the applicant's condition prevent him or her from performing "*any other kind of substantial gainful work which exists in the national economy?,*" meaning work that "*exists in significant numbers*" either in the region of residence or in several regions of the country.[4] If yes, the application is accepted and benefits are awarded. If not, the application is denied. In this step, the residual functional capacity is applied against a vocational grid that considers the individual's age, education, and the transferability of previously learned and exercised skills to other jobs. The vocational grid directs an allowance or denial of benefits.

The elements of ORS are designed with the needs of Steps 4 and 5 of disability adjudication in mind. As noted earlier, there are four different categories of information that are collected. Educational requirements include whether literacy is needed, degrees required with respect to formal education, and certifications, licenses, and training. These elements, in turn, are used to calculate specific vocational preparation. Mental and cognitive elements include task complexity, work control, and interaction with regular contacts.[5] A wide range of physical demands is asked about, including hearing, use of keyboarding, visual acuity, sitting, standing, stooping, kneeling, crawling, crouching, pushing, pulling, reaching, strength, climbing, and manipulation. Finally, environmental conditions comprise such elements as the temperature, exposure to fumes, humidity, and wetness. Appendix table 5A.1 contains a full list of data elements.

Despite the fact that ORS is designed for disability adjudication, as noted in the first section, that does not mean it cannot be put to more general research purposes. In section 5.3, we discuss links between a classification of jobs based on ORS elements and the influential job categorization scheme of Autor, Levy, and Murnane (2003).

5.2.2   ORS Procedures and Sampling

The goal of ORS is to collect and publish occupational information that meets the needs of SSA at the level of the eight-digit standard occupational classification (SOC) that is used by the Occupational Information Network (O*NET).[6] The ORS data are collected under the umbrella of

---

4. Quotations are from the Social Security Act Section 223(d)(2).
5. The wording of the mental and cognitive elements have been changed for production. A sample of the collection form is available at http://www.bls.gov/ncs/ors/occupational_requirements_survey_elements_private.pdf.
6. The occupational classification system most typically used by BLS is the six-digit SOC (https://www.bls.gov/soc/), generally referred to as "detailed occupations." O*NET uses a more detailed occupational taxonomy (https://www.onetcenter.org/taxonomy.html), classifying occupations at eight digits and referring to these as "O*NET-SOC 2010 occupations." There are 840 six-digit SOCs and 1,110 eight-digit SOCs.

the Bureau of Labor Statistics National Compensation Survey (NCS)[7] program. The NCS is an establishment-based survey that provides measures of (a) employer costs for employee compensation (ECEC), (b) compensation trends (Employment Cost Index, or ECI), (c) the incidence of employer-provided benefits among workers, and (d) provisions of selected employer-provided benefit plans. The NCS uses field economists (FEs) to collect data, rather than, for instance, mailing out questionnaires. The FEs are well suited for ORS data collection as their training focuses on identifying the appropriate respondent, probing the respondent to clarify apparent inconsistencies in responses, and following up with respondents to ensure data are complete and accurate. The FEs generally collect data elements through either a personal visit to the establishment or remotely via telephone, email, mail, or a combination of modes.

The ORS preproduction sample was drawn from the same frame as the NCS—the Quarterly Census of Employment and Wages, which includes all establishments covered by state unemployment insurance laws, and a supplementary file of railroads. The frame contains virtually all establishments in the fifty United States and the District of Columbia in the private sector (excluding agriculture, forestry and fishing, and private households) and in state and local governments.[8] The preproduction ORS sample contains 2,549 establishments. Approximately 15 percent of these units are government owned and 85 percent privately owned. Roughly one-third of the ORS preproduction sample consists of establishments that are also in the NCS sample. This overlap is notable because, as we discuss in greater detail in section 5.3, for this portion of the sample it is possible to obtain wage and other data to match with the ORS elements.

Of the 2,549 establishments contacted by field economists, 1,851 of them provided usable data, indicating a usable establishment response rate of 73 percent. Some 6 percent of the initial sample was either out of business, out of scope, or had no jobs that were within scope, with the remaining 21 percent constituting refusals.

For each establishment in the ORS sample, jobs were selected for inclusion in the survey with probability proportional to incumbent employment; these jobs are referred to as "quotes." The number of jobs selected within a private establishment varies from four to eight, based on establishment size, and, in government, the number of jobs ranges from four to twenty. It is common for multiple individuals within an establishment to have the same job (e.g., elementary school teachers within a school/school district), which can result in fewer individual quotes for that establishment. Because the quote-level information is tied to the job, not the individual, sampling a certain number of jobs within an establishment is not equivalent to sampling a certain number of workers within an establishment.

---

7. For details on the NCS, see http://www.bls.gov/ncs/.
8. Federal government workers are out of scope for ORS.

The ORS preproduction data collection began in October 2014 and continued until May 2015. At the close of the data-review process, information on 7,109 quotes or jobs had been collected from the 1,851 establishments, slightly fewer than four jobs per establishment. These jobs spanned all twenty-two unique two-digit SOCs in scope for ORS and 704 unique eight-digit SOCs.[9] The 704 eight-digit SOCs represent 63.4 percent of the 1,110 unique eight-digit SOCs. In order to be able to present estimates that cover the economy as a whole and not overload the reader with numbers, most of the occupational estimates we present in the next section are at the more aggregate level of nine major occupations. We also present estimates for eleven major industries.

5.3   Occupational Requirements: Evidence from the ORS Preproduction Sample

5.3.1   Educational Requirements

We now turn to actual estimates of job requirements from the ORS preproduction sample, starting with the category of educational requirements. It is important to note that these are "research" estimates only. Due to alternative categorizations of certain data elements and different approaches to calculating standard errors, estimates presented in this chapter may not match any official estimates from the preproduction data released by BLS.

Spurred in part by the rise in returns to a college education—for instance, between 1979 and 2013, the wage premium earned by college graduates relative to high school graduates widened from 24.95 percent to 50.18 percent for women and from 20.18 percent to 48.44 percent for men[10]—growing attention is being paid in the political arena to boosting attendance at college, in part by making it more affordable. According to the Obama administration, "Earning a postsecondary degree or credential is no longer just a pathway to opportunity for a talented few; rather, it is a prerequisite for the growing jobs of the new economy."[11] With this in mind, the administration asserted that everyone should obtain at least one year of higher education or postsecondary training. In this context, it is interesting to note that, according to ORS estimates shown in table 5.1, an associate's degree is required in 4 percent of jobs, a bachelor's degree in 18 percent, and a graduate or professional degree in 5 percent. Thus, according to ORS, only about one-quarter of

---

9. There are twenty-three two-digit SOCs in the classification system, but military (SOC 55) is out of scope for ORS.

10. These estimates are from EPI analysis of Current Population Survey Outgoing Rotation Group microdata. The college wage premium is the percent by which wages of college graduates exceed those of otherwise equivalent high school graduates, regression adjusted (http://www.epi.org/chart/swa-wages-figure-4n-college-wage-premium-2/).

11. https://obamawhitehouse.archives.gov/issues/education/higher-education.

Table 5.1   Educational requirements, ORS and O*NET

ORS educational category	Percent	O*NET educational category	Percent
No literacy	2.6		
Literacy, no degree	28.1	Less than high school	13.6
High school diploma	43.2	High school diploma	34.9
		Postsecondary certification	8.4
		Some college	7.7
Associate's degree	4.0	Associate's degree	7.6
Baccalaureate degree	17.7	Baccalaureate degree	17.8
		Postbaccalaureate certificate	1.2
Postbaccalaureate degree	4.5	Postbaccalaureate degree	8.7

employment requires any type of college education. A high school degree is, however, required for 43 percent of jobs. No degree is required in 31 percent of employment, with 2.6 percent of all jobs said to not require any literacy whatsoever.

How do these results compare to those from other sources that have tried to measure the same concept? O*NET also assesses the education requirements of occupations, though, because it does not publish economy-wide estimates, we calculated them by averaging estimates at the detailed occupation level using weights obtained from BLS's Occupational Employment Statistics program. The categories used by O*NET, in part because they involve certifications, are somewhat different than those used by ORS, but some comparisons can still be made.

Whereas ORS indicates no degree is required in 31 percent of the jobs, in O*NET the category for less than high school contains only 14 percent of employment.[12] The ORS data indicate that 43 percent of jobs require a high school degree, which is roughly the same as the proportion in the O*NET categories high school or high school plus certification. O*NET, however, has 15 percent of employment in the categories for individuals either with some college or an associate's degree, while only 4 percent of jobs is in the associate's degree category in ORS. The percentages requiring a bachelor's degree are similar across the two sources, but O*NET has a higher proportion in the postbaccalaureate category (10 percent versus 5 percent), which in O*NET includes everything ranging from postbaccalaureate certification to postdoctoral training.

The ORS education requirements estimates can also be compared to a relatively recent source of nationally representative data that has a number of elements in common with ORS, Michael Handel's Survey of Workplace Skills, Technology and Management Practices (STAMP). STAMP's estimates are based on self-reports of job incumbents and its first wave (of two)

12. O*NET estimates used are from version 19.

was conducted between October 2004 and January 2006, with a sample of 2,304 respondents. The data are not publicly available but some comparisons can be made with ORS on the basis of results presented in Handel (2015c). Instead of inquiring directly about literacy, STAMP asked whether any reading was required on the job. According to STAMP, some reading was required of 96 percent of the workforce, compared to the estimate in ORS that 97.4 percent of jobs required literacy. STAMP divided occupations into five groups: upper white collar (management, professional, technical occupations), lower white collar (clerical, sales), upper blue collar (craft and repair workers—e.g., construction trades, mechanics), lower blue collar (factory workers, truck drivers, etc.) and service (e.g., food service workers, home health care aides, childcare, janitors, police and firefighters). The percentage where reading is required ranged from 91 percent for the two blue-collar groups up to 99 percent for the upper white-collar one.

Handel (2015c) also provides information for the educational requirements of jobs. The numbers are fairly close to those from ORS in terms of the shares requiring a bachelor's degree or beyond. According to STAMP, a graduate degree was required in 6.3 percent of the jobs, versus 5 percent in ORS, with a bachelor's degree needed in an additional 20.8 percent (18 percent in ORS) of the jobs. Some college but less than a bachelor's degree was required in 16.5 percent of the jobs, much greater than in ORS. A high school degree by itself was required in 42.6 percent of the jobs and a high school degree plus vocational training in an additional 6.3 percent of the jobs. The remaining 7.6 percent required less than a high school degree.

5.3.2 Specific Vocational Preparation

Aside from formal education requirements,[13] ORS also asked about prior experience, postemployment training, and certificates and licenses. The duration associated with all of these are used to calculate SVP, which, as noted above, is the amount of time needed for an individual to get to an average level of performance. Specific vocational preparation totals time spent both in formal education and certification and training programs that prepared the individual for the job (preemployment training), required prior work experience in related jobs, and the time needed in the job itself to get to average performance (postemployment training). It is important to keep in mind that SVP could be high both because a long period of specialized on-the-job training is needed and because much time must be spent in special-

13. For the purposes of SVP, formal education focuses on the "vocational" component of the education. High school, for example, is not included in formal education, except in the rare case that an individual spent time in a vocational high school program. Generally, a four-year college degree will have two years of general education requirements, which means only two years count toward SVP. Postbaccalaureate degrees tend to be entirely vocational in nature, in which case the entire length of the postbaccalaureate degree is included in the SVP measure as well as two years of college education.

Table 5.2   **Specific vocational preparation by occupation and industry (percent)**

	Short demo/1 month	More than 1 month up to 1 year	More than 1 year up to 4 years	Over 4 years
All workers	33	17	32	18
Occupation				
Management, business, financial	—	—	32	65
Professional and related	4	4	57	35
Service	61	22	15	2
Sales and related	54	13	25	9
Office and admin.	26	25	40	8
Construction and extraction	—	—	25	30
Installation, maintenance, repair	10	2	50	24
Production	41	24	29	6
Transport. and material moving	57	27		
Industry				
Construction	15	25	27	34
Manufacturing	32	21	32	15
Wholesale trade	36	16	30	18
Retail trade	62	15	21	3
Transport and warehousing	45	—	26	—
Financial activities	—	—	48	27
Professional and business services	23	14	35	28
Education and health Services	22	14	44	21
Leisure and hospitality	68	16	11	5
Other services	30	34	25	11
Public admin.	11	22	45	21

Note: Dash indicates no workers in this category or data did not meet publication criteria.

ized formal schooling. Specific vocational preparation is measured in days and then grouped into nine categories ranging from "short demonstration" to over ten years. Owing to the sparseness of responses for some categories,[14] particularly for estimates by industry and occupation, we collapse these nine categories into four: one month or below; more than one month up to and including one year; more than one year up to and including four years; and more than four years.

As shown at the top of table 5.2, across all workers, according to ORS respondents, about one-third of jobs can be learned within one month's time. At the other end of the spectrum, a bit more than one-sixth of jobs require over four years to get to average performance. Looked at differently, roughly half of employment requires less than one year of SVP, and the other half needs more.

14. Estimates are not shown on the tables if their relative standard errors (RSEs) exceed 0.3. In addition, when the sum of a group of estimates is equal to one, a suppression for RSE reasons generally necessitates a secondary suppression, given that it would be possible to deduce the suppressed estimate's value from the values of the other estimates.

We now examine SVP by major occupation (nine categories) and major industry (eleven categories) to get a better understanding of what is behind the distribution for the economy as a whole. As occupation is what one does, while industry is where one does it, in general, one would expect there to be larger differences by occupation than industry in education requirements, skills demanded, and tasks performed. Support for this supposition can be found in the fact that occupations have more explanatory power than industries with respect to other measures related to the labor market, such as wages (e.g., see Pierce 1999). Though a given occupation may differ across industries, much of the differences we will note across industries are a result of their differing occupational compositions.

As table 5.2 shows, there is substantial variation by major occupation in SVP. Both management, business, and financial occupations and professional and related occupations have more than 90 percent of employment in categories where the SVP exceeds one year. In contrast, service, sales and related, and transportation and material-moving occupations all have a majority of employment where SVP is one month or lower.

Examining SVP by major industry, one sees less variation than by occupation, with a few of the industry SVP distributions being fairly close to that of the economy as a whole. There are notable exceptions, though. On the low SVP side are those industries where SVP is less than a year for substantially more than half of employment, which include retail trade, transport and warehousing, leisure and hospitality, and other services. On the high SVP side, where SVP is substantially greater than one year for much more than 50 percent of employment, are the following industries: financial activities, professional and business services, education and health services, and public administration.

As previously mentioned, the value of SVP can be driven by requirements of formal education, preemployment training, prior work experience, or postemployment training (see figure 5.1). Across all workers, the largest shares of SVP are postemployment training (37 percent) and prior work experience (39 percent). This varies markedly by SVP categories. For those in jobs requiring little preparation, nearly all of the SVP component is captured in postemployment training. At the other extreme, jobs with the highest levels of SVP have nearly all vocational preparation captured by required formal education (29 percent) and prior work experience (62 percent).

5.3.3   Mental and Cognitive Demands

We now turn to the second category of data collected by ORS, mental and cognitive demands, and begin with the element of task complexity. In response to the question "how complex are tasks in this occupation?" respondents were able to choose from five different categories: very complex, complex, moderate, simple, and very simple. Once again, we collapse categories (complex and very complex, moderate, simple and very simple) to obtain

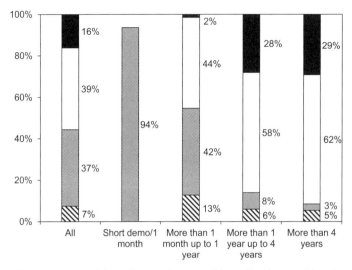

Fig. 5.1   Components of specific vocational preparation
Note: The bar for short demo/one-month duration shows only postemployment training, due to the percentages in the remaining categories not meeting publication criteria.

more reliable estimates. About one-half (51 percent) of jobs were rated in the simplest category, around one-third (34 percent) as moderate, with the remaining 15 percent in jobs rated in the most complex category. These shares show large differences across major occupations. Management, business, and financial occupations (56 percent), and professional and related occupations (36 percent) are the only occupation groups where the share of the most complex category exceeds that for the economy as a whole, with the next highest occupation having a share of only 14 percent. Examined from the other end of the complexity spectrum, transportation and material-moving occupations (85 percent) and service occupations (81 percent) have the highest shares of the simplest jobs, with sales and related, office and administration and production also having more than a majority share in this category (see table 5.3).

Are there major differences by industry in terms of the distribution of task complexity? Such differences are, once again, less notable than those for occupation, though still present. For instance, leisure and hospitality (83 percent), transport and warehousing (77 percent), and retail trade (74 percent) have higher than average shares of the simplest jobs, while public administration (26 percent), professional and business services (23 percent), financial activities (21 percent), and education services (20 percent) have above average shares of the most complex jobs.

A second dimension of cognitive demands is how closely controlled an

Table 5.3   Cognitive elements by occupation and industry (percent)

	Task complexity			Work control			Regular contacts		
	Complex/ very complex	Moderate	Simple/ very simple	Closely/ very closely	Moderate	Loose/ very loose	Structured/ very structured	Semi- structured	Unstructured/ very unstructured
All workers	15	34	51	58	29	13	72	22	6
Occupation									
Management, business, financial	56	—	—	14	38	48	28	48	24
Professional and related	36	53	10	20	53	28	41	48	11
Service	2	17	81	84	13	3	90	8	2
Sales and related	5	29	66	69	22	9	71	21	8
Office and admin.	2	36	62	70	27	3	85	—	—
Construction and extraction	12	41	47	59	—	—	78	—	—
Installation, maintenance, repair	—	62	—	—	52	—	85	—	—
Production	2	27	71	79	—	—	95	—	—
Transport. and material moving	—	—	85	85	—	—	96	—	—
Industry									
Construction	15	45	40	52	37	11	78	—	—
Manufacturing	11	35	54	62	27	11	81	16	3
Wholesale trade	15	38	47	53	29	18	65	—	—
Retail trade	3	23	74	79	18	3	86	—	—
Transport and warehousing	—	—	77	77	—	—	87	—	—
Financial activities	21	51	29	48	34	18	56	32	13
Professional and business services	23	33	44	48	33	19	62	29	9
Education and health services	20	41	39	47	38	16	63	31	6
Leisure and hospitality	3	14	83	85	11	4	91	—	—
Other services	—	—	57	62	26	12	78	—	—
Public admin.	26	45	29	36	44	20	57	27	16

occupation's work is. We collapse five categories for work control (very loosely, loosely, moderately, closely, and very closely) to three (closely and very closely, moderately, and loosely and very loosely) for reasons of reliability. Nearly three-fifths of employment was rated as being closely or very closely controlled, with a further 29 percent moderately controlled, and 13 percent loosely or very loosely controlled. There is similar variability across major occupations, as with task complexity. Management, business, and financial occupations (48 percent) and professional and related occupations (28 percent) are the only occupation groups where the share of loosely or very loosely controlled jobs surpasses the economy-wide average. Service and transportation and material-moving occupations have about 85 percent of employment in closely or very closely controlled jobs, with production occupations not far behind at 79 percent.

Major industries with a much higher than average proportion of closely or very closely controlled jobs include leisure and hospitality (85 percent), retail trade (79 percent), and transport and warehousing (77 percent). Public administration (20 percent), professional and business services (19 percent), and financial activities (18 percent) rank highest in terms of the share in the loosely or very loosely controlled category.

The final cognitive element we will consider involves responses to the question, "What type of work-related interactions does the occupation have with regular contacts?" As with the other two cognitive elements, five categories have been collapsed into three (structured and very structured, semistructured, unstructured and very unstructured).[15] For the economy as a whole, structured or very structured contacts predominate, being the case in nearly three-quarters of employment (72 percent). Semistructured contacts account for about one-fifth of employment (22 percent), with the remaining 6 percent in unstructured or very unstructured contacts. Those in management, business, and financial occupations are much less likely to have unstructured or very unstructured contacts (24 percent), while the contacts of those in transportation and material moving (96 percent), production (95 percent), service (91 percent), office and administration (85 percent), and installation, maintenance, and repair (85 percent) are more likely to be structured or very structured.

By industry, once again, there is less variability than by occupation, though leisure and hospitality (91 percent), transport and warehousing (87 percent), and retail trade (86 percent) stand out as sectors where contacts are particularly structured.

Thus far, we have been examining education requirements and cognitive

15. Very structured is defined as exchanging straightforward, factual information; structured involves coordinating and routine problem solving; semistructured includes problem solving, discussing, soft selling; unstructured includes influencing, persuading, hard selling; and very unstructured includes defending, negotiating, and resolving controversial or long-term issues.

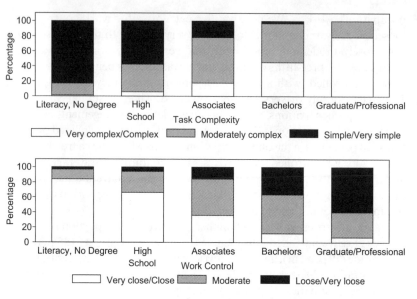

Fig. 5.2   Mental and cognitive elements by educational requirements

demands independently, but it is also of interest to see how they are inter-related. For instance, how do cognitive demands vary by education require-ment? Figure 5.2 makes it apparent that both task complexity and work control are strongly ordered by the amount of education required.

As one would expect, as education requirements increase, the share of simple and very simple jobs decreases and the proportion of complex and very complex jobs increases. Work control is related in a similar fashion, as higher educational requirements are associated with jobs that are controlled more loosely. Figure 5.3 is similar to figure 5.2, except cognitive demands are arrayed against four (collapsed) levels of specific vocational preparation instead of against degrees required. As with education, as the level of SVP rises, task complexity rises, while jobs become more loosely controlled.

Indirectly apparent in figures 5.2 and 5.3 is the relationship between task complexity and work control. Looking at the lowest level of educational attainment depicted (literacy, no high school degree) or at the lowest level of SVP (short demo/one month) shows that these jobs are characterized by simple/very simple tasks and are closely/very closely controlled. A direct comparison between complexity and control is presented in figure 5.4.

The graph depicts the joint probabilities of the categories of task com-plexity and work controls. Roughly 48 percent of jobs in the economy can be classified as simple/very simple and closely/very closely controlled. As the complexity level rises, the level of control decreases—the diagonal joint probabilities (from lower left to upper right) have the highest density. The lower-right corner, jobs that are simple/very simple and very closely/closely

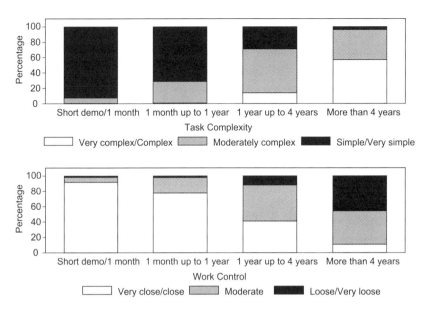

Fig. 5.3   Mental and cognitive elements by specific vocational preparation

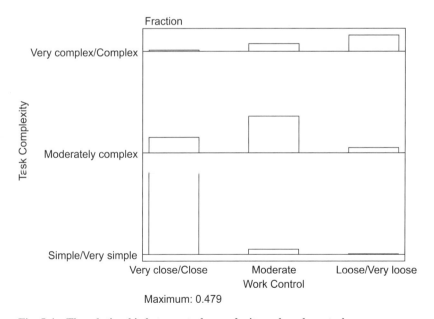

Fig. 5.4   The relationship between task complexity and work control

controlled, include occupations such as cashiers and laborers and freight, stock, and material movers. Moving up diagonally, jobs that are both moderately complex and moderately controlled include teaching occupations and very complex/complex and loose/very loosely controlled jobs include specialized nurses and software designers. Simple jobs that are moderately controlled include jobs with low barriers to entry that are typically performed off-site from one's direct employer, such as landscapers and personal care aides. Complex jobs that are moderately controlled include accountants.

Not surprising, but notable, is the very small percentage of jobs that are both simple and loosely controlled. This intersection represents a key set of job alternatives for individuals with certain types of cognitive impairments.

Autor, Levy, and Murnane (2003) developed a task model to predict the impact of computerization on different kinds of jobs. They divided occupations into a 2 × 2 grid, with one dimension defined by whether the tasks in the occupations are routine or nonroutine, and the other defined on the basis of whether the tasks are manual or analytical. They hypothesize substantial computer substitution for routine tasks, whether manual or analytical. For nonroutine tasks, they hypothesize strong possibilities for complementarities for the analytical occupations, but limited possibilities for substitution or complementarities for the manual occupations.

While ORS does not contain the same variables as Autor, Levy, and Murnane, one can compare the jobs in our 3 × 3 grid in figure 5.4 to those in Autor, Levy, and Murnane's 2 × 2 grid. The closely controlled/simple cell in figure 5.4 appears to contain jobs similar to those in Autor, Levy, and Murnane's routine/manual category (picking and sorting, repetitive assembly). Their nonroutine/analytical box (e.g., medical diagnosis and legal writing) also has much in common with the four categories in figure 5.4 having moderate or greater complexity and moderate or less control.

5.3.4   Strength Requirements

We now turn to physical demands and examine a variable called strength, which is a key element in SSA's disability process. The variable captures a number of different dimensions of physical demands and is used to categorize work as either sedentary, light, medium, heavy, or very heavy. For instance, sedentary work is where the job requirements are as follows: standing for no more than 3/8 of the day; lifting of up to ten pounds occasionally; lifting a negligible weight frequently; lifting no weight constantly; no pushing with arms/hands; no pushing with legs/feet; and no pulling with feet only. At the other end of the spectrum, a heavy job requires the incumbent to lift more than 100 pounds occasionally, lift more than fifty pounds frequently, and lift more than ten pounds constantly. As before, we show estimates with the categories collapsed into three (sedentary and light, medium, heavy and very heavy).

As shown in table 5.4, some 70 percent of employment is estimated to be in the sedentary and light, 22 percent in the medium, and the remaining

Table 5.4   **Strength by occupation and industry (percent)**

	Light/sedentary	Medium	Heavy/very heavy
All workers	70	22	8
Occupation			
Management, business, financial	90	—	—
Professional and related	84	12	4
Service	65	28	6
Sales and related	69	—	—
Office and admin.	88	9	3
Construction and extraction	28	42	30
Installation, maintenance, repair	32	45	23
Production	45	35	20
Transport. and material moving	52	28	20
Industry			
Construction	32	42	26
Manufacturing	51	33	16
Wholesale trade	61	—	—
Retail trade	58	35	7
Transport and warehousing	59	—	—
Financial activities	90	—	—
Professional and business services	84	—	—
Education and health services	79	16	5
Leisure and hospitality	70	—	—
Other services	81	—	—
Public admin.	63	26	11

Note: Dash indicates no workers in this category or data did not meet publication criteria.

8 percent in the heavy categories. Around 85 to 90 percent of employment in management, business, and financial occupations, professional and related occupations, and office and administration occupations is in the sedentary and light category. Major occupations with smaller proportions of sedentary and light work include construction and extraction (28 percent), installation, maintenance and repair (32 percent), production (45 percent), and transportation and material moving (52 percent). By industry, financial activities (90 percent) and other services (81 percent) have the highest proportions of sedentary and light work.

5.4   Occupational Requirements and Wages

In this section, we explore the relationship between various ORS elements and wages, measuring the returns associated with various skills and illustrating the use of ORS data for labor market analysis. Because ORS itself does not measure wages, we take the 2,106 ORS quotes that overlap with the NCS sample and are able to obtain average hourly wage measures for 1,523 of these from the fourth quarter of 2014. It is rare that one has measures of skill and pay for the same job, as most of the research on pay and skills, at least in the United States, relies on merging in occupation-level measures from the *DOT* or O*NET onto data sets with measures of pay.[16]

Before turning to regressions containing ORS elements, it may be useful to say more about the dependent variable, average hourly wages, which comes from the Employer Costs for Employee Compensation (ECEC) portion of the National Compensation Survey. In the ECEC, earnings are defined to include incentive pay but exclude premium pay for overtime, holiday, and weekend work; shift differentials; bonuses not directly tied to production; payments by third parties such as tips; and payment in kind such as room and board. The ECEC data are converted to a cost per hour worked using work schedule information common to all workers and averaged over the incumbents within a job. Wage data from the ECEC or related components of the NCS have been used in a number of different studies, including ones on public-private compensation differentials (Gittleman and Pierce 2012, Munnell et al. 2011), inequality (Pierce 2010), and interindustry wage differentials (Gittleman and Pierce 2011).

These average hourly wage data can be linked to the ORS data by job. While the fact that these data are averages over incumbents is, in certain circumstances, a disadvantage relative to having data on each individual worker, the ORS data elements apply to each incumbent so there is a match between the level of aggregation of the wage data and that of the ORS elements. The key advantage to this approach is that the data on earnings and requirements are directly linked. Most studies that examine the returns to job attributes rely on linking microdata on individuals (typically the Current Population Survey or census public-use microdata) with jobs (from *DOT* or O*NET) by aligning occupation codes and merging in occupational averages (e.g., Autor, Levy, and Murnane 2003; Abraham and Spletzer 2009; Ingram and Neumann 2006). As Abraham and Spletzer acknowledge, inaccurate detailed occupation coding in the CPS and census raise data-quality concerns when data sets are matched based on occupation. As the NCS and ORS data are collected based on the same "quote" or job at the establishment, the linkage between pay and ORS elements should be accurate.[17]

We first present the average hourly wage (and associated 95 percent confidence interval) for categories defined by the key variables of interest—SVP, education, task complexity, regular contacts, and strength. With the possible exception of strength, the mean wage associated with the different categories for each of these variables follows a predictable pattern, as is evident in figures 5.5, 5.6, and 5.7.[18]

---

16. See Autor and Handel (2013) for an exception and further discussion.

17. The fact that ORS has data by the job rather than averages for the occupation as a whole means that it should be possible to use ORS elements to explain within-occupation wage variation. Such an undertaking will have to wait, however, until ORS is a full-scale survey (with a larger sample) and is dependent on funding to collect wages along with ORS elements.

18. Average wages for jobs with no literacy requirement are not provided, though this category is included in the regression models.

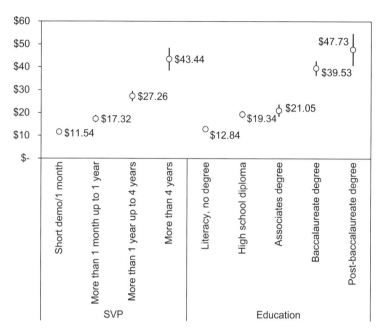

Fig. 5.5   Average hourly wages by SVP and education categories

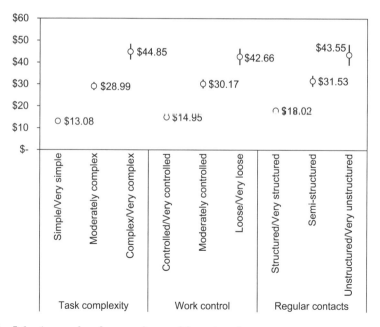

Fig. 5.6   Average hourly wages by cognitive categories

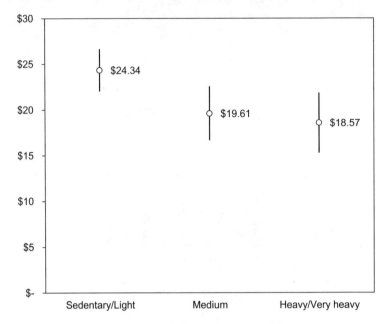

Fig. 5.7   Average hourly wages by strength category

Turning to the multivariate analysis, all models regress the natural log of the wage on a set of NCS establishment (size, industry, and private/public sector) and job characteristics (full-time/part-time and union/nonunion). Establishment size is captured by four categories: 0–49 (the reference group), 50–99, 100–499, and 500 or more workers. Controls for industry are made at the broad NAICS grouping: mining and utilities (reference group), construction, manufacturing, wholesale trade, retail trade, transportation and warehousing, information, financial services, professional and business services, education and health services, leisure and hospitality, other services, and public administration. Ownership is controlled for with a dummy variable for private sector (state and local government is the reference group).

Four models are estimated. Model 1 includes only additional controls for education and Model 2 expands this to include cognitive elements and strength. Models 3 and 4 are similarly structured, but include SVP rather than education. Consistent with past research, the establishment variables indicate the presence of establishment-size effects (Brown and Medoff 1989) and interindustry differentials (Gittleman and Pierce 2011), and little difference by whether employment is in the private sector or in state and local government (Gittleman and Pierce 2012). In terms of job characteristics, there are premiums for union status (Gittleman and Pierce 2007) and full-time status (Lettau 1997). These and all our estimation results are presented in appendix table 5A.2.

We first consider education requirements, where the omitted group is jobs

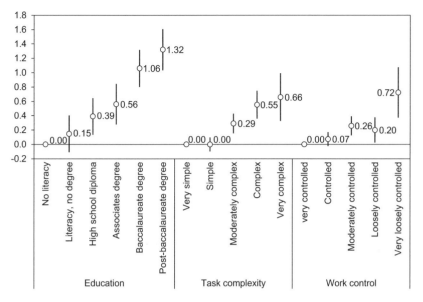

Fig. 5.8   Returns to education, task complexity, and work control

where no literacy is required. The point estimates are ordered in terms of increasing education, but the standard errors are large, in part because of item nonresponse. Nonetheless, there is support for the hypothesis that those jobs requiring bachelor's degrees or higher have greater earnings than other jobs. The R-squared, including establishment and job characteristics, is 0.67, high compared to what one would get in a comparable regression using household data. With just establishment and job controls, the R-squared is 0.43. The magnitude of the return from an associate's degree relative to a high school diploma is similar to that in Card (1999) and Carneiro, Heckman, and Vytlacil (2011), who find returns to additional years of education post–high school on the order of 6–11 percent per year (depending on model specification). The return to a college degree from our model is generally larger in magnitude than in the literature, though the overall ordering of the returns to education follow a sensible pattern when taken as a whole.

Figure 5.8 presents coefficients on education, task complexity, and work control from Model 2. Adding cognitive variables to the model decreases the returns to education considerably—roughly halving them for most categories. This is similar to analysis of the PIAAC, which finds that the returns to education decrease by approximately one-third when skills variables are included in the model (OECD 2013). The R-squared for the model with a full set of controls for work requirements is 0.77.

It may be worth highlighting again the distinction between tasks (a unit of work activity to produce output) and skills (a worker's endowment of capabilities for performing various tasks). As Autor and Handel (2013) note, in the Mincer earnings model, skills, as proxied by education and experience, have an economy-wide price. But because of the ongoing self-selection of workers into tasks and the bundling of task demands within jobs, these authors view the Roy model as a more appropriate one for analyzing returns to tasks. One implication of this model is that a Mincer-type regression will not generally recover the average returns to the tasks. While the cognitive demands that we use in the regression analysis may not fit neatly into the skill-task distinction, we nonetheless view this as a useful exploratory exercise.

The first cognitive demand we consider in the regression analysis is task complexity. Wages are ordered by the levels of this element. Those in very complex jobs earn 0.66 log points more than those in simple ones and those in the moderately complex jobs earn 0.29 log points more than those in the very simplest (see figure 5.8). The positive relationship between cognitive tasks and wages corroborates Autor and Handel (2013), though the magnitude of the relationship cannot be easily compared with theirs. Additionally, if we consider task complexity as roughly synonymous with analytical content, then these results also roughly align with those of Abraham and Spletzer (2009) and Ingram and Neumann (2006).

The results from work control are similar to those from task complexity in that wages are increasing in how loosely the job is controlled. In the final cognitive demand that we consider, type of regular contacts is not significant in the model, which is consistent with the finding by Pierce (1999) that coefficients on contacts variables tend to be small and imprecisely estimated in log-wage regressions using NCS wage data. The controls for strength are also jointly insignificant. There is no consensus in the literature on the empirical relationship between physical demand and wages. Abraham and Spletzer (2009) and Autor and Handel (2013) find negative returns to jobs requiring physical skills, while Ingram and Neumann (2006) estimate positive returns to jobs requiring physical effort (though they also include education controls in their models).

The results for specific vocational preparation (SVP) are similar to those for education requirements. All the coefficients are significantly positive relative to the omitted group of short demonstration. The R-squared for Model 3, which controls for SVP, establishment and job characteristics, is similar to that for education requirements at 0.66. Much like the Model 2 results, including controls for the cognitive and strength requirements roughly halves the coefficients on SVP (Model 4 in appendix table 5A.2). Owing to relatively large standard errors, the adjacent categories are not significantly different at the 10 percent level.

As seen in figure 5.9, there are substantial returns to task complexity and work control requirements after controlling for SVP. The pattern of returns to these cognitive skills is similar to those in the education model (figure 5.8). Also similar to that model, the returns to regular contacts and strength are not statistically significant.

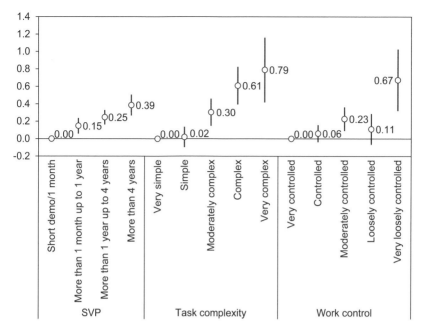

Fig. 5.9   Returns to SVP, task complexity, and work control

5.5   Occupational Requirements and Safety Outcomes

How do the physical demands and environmental conditions measured by ORS affect injury and illness rates? In this section, we present a second type of illustrative multivariate analysis, examining the relationship between the risk of an occupation, as measured by various ORS elements, and the outcomes of that risk, as captured by the injury and illness rate. Ideally, we would take the approach we did with the wage data, and match a job's ORS elements to its own injury and illness rate from the Survey of Occupational Injuries and Illnesses (SOII).[19] If we were to do this, however, the resulting sample would be both very small and unrepresentative, given that there is likely to be little overlap between the ORS sample and the SOII sample. Instead, the approach we use here is to aggregate both the risk and the injury and illness data by three-digit SOC, as this seems to be the lowest level of aggregation we can use where there is enough sample per occupation to adequately measure risk in the ORS data.

Similar research in this area has used the O*NET to calculate occupational risk and used panel data on a worker's occupational history to calculate the impact of accumulated risk on chronic diseases later in life (Dembe et al. 2014). In contrast, our work here focuses on the impact of occupational risks on occupational injuries and illnesses. Without the occupational his-

19. For more information on the SOII, see http://www.bls.gov/iif/.

tories used in earlier work, we are unable to calculate the cumulative effect of exposure to risk over long time periods, and therefore focus primarily on traumatic injuries. One advantage of the ORS data is that, in addition to knowing the mean level of risk for the occupation, we also have information on the distribution of risk within the occupation. This additional information allows us to focus on elements of risk that are more closely associated with specific occupations.

While the ORS sample contains ninety-two unique three-digit SOCs, to get reliable measures we require twenty observations for each of these, which reduces the number to sixty. An additional eleven three-digit SOCs were dropped because of item-level nonresponse (if there were fewer than twenty responses per item), and for one three-digit SOC no injury and illness rate was available from the SOII. Thus, forty-eight three-digit SOCs remained for analysis.[20]

Risk of injury and illness in the ORS is captured by many variables, with most of these in the categories environmental conditions and physical demands. We have both dichotomous measures of the presence of risk, as well as measures of the percentage of time at job with risk. While the latter is potentially a better measure, it is often not available for a large fraction of the sample. Thus, we are more likely to use the dichotomous variables, which can make use of cases where the respondent answered that the risk was present, but the duration was unknown.

Having forty-eight three-digit SOCs for analysis leaves us with a relatively small number of degrees of freedom compared to the number of ORS elements that can potentially explain injury rates. Because we are running regressions at the three-digit SOC level, we are, moreover, interested in restricting ourselves to those ORS elements where occupation has considerable explanatory power. To address both considerations, we regress each ORS element individually on three-digit SOC dummy variables. We choose as regressors those ten elements where occupation has the most explanatory power, in all cases with an R-squared exceeding 0.35.

Eight of the ten elements are environmental conditions or physical demands that may affect risk directly. They are traditional keyboarding, encountering wetness, sitting (percentage of hours), working near moving mechanical parts, working in high exposed places, driving required, amount ever lifting/carrying and gross manipulation (percentage of hours).[21] The other two elements are cognitive demands considered above—task complexity and work control—which may capture other dimensions of occupations that affect risk.

In table 5.5, we examine injury and illness rates at the occupation level

---

20. We estimate that the dropped three-digit SOCs account for less than 15 percent of total employment.
21. Unless otherwise indicated, element indicates presence or absence.

Table 5.5   Occupational requirements and injury/illness rates

Variables	Total incidence rate	Event				Nature		Source		
		Transportation incidents	Falls to lower level	Struck by object or equipment	Sprains, strains, tears	Soreness, pain	Carpal Tunnel Syndrome	Parts and materials	Ladder	Vehicles
Keyboarding: traditional	-28.54	-13.25**	-1.14	-0.04	5.60	-11.59	-0.32	4.78	4.17**	-15.10
	(50.23)	(6.50)	(3.60)	(8.45)	(21.73)	(10.65)	(0.78)	(7.21)	(1.66)	(10.84)
Sitting (%)	98.98**	24.76***	4.61	2.16	41.73**	24.36**	1.47**	4.87	-6.59***	43.29***
	(46.74)	(6.05)	(3.35)	(7.86)	(20.22)	(9.91)	(0.72)	(6.70)	(1.60)	(10.09)
Wetness	-15.51	-6.63	-4.34	-8.72	-3.30	0.35	0.24	-14.49***	-1.53	-13.49*
	(35.97)	(4.65)	(2.58)	(6.05)	(-5.56)	(7.63)	(0.61)	(5.14)	(1.21)	(7.76)
Exposed places	-5.18	-21.29***	25.41***	8.56	-0.42	5.28	-2.79***	19.02**	25.00***	-40.72***
	(58.20)	(7.53)	(4.17)	(9.79)	(25.18)	(12.34)	(0.87)	(8.30)	(1.91)	(12.56)
Moving parts	-34.56	-20.20***	-10.95***	24.48***	-37.50*	-21.65**	4.79***	30.51***	-4.32***	-23.61**
	(46.86)	(6.06)	(3.36)	(7.88)	(20.27)	(9.94)	(0.72)	(6.69)	(1.54)	(10.11)
Driving	9.81	15.07***	3.79	-6.51	14.34	6.38	-1.04	-2.13	2.59*	24.17***
	(38.60)	(4.99)	(2.76)	(6.49)	(16.70)	(8.18)	(0.62)	(5.51)	(1.41)	(8.33)
Gross manipulation (%)	217.63*	21.63	26.15***	65.75***	94.85*	26.19	-2.02	53.50***	8.38**	71.99***
	(109.20)	(14.12)	(7.82)	(18.36)	(47.25)	(23.15)	(2.01)	(15.67)	(3.60)	(23.57)
Lift/carry, ever amount	3.62***	0.45***	0.02	0.00	1.64***	0.72***	0.01	-0.05	-0.11***	0.57***
	(0.73)	(0.09)	(0.05)	(0.12)	(0.32)	(0.16)	(0.01)	(0.10)	(0.02)	(0.16)
Work control	18.23	4.93	1.91	4.18	-3.40	3.57	-0.88	4.44	-1.76	6.56
	(46.78)	(6.05)	(3.35)	(7.87)	(20.24)	(9.92)	(0.83)	(6.68)	(1.54)	(10.10)
Task complexity	-45.38	-7.37	-2.27	-0.80	-12.39	-12.55	-0.27	-0.80	1.13	-11.37
	(50.97)	(6.59)	(3.65)	(8.57)	(22.05)	(10.81)	(0.86)	(7.27)	(1.72)	(11.00)
Observations	48	48	48	48	48	48	40	47	40	48
R-squared	0.86	0.83	0.85	0.86	0.84	0.82	0.70	0.87	0.92	0.84

Note: Numbers are coefficients from regressions of injury/illness rates on occupational requirements variables. Standard errors in parentheses.

***Significant at the 1 percent level.

**Significant at the 5 percent level.

*Significant at the 10 percent level.

from the SOII, overall, and then by the event causing the injury, nature of the injury, and source of the injury. The rates are measured as cases per 10,000 full-time equivalent workers. Looking at the overall incidence rate, we see that those occupations where there are higher levels of lifting and carrying have a higher injury and illness rate, which is also true of those occupations with more sitting and more gross manipulation. While sitting may not seem to be a risky activity, we will see when we turn to events, nature, and sources why it is associated with higher incidence rates.

First to be examined is event causing an injury. It is no surprise that those occupations with a high rate of injuries caused by transportation incidents are strongly positively associated with driving and lifting/carrying, as these are activities associated with the jobs of transportation workers. Standard keyboarding, working in high exposed places and near mechanical moving parts have a negative relationship with transportation injury rates, presumably because these elements are less common among transportation workers.

Falls to lower levels are, quite sensibly, positively associated with working in high exposed places. They also have a positive relationship with gross manipulation, while being negatively related to working with mechanical moving parts. The final event we consider, struck by object or equipment, is, appropriately, positively associated with working with mechanical moving parts, as well as with gross manipulation.

Some interesting relationships are also evident in our examination of the nature of the injuries. Both strains, sprains, and tears and soreness and pain are positively related to gross manipulation and lifting/carrying. Carpal tunnel syndrome, in contrast, is more likely to be found in jobs where there are mechanical moving parts and where there is a relatively high amount of sitting.

Finally, we consider the source of the injury. Parts and materials injuries are more likely to come in jobs with gross manipulation and mechanical moving parts, but less likely when there is exposure to wetness. Injuries where the ladder is the source are more common in jobs where workers are in high exposed places. They also have positive relationships with gross manipulation and standard keyboarding, and negative relationships with large amounts of sitting, working near mechanical moving parts, and the amount of ever lifting and carrying. Injuries where vehicles are the source are somewhat similar to transportation incidents in that both are more apt to be present in jobs where there is driving, lifting/carrying, much sitting, and where there isn't work in high exposed places or with mechanical moving parts.

5.6   Conclusion: The Potential of ORS for Research

Employing information from the preproduction version of ORS, we have presented a set of estimates of some key occupational requirements for all workers, as well as by broad occupation and industry categories. We have also illustrated how ORS data can be used in analysis, focusing on wage determination and the role of job requirements in injury and illness rates.

As BLS moves into production collection, ORS data will be collected annually on a substantially larger sample (roughly 6,000 establishments planned in year 1). The ORS is currently approved for an initial three years of collection with the goal of having reliable estimates for the vast majority of the data elements at the eight-digit O*NET SOC level at the end of the period. While the data elements and collection procedures are intended to support SSA in disability adjudication, these data elements will likely also be useful for a variety of other stakeholders, including researchers.

In addition to the research questions discussed in the introduction, we propose some other areas of research in which ORS data may prove useful. First, our initial analysis linking ORS estimates of job requirements, particularly the physical requirements, to safety outcomes suggests that ORS may be a valuable data set for occupational safety and health researchers. As ORS will ultimately have full sets of estimates on the types of physical and environmental conditions required at a detailed occupation level, it can be used in research focused on a particular occupation (truck driving, for example) or focused on a specific injury that may occur across occupations, linked to underlying physical requirements (such as the relationship between reaching and musculoskeletal injuries).

In addition to considering the direct links between more obviously "risky" job requirements and injuries, ORS data may inform studies of the role of job requirements and illness. Occupational illnesses are typically less well understood than injuries since they tend to result from longer-term exposure to risk factors. Recent research focuses on the relationship between sedentary behavior (including prolonged periods of sitting while at work) and a variety of long-term adverse health outcomes including obesity, type II diabetes, and cardiovascular disease (Dunstan et al. 2012; van der Ploeg et al. 2012; Proper et al. 2011). The ORS data can be used to identify the sets of occupations where workers sit most and the duration/percent of time of sitting, as well as the ability of workers to alternate between sitting and standing at will.

Finally, the current financial strain on the SSI and SSDI programs has led to a great deal of research regarding the barriers involved in getting persons with disabilities to return to the workforce. Extensive research exists that documents the negative relationship between SSDI receipt and labor force participation (Autor and Duggan 2006; Maestas, Mullen, and Strand 2013; von Wachter, Song, and Manchester 2011). The ORS does not ask respondents about accommodations for workers with disabilities; however, disability researchers and advocates may be able to use the ORS data on physical requirements to identify jobs in which specific accommodations may result in more employment opportunities for individuals with disabilities. For example, understanding the characteristics of establishments where some production workers are able to sit or stand at will may lead to recommendations for translating this flexibility into other sectors.

Similarly, identifying jobs that are moderately or loosely controlled but require relatively low levels of SVP provides opportunities to identify the training programs necessary to place individuals with cognitive impairments and relatively low levels of education in such jobs. Recent research has found that participation in state workforce programs increases the likelihood of return to work among SSDI beneficiaries. Information on the amount of training needed to perform certain jobs may help workforce boards target their programs to such workers.

Appendix

Table 5A.1   List of ORS elements

Specific vocational preparation—Four elements	
Minimum formal education or literacy required	Prior work experience
Preemployment training (license, certification, other)	Postemployment training

Mental and cognitive demands—Nine elements	
Closeness of job-control level	Frequency of verbal work-related interaction with other contacts
Complexity of task level	
Frequency of deviations from normal work location	Frequency of verbal work-related interaction with regular contacts
Frequency of deviations from normal work schedule	Type of work-related interactions with other contacts
Frequency of deviations from normal work tasks	Type of work-related interactions with regular contacts

Auditory/Vision—Ten elements	
Driving, type of vehicle	Hearing: Other sounds
Communicating verbally	Passage of hearing test
Hearing: One on one	Far visual acuity
Hearing: Group	Near visual acuity
Hearing: Telephone	Peripheral vision

Environmental conditions—Eleven elements	
Extreme cold	Noise-intensity level
Extreme heat	Outdoors
Fumes, noxious odors, dusts, gases	Proximity to moving mechanical parts
Heavy vibration	Toxic, caustic chemicals
High, exposed places	Wetness
Humidity	

Table 5A.1   (continued)

Physical demands, exertion—Fourteen elements

Most weight lifted/carried ever	Standing and walking
Push/pull with feet only: One or both	Weight lifted/carried 2/3 of the time or more
Push/pull with foot/leg: One or both	(range)
Push/pull with hand/arm: One or both	Weight lifted/carried 1/3 up to 2/3 of the
Pushing/pulling with feet only	time (range)
Pushing/pulling with foot/leg	Weight lifted/carried from 2 percent up to
Pushing/pulling with hand/arm	1/3 of the time (range)
Sitting	Weight lifted/carried up to 2 percent of the
Sitting versus standing at will	time (range)

Physical demands, reaching/manipulation—Fourteen elements

Overhead reaching	Gross manipulation: One hand or both
Overhead reaching: One or both	Foot/leg controls
At/below shoulder reaching	Foot/leg controls: One or both
At/below shoulder reaching: One or both	Keyboarding: Ten key
Fine manipulation	Keyboarding: Other
Fine manipulation: One hand or both	Keyboarding: Touch screen
Gross manipulation	Keyboarding: Traditional

Physical demands, postural—Seven elements

Climbing ladders/ropes/scaffolds	Crouching
Climbing ramps/stairs: Structural only	Kneeling
Climbing ramps/stairs: Work related	Stooping
Crawling	

Table 5A.2   Full-estimation results

Variable grouping	Variable	Education only (1)	Ed., cognitive, strength (2)	SVP only (3)	SVP, cognitive, strength (4)
Education	Literacy, no degree	0.147	0.133		
		(0.131)	(0.104)		
	High school diploma	0.390***	0.211*		
		(0.131)	(0.108)		
	Associate's degree	0.561***	0.275**		
		(0.145)	(0.123)		
	Baccalaureate degree	1.060***	0.511***		
		(0.132)	(0.116)		
	Postbacc. degree	1.317***	0.632***		
		(0.147)	(0.127)		
SVP	More than 1 month, up to 1 year			0.229***	0.147***
				(0.0495)	(0.0461)
	More than 1 year, up to 4 years			0.551***	0.246***
				(0.0454)	(0.0431)
	More than 4 years			1.015***	0.386***
				(0.0612)	(0.0623)

(continued)

Table 5A.2   (continued)

Variable grouping	Variable	Education only (1)	Ed., cognitive, strength (2)	SVP only (3)	SVP, cognitive, strength (4)
Task complexity	Very complex		0.658***		0.790***
			(0.171)		(0.191)
	Complex		0.553***		0.607***
			(0.104)		(0.110)
	Moderately complex		0.290***		0.304***
			(0.0702)		(0.0779)
	Simple		−0.000783		0.0190
			(0.0514)		(0.0581)
Work control	Controlled		0.0735		0.0599
			(0.0525)		(0.0485)
	Moderately controlled		0.258***		0.227***
			(0.0684)		(0.0650)
	Loosely controlled		0.200**		0.109
			(0.0902)		(0.0907)
	Very loosely controlled		0.723***		0.674***
			(0.179)		(0.178)
Regular contacts	Structured contacts		0.00759		0.00709
			(0.0387)		(0.0389)
	Semi-structured contacts		0.00543		0.0576
			(0.0485)		(0.0525)
	Unstructured contacts		0.0441		0.105
			(0.100)		(0.0986)
	Very unstructured contacts		−0.0794		−0.0413
			(0.0584)		(0.0761)
Strength	Light		0.0401		0.0386
			(0.0439)		(0.0467)
	Medium		0.00860		−0.0301
			(0.0489)		(0.0506)
	Heavy		0.00523		0.0122
			(0.0818)		(0.0903)
	Very heavy		0.0496		0.0394
			(0.0667)		(0.0700)
Establishment size	50–99 employees	0.0319	−0.0265	0.0375	−0.0409
		(0.0594)	(0.0527)	(0.0586)	(0.0523)
	100–499 employees	0.0918*	0.104**	0.0918*	0.110**
		(0.0484)	(0.0507)	(0.0522)	(0.0533)
	500 or more employees	0.229***	0.166***	0.169***	0.144**
		(0.0601)	(0.0585)	(0.0635)	(0.0588)
Sector	Private sector	0.0318	−0.104**	−0.0578	−0.157***
		(0.0540)	(0.0511)	(0.0626)	(0.0533)
Union coverage	Union	0.268***	0.289***	0.367***	0.316***
		(0.0380)	(0.0418)	(0.0499)	(0.0474)
Time base	Full time	0.264***	0.167***	0.182***	0.140***
		(0.0360)	(0.0373)	(0.0453)	(0.0393)
Industry	Construction	−0.195	0.250**	−0.162	0.297***
		(0.129)	(0.107)	(0.154)	(0.104)

Table 5A.2   (continued)

Variable grouping	Variable	Education only (1)	Ed., cognitive, strength (2)	SVP only (3)	SVP, cognitive, strength (4)
	Manufacturing	−0.455***	0.0510	−0.208	0.153
		(0.103)	(0.111)	(0.150)	(0.0944)
	Wholesale trade	−0.523***	0.0312	−0.333**	0.157*
		(0.133)	(0.115)	(0.142)	(0.0943)
	Retail trade	−0.548***	−0.0889	−0.282**	0.000860
		(0.103)	(0.0995)	(0.137)	(0.0930)
	Transportation and	−0.375***	0.117	−0.230	0.139
	warehousing	(0.114)	(0.0819)	(0.164)	(0.0985)
	Information	−0.357***	−0.112	−0.0399	−0.00751
		(0.111)	(0.195)	(0.164)	(0.212)
	Financial activities	−0.434***	−0.00682	−0.0931	0.134
		(0.119)	(0.0986)	(0.148)	(0.0907)
	Professional and	−0.426***	0.0599	−0.0706	0.177
	business services	(0.125)	(0.114)	(0.155)	(0.111)
	Education and health	−0.658***	−0.170**	−0.263**	−0.0366
	services	(0.0926)	(0.0719)	(0.134)	(0.0632)
	Leisure and	−0.835***	−0.338***	−0.571***	−0.265***
	hospitality	(0.102)	(0.0960)	(0.139)	(0.0911)
	Other services	−0.494***	−0.0132	−0.339**	0.0398
		(0.130)	(0.123)	(0.152)	(0.116)
	Public administration	−0.528***	−0.264***	−0.306**	−0.225***
		(0.103)	(0.0705)	(0.141)	(0.0585)
	Constant	2.572***	2.180***	2.518***	2.239***
		(0.173)	(0.145)	(0.151)	(0.0983)
	R-squared	0.673	0.769	0.662	0.766

Note: Standard errors in parentheses.
***Significant at the 1 percent level.
**Significant at the 5 percent level.
*Significant at the 10 percent level.

References

Abraham, Katharine G., and James R. Spletzer. 2009. "New Evidence on the Returns to Job Skills." *American Economic Review* 99 (2): 52–57.

Acemoglu, Daron, and David H. Autor. 2011. "Skills, Tasks and Technologies: Implications for Employment and Earnings." In *Handbook of Labor Economics*, vol. 4B, edited by David Card and Orley Ashenfelter. Amsterdam: North Holland.

Autor, David H. 2015. "Why Are There Still So Many Jobs? The History and Future of Workplace Automation." *Journal of Economic Perspectives* 29 (3): 3–30.

Autor, David H., and Mark Duggan. 2006. "The Growth in the Social Security Disability Rolls: A Fiscal Crisis Unfolding." NBER Working Paper no. 12436, Cambridge, MA.

Autor, David H., and Michael J. Handel. 2013. "Putting Tasks to the Test: Human Capital, Job Tasks and Wages." *Journal of Labor Economics* 31 (2, pt. 2): S59–96.

Autor, David H., Frank Levy, and Richard J. Murnane. 2003. "The Skill Content of Recent Technological Change: An Empirical Exploration." *Quarterly Journal of Economics* 118 (4): 1279–333.

Blinder, Alan S. 2009. "How Many U.S. Jobs Might Be Offshorable?" *World Economics* 10 (2): 41–78.

Borghans, Lex, Bas ter Weel, and Bruce A. Weinberg. 2014. "People Skills and the Labor-Market Outcomes of Underrepresented Groups." *Industrial and Labor Relations Review* 67 (2): 287–334.

Bound, John, and George E. Johnson. 1992. "Changes in the Structure of Wages in the 1980s: An Evaluation of Alternative Explanations." *American Economic Review* 82 (3): 371–92.

Brown, Charles, and James Medoff. 1989. "The Employer Size-Wage Effect." *Journal of Political Economy* 97 (5): 1027–59.

Card, David. 1999. "The Causal Effect of Education on Earnings." *Handbook of Labor Economics* 3:1801–63.

Carneiro, Pedro, James J. Heckman, and Edward J. Vytlacil. 2011. "Estimating Marginal Returns to Education." *American Economic Review* 101 (6): 2754–81.

Dembe, Allard E., Xiaoxi Yao, Thomas M. Wickizer, Abigail B. Shoben, and Xiuwen Sue Dong. 2014. "Using O*NET to Estimate the Association between Work Exposures and Chronic Diseases." *American Journal of Industrial Medicine* 57 (9): 1022–31.

Deming, David J. 2015. "The Growing Importance of Social Skills in the Labor Market." NBER Working Paper no. 21473, Cambridge, MA.

Dunstan, David W., Bethany Howard, Genevieve N. Healy, and Neville Owen. 2012. "Too Much Sitting—A Health Hazard." *Diabetes Research and Clinical Practice* 97 (3): 368–76.

Firpo, Sergio, Nicole M. Fortin, and Thomas Lemieux. 2011. "Occupational Tasks and Changes in the Wage Structure." IZA Discussion Paper no. 5542, Institute of Labor Economics. February.

Gittleman, Maury, and Brooks Pierce. 2007. "New Estimates of Union Wage Effects." *Economics Letters* 95 (2): 198–202.

———. 2011. "Inter-Industry Wage Differentials, Job Content and Unobserved Ability." *Industrial and Labor Relations Review* 64 (2): 356–74.

———. 2012. "Compensation for State and Local Government Workers." *Journal of Economic Perspectives* 26 (1): 217–42.

Handel, Michael J. 2015a. "Occupational Requirements Reliability and Validity Literature Research." Report to Bureau of Labor Statistics. http://www.bls.gov/ncs/ors/handel_report_feb15.pdf.

———. 2015b. "O*NET: Strengths and Limitations." Unpublished manuscript, Northeastern University.

———. 2015c. "What Do People Do at Work? A Profile of U.S. Jobs from the Survey of Workplace Skills, Technology, and Management Practices (STAMP)." Unpublished manuscript, Northeastern University.

Ingram, Beth F., and George R. Neumann. 2006. "The Returns to Skill." *Labour Economics* 13 (1): 35–59.

Jensen, J. Bradford, and Lori G. Kletzer. 2010. "Measuring Tradable Services and the Task Content of Offshorable Services Jobs." In *Labor in the New Economy*, edited by Katharine G. Abraham, James R. Spletzer, and Michael Harper. Chicago: University of Chicago Press.

Katz, Lawrence F., and Kevin M. Murphy. 1992. "Changes in Relative Wages: Supply and Demand Factors." *Quarterly Journal of Economics* 107:35–78.

Lettau, Michael K. 1997. "Compensation in Part-Time Jobs versus Full-Time Jobs. What If the Job Is the Same?" *Economics Letters* 56 (1): 101–6.

Maestas, Nicole, Kathleen J. Mullen, and Alexander Strand. 2013. "Does Disability Insurance Receipt Discourage Work? Using Examiner Assignment to Estimate Causal Effects of SSDI Receipt." *American Economic Review* 103 (5): 1797–829.

Munnell, Alicia, Jean-Pierre Aubry, Josh Hurwitz, and Laura Quinby. 2011. "Comparing Compensation: State-Local versus Private Sector Workers." Research Study, Center for State & Local Government Excellence. https://slge.org/publications /comparing-compensation-state-local-versus-private-sector-workers.

Murnane, Richard, John B. Willett, and Frank Levy. 1995. "The Growing Importance of Cognitive Skills in Wage Determination." *Review of Economics and Statistics* 77 (2): 251–66.

Oldenski, Lindsay. 2014. "Offshoring and the Polarization of the U.S. Labor Market." *Industrial and Labor Relations Review* 67 (3, suppl.): 734–61.

Organisation for Economic Co-operation and Development (OECD). 2013. *OECD Skills Outlook 2013: First Results from the Survey of Adult Skills.* Paris: OECD Publishing. http://dx.doi.org/10.1787/9789264204256-en.

Pierce, Brooks. 1999. "Using the National Compensation Survey to Predict Wage Rates." *Compensation and Working Conditions* Winter: 8–16.

———. 2010. "Recent Trends in Compensation Inequality." In *Labor in the New Economy*, edited by Katharine G. Abraham, James R. Spletzer, and Michael Harper. Chicago: University of Chicago Press.

Proper, Karin I., Amika S. Singh, Willem Van Mechelen, and Mai J. M. Chinapaw. 2011. "Sedentary Behaviors and Health Outcomes among Adults: A Systematic Review of Prospective Studies." *American Journal of Preventive Medicine* 40 (2): 174–82.

van der Ploeg, Hidde P., Tien Chey, Rosemary J. Korda, Emily Banks, and Adrian Bauman. 2012. "Sitting Time and All-Cause Mortality Risk in 222,497 Australian Adults." *Archives of Internal Medicine* 172 (6): 494–500. https://doi.org/10.1001 /archinternmed.2011.2174.

Von Wachter, Till, Jae Song, and Joyce Manchester. 2011. "Trends in Employment and Earnings of Allowed and Rejected Applicants to the Social Security Disability Insurance Program." *American Economic Review* 101 (7): 3308–29.

Weinberger, Catherine J. 2014. "The Increasing Complementarity between Cognitive and Social Skills." *Review of Economics and Statistics* 96 (5): 849–61.

III

Skills, Inequality, and Polarization

6

Noncognitive Skills as
Human Capital

Shelly Lundberg

6.1 Introduction

Human capital plays a central role in all analyses of economic growth. In empirical growth models, the standard proxy for human capital is educational attainment, but this is an indirect and very imperfect measure of labor skills. Educational attainment is also a skill measure that is not comparable across nations (or over time) due to variation in educational quality. Hanushek and Kimko (2000) found that scores on international examinations are more important than years of educational attainment for economic growth, and a robust literature concerning the role of cognitive skills in economic development has emerged (Hanushek and Woessmann 2008). As evidence grows that other, so-called noncognitive, skills have large and significant impacts on individual earnings and other economic outcomes, the research on growth may need to incorporate these additional dimensions of human capital. We are far, however, from a clear understanding of how to define and measure noncognitive skills in a way that would allow for meaningful cross-country analysis.

The idea that noncognitive skills are both important outcomes of the educational process and inputs to human capital production has a long history in labor economics. Bowles and Gintis (1976), in their classic study of

Shelly Lundberg is the Leonard Broom Professor of Demography at the University of California, Santa Barbara, adjunct professor at the University of Bergen, and a research fellow of the IZA.

I appreciate helpful comments from David Deming, Charles Hulten, Valerie Ramey, and Jenna Stearns, and the excellent assistance I have received from Sarah Bana. For acknowledgments, sources of research support, and disclosure of the author's material financial relationships, if any, please see http://www.nber.org/chapters/c13701.ack.

the American education system, assert that "employer-valued attributes," including perseverance and punctuality, are important products of schooling. Weiss (1988) shows that nearly all of the relationship between high school graduation and earnings can be explained by the lower quit propensities and lower rates of absenteeism of high school graduates compared to high school dropouts. Heckman and a number of collaborators have worked to incorporate noncognitive skills into the economic analysis of individual achievement, noting that "personality, persistence, motivation, and charm matter for success in life" (Heckman, Stixrud, and Urzua 2006). There is now considerable evidence that these traits, in addition to cognitive ability and academic achievement, are important determinants of economic success. In particular, socioeconomic gaps in noncognitive traits at early ages are implicated in the intergenerational transmission of inequality. This represents an important shift in economists' conception of human capital, moving beyond brains and brawn to incorporate a broad set of psychosocial capabilities.

In a very short period of time, a substantial literature has appeared on noncognitive skills—their economic payoffs, the sources of socioeconomic disparities in skill levels, and the possible role of early investments in augmenting noncognitive skills and reducing these disparities. A recent Organisation for Economic Co-operation and Development (OECD) report by Kautz et al. (2015, 7) reviews much of this literature, with a particular focus on the outcomes of early interventions, and reaches the following conclusions: (a) noncognitive skills are valuable in school and in the labor market, (b) reliable measures of noncognitive skills are available, and (c) individual skills are stable at a point in time, but can be shaped in the early years of life.

The first of these conclusions is undoubtedly true, and the evidence for the third is accumulating rapidly. The second conclusion is perhaps premature—some serious issues persist with respect to the measurement of noncognitive skills, and especially the estimation of skill disparities between groups. One issue is a lack of consensus about what noncognitive skills are, and the absence of a consistent set of metrics that can be applied across studies. In Kautz et al., noncognitive skills are defined as "personality traits, goals, character, motivations, and preferences that are valued in the labour market, in school, and in many other domains," which is an astonishingly broad characterization. A second issue is the widespread use of behavior as, de facto, a pure indicator of skill, rather than an outcome that also depends on incentives, beliefs, and situation. The comparability of such measures across population groups defined by gender, ethnicity, or socioeconomic status or across nations is highly suspect.

The label "noncognitive" is a controversial one and psychologists disapprove, informally, of its popularity among economists. Alternative terms have been used, including socioemotional skills, soft skills, personality skills and, most recently, character, but I will use "noncognitive" consistently because it is familiar and a clearly superior alternative has not emerged. Indices of children's noncognitive skills are usually based on teacher and parent reports of the child's behavior, including their ability to focus attention on tasks, social skills, and externalizing (disruptive or aggressive) behavior. Measures of adult skills are sometimes based on behavioral assessments (or administrative records such as criminal histories) but more commonly rely on self-reports of the individual's behavioral tendencies, feelings, or beliefs, including assessments of self-esteem, conscientiousness, and persistence.

In this chapter, I review some of the recent literature on the association between noncognitive skill metrics and important economic outcomes such as educational attainment and earnings. Some characteristic patterns of effects are illustrated using two longitudinal surveys that track recent cohorts from adolescence to young adulthood, but have not been extensively used in previous studies of noncognitive skills. I find that some measures of social and emotional problems in early adolescence have strong negative associations with educational attainment, while others do not. All skill proxies have weak effects on earnings conditional on education. Parental and youth reports of the same behaviors have independent influences on education outcomes. Though this is a standard empirical exercise in this literature, the results are not easy to interpret. They do suggest that adolescent noncognitive skills may be particularly important in navigating the path through school, rather than having independent influences on labor productivity. I also show, using an example involving impulsivity and crime, that measurement and endogeneity problems make one common empirical exercise—the documentation of skill gaps between groups and assessments of the contribution of these gaps to inequality—extremely problematic.

The research agenda on incorporating noncognitive skills into economic growth models is rather daunting. First, we need some agreement on a standard battery of noncognitive skill assessments at different stages of human development. The early childhood intervention literature has been able to rely on measures used by developmental psychologists, but as we move through childhood to adolescence and adulthood, the situation becomes rather chaotic since there are too many behavioral domains and psychological inventories to choose from. Second, we need research that disentangles the effect of skills on economic outcomes from impacts that occur through other channels, parental and environmental, that have helped to shape these skills. This standard identification problem has been inadequately addressed in the current literature. Finally, evidence is emerging that the returns to traits that have been labeled noncognitive skills are highly heterogeneous—traits that are useful in some social, economic, and cultural environments may be harmful in others. This complicates international comparisons in a way that does not arise with cognitive skills.

Despite these difficulties, broadening the economic concept of human capital is an important exercise. Research in neurobiology and developmental psychology indicates that noncognitive skills emerge from the same developmental processes as conventionally measured cognitive abilities. Early interventions that enrich children's environments and reduce stress can lead to improvements in executive functioning that foster the ability to regulate emotions and attention, as well as to acquire vocabulary. These skills are strongly predictive of educational outcomes and attainment, and may be leveraged by complementarities between sets of skills in the human capital acquisition process. Early investments in noncognitive skills may have important positive effects on growth by increasing the returns to other educational inputs. Finally, as technological change transforms the labor market and the task requirements of jobs, the returns to skills that foster effective human interaction seem likely to continue to rise (Deming 2017).

6.2 Noncognitive Skills Enter the Human Capital Literature

New studies that document the returns to psychosocial traits and behavioral tendencies, or the impact of early treatments on these traits, emerge almost daily. Researchers have found that a variety of such indicators are significant predictors of economic outcomes including wages, earnings, health, crime, and relationship stability. One of the key features of this literature is the bewildering array of personal traits and actions that the "noncognitive skill" label has been applied to, including teacher assessments of social skills, parental reports of toddler temperament, self-reported beliefs about personal control, and administrative records of school suspensions. In general, these are measures of convenience, adopted by researchers because they happen to be available on surveys or administrative registers and turn out to be correlated with interesting outcomes. These noncognitive metrics can be sorted into three broad categories:

1. Self-assessments. These instruments ask individuals to respond to questions that indicate "This is what I am like" or "This is what I believe." Personality traits are perhaps the most commonly used self-assessments in the economics literature. For example, a positive response to "I sympathize with others' feelings" is one component of the Big Five personality trait, agreeableness, while agreeing with "When I make plans, I am almost certain that I can make them work" is indicative of an internal locus of control (or high self-efficacy).

2. Parent/teacher reports of a child's behavior, tendencies, or abilities. Behavior problem indices that include measures of externalizing and internalizing behavior, as well as reports of persistence, ability to focus, and social skills, have been extensively used by psychologists and education researchers, and are available in many large-scale data sets.

3. Administrative records. Registers of school disciplinary actions, crimi-

nal justice contacts, or military service can sometimes be linked to subsequent economic outcomes.

The more recent economics literature on noncognitive skills (including the controversial label) came into prominence with two studies by James Heckman and coauthors. One of these relied on behavioral indicators of skills, while the other used self-assessments. Heckman and Rubinstein (2001) find that General Education Development/Diploma (GED) recipients are more likely to engage in drug use and to commit minor crimes than conventional high school graduates. They infer that the absence of a positive economic return to GED recipiency is due to a shortfall in noncognitive skills among those who receive this credential. Heckman, Stixrud, and Urzua (2006), using adolescent measures of self-efficacy and self-esteem in the National Longitudinal Survey of Youth 1979 as indicators of noncognitive ability, find that noncognitive and cognitive skills are equally important in determining a variety of economic and social outcomes. Both of these papers have been influential and have alerted economists to the potential significance of traits other than cognitive ability that contribute to economic success.

6.2.1   Personality, Self-Control, and Social Skills

In the first decade of the century, many researchers took advantage of newly available (self-assessed) personality inventories included in large longitudinal surveys, including the British Household Panel Study (BHPS), the German Socio-Economic Panel Study (SOEP), and the Household, Income and Labour Dynamics in Australia (HILDA) Survey. Most surveys included a fifteen-item short form of the "Big Five" personality inventory, which consists of the traits openness to experience, conscientiousness, extraversion, agreeableness, and neuroticism/emotional stability. The Big Five was developed and extensively evaluated by psychologists, and is broadly accepted as a meaningful and consistent construct for describing human differences (Goldberg 1981).

Economic studies of personality focused initially on the determinants of earnings and other labor market outcomes. In general, high emotional stability and low agreeableness have been found to be positively associated with earnings for men, and in some cases for women (Mueller and Plug 2006; Heineck 2011; Nyhus and Pons 2005). Personality traits also influence the sorting of workers across occupations, and this can be interpreted as the result of either varying preferences over job attributes or occupation-specific determinants of productivity (Filer 1986; Krueger and Schkade 2008). Nandi and Nicoletti (2014) decompose the pay gaps between personality groups in the BHPS data into components that can be explained by personality-based differences in occupation, education, work experience, and unexplained components. They find that the observed pay premium for openness can be explained by higher education and by sorting into higher-paid occupations, but that the pay premium for extraversion and the penalties for neuroticism and agreeableness cannot. Another personality construct, self-efficacy or locus of control, has also been found to be positively related to a variety of labor market outcomes (Heineck and Anger 2010; Cobb-Clark, Caliendo, and Uhlendorff 2015; Cobb-Clark 2015). Personality and other socioemotional traits also have important associations with the propensity to marry and with relationship stability (Lundberg 2012, 2015).

Even though the study of personality originated as an attempt to understand why some highly intelligent individuals perform well in school and in later life while others do not, the relationship between personality and education has not received as much attention from economists as have personality effects on earnings. Pioneers in the development of intelligence quotient (IQ) tests, such as Binet and Terman, were aware of the significance of qualities other than cognitive ability in determining success, and identified the key features of this dimension of "character" as perseverance and attentiveness—aspects of the Big Five trait, conscientiousness (Almlund et al. 2011). A large literature in psychology and education finds that conscientiousness and behaviors related to conscientiousness, such as persistence and self-control, are strongly predictive of grades in school and other measures of educational success.

Measuring noncognitive skills via self-assessments such as personality inventories cannot begin before middle childhood at the earliest. Assessments of younger children rely on behavioral measures, and the "marshmallow studies" have produced the best known of these. Beginning in the late 1960s, psychologist Walter Mischel led a series of studies that showed a strong association between the ability to delay gratification as a four-year-old and later test scores, educational attainment, and health (Mischel, Ebbesen, and Raskoff Zeiss 1972). Larger studies have used observational measures such as parent and teacher reports of externalizing behavior—arguing, fighting, acting impulsively or disruptively—and social skills. Children from disadvantaged backgrounds begin school well behind their peers in the ability to focus their attention and control their impulses, and these gaps tend to persist as they progress through school. The predictive power of early assessments vary: teacher evaluations of eighth grade misbehavior are correlated with educational and labor market outcomes (Segal 2013), but some studies fail to find any relationship between school-entry skills such as attention and later outcomes (Duncan and Magnuson 2011).

Recent years have seen many creative uses of administrative and survey data to infer noncognitive skills and link them to later outcomes. For example, a psychologist's assessment of the suitability of a young man for military service predicts his suitability for other jobs as well (Lindqvist and Vestman 2011), and interviewer reports of survey respondent fidgeting are correlated with later economic outcomes (Cadena and Keys 2015).

6.2.2   What Are Noncognitive Skills and Where Do They Come From?

Critics have objected to the use of the label "noncognitive" skills to describe any productive characteristic that is not measured in standard cognitive batteries and academic achievement tests. This is because behaviors such as task persistence and effective social interaction require cognitive input in a way that is not clearly distinct from the cognitive demands of completing a Raven's Matrices test. The unifying principle in this view of human skills is the psychological concept of executive functioning, an umbrella term for the management of cognitive processes. A recent World Bank report on early development links cognitive and noncognitive skills through the developmental process:

> The cognitive components of self-regulation, referred to as executive function, include the ability to direct attention, shift perspective, and adapt flexibly to changes (cognitive flexibility); retain information (working memory); and inhibit automatic or impulsive responses in order to achieve a goal such as problem solving (impulse control). . . . Self-regulation also includes emotional components such as regulating one's emotions, exhibiting self-control, and delaying gratification to enjoy a future reward. (World Bank 2015, 100)

Behavioral inhibition or self-regulation is at the core of most identified noncognitive, as well as cognitive, capabilities. The ability to focus on schoolwork, get along with classmates, abstain from drugs, and persevere on tasks is a set of skills with the same developmental origins as the ability to read well and solve math problems. The role of executive function in regulating behavior will vary depending on circumstances and developmental stages, but the consistent importance of cognitive control in shaping a broad range of capabilities highlights the inaptness of the term "noncognitive."

The case for treating noncognitive skills as a type of human capital is that many dimensions, such as self-control, appear to be relatively stable, but augmentable, traits that enhance task performance, increase labor productivity, and contribute to positive economic outcomes. The question "where do they come from?" is only beginning to be answered. Personality traits are strongly heritable, and twin studies find that 40–60 percent of variation in personality is genetic (Bouchard and Loehlin 2001; Anger 2012). Advances in neuroscience, molecular biology, developmental psychology, and economics are beginning to link deficits in a broad range of behavioral, health, and cognitive abilities to early experiences and environmental conditions, including toxic stress and pollution (Shonkoff et al. 2012; Currie 2011). The implication is that the mental regulatory skills represented by the term executive functioning are affected by early (including prenatal) conditions.

Kautz et al. (2015) provide a very comprehensive survey of interventions designed to improve cognitive and noncognitive skills at a variety of ages, from infancy through adolescence. For most programs, the evidence for a treatment effect on noncognitive skills is inferential: the intervention has no measurable lasting impact on cognitive or academic abilities, but does have a long-term positive effect on education, employment, or crime. The best-known set of results is perhaps the impact of the Perry Preschool Project, an intensive program for three- to four-year-old low-income children with treatment and control groups that had long-term impacts on test scores, adult crime, and male income, though no lasting effect on IQ. A recent paper bolsters the argument that these effects were due to a noncognitive skill increase by showing that there were intermediate effects on indices of externalizing behavior and (female) academic motivation (Heckman, Pinto, and Savelyev 2013).[1] The Jamaican Supplementation Study provided two years of nutritional supplements and a parenting intervention that encouraged stimulation of stunted children age nine to twenty-four months at the beginning of the program. The stimulation treatment outperformed the nutritional treatment, with substantial effects on adult earnings and on cognitive and psychosocial skills in late adolescence (Gertler et al. 2014). There are few examples of interventions at later ages with long-term follow-up, but Project Star, in which some children were randomly assigned to smaller kindergarten classes, had no lasting effect on test scores but appeared to lead to higher earnings in early adulthood (Chetty et al. 2011). Following the success of the Jamaican study, many recent interventions have focused on improving parenting as a way to reach children very early in life. These include programs that encourage parents to interact with children in developmentally appropriate ways and others that directly target maternal stress and mental health issues that may impact parenting quality.[2]

Treating noncognitive skills as a form of human capital raises one rather confusing issue: Is it more appropriate to think of the varied indicators that have appeared in the recent economics literature as skills, or as preferences? Referring to psychological traits as "skills" is an attempt to maintain the economic distinction between preferences and constraints, but in fact, the line is rather blurred. For example, the personality trait "extraversion" reflects both social skills and an orientation toward social interaction. In their analysis of intergenerational mobility, Bowles, Gintis, and Osborne (2001) emphasize the role of parents and schools in passing on "incentive-enhancing preferences" (such as patience and self-control) as an important mechanism for transmitting economic privilege across generations. Intuitively, the self-regulation that leads to deferred gratification in the marshmallow test must be closely allied with our concept of time preference. Yet,

1. Two older programs (Perry Preschool and Abecedarian) are positive outliers among the large set of early childhood education programs in their impacts on later human capital, and we know little about the connections between program components and particular sets of skills (Duncan and Magnuson 2013).
2. See the review in World Bank (2015, chapter 5).

the empirical associations between personality and economic preference parameters are very weak (Almlund et al. 2011; Rustichini et al. 2012) and one study finds that personality and preference indicators have largely independent effects on a large set of outcomes, including health, life satisfaction, wage, unemployment, and education (Becker et al. 2012). For noncognitive skills, we have no conceptual framework comparable to the choice theory that defines preference parameters, and this impedes any effort to move beyond a piecemeal approach to noncognitive skills and develop a standardized set of instruments.

6.3   Noncognitive Skills and Adult Outcomes in NLSY97 and Add Health

To illustrate some of the characteristics of early noncognitive skill measures as predictors of future educational attainment, wages, and employment, I use data from the National Longitudinal Survey of Youth 1997 (NLSY97) and the National Longitudinal Study of Adolescent to Adult Health (Add Health), which follow similar recent cohorts from early adolescence to young adulthood. The first wave of each study includes a set of noncognitive skill indicators, ranging from skimpy in NLSY97 to abundant in Add Health, that has been relatively unutilized by economists. The purpose of this exercise is to choose, a priori, a promising and typical set of indicators of adolescent angst, confidence, and behavioral difficulties, to see whether they predict later educational attainment and labor market outcomes, and to report all the results transparently and comprehensively. I find that some plausible adolescent noncognitive skill indicators are significant predictors of educational attainment while others, equally promising, are not, and that all are weak predictors of earnings and wages. For simplicity, I report only the results for the male subsamples, though the patterns in the female models are very similar.

6.3.1   National Longitudinal Survey of Youth 1997 (NLSY97)

The NLSY97 began with a nationally representative sample of 9,000 youths who were twelve to sixteen years old at the first wave and twenty-six to thirty-two when they were interviewed in 2011–2012. In Round 1, a version of the Armed Services Vocational Aptitude Battery (ASVAB) was administered, so we have a measure of academic skills and knowledge of the sort that is widely used as a measure of "cognitive skills," and also several indicators of noncognitive skills. This is in no sense a remarkable set of skill measures, but it does include a set of noncognitive indicators that are asked of both parents and children, which is relatively rare in large surveys. Also, the survey subjects are old enough in the last round that completed education and usable labor market information is available for almost all of them.

To measure behavioral and emotional problems in the first wave of the

NLSY97, a set of six items that were developed as indicators of children's mental health for the National Health Interview Survey (NHIS) were used. These items were, in turn, used as part of the Child Behavior Checklist (Achenbach and Edelbrock 1981). The items selected for the NHIS were those that provided the best discrimination between children who were referred or not referred for mental health services, by age category and gender. The NLSY97 uses items selected for boys and girls age twelve to seventeen, and each is asked of the parent as well as the youth. The four items that are asked of boys are whether he (a) has trouble concentrating or paying attention, (b) doesn't get along with other kids, (c) lies and cheats, and (d) is unhappy, sad, or depressed. These Achenbach index items are coded here as binary with "sometimes/somewhat true" combined with "often true" (a rare response). Factor analysis indicates that these measures cannot be combined into a mental health index, and so they are entered into the education and labor market outcome models separately. There is a general tendency for these reports of problem behaviors to fall with mother's education, though there are exceptions (e.g., mother reports that sons are depressed). The mean ASVAB percentile is strongly increasing in mother's education, as is an optimism index (constructed from four items such as "In uncertain times, I usually expect the best"). Youths report substantially higher rates of problem behaviors than do parents, on average, and the correlation between parent and youth responses is relatively low for most items (.19 to .30).

Table 6.1 reports the results for ordered probit models of educational attainment (defined in six levels from less than high school through postgraduate degree) and linear probability models of college graduation for men, where the independent variables include youth and parent-reported behavior problems, optimism, cognitive ability, and maternal characteristics. One self-reported noncognitive measure is significantly associated with educational attainment (trouble paying attention) as are two parent-reported items (lies or cheats and depressed). If both parent and youth reports are included in the model, the significance levels and magnitude of these coefficients change very little. These associations are substantial—a self-report of "trouble paying attention" by a teenager is equivalent to a decrease of 10 ASVAB percentiles in the categorical education model. The optimism index is never significantly associated with education (or with other outcomes).

Table 6.2 shows that, for this particular set of noncognitive indicators, there is little direct influence on wages and employment[3] once educational attainment is controlled for. Personality studies usually find significant

---

3. Employment is defined as positive earnings and twenty-five or more hours of work per week.

Table 6.1   Educational attainment, men (National Longitudinal Survey of Youth 1997)

	Educational attainment (0–5)		Bachelor's degree or above	
	(1)	(2)	(3)	(4)
Achenbach Child Behavior Checklist— Self-report				
Trouble paying attention	−0.2360***		−0.0985***	
	(0.0695)		(0.0228)	
Does not get along well with others	−0.0327		0.0033	
	(0.0681)		(0.0224)	
Lies or cheats	−0.0396		−0.0256	
	(0.0666)		(0.0219)	
Unhappy, sad, or depressed	0.0170		0.00218	
	(0.0681)		(0.0224)	
Achenbach Child Behavior Checklist— Parent report				
Trouble paying attention		−0.0530		−0.0221
		(0.0709)		(0.0234)
Does not get along well with others		−0.0173		−0.0079
		(0.0798)		(0.0263)
Lies or cheats		−0.1550*		−0.0305
		(0.0711)		(0.0233)
Unhappy, sad, or depressed		−0.2540**		−0.0744**
		(0.0776)		(0.0255)
Optimism index	0.0188	0.0228	0.0062	0.0085
	(0.0197)	(0.0194)	(0.00649)	(0.00640)
ASVAB age-normed percentile	0.0164***	0.0161***	0.0045***	0.0045***
	(0.00140)	(0.00144)	(0.000441)	(0.000459)
Observations	1,178	1,178	1,178	1,178
Adjusted R^2			0.283	0.280

Notes: Ordered probit and linear probability models. Standard errors in parentheses. Model also includes controls for mother's education, race, ethnicity, and region.

***Significant at the 0.1 percent level.

*Significant at the 1 percent level.

Significant at the 5 percent level.

direct impacts of personality traits on earnings, conditional on education, but it is not uncommon for noncognitive indicators based on early reports of emotional and behavioral problems to primarily affect the education process and have little direct association with later outcomes.[4] In results not reported here, there are significant interactions between cognitive skills and

4. In fact, Papageorge, Ronda, and Zheng (2017) find that childhood externalizing behavior, though it reduces educational attainment, has a positive association with adult earnings.

Table 6.2   Employment and wages, men (National Longitudinal Survey of
Youth 1997)

	Employment[a] (1)	Log wage (2)
GED	−0.0154	0.0942
	(0.0593)	(0.148)
HS diploma	**0.133****	**0.369****
	(0.0509)	(0.124)
Associate's degree	0.0964	**0.496****
	(0.0678)	(0.153)
Bachelor's degree	**0.243*****	**0.505*****
	(0.0631)	(0.145)
Graduate degree	**0.164***	**0.690*****
	(0.0829)	(0.183)
Achenbach Child Behavior Checklist—Parent report		
Trouble paying attention	−0.0170	−0.0164
	(0.0299)	(0.0645)
Does not get along well with others	**−0.0720***	−0.102
	(0.0332)	(0.0749)
Lies or cheats	−0.0227	0.0302
	(0.0298)	(0.0645)
Unhappy, sad, or depressed	−0.0073	−0.1300
	(0.0323)	(0.0704)
Optimism index	−0.0094	0.0136
	(0.0082)	(0.0183)
ASVAB age-normed percentile	−0.0002	0.0010
	(0.0006)	(0.0013)
Observations	1,006	772
Adjusted R^2	0.053	0.107

Notes: Standard errors in parentheses. Model also includes controls for mother's education, race, ethnicity, and region.
[a] Positive earnings and twenty-five hours or more of work a week.
***Significant at the 0.1 percent level.
**Significant at the 1 percent level.
*Significant at the 5 percent level.

some of the noncognitive measures—the effect of the ASVAB percentile on college graduation is substantially attenuated for men whose parents reported that they "did not get along well with others" when young. One possible interpretation of this result is that social skills and self-control alter the human capital production function by enhancing the learning environment, but the potential endogeneity of these measures is worth pointing out: parental reports that their child has poor social skills may be a signal of parental characteristics that affect school success rather than a valid measure of the child's noncognitive skills (Datta Gupta, Lausten, and Pozzoli 2012).

6.3.2   National Longitudinal Study of Adolescent to Adult Health
(Add Health)[5]

The Add Health study is a good companion to the NLSY97, since it surveys almost the same birth cohorts and follows them for a similar period, but provides a very different set of noncognitive skill indicators in Wave I. The study began in 1994–1995 with a nationally representative, school-based survey of more than 90,000 students in grades 7 through 12. About 20,000 respondents were followed in subsequent surveys, the last of which (Wave IV) was conducted in 2007–2008 when the respondents were between twenty-four and thirty-two years of age. To increase comparability with the NLSY97 results, I restrict the sample to men. By Wave IV most, though not all, of these young men will have completed their formal education and acquired some work experience.

The Add Health data is very rich, and Wave I contains a wealth of questions about the adolescents' attitudes, beliefs, and behaviors that could be used to construct noncognitive skill measures. I have chosen to include fairly standard indices of self-esteem and depression, and constructed a school problems index from youth reports of problems experienced with classmates, teachers, or homework. Finally, I have included a positive response to the question "When making decisions, you usually go with your 'gut feeling' without thinking too much about the consequences of each alternative" as a measure of impulsivity. Cognitive skills are measured with a computer-assisted version of the Peabody Picture Vocabulary Test administered in Wave I.

Table 6.3 shows that depression and self-esteem have small and generally insignificant associations with educational attainment (or the probability of graduating from college), but the school problems index and impulsivity have large and significant associations, ranging from one-third to one-half of the magnitude of cognitive ability (all measures are standardized). As with the NLSY97 measures, there are no significant effects of noncognitive skills measured in early and mid-adolescence on labor market outcomes (in this case log earnings) once educational attainment has been controlled for. These results highlight the context-specificity of many measures of noncognitive skills—the emotional states and behavior problems of adolescents clearly flag educational difficulties, but are less predictive of longer-term capabilities.

5. Add Health is a program project directed by Kathleen Mullan Harris and designed by J. Richard Udry, Peter S. Bearman, and Kathleen Mullan Harris at the University of North Carolina at Chapel Hill, and funded by grant P01-HD31921 from the Eunice Kennedy Shriver National Institute of Child Health and Human Development, with cooperative funding from twenty-three other federal agencies and foundations. Special acknowledgment is due Ronald R. Rindfuss and Barbara Entwisle for assistance in the original design. Information on how to obtain the Add Health data files is available on the Add Health website (http://www.cpc.unc.edu/addhealth). No direct support was received from grant P01-HD31921 for this analysis.

Table 6.3   Educational attainment and earnings, men (National Longitudinal Study of Adolescent to Adult Health)

	Educational attainment (1)	Bachelor's degree or above (2)	Log earnings (3)
HS diploma			0.2692**
			(0.0928)
Some college			0.3839***
			(0.1049)
Associate's degree			0.5020***
			(0.1088)
Bachelor's degree			0.6190***
			(0.1016)
Graduate degree			0.6392***
			(0.1209)
School problems index	−0.2072***	−0.0497***	0.0024
	(0.0214)	(0.0070)	(0.0238)
Depression index	0.0175	0.0006	−0.0230
	(0.0219)	(0.0075)	(0.0262)
Self-esteem index	0.0331	0.0175*	0.0062
	(0.0214)	(0.0076)	(0.0238)
Impulsivity	−0.1169***	−0.0377***	−0.0168
	(0.0200)	(0.0073)	(0.0212)
Cognitive ability (AH Picture Vocabulary Test)	0.3561***	0.0920***	0.0867**
	(0.0230)	(0.0082)	(0.0329)
Observations	5,743	5,743	5,373
Adjusted R^2		0.203	0.094

Notes: Standard errors in parentheses. Model also includes controls for mother's education, race, and ethnicity.
***Significant at the 0.1 percent level.
**Significant at the 1 percent level.
*Significant at the 5 percent level.

In many ways, this is a typical set of nonexperimental, noncognitive skill results—we can show that some characteristics and behavioral tendencies measured relatively early in life have significant associations with later outcomes, particularly educational attainment. The interpretation of the results is difficult—clearly no causal statements would be appropriate. Problems in school can reflect deficiencies in parenting or an adverse school environment as well as adolescent skills, and it is unlikely that we could control for school and parent characteristics well enough to eliminate omitted characteristics. It is not surprising, perhaps, that reported behavior now may be strongly predictive of behavior in the future, but that association may reflect continuity in either characteristics or in circumstances.

6.4   Measurement

To date, the economics literature on noncognitive skills has made major contributions to our understanding of the production and the productivity of human capital. It has broadened our understanding of human capabilities and the multidimensional nature of productive skills, and has focused attention on the early stages of life, when executive functioning and the regulatory capacities that flow from it can be degraded or enhanced. Considerable progress has been made in modeling the production of multidimensional forms of human capital (Cunha and Heckman 2008). The development of a coherent body of empirical knowledge, however, has been hampered by the absence of a broadly accepted conceptual framework that maps developmental stages into identifiable skills and by the pursuit of an opportunistic approach to measurement. Summarizing the literature is difficult given the astonishing variety of skill proxies that economists, tapping existing data, have used. There are also conceptual problems that arise when we interpret the coefficients in tables 6.1–6.3 as estimates of the returns to noncognitive skills.

The first issue an obvious one: skills, including noncognitive ones, are endogenous. They are likely to be correlated with parental resources, environmental influences, and other skills that we don't happen to have measured, and so any causal interpretation of their apparent effects is inappropriate. The skepticism that we as a profession bring to interpreting a coefficient on a measure of IQ in an education or earnings equation seems to desert us occasionally when we are faced with a novel measure of noncognitive skill. The link between the self-control exercised by the patient children in the marshmallow experiment and their later successes may reflect not the actual return to developing patience early in life, but rather the quality of their parenting by other pathways. An interesting concrete example of this conflation can be found in Dohmen et al. (2010), who find substantial bias in the estimated "effects" of cognitive ability, risk aversion, and patience on key adult outcomes when all three measures are not included in the model.

Second, observed or reported behavior, while it may be reflective of noncognitive skills, also depends on other traits, incentives, beliefs, and situational factors, which we are unlikely to be able to control for. In the framework of Kautz et al. (2015), skills are measured based on task performance, which in turn depends on multiple skills and effort. They argue in favor of using behaviors as measures of skill, and attempting to control for other factors that influence performance, in order to avoid the reference bias that is likely to influence self-reported psychological scales.[6]

---

6. They show that average levels of conscientiousness across countries are not positively related to work hours, though there is a strong within-country correlation.

We now know that this identification problem affects standard measures of cognitive ability, since they depend on test performance. IQ test scores, far from being pure indicators of intellectual ability, are influenced by personality and motivation. Borghans, Meijers, and ter Weel (2008) find that substantial portions of variance in achievement test scores depend on personality, not cognitive ability, and Segal (2012) shows that incentives increase performance on low-stakes cognitive tests. Invoking racial stereotypes can affect test performance (Steele and Aronson 2005). Measures of children's noncognitive skills that are based on teacher and parent reports of externalizing behavior, lying, or the child's ability to maintain focus on an assigned task are likely to be much more sensitive than cognitive test results to incentives, expectations, and peer effects. Particularly problematic is the interpretation of differences in test scores or behavior between children from high- and low-income families as pure differences in skills, when their environments are likely to vary substantially.

Borghans et al. (2011) focus on the problem of identifying traits from observed behavior, noting that behavior is influenced by incentives and by multiple traits. Incentives, in particular, may vary systematically by groups in the population defined over income, race, or gender. The task performance of individual i in group j, Y_{ij}, will depend on their level of skill, θ_j, and their chosen level of effort, e_i. Measuring skills on the basis of task performance requires that we control for effort, which is usually unobservable. There are a couple of ways that group membership can enter this process of inferring skills from observed performance. One possible source of group dependence is that the mapping of skill and effort into performance, φ_j, may vary by groups if, for example, teacher assessments are biased. Alternatively, the choice of effort will depend both on an individual effort endowment ($\bar{e}_i$) and incentives (p_{ij}) that may have a group-specific component (such as social sanctions against behavior that does not conform to gender norms)

$$Y_{ij} = \varphi_j(\theta_i, e_i)$$

$$e_i = f(\bar{e}_i, p_{ij}).$$

There may also be important environmental drivers of task performance, such as the intensity of other demands on a person's capabilities. An individual's reserves of self-control can be depleted by exertions of control (Muraven and Baumeister 2000). Experiments have shown that resisting temptation leads to a weakened ability to resist subsequent temptations, and individuals who have to cope with stressors such as noise and crowding are less able to delay gratification. Mani et al. (2013) find that poverty appears to degrade cognitive functioning. The farmers in their study exhibit diminished cognitive functioning before the harvest, when they are poor, compared to after the harvest, when they are rich. The differences are not accounted for by nutrition or work effort, and appear to be due to poverty-related demands on mental resources. Poor children, who are likely to face more chaotic and stressful conditions at home, may be less able to muster the resources to maintain focus and control at school, even if their fundamental capabilities are identical to those of other children.

Children with identical levels of a trait such as self-control may also have different expectations about the payoffs to exerting control, and in fact these payoffs may be dependent on context. In a variant of the marshmallow test, researchers preceded the classic test with two sessions in which randomly assigned children were primed to believe that their environment was reliable or unreliable (promised art supplies either did or did not show up). Children who had been exposed to the unreliability of the experimenters' promises scored substantially worse on the marshmallow gratification delay test (Kidd, Palmeri, and Aslin 2013). The researchers conclude that differences in performance on the marshmallow test may be due, not just to differences in self-control capabilities, but also to experiences about the reliability of the children's environments.

The return to noncognitive skills, in particular, seems to be highly context-dependent, and evidence of heterogeneity in returns is beginning to emerge. The positive association between a child's externalizing behavior and adult earnings that Papageorge, Ronda, and Zheng (2017) report does not extend to individuals from disadvantaged backgrounds.[7] Lundberg (2013) finds that the relationship between personality traits and college graduation in the United States varies by socioeconomic status, with conscientiousness having a substantial payoff only for youth with highly educated mothers. Such heterogeneity in returns should affect investments in skills as individuals set marginal costs equal to expected marginal returns. In environments such as the unreliable marshmallow test, developing impulse control may not make much sense—when such skills are not rewarded, they are not likely to be reinforced.

If observed behaviors depend not just on skills, but also on context—via perceived payoffs, distractions, peer effects, or supportive surroundings—then difficulties arise in comparing noncognitive skills that rely on behavioral assessments across groups. Early behavior can predict later behavior either because of persistent traits/skills or because of correlated circumstances. On the other hand, as Kautz et al. (2015) point out, group disparities based on self-reports about behavioral tendencies and beliefs such as personality can be affected by reference bias, in that how you assess yourself and your behavior may depend on peer behavior or cultural norms. One way to proceed is to compare alternative indicators of the same underlying skill.

---

7. Note that there are two ways to interpret this result: one, as true heterogeneity in the results to skill, or two, as instability in the mapping from skills to behavior across socioeconomic groups.

6.4.1   Male Impulsivity and Crime

Self-control is fundamental to many conceptualizations of noncognitive skill, as the marshmallow tests illustrate, and crime is thought to be strongly associated with deficits in self-control. The criminology literature links early difficulties in self-regulation and a failure to consider long-term consequences with later criminal behavior (Gottfredson and Hirschi 1990; Wright et al. 1999). The Add Health data includes several early indicators of impulsivity or low self-control that permit us to compare how well different measures predict later criminal behavior.

In the first wave of the study, when the Add Health subjects are in middle school or the early years of high school, three possible measures of impulsivity are collected that correspond to three of the basic types of noncognitive skill data: self-assessment, administrative records, and observed behavior:

- Self-assessment: "When making decisions, you usually go with your 'gut feeling' without thinking too much about the consequences of each alternative." The youth is classified as impulsive if he or she responds "agree" or "strongly agree" to this question.
- (Potential) administrative data: "Have you ever received an out-of-school suspension from school?" Since the majority of school suspensions are reported to be due to either disobedience or disruptive/disrespectful behavior, suspensions are likely to be strongly driven by individual impulsivity.
- Interviewer remarks: "Did the respondent ever seem bored or impatient during the interview?"

These three measures of impulsivity are positively, but not very strongly, correlated, with the strongest correlation being 0.12 between the self-assessment and report of school suspensions.

In Wave IV, when the subjects are age twenty-six to thirty-two, several measures of criminal activity and criminal justice contact are collected. These include an indicator for ever having been arrested, and reports of whether, in the past twelve months, the individual has deliberately damaged property, gotten involved in a physical fight, used or threatened to use a weapon, hurt someone so badly they needed medical care, or used a weapon or engaged in any other crime, including theft and selling drugs. Means of the impulsivity and crime measures for the male respondents are reported in table 6.4.

Predictive power is often used in noncognitive skill studies as evidence in support of the interpretation of a behavioral outcome as a valid skill measure. Kautz et al. (2015), for example, cite studies showing that behavioral measures are at least as good at predicting crime as self-reported psychologi-

Table 6.4   **Means of early impulsivity and later crime indicators, men (National Longitudinal Study of Adolescent to Adult Health)**

Impulsivity measures, Wave I	
Self-reported impulsivity	0.40
School suspension	0.35
Interviewer report	0.14
Self-reported crime and arrests, Wave IV	
Crime (in past 12 months)	0.31
Ever arrested	0.41

cal scales and conclude that behaviors can be used to infer a skill "as long as the measurement accounts for other skills and aspect of the situation." Table 6.5 reports results for linear probability models of impulsivity effects on crime and arrests (the patterns are similar if we use indicators of specific categories of crime). All three impulsivity indicators predict crime and arrests, with school suspensions having the strongest effect. When all impulsivity measures are included in the models (columns [4] and [8]), interviewer reports of impatience no longer has a significant association with crime. The inclusion of family background variables such as mother's education and family structure reduce the impulsivity coefficients by about 9 percent. Since school suspensions are most often triggered by disruptive behavior that suggests low self-regulation, it is plausible that they will be strongly predictive of future crime and criminal justice system contact.

Does this mean that a record of school suspensions is the best measure of crime-related impulsivity that is available in the Add Health study? What we should be looking for is a measure of capabilities that is not also a proxy for other factors driving behavior (such as incentives). In this respect, suspensions are a problematic measure of impulsivity. Table 6.6 reports the results from regressions that use other measures of noncognitive skills, family background, and race to predict the three measure of impulsivity. Both suspensions and the self-report are correlated with personality traits, but only suspensions are strongly related to mother's education. Most striking is the result that being black increases school suspensions by 50 percent, but does not change self-reported impulsivity and has a modest positive impact on interviewer reports of restlessness. In the racial dimension, other factors that drive behavior or school discipline are clearly relevant—school quality, racial bias in teacher and school responses to behavior, or even different expectations about the rewards of restraint in school are likely to be relevant. Clearly, race is an "aspect of the situation" that can be controlled for, but we are unlikely to be able to control consistently for home and neighborhood characteristics that affect behavior and drive this group discrepancy. Behavioral outcomes that depend on expected rewards, beliefs, other demands on a student's capabilities, or differential treatment by teachers and other

Table 6.5   Effects of Wave I impulsivity on Wave IV crime and arrests, men (National Longitudinal Study of Adolescent to Adult Health)

	Crime				Ever arrested			
	(1)	(2)	(3)	(4)	(5)	(6)	(7)	(8)
Self-reported impulsivity	.0477***				.1074***			
	(0.0116)				(0.0123)			
+ family background	.0443***			.0354***	.0971***			.0704***
	(0.0116)			(0.0117)	(0.0122)			(0.0120)
School suspension		.0970***				.2710***		
		(0.0119)				(0.0123)		
+ family background		.0879***		.0842***		.2497***		.2389***
		(0.0122)		(0.0123)		(0.0126)		(0.0127)
Interviewer report			.0373**				.0738**	
			(0.0165)				(0.0177)	
+ family background			.0357**	.0255			.0692**	.0433**
			(0.0165)	(0.0166)			(0.0175)	(0.0171)

Note: Family background variables include mother's education and a dummy variable for living with both parents in Wave I.

***Significant at the 0.1 percent level.

**Significant at the 1 percent level.

*Significant at the 5 percent level.

Table 6.6   **Predicting Wave I measures of impulsivity, men (National Longitudinal Study of Adolescent to Adult Health)**

	Self-reported impulsivity (1)	School suspension (2)	Interviewer report (3)
African American	.0088	.1748***	.0230**
	(0.0153)	(0.0143)	(0.0108)
Lived with both parents	−.0390***	−.1238***	−.0030
	(0.0124)	(0.0116)	(0.0087)
Mother high school	.0133	−.0645***	−.0187
	(0.0177)	(0.0165)	(0.0125)
Mother some college	−.0110	−.0729***	−.0193
	(0.0204)	(0.0191)	(0.0144)
Mother college graduate	−.0311	−.1664***	−.0320**
	(0.0196)	(0.0183)	(0.0138)
Personality			
Openness	−.0465***	−.0219***	−.0070
	(0.0064)	(0.0060)	(0.0045)
Conscientiousness	−.0084	−.0104*	.0036
	(0.0063)	(0.0059)	(0.0045)
Extraversion	.0272***	0.309***	.0090*
	(0.0063)	(0.0059)	(0.0045)
Agreeableness	−.0181***	−.0340***	−.0180***
	(0.0064)	(0.0060)	(0.0046)
Neuroticism	.0360***	0.0533***	.0022
	(0.0067)	(0.0061)	(0.0046)
Observations	6,577	6,599	6,605
Adjusted R^2	0.024	0.091	0.004

Note: Standard errors in parentheses.
***Significant at the 0.1 percent level.
**Significant at the 1 percent level.
*Significant at the 5 percent level.

authorities are going to generate flawed measures of skill disparities across socioeconomic groups.

6.5   Noncognitive Human Capital and Growth

The case for broadening the concept of human capital to include noncognitive skills is a strong one. Many studies have shown that enriched environments in early childhood lead to positive outcomes later in life beyond their influence on measured cognitive skills, but evidence of the impact of education (or educational quality) on noncognitive skills is only beginning to emerge. Some personality traits are associated with positive outcomes in education and the labor market, though returns appear to vary by socioeconomic status (education) and occupation (earnings). Measures of adolescent emotional and behavioral problems, though they are not strongly predictive of labor market outcomes conditional on education, do have strong associations with educational attainment.

Some interesting issues to explore in future research on noncognitive skills concern possible complementarities between skills in educational and production processes. Noncognitive skills such as attention and self-control can increase the productivity of educational investments. Disruptive behavior and crime impose negative externalities in schools and communities that increased levels of some noncognitive skills could ameliorate. Aizer (2008) shows that diagnosis and treatment of attention deficit disorder (ADD) improves classroom peer behavior, which in turn increases student achievement. To indulge in pure speculation, it may be that broad improvements in noncognitive skills could have positive effects on technological innovation if these skills improve institutional quality and levels of cooperation within institutions.

To date, however, the state of our knowledge about the production of and returns to noncognitive skills is rather rudimentary. We lack a conceptual framework that would enable us to consistently define multidimensional noncognitive skills, and our reliance on observed or reported behavior as measures of skill make it impossible to reliably compare skills across groups that face different environments. Finally, there is increasing evidence that the returns to noncognitive skills may be highly context-dependent, a factor that limits our ability to extract policy recommendations from the existing literature.

References

Achenbach, Thomas M., and C. Edelbrock. 1981. *Child Behavior Checklist*. Burlington: University of Vermont.
Aizer, Anna. 2008. "Peer Effects and Human Capital Accumulation: The Externalities of ADD." NBER Working Paper no. 14354, Cambridge, MA.
Almlund, Mathilde, Angela Lee Duckworth, James J. Heckman, and Tim D. Kautz. 2011. "Personality Psychology and Economics." In *Handbook of the Economics of Education*, vol. 4, edited by Eric A. Hanushek, Stephen Machin, and Ludger Woessmann, 1–182. Amsterdam: Elsevier Science.
Anger, Silke. 2012. "Intergenerational Transmission of Cognitive and Noncognitive Skills." In *From Parents to Children: The Intergenerational Transmission of Advantage*, edited by John Ermisch, Markus Jäntti, and Timothy Smeeding, 393–421. New York: Russell Sage Foundation.
Becker, Anke, Thomas Deckers, Thomas Dohmen, Armin Falk, and Fabian Kosse. 2012. "The Relationship between Economic Preferences and Psychological Personality Measures." *Annual Review of Economics* 4:453–78.
Borghans, Lex, Bart H. H. Golsteyn, James Heckman, and John Eric Humphries.

2011. "Identification Problems in Personality Psychology." *Personality and Individual Differences* 51 (3): 315–20.

Borghans, Lex, Huub Meijers, and Bas ter Weel. 2008. "The Role of Noncognitive Skills in Explaining Cognitive Test Scores." *Economic Inquiry* 46 (1): 2–12.

Bouchard Jr., Thomas J., and John C. Loehlin. 2001. "Genes, Evolution, and Personality." *Behavior Genetics* 31 (3): 243–73.

Bowles, Samuel, and Herbert Gintis. 1976. *Schooling in Capitalist America: Educational Reform and the Contradictions of American Life*. New York: Basic Books, Inc.

Bowles, Samuel, Herbert Gintis, and Melissa Osborne. 2001. "The Determinants of Earnings: A Behavioral Approach." *Journal of Economic Literature* 39 (4): 1137–76.

Cadena, Brian, and Benjamin Keys. 2015. "Human Capital and the Lifetime Costs of Impatience." *American Economic Journal: Economic Policy* 7 (3): 126–53.

Chetty, Raj, John N. Friedman, Nathaniel Hilger, Emmanuel Saez, Diane Whitmore Schanzenbach, and Danny Yagan. 2011. "How Does Your Kindergarten Classroom Affect Your Earnings? Evidence from Project Star." *Quarterly Journal of Economics* 126 (4): 1593–660.

Cobb-Clark, Deborah. 2015. "Locus of Control and the Labor Market." IZA Discussion Paper no. 8678, Institute for the Study of Labor.

Cobb-Clark, Deborah, Marco Caliendo, and Arne Uhlendorff. 2015. "Locus of Control and Job Search Strategies." *Review of Economics and Statistics* 97 (1): 88–103.

Cunha, Flavio, and James J. Heckman. 2008. "Formulating, Identifying and Estimating the Technology of Cognitive and Noncognitive Skill Formation." *Journal of Human Resources* 43 (4): 738–82.

Currie, Janet. 2011. "Inequality at Birth: Some Causes and Consequences." *American Economic Review* 101 (3): 1–22.

Datta Gupta, Nabanita, Mette Lausten, and Dario Pozzoli. 2012. "Does Mother Know Best? Parental Discrepancies in Assessing Child Functioning." IZA Discussion Paper no. 6962, Institute for the Study of Labor.

Deming, David J. 2017. "The Growing Importance of Social Skills in the Labor Market." *Quarterly Journal of Economics* 132 (4): 1593–640.

Dohmen, Thomas, Armin Falk, David Huffman, and Uwe Sunde. 2010. "Are Risk Aversion and Impatience Related to Cognitive Ability?" *American Economic Review* 100 (3): 1238–60.

Duncan, Greg J., and Katherine Magnuson. 2011. "The Nature and Impact of Early Achievement Skills, Attention Skills, and Behavior Problems." In *Whither Opportunity? Rising Inequality, Schools, and Children's Life Chances*, edited by Greg J. Duncan and Richard J. Murnane. New York: Russell Sage Foundation.

———. 2013. "Investing in Preschool Programs." *Journal of Economic Perspectives* 27 (2): 109–32.

Filer, Randall K. 1986. "The Role of Personality and Tastes in Determining Occupational Structure." *Industrial & Labor Relations Review* 39 (3): 412–24.

Gertler, Paul, James Heckman, Rodrigo Pinto, Arianna Zanolini, Christel Vermeersch, Susan Walker, Susan M. Chang, and Sally Grantham-McGregor. 2014. "Labor Market Returns to an Early Childhood Stimulation Intervention in Jamaica." *Science* 344 (6187): 998–1001.

Goldberg, Lewis R. 1981. "Language and Individual Differences: The Search for Universals in Personality Lexicons." In *Review of Personality and Social Psychology*, vol. 2, edited by Ladd Wheeler, 141–65. Thousand Oaks, CA: Sage.

Gottfredson, Michael R., and Travis Hirschi. 1990. *A General Theory of Crime.* Stanford, CA: Stanford University Press.

Hanushek, Eric A., and Dennis D. Kimko. 2000. "Schooling, Labor-Force Quality, and the Growth of Nations." *American Economic Review* 90 (5): 1184–208.

Hanushek, Eric A., and Ludger Woessmann. 2008. "The Role of Cognitive Skills in Economic Development." *Journal of Economic Literature* 46 (3): 607–68.

Heckman, James J., Rodrigo Pinto, and Peter Savelyev. 2013. "Understanding the Mechanisms through Which an Influential Early Childhood Program Boosted Adult Outcomes." *American Economic Review* 103 (6): 1–35.

Heckman, James J., and Yona Rubinstein. 2001. "The Importance of Noncognitive Skills: Lessons from the GED Testing Program." *American Economic Review* 91 (2): 145–49.

Heckman, James J., Jora Stixrud, and Sergio Urzua. 2006. "The Effects of Cognitive and Noncognitive Abilities on Labor Market Outcomes and Social Behavior." *Journal of Labor Economics* 24 (3): 411–82.

Heineck, Guido. 2011. "Does It Pay to Be Nice? Personality and Earnings in the United Kingdom." *Industrial & Labor Relations Review* 64 (5): 1020–38.

Heineck, Guido, and Silke Anger. 2010. "The Returns to Cognitive Abilities and Personality Traits in Germany." *Labour Economics* 17 (3): 535–46.

Kautz, Tim, James J. Heckman, Ron Diris, Bas ter Weel, and Lex Borghans. 2015. "Fostering and Measuring Skills: Improving Cognitive and Non-Cognitive Skills to Promote Lifetime Success." OECD Education Working Paper no. 110, OECD Publishing.

Kidd, Celeste, Holly Palmeri, and Richard N. Aslin. 2013. "Rational Snacking: Young Children's Decision-Making on the Marshmallow Task Is Moderated by Beliefs about Environmental Reliability." *Cognition* 126 (1): 109–14.

Krueger, Alan B., and David Schkade. 2008. "Sorting in the Labor Market: Do Gregarious Workers Flock to Interactive Jobs?" *Journal of Human Resources* 43 (4): 859–83.

Lindqvist, Erik, and Roine Vestman. 2011. "The Labor Market Returns to Cognitive and Noncognitive Ability: Evidence from the Swedish Enlistment." *American Economic Journal: Applied Economics* 3 (1): 101–28.

Lundberg, Shelly. 2012. "Personality and Marital Surplus." *Journal of Labor Economics* 1 (1): 1–21.

———. 2013. "The College Type: Personality and Educational Inequality." *Journal of Labor Economics* 31 (3): 421–41.

———. 2015. "Skill Disparities and Unequal Family Outcomes." *Research in Labor Economics* 41:177–212.

Mani, Anandi, Sendhil Mullainathan, Eldar Shafir, and Jiaying Zhao. 2013. "Poverty Impedes Cognitive Function." *Science* 341 (6149): 976–80.

Mischel, Walter, Ebbe B. Ebbesen, and Antonette Raskoff Zeiss. 1972. "Cognitive and Attentional Mechanisms in Delay of Gratification." *Journal of Personality and Social Psychology* 21 (2): 204–18.

Mueller, Gerrit, and Erik Plug. 2006. "Estimating the Effect of Personality on Male and Female Earnings." *Industrial & Labor Relations Review* 60 (1): 3–22.

Muraven, Mark, and Roy F. Baumeister. 2000. "Self-Regulation and Depletion of Limited Resources: Does Self-Control Resemble a Muscle?" *Psychological Bulletin* 126 (2): 247–59.

Nandi, Alita, and Cheti Nicoletti. 2014. "Explaining Personality Pay Gaps in the UK." *Applied Economics* 46 (26): 3131–50.

Nyhus, Ellen K., and Empar Pons. 2005. "The Effects of Personality on Earnings." *Journal of Economic Psychology* 26 (3): 363–84.

Papageorge, Nicholas W., Victor Ronda, and Yu Zheng. 2017. "The Economic Value of Breaking Bad: Misbehavior, Schooling and the Labor Market." IZA Discussion Paper no. 10822, Institute for the Study of Labor.

Rustichini, Aldo, Colin G. DeYoung, Jon C. Anderson, and Stephen V. Burks. 2012. "Toward the Integration of Personality Theory and Decision Theory in the Explanation of Economic and Health Behavior." IZA Discussion Paper no. 6750, Institute for the Study of Labor.

Segal, Carmit. 2012. "Working When No One Is Watching: Motivation, Test Scores, and Economic Success." *Management Science* 58 (8): 1438–57.

———. 2013. "Misbehavior, Education, and Labor Market Outcomes." *Journal of the European Economic Association* 11 (4): 743–79.

Shonkoff, Jack P., Andrew S. Garner, Benjamin S. Siegel, Mary I. Dobbins, Marian F. Earls, Laura McGuinn, John Pascoe, and David L. Wood. 2012. "The Lifelong Effects of Early Childhood Adversity and Toxic Stress." *Pediatrics* 129 (1): e232–46.

Steele, Claude M., and Joshua Aronson. 2005. "Stereotypes and the Fragility of Academic Competence, Motivation, and Self-Concept." In *Handbook of Competence and Motivation*, edited by Andrew J. Elliot and Carol S. Dweck, 436–55. New York: Guilford Press.

Weiss, Andrew. 1988. "High School Graduation, Performance, and Wages." *Journal of Political Economy* 96 (4): 785–820.

World Bank. 2015. "World Development Report 2015: Mind, Society, and Behavior." Washington, DC, World Bank. https://doi.org/10.1596/978-1-4648-0342-0.

Wright, Bradley R. Entner, Avshalom Caspi, Terrie E. Moffitt, and Phil A. Silva. 1999. "Low Self-Control, Social Bonds, and Crime: Social Causation, Social Selection, or Both?" *Criminology* 37 (3): 479–514.

Comment David J. Deming

Shelly Lundberg has written an important chapter about the rapidly growing study of "noncognitive" skills in economics. This chapter should be required reading for social scientists who seek to use measures of noncognitive skills in schools and other educational settings to make important policy decisions. I largely agree with her conclusions about the state of the literature, which I summarize crudely as follows. Although the evidence is overwhelming that so-called noncognitive skills are important predictors of many important life outcomes, we do not really agree on what they are (and importantly, what they are not). Thus we have very little idea of how to measure noncognitive skills well, and even less idea of how to use measures of noncognitive skills to make high-stakes policy decisions.

David J. Deming is professor of public policy at the Harvard Kennedy School, professor of education and economics at the Harvard Graduate School of Education, and a research associate of the National Bureau of Economic Research.

For acknowledgments, sources of research support, and disclosure of the author's material financial relationships, if any, please see http://www.nber.org/chapters/c13702.ack.

In my view, *measurement* is the fundamental challenge for social scientists who want to study noncognitive skills. I would characterize existing measures of noncognitive skills as having one of two problems. First, self-assessment measures such as personality inventories (e.g., the "Big Five") are arguably valid but unreliable across contexts, often in ways that make them difficult to use for any practical purpose. On the other hand, administrative records of behavior are reliable (in a statistical sense) and predictive—but possibly invalid—measures of the underlying skill. All of the measures of noncognitive skills that I have seen used in research have—to varying degrees—one of these two problems.

While no measure is perfect, cognitive skills are much better measured than noncognitive skills in terms of both validity and reliability. One might conclude from this that the construct of cognitive skill is *inherently* more valid. However, this ignores the history of measurement. Psychologists—and the testing industry—spent several decades and millions of dollars systematically improving and refining the measurement of cognitive skills. I conclude by advocating for an equally careful and rigorous approach to the theoretical refinement and measurement of noncognitive skills.

Reliability and Self-Assessment

As Lundberg points out, most self-assessment measures ask individuals to answer questions that indicate "what I am like" or "this is what I believe." An example is the Big Five personality inventory, a rigorously developed psychological model that distills human personality into five factors—extraversion, conscientiousness, agreeableness, neuroticism, and openness to experience.

The five factors were originally derived from a statistical factor analysis of a much larger number of potential personality traits (see John and Srivastava [1999] for an overview and history of the Big Five). Thus, in a sense they are statistical rather than theoretical constructs, chosen because they are distinct and orthogonal to one another rather than for higher-minded reasons. Still, the existence of these five distinct and mostly comprehensive set of personality factors has been replicated by psychologists in many other settings spanning geography, culture, and time. Agreement is hardly unanimous and criticisms of the Big Five abound, yet it probably represents the best case scenario for "noncognitive" skill measures that are based on self-assessment. Moreover, Big Five personality measures—especially conscientiousness—are strongly positively correlated with educational attainment, labor market earnings, and other important life outcomes (e.g., Heckman and Kautz 2012).

Are self-assessments such as the Big Five *reliable*? That depends on what you mean by reliable. The most basic definition is *test-retest* reliability, where one administers the same assessment to the same person under the same conditions over a very short period of time, and estimates the correlation between assessments. Of course a perfect replication of the test environment is never possible, but under ideal conditions the test-retest reliability of the Big Five is extremely high. The correlation between assessments ranges from 0.8 to 0.9, depending on the length of the test instrument and the specific factor being studied (John and Srivastava 1999). This is very similar to the test-retest reliability for IQ, for example.

However, policy is not made in a lab, and the evidence for the reliability of self-assessments in the field and across contexts is much less reassuring. Schmitt et al. (2007) administer the Big Five personality questionnaire in a number of OECD countries and show that the correlation between conscientiousness (the tendency to work hard and be persistent) and average hours worked is *negative*. This is particularly striking in the case of respondents in France and South Korea. South Koreans report working nearly 2,500 hours per year, compared to around 1,500 hours for their French counterparts. Yet France places fourth and South Korea places twenty-fifth out of twenty-six when respondents are asked to self-assess their conscientiousness. West et al. (2015) find that students who are randomly assigned to a set of schools known for their emphasis on character building and hard work (so-called No Excuses charter schools) self-report *lower* levels of conscientiousness, self-control, and "grit." In both cases, respondents are comparing themselves to those around them. This makes it difficult or impossible to compare measures of "noncognitive" skills across very different contexts.

Noncognitive skill measures that are sensitive to context are particularly problematic in high-stakes settings. Put bluntly, personality assessments can be easily gamed if one knows what the "right answer" is supposed to be. For example, personality tests are often administered by large retail companies as part of the job applicant screening process. A cursory web search for "job application test answers" reveals that there is a robust market in teaching people how to successfully game personality assessments.

Notably, gaming is possible even without access to specific test items. Conscientiousness is among the best predictors of job performance, and so employers would like to screen for this personality trait. Big Five question items that measure conscientiousness include Likert scale items (1 to 5 numerical responses that range from strongly disagree to strongly agree) such as "I see myself as someone who does a thorough job" or "I am always prepared." It is not hard to foresee that placing a high weight on conscientiousness in hiring will lead to a sudden and dramatic increase in the self-reported persistence and diligence of the average applicant!

Using Behaviors to Measure "Noncognitive" Skills

Kautz et al. (2015) discuss this problem of "reference bias" in self-assessed measures. They propose using behaviors as alternative measures of skills:

all tasks or behaviors can be used to infer a skill as long as the measurement accounts for other skills and aspects of a situation. . . . Self-reported scales should not be assumed to be more reliable than behaviors, although personality psychologists often assume so. The question is which measurements are most predictive and which can be implemented in practice. The literature suggests that there are objective measurements of noncognitive skills that are not plagued by reference bias. (Kautz et al. 2015, 17–18)

In other words, behavioral measures of noncognitive skills might be better than self-assessments if they are predictive and reliable across contexts (e.g., not plagued by reference bias).

Lundberg points out that using observed behaviors to measure skills is potentially problematic if behavior also depends on social context. Using the Add Health data, she shows that (a) self-reported impulsivity is correlated with school suspensions and with crime, (b) African Americans are much more likely to be suspended from school, and (c) there are no racial differences in self-reported impulsivity. Thus it is problematic to use school suspensions as a behavioral measure of impulsivity, since suspensions are also determined by school context, racial discrimination, and other unknown factors.

I think this critique is extremely important, and it points out deeper issues with the measurement of noncognitive skills. Sometimes measures are too predictive—or alternatively, they are predictive *because* the underlying construct is invalid. School suspensions capture some measure of the student's impulsivity, but also what type of school they attend, their gender and race, and many other things. In these situations, one's confidence in the ability to use the behavior as a proxy for skills hinges on one's ability to control for everything else that is important. This is a classic omitted variables bias problem—the behavior (school suspensions) captures the underlying skill, but also many other things.

I must note that Kautz et al. (2015) includes a very careful discussion of the pros and cons of these issues, and Borghans et al. (2011) go into even more detail on identification issues in the use of behavior measures. So the authors are not unaware of these concerns.

Nonetheless, I think the issue of construct validity is mostly underappreciated in the literature. There is simply no substitute for careful development of a theoretically sound underlying construct. We will never be able to measure noncognitive skills well if we do not understand what we are measuring.

The Way Forward

One pessimistic response is that we will never be able to measure noncognitive skills well because noncognitive skills do not exist. The most reductive view, which we have all seen from time to time, is that IQ is everything. While it is easy to reject this extreme form of the argument, it is not so easy to reject a weaker form that cognitive skills are more important predictors of life outcomes than noncognitive skills. This argument starts with the observation that cognitive tests are both more predictive and more reliable than noncognitive measures, whether self-assessed or behavioral.

However, if measurement error in skills is classical, then the coefficient on skills in a regression with an outcome such as log wages will be attenuated toward zero, with the degree of attenuation decreasing in the reliability of the measure. Thus, if noncognitive skills are measured more poorly than cognitive skills, we will tend to underestimate their importance.

More broadly, we must recognize that measures of cognitive and noncognitive skills do not just appear from nowhere. Rather, they are developed over many years and by many different researchers, often for an initially narrow purpose. The modern IQ test was created as a means to diagnose mental retardation in schoolchildren, with lower scores simply indicating that children were unable to perform tasks that were "typical" for their same-age peers. The later reification of "g" as general intelligence was based on the observation that children's grades and test scores can be statistically best explained by a single common factor.

All this is to say that the scholarly consensus about the importance of different human capacities is often driven by how well these capacities can be measured. For example, if we could develop reliable and context-invariant tests of important noncognitive capacities such as self-control and social intelligence, I would not be surprised if they ended up being better predictors than IQ of labor market outcomes.

Here I am optimistic that we can more fruitfully exploit comparative advantage between psychologists and economists. Psychologists have carefully developed measures that map cleanly to underlying constructs, but they have (for the most part) not subjected these measures to rigorous testing in a variety of field settings. Economists, on the other hand, have gleefully used convenient, off-the-shelf measures of questionable validity (n.b., I am as guilty as anyone in this regard) to make broad generalizations about the importance of noncognitive skills, with an exact definition of these skills to be determined. When it comes to noncognitive skills, we economists are the proverbial drunk searching under the street lamp for his keys, because that is where the light is located.

I will close with a specific example of this possible complementarity across disciplines. The Reading the Mind in the Eyes Test (RMET) is a test of emotion recognition or social sensitivity developed by Simon Baron-Cohen and colleagues (e.g., Baron-Cohen et al. 2001). The RMET was originally created for a narrow purpose—to diagnose so-called theory of mind deficits such as Asperger Syndrome and high-functioning autism in otherwise well-functioning adults. However, much like IQ, psychologists have discovered that the RMET has predictive power for a wide variety of outcomes within a general population. Woolley et al. (2010) randomly assign participants to teams and find that the team's average score on the RMET predicts task performance after controlling for group average IQ.

While the RMET is not perfect, it is superior to many other measures of noncognitive skills in at least two respects. First, the RMET overcomes some of the limitations of self-assessment because there is a correct answer to the question items. This prevents reference group bias as well as strategic responses in high-stakes settings. Second, there is a well-grounded theory of how the underlying capacity (theory of mind) relates to task performance (emotion recognition in human faces), and in turn how task performance relates to outcomes (see Deming [2017] for a more thorough discussion of the connection between social skills and labor market success). This helps with the concern that a poorly defined construct measures "too much."

There are many studies in psychology journals that probe the validity and reliability of the RMET and other measures of social and emotional intelligence across settings, samples, and cultures. A recent meta-analysis finds a modest positive correlation of about 0.25 between IQ and the RMET (Baker et al. 2014). Most of the studies in this meta-analysis rely on small convenience samples.

What we do not have—and what I am hoping economists can provide—is a sense of how the RMET or other measures of social intelligence vary in a broader population. What is the correlation between social intelligence and measures of socioeconomic status such as income and parental education? Does the RMET predict life outcomes at all, and is it differentially predictive for key subgroups? These are only initial questions in what I hope is an emerging paradigm—improving the theory and measurement of noncognitive skills.

References

Baker, C. A., E. Peterson, S. Pulos, and R. A. Kirkland. 2014. "Eyes and IQ: A Meta-Analysis of the Relationship between Intelligence and Reading the Mind in the Eyes." *Intelligence* 44:78–92.

Baron-Cohen, S., S. Wheelwright, J. Hill, Y. Raste, and I. Plumb. 2001. "The Reading the Mind in the Eyes Test Revised Version: A Study with Normal Adults, and Adults with Asperger Syndrome or High-Functioning Autism." *Journal of Child Psychology and Psychiatry* 42 (2): 241–51.

Borghans, Lex, Bart H. H. Golsteyn, James Heckman, and John Eric Humphries. 2011. "Identification Problems in Personality Psychology." *Personality and Individual Differences* 51 (3): 315–20.

Deming, David J. 2017. "The Growing Importance of Social Skills in the Labor Market." *Quarterly Journal of Economics* 132 (4): 1593–640.

Heckman, James J., and Tim Kautz. 2012. "Hard Evidence on Soft Skills." *Labour Economics* 19 (4): 451–64.

John, Oliver P., and Sanjay Srivastava. 1999. "The Big Five Trait Taxonomy: History, Measurement, and Theoretical Perspectives." In *Handbook of Personality: Theory and Research*, vol. 2, edited by L. A. Pervin and O. P. John, 102–38. New York: Guilford Press.

Kautz, Tim, James J. Heckman, Ron Diris, Bas ter Weel, and Lex Borghans. 2015. "Fostering and Measuring Skills: Improving Cognitive and Non-Cognitive Skills to Promote Lifetime Success." OECD Education Working Paper no. 110, OECD Publishing.

Schmitt, D. P., J. Allik, R. R. McCrae, and V. Benet-Martínez. 2007. "The Geographic Distribution of Big Five Personality Traits: Patterns and Profiles of Human Self-Description across 56 Nations." *Journal of Cross-Cultural Psychology* 38 (2): 173–212.

West, Martin R., M. A. Kraft, A. S. Finn, R. E. Martin, A. L. Duckworth, C. F. O. Gabrieli, and J. D. E. Gabrieli. 2015. "Promise and Paradox Measuring Students' Non-Cognitive Skills and the Impact of Schooling." *Educational Evaluation and Policy Analysis.* https://doi.org/10.3102%2F0162373715597298.

Woolley, A. W., C. F. Chabris, A. Pentland, N. Hashmi, and T. W. Malone. 2010. "Evidence for a Collective Intelligence Factor in the Performance of Human Groups." *Science* 330 (6004): 686–88.

7

Wage Inequality and Cognitive Skills
Reopening the Debate

Stijn Broecke, Glenda Quintini,
and Marieke Vandeweyer

7.1 Background and Objectives

In the late 1990s and early into the twenty-first century, a brief debate raged on the importance of cognitive skills in explaining international differences in wage inequality—a debate that was never really settled. On the one hand, Blau and Kahn (1996, 2005) and Devroye and Freeman (2001) argued that differences in cognitive skills played a relatively minor role in explaining differences in wage inequality between the United States and other advanced economies while, on the other hand, Leuven, Oosterbeek, and van Ophem (2004) claimed that around one-third of the variation in relative wages between skill groups across countries could be explained by differences in the net supply of skills.

While these papers used different methodologies and, in fact, addressed slightly different issues (wage inequality versus skills wage premiums), what was really at stake was the role of the market (demand and supply) as an explanation for differences in the returns to skill versus an alternative explanation that attributes skill prices to differences in institutional setups, like

Stijn Broecke is a senior economist in the Skills and Employability division of the Directorate of Employment, Labour and Social Affairs at the Organisation for Economic Co-operation and Development (OECD). Glenda Quintini is a senior economist in the Skills and Employability division of the Directorate of Employment, Labour and Social Affairs at the OECD. Marieke Vandeweyer is a researcher at the University of Leuven and a labor market economist in the Skills and Employability division of the Directorate of Employment, Labour and Social Affairs at the OECD.

We are grateful for useful comments from the editors and Frank Levy on an earlier draft of this chapter. This work should not be reported as representing the official views of the OECD or its member countries. The opinions expressed and remaining errors are those of the authors. For acknowledgments, sources of research support, and disclosure of the authors' material financial relationships, if any, please see http://www.nber.org/chapters/c13703.ack.

the minimum wage and unionization. This mirrors a wider debate in the economic literature that has pitched the market (including the role of technological change and international trade) against institutions in explaining wage dispersion. As argued by Salverda and Checchi (2014), this literature really consists of two separate strands that, despite not being mutually exclusive, have developed in parallel with very little interaction between the two.

Since the publication of these papers, the debate on the importance of cognitive skills in explaining international differences in wage inequality has been left largely untouched. During this period, however, inequality has continued to rise. In the United States, the P90/P10 earnings ratio rose from 3.75 in 1975 to 4.59 in 1995 and to 5.22 in 2012.[1] At the same time, a growing body of evidence has demonstrated that inequality has high social costs (Krueger 2012; Pickett and Wilkinson 2011; Stiglitz 2012), and there also appears to be a growing consensus that inequality may be bad for economic growth (Ostry, Berg, and Tsangarides 2014; Cingano 2014).

Recently, with the availability of new data (the Survey of Adult Skills—PIAAC),[2] researchers have started looking again at the relationship between cognitive skills and wage inequality. Using decomposition methods identical or similar to Blau and Kahn (2005), Paccagnella (2015) and Pena (2014) also find that skills contribute very little to international differences in wage inequality, and that skills prices play a far more important role. From this, these authors conclude that differences in inequality must be driven primarily by differences in institutions—a view echoed by another recent paper (Jovicic 2015). However, neither of these studies considers the early criticisms made by Leuven, Oosterbeek, and van Ophem (2004) of the Blau and Kahn (2005) work. In particular, Leuven, Oosterbeek, and van Ophem (2004) argued that skills prices will not only reflect institutional setups but also basic market forces, and that the decomposition approach taken by Blau and Kahn (2005) ignores important dynamic aspects of the relationship between skills supply and demand that determine both the returns to skill and wage inequality.

In this chapter, we reconsider both sides of the argument, and conclude that the new wave of studies based on the PIAAC data (Jovicic 2015; Paccagnella 2015; Pena 2014) may have been too quick in dismissing the importance of cognitive skills in explaining international differences in wage inequality. First, we simulate alternative wage distributions for the United States using the methods proposed by DiNardo, Fortin, and Lemieux (1996) and Lemieux (2002, 2010) to see what would happen to wage inequality in the United States if it had (a) the skills endowments and (b) the skills prices of other PIAAC countries. Consistent with the aforementioned studies, this exercise leads us to conclude (a) that differences in skills endowments can-

1. These figures are taken from the OECD earnings database and are estimated using gross usual weekly earnings of full-time workers age sixteen and over from the Current Population Survey.

2. PIAAC stands for the Programme for the International Assessment of Adult Competencies.

not explain much of the higher wage inequality observed in the United States, and (b) that higher skills prices in the United States account for a much larger share (nearly one-third on average) of the difference in wage inequality.

However, as argued by Leuven, Oosterbeek, and van Ophem (2004), this price effect will not just reflect differences in institutions. Indeed, the higher price of skills in the United States will reflect at least two factors: (a) differences in institutions, but also (b) differences in the relative supply of, and demand for, skills. To evaluate the importance of the latter, we follow Leuven, Oosterbeek, and van Ophem (2004) and use Katz and Murphy's (1992) demand and supply model to study the relationship between the net supply of skills, on the one hand, and wage inequality, on the other. While tentative, this analysis shows that market forces do indeed matter, and that differences in the relative net supply of high- versus medium-skilled workers can account for 29 percent of the higher P90/P50 wage ratio in the United States (although the net supply of skills explains little of the higher wage inequality at the bottom of the wage distribution). We show that these findings are robust to the inclusion of labor market institutions in the set of control variables of the regression.

We also explore the extent to which higher wage inequality in the United States might be compensated for by relatively higher employment rates among the low skilled. Contrary to this "wage compression" hypothesis, and consistent with findings from Freeman and Schettkat (2001) and Jovicic (2015), we find that the employment (unemployment) rates of the low skilled are not much higher (lower) in the United States relative to those of the high skilled than they are in other countries. We also find that the ratio between the average skills levels of the employed and the unemployed is quite high in the United States which, once again, is inconsistent with the idea that higher wage inequality is the price paid for better employment outcomes for the low skilled.

The next section of this chapter describes the PIAAC data we use in our analysis, and provides a descriptive overview of wage inequality, skills endowments, and prices in the twenty-two Organisation for Economic Co-operation and Development (OECD) countries included in our sample. Section 7.3 introduces the method we employ for analyzing international differences in wage inequality and presents the results obtained. Section 7.4 covers the demand and supply analysis, and section 7.5 tests the robustness of these findings to the inclusion of labor market institutions. Section 7.6 explores the wage compression hypothesis, while section 7.7 concludes and offers some pointers for future research.

7.2   Data

The data collected by the OECD's 2012 Survey of Adult Skills (PIAAC) offers an unparalleled opportunity to investigate the relationship between cognitive skills and wage inequality. The survey directly assessed the proficiency of around 166,000 adults (age sixteen to sixty-five) from twenty-four countries[3] in literacy, numeracy, and problem solving in technology-rich environments. In addition, the survey collected information on individuals' skills use in the workplace, as well as on their labor market status, wages, education, experience, and a range of demographic characteristics. The achieved samples range from around 4,500 in Sweden to nearly 27,300 in Canada. In this chapter, the focus is on the twenty-two OECD countries in the sample (i.e., excluding the Russian Federation and Cyprus).

The direct assessment of cognitive skills in PIAAC represents a significant improvement over the more traditional skills proxies (such as years of education, qualification levels, and experience) used in many other surveys and research. Such direct measures are particularly important when doing international comparisons because a year of education, for example, will mean something very different from one country to another, partly because there are important differences in the quality of educational systems between countries. By contrast, the PIAAC assessments were deliberately designed to provide reliable measures of skills proficiency that can be compared across countries, languages, and cultures. There is also a growing body of research that has highlighted the importance of cognitive skills in determining a range of labor market outcomes, including employment and wages (e.g., OECD 2014; Hanushek et al. 2015).

It is important to point out that cognitive skills are not the same as the task-based definition of skill emerging from the literature on routine-biased technological change (see, e.g., Autor, Levy, and Murnane 2003; Autor, Katz, and Kearney 2006; Autor and Dorn 2013). While cognitive skills can be seen as characteristics of the worker and reflect his or her education and personal background, as well as a number of other factors, tasks focus on the content of occupations. There is not necessarily a one-to-one mapping between the two, and any worker with a particular skills set can perform a variety of tasks. In addition, the set of tasks performed by a worker can change in response to changes in the labor market, which are driven by technological progress, globalization, and other such trends.

The two skills concepts are nonetheless closely related. According to the routine-biased technological change hypothesis, routine tasks (i.e., those that can easily be automated) are disappearing (and with it the demand for routine skills), while the demand for nonroutine tasks and skills is rising. The concept of nonroutine skills encompasses a wide array of skills, but cognitive skills (or "key information-processing skills" as they are sometimes

---

3. Twenty-two OECD countries/regions: Australia, Austria, Canada, the Czech Republic, Denmark, Estonia, Finland, Flanders (Belgium), France, Germany, Ireland, Italy, Japan, Korea, the Netherlands, Norway, Poland, the Slovak Republic, Spain, Sweden, the United Kingdom (England and Northern Ireland), and the United States; one region; as well as two non-OECD countries: Cyprus and the Russian Federation.

referred to) form an essential part of them. These skills provide a fundamental basis for the development of other, higher-order skills, and are necessary in a broad range of contexts, including work. The close relationship between the two concepts is borne out by the data: just as there has been an increase in employment in nonroutine occupations, there has been growth in the share of employment in occupations associated with the highest levels of key information-processing skills (OECD 2013).

It is also important to point out that cognitive skills are assessed in PIAAC by focusing on the ability of individuals to perform certain tasks. For example, numeracy skills in PIAAC are defined as the ability to "access, use, interpret and communicate mathematical information and ideas in order to engage in and manage the mathematical demands of a range of situations in adult life." To this end, numeracy involves managing a situation or solving a problem in a real context, by responding to mathematical content/information/ideas represented in multiple ways (OECD 2013). Literacy and problem solving in technology-rich environments are assessed in a similar way.

Finally, while PIAAC collected information on three different cognitive skills, only numeracy skills will be used in the present chapter. This is because the three measures are highly correlated and the conclusions reached do not depend on the choice of measure.

A second strength of the present chapter is its ability to draw on detailed (and continuous) wage data for the twenty-two OECD countries/regions that are covered by PIAAC. In contrast, Leuven, Oosterbeek, and van Ophem (2004) could use only fifteen (out of twenty) countries that participated in the International Adult Literacy Survey (IALS—a predecessor of PIAAC), because wage information was only available in quintiles for the other five countries. Similarly, Blau and Kahn (2005) cite wage-data restrictions as a primary reason for focusing on just nine of the advanced countries included in IALS, while Devroye and Freeman (2001) use eleven. Even among the fifteen countries covered by Leuven, Oosterbeek, and van Ophem (2004), wage data were only available in twenty intervals for three of them (Germany, the Netherlands, and Switzerland), while it was impossible to calculate hourly wages in the case of Sweden. Finally, the more recent research using PIAAC data also suffers from similar problems. In the data used by Pena (2014), for example, continuous wage data is missing for five of the countries (including the United States), while Jovicic (2015) does not have access to continuous wage data for Austria, Canada, and Sweden.

Table 7.1 offers some basic descriptive statistics on the number of observations in PIAAC with valid wage observations, as well as on the level and dispersion of both skills and wages. The table shows that the United States combines one of the lowest levels of skill (only Spain and Italy do worse) with the highest skill dispersion (both at the top and at the bottom of the distribution). Gross hourly wages (which are expressed in PPP-corrected USD) are among the highest in the United States (although they are higher still in Ireland, Flanders,

Table 7.1   Descriptive statistics: Skills and wages by country

		Skill				Wages			
	N	Mean	P90/P10	P90/P50	P50/P10	Mean	P90/P10	P90/P50	P50/P10
Australia	4,371	276	1.60	1.21	1.32	18.90	3.14	1.90	1.65
Austria	2,943	279	1.54	1.19	1.29	19.06	3.05	1.83	1.67
Canada	16,116	271	1.66	1.22	1.35	20.37	3.94	1.94	2.03
Czech Republic	2,630	279	1.49	1.18	1.26	8.96	2.88	1.68	1.71
Denmark	4,448	286	1.52	1.19	1.28	23.84	2.58	1.55	1.66
England/N. Ireland (UK)	4,801	271	1.63	1.23	1.33	18.40	3.53	2.07	1.71
Estonia	3,972	277	1.51	1.19	1.26	9.64	4.71	2.24	2.10
Finland	3,251	292	1.51	1.19	1.26	19.30	2.54	1.70	1.50
Flanders (B)	2,736	287	1.54	1.19	1.30	22.23	2.61	1.67	1.56
France	3,696	261	1.73	1.23	1.40	15.58	2.56	1.77	1.45
Germany	3,382	278	1.60	1.20	1.33	18.82	4.22	1.88	2.25
Ireland	2,784	265	1.61	1.22	1.32	21.57	3.57	2.08	1.71
Italy	1,815	255	1.66	1.22	1.36	16.14	3.42	1.99	1.72
Japan	3,262	292	1.46	1.17	1.25	16.09	4.08	2.32	1.76
Korea	3,097	268	1.52	1.18	1.29	17.84	5.83	2.68	2.18
Netherlands	3,162	287	1.51	1.18	1.28	21.47	3.24	1.79	1.81
Norway	3,553	286	1.55	1.19	1.30	24.32	2.44	1.60	1.52
Poland	3,908	267	1.59	1.22	1.31	9.27	3.89	2.15	1.81
Slovak Republic	2,505	285	1.44	1.17	1.24	8.90	4.01	2.15	1.87
Spain	2,456	258	1.61	1.20	1.34	14.96	3.60	2.05	1.75
Sweden	2,888	287	1.55	1.19	1.30	18.68	2.18	1.59	1.37
United States	2,793	261	1.75	1.24	1.41	21.52	4.81	2.40	2.01

Notes: Skills refer to proficiency in numeracy and are expressed in score points (1 = minimum and 500 = maximum). Wage data are trimmed, by country, at the top and bottom percentiles. Wages are hourly, include bonuses, and are expressed in PPP-corrected USD.

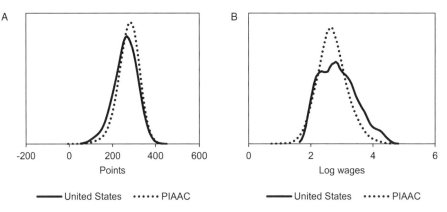

Fig. 7.1   **Skills and wage distributions, United States and PIAAC average. *A*, skill distribution; *B*, wage distribution.**
Note: Obtained by kernel density estimation.

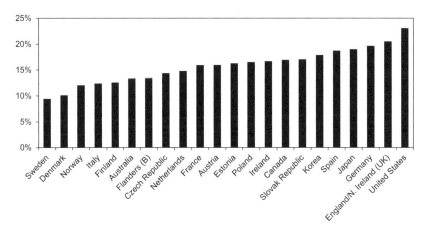

Fig. 7.2   The return to skill, United States and other PIAAC countries
Note: The figure shows the coefficient on skill from a regression of log hourly wages (including bonuses) for wage and salary earners (in PPP-corrected USD) on standardized numeracy scores and a quartic of experience.

Denmark, and Norway). Wage inequality in the United States (as measured by the P90/P10 wage ratio) is second only to Korea, and is particularly high at the top of the distribution. In contrast, Canada, Estonia, Korea, and Germany all have P50/P10 wage ratios higher than that observed in the United States. Figure 7.1 shows the full skill and wage distributions of the United States in comparison to the PIAAC average. The shapes and positions of these curves confirm the higher skills and wage inequality in the United States, as well as the lower average skill level of the employed population.

To conclude this section, figure 7.2 shows the results of a simple Mincer-type regression of log wages on skills, experience, and experience squared, and confirms that the higher return to skill in the United States might be one of the key reasons why wage inequality is so much higher. Indeed, among the twenty-two countries shown in figure 7.2, the United States is the country with the highest return to skill (more than twice as high as in Sweden and Denmark). As will be argued throughout this chapter, this higher return to skill in the United States will reflect a combination of differences in (a) the demand for and supply of skill, and (b) labor market institutions, policies, and practices.

7.3   The Role of Skills and Skills Prices

In this section, we estimate the extent to which higher wage inequality in the United States is associated with differences in (a) skills endowments, and (b) skills prices. Our method differs from those used in the previous research on wage inequality and cognitive skills, and brings a number of improvements. Both Devroye and Freeman (2001) and Jovicic (2015) use a simple variance decomposition method, which cannot account for the full distributional aspects of both wages and skills. Blau and Kahn (2005) and Pena (2014) use the Juhn, Murphy, and Pierce (1993) decomposition—but this method has become the subject of a number of criticisms over time (Yun 2009; Suen 1997; Fortin, Lemieux, and Firpo 2010).[4] Finally, Paccagnella (2015) resorts to unconditional quantile regressions (Fortin, Lemieux, and Firpo 2010), but his application of the method only allows an analysis of the effect of overall, average skill levels (and not the entire skills distribution) on wage inequality. Instead, we draw on DiNardo, Fortin, and Lemieux (1996) and Lemieux (2002, 2010) and simulate counterfactual wage distributions using reweighting techniques. As will be shown below, an important attraction of this method lies in its simplicity and the visual inspection of alternative wage distributions that it permits.

While we believe that our approach offers some improvement over previous methods used in the literature, the conclusions we reach in this section are essentially the same as those reached by other authors—that is, that differences in skills endowments across countries can account for little of the difference in wage inequality, while differences in skills prices (or how skills are rewarded) appear to play a far more important role. We begin this section by explaining our methodology in some more detail, and then present the results.

---

4. One of the main criticisms of the Juhn, Murphy, and Pierce decomposition concerns the "residual imputation" step. In this step, the residuals of the base country are replaced with the similarly ranked residuals of the comparator country. However, a key assumption behind this approach is that these residuals (from a regression of wages on skills) are independent of skills, which is clearly unrealistic. For further detail, see Fortin, Lemieux, and Firpo (2010).

7.3.1   Simulating Counterfactual Wage Distributions

To estimate the contributions of skills prices and skills endowments to higher wage inequality in the United States, we will estimate two sets of alternative wage distributions. In the first, we impose the skills distributions of the other PIAAC countries onto the United States (holding skills prices constant). In the second, we impose the skills prices of the other PIAAC countries onto the United States (this time holding skills endowments constant).

7.3.2   The Effect of Skill Endowments

To see what would happen to wage inequality in the United States if it had the same skills distribution as the other PIAAC countries, we reweight the United States data to make the skills profile of its workforce resemble that of the comparator country. We then estimate the difference this makes to wage inequality. Intuitively, if the comparator country has more skilled workers, then the reweighting method will give more weight to skilled workers in the United States, while reducing the weight given to less skilled ones. Because the other characteristics of the individuals are left unchanged (including their wages), this results in an alternative wage distribution. This alternative wage distribution can then be used to calculate standard measures of wage inequality that can be compared to those estimated on the original wage distribution. The difference between the two measures of wage inequality can be attributed to the difference in skills endowments.

More formally, assume one is interested in seeing what would happen to the wage distribution of the United States (US) if it had the same skills distribution as country x. Then, taking an individual i in the United States, the original sample weights $\omega_{i,\text{US}}$ for that individual are replaced by a counterfactual weight $\omega'_{i,\text{US}} = \omega_{i,\text{US}}\Psi_i$ where Ψ_i represents the reweighting factor. While DiNardo, Fortin, and Lemieux (1996) suggest regression methods to compute the reweighting factor Ψ_i, the latter may be obtained more simply and nonparametrically if the data can be divided up in a finite number of cells (Lemieux 2002). In the case of skills, this is indeed possible.

In practice, the procedure is implemented as follows. The data for the United States and the comparator country are divided into skill cells/intervals s of 5 points each,[5] and the shares of the total workforce employed in each cell, $\theta_{s,\text{US}}$ and $\theta_{s,x}$, are calculated. One can then reweight the US data to approximate the skills distribution of the comparator country by simply using the following reweighting factor:

$$\Psi_i = \frac{\theta_{s,x}}{\theta_{s,\text{US}}}.$$

---

5. Except for individuals at the top (more than 355 points) and bottom (fewer than 180 points) of the distribution. These are put into two separate groups.

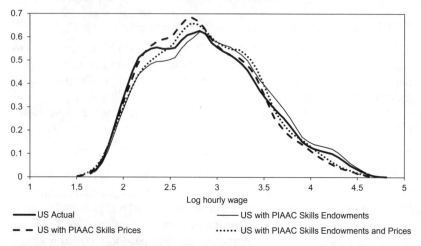

——— US Actual   ——— US with PIAAC Skills Endowments

– – US with PIAAC Skills Prices   ⋯⋯⋯ US with PIAAC Skills Endowments and Prices

Fig. 7.3   Simulating alternative wage distributions in the United States based on PIAAC skills endowments and prices

7.3.3   The Effect of Skill Prices

The price effect simulations are inspired by a method proposed by Lemieux (2002). Intuitively, we give individuals with a certain skill level in the United States the same return to skill as individuals with that skill level would obtain in country x. More formally: assuming that the data can be divided up in a finite number of cells (e.g., intervals s of 5 numeracy points each), then changes in skill prices can be simulated by comparing the mean of (log) wages of skill group s in the United States, $y_{s,\mathrm{US}}$, with the mean of (log) wages in skill group s in country x, $y_{s,x}$. The new (log) wage for each individual i in the United States, $y'_{i,\mathrm{US}}$, can then be calculated by adding the difference between country x's average (log) wage for skill group s and the average (log) wage for skill group s in the United States:

$$y'_{i,\mathrm{US}} = y_{i,\mathrm{US}} + (y_{s,x} - y_{s,\mathrm{US}}).$$

Price and quantity effects may of course be applied simultaneously to obtain a joint effect on the wage distribution. The order in which these effects are calculated does not affect the outcome, since both are calculated within the same skill cell.

Figure 7.3 illustrates the effect on the United States wage distribution of (a) adopting the skills distribution of the average PIAAC country,[6]

6. The average PIAAC country is constructed on the basis of all PIAAC observations. However, because countries with larger populations would have a greater weight and, therefore, a disproportionate influence on the distribution, the survey weights are rescaled so that the sum of each country's weights is equal to one. In essence, this is equivalent to taking an unweighted average across countries. In addition, because wage levels differ significantly across countries,

(b) adopting the skills prices of the average PIAAC country, and (c) adopting both the skills distribution and prices of the average PIAAC country simultaneously. As the figure shows, imposing the skills distribution of the average PIAAC country onto the United States would change the wage distribution somewhat, but would have relatively little effect on wage inequality (as indicated by the height of the distribution). Imposing skills prices of the average PIAAC country would, however, have a more important compressing effect on the wage distribution. Similarly, imposing both the skills distribution and prices of the average PIAAC country onto the United States would lead to a fall in wage inequality.

Table 7.2 contains the full set of results from our analysis.[7] The first set of columns shows the impact on wage inequality in the United States if it adopted the skills distribution of the comparator country. It essentially confirms the findings of earlier papers (e.g., Blau and Kahn 2005) that the contribution of cognitive skills to explaining higher wage inequality in the United States is small. One difference is that the earlier analysis had found that the contribution of skills was positive (ranging from 3 percent to 13 percent on average), while table 7.2 indicates that, in most cases, the contribution is actually negative—that is, that the P90/P10 wage ratio in the United States would increase if it had the skills distribution of the comparator country (the estimates suggest that it would be around 10 percent higher on average). Only if the United States had the skills distribution of France, Poland, Ireland, Italy, and Spain would wage inequality fall.

While surprising, these results are consistent with the recent findings of Paccagnella (2015), who finds that average skills levels in the United States can account for −4 percent, on average, of the higher P90/P10 wage ratio in the United States (although the author controls for educational attainment in addition to skills, which is likely to explain the lower estimate). Again, similar to Paccagnella (2015), table 7.2 suggests that these negative effects are driven primarily by the P50/P10 wage ratio (i.e., the bottom of the wage distribution).[8] These counterintuitive results can be explained by the skills profile of wages in the United States, which is significantly steeper in the top half of the skills distribution. Because skills prices are held constant in the analysis, increasing the number of skilled workers in the United States mechanically results in higher wage inequality as the wages of those at the P50 of the wage distribution would increase faster than the wages of those at the P10.

---

they need to be adjusted before being combined into a single PIAAC distribution (which would otherwise be too wide). Wages are therefore demeaned by country, and all the analysis is carried out on these country-specific deviations from the mean.

7. The full set of figures associated with these simulations can be found in appendix figure 7A.1.

8. Blau and Kahn (2005) also find some negative effects, but these are at the top of the wage distribution (P90/P50), and for males.

Table 7.2   The role of skills endowments and skills prices in explaining higher wage inequality in the United States

	(1) Skills endowments			(2) Skills prices			(3) Skills endowments and prices		
	P90/P10	P90/P50	P50/P10	P90/P10	P90/P50	P50/P10	P90/P10	P90/P50	P50/P10
Australia	-9.5	4.9	-24.8	35.0	37.8	26.3	23.7	36.1	4.3
Austria	-5.6	12.4	-30.8	25.9	27.3	18.8	15.9	29.2	-7.3
Canada	-19.7	0.7	397.1	55.3	33.5	-420.3	40.4	32.6	-132.6
Czech Republic	-1.5	12.9	-31.3	18.1	22.6	3.7	16.5	27.0	-10.6
Denmark	-11.4	3.9	-39.2	30.0	26.7	28.7	15.1	23.4	-8.1
England/N. Ireland (UK)	-10.5	3.5	-22.0	18.1	21.3	13.1	11.4	27.2	-5.0
Estonia	-33.8	55.1	97.6	309.1	68.6	-53.4	260.6	96.6	12.7
Finland	-18.6	-0.5	-34.0	27.5	32.2	15.4	13.0	29.4	-10.6
Flanders (B)	-13.1	2.7	-30.7	27.6	29.3	18.5	15.4	27.6	-6.5
France	2.9	4.2	1.0	25.7	29.1	17.0	26.4	30.7	16.8
Germany	-27.9	7.9	44.2	63.7	25.1	-21.5	30.8	23.1	10.7
Ireland	8.3	16.1	0.4	30.1	46.2	12.6	32.8	55.0	9.3
Italy	14.8	17.5	9.2	36.3	42.1	24.8	49.1	56.7	34.8
Japan	-50.7	31.7	-70.0	43.7	149.8	17.1	-11.7	113.2	-43.6
Korea	-13.2	-39.4	22.5	-31.0	-43.1	-18.9	-42.4	-69.9	-9.9
Netherlands	-17.9	6.2	-74.8	34.5	31.4	36.4	15.4	29.9	-26.7
Norway	-12.9	1.1	-28.1	25.9	25.7	19.2	11.5	20.8	-5.6
Poland	7.7	27.2	-13.8	41.8	53.7	26.6	41.9	64.8	14.7
Slovak Republic	-11.9	40.4	-91.9	27.5	38.0	9.7	13.5	58.9	-58.7
Spain	19.9	26.9	9.5	24.9	30.3	16.0	40.2	52.1	22.8
Sweden	-12.4	2.1	-23.9	28.0	30.2	18.0	15.7	27.3	-2.1

Notes: The table shows the proportion of higher wage inequality in the United States that can be attributed to differences in (1) the skills distribution, (2) skills prices, and (3) the skills distribution and skills prices together. For example, 4.9 percent of the difference in the P90/P50 ratio between the United States and Australia can be explained by differences in the skills distribution between the two countries. Negative values indicate that the difference in wage inequality between the United States and the comparator country would increase if the United States had the characteristics of the comparator country.

The difference between our results and those of Blau and Kahn (2005) may be driven by the different methodology that we use. When we apply Blau and Kahn's (2005) methodology to the PIAAC data, we still find that the contribution of skills is small and negative on average (−7.9 percent)—however, this result is driven primarily by a large negative effect for Estonia.[9] Excluding this country, we find that the contribution of skills is still small, but positive (4.5 percent, on average)—as in Blau and Kahn (2005). However, to confirm that this difference in results is truly driven by the difference in methodology, one would also want to run the experiment the other way around, and use our methodology on the IALS data. Unfortunately this was not possible because access to the detailed IALS wage data is restricted for the United States and we were unable to obtain access to these. One cannot rule out, therefore, that some of the difference between our results and those of Blau and Kahn (2005) is also driven by (a) a real change over time in the role that skills play in explaining higher wage inequality in the United States, and (b) the country coverage in PIAAC, which is different from the one of IALS.

On the whole, however, the most important conclusion that emerges from the above analysis is that, despite our different (and, we believe, improved) methodology, our findings are largely consistent with those of Blau and Kahn (2005)—that is, differences in skills endowments across countries cannot account for much of the differences in wage inequality.

The second set of findings presented in table 7.2 are also consistent with both Blau and Kahn (2005) and Paccagnella (2015)—that is, skills prices can account for a significantly larger share of higher wage inequality in the United States than can skills endowments. The contribution of skills prices ranges from 18 percent in the Czech Republic to nearly 64 percent in Germany, and can explain nearly one-third, on average, of the higher wage inequality in the United States (excluding both Estonia and Korea, two clear outliers). Skills prices also tend to play a slightly more important role in explaining wage inequality at the top than at the bottom of the wage distribution: this is the case in eighteen of the twenty-one country comparisons shown in table 7.2.

While Blau and Kahn (2005) at least acknowledged the possibility that higher skills prices could reflect market forces as well as differences in institutions, the more recent research using PIAAC simply ignores this argument. Paccagnella (2015) concludes that the greater contribution of skills prices to wage inequality "suggests that economic institutions [. . .] are the main determinants of wage inequality," but without actually proving this point. Similarly, Pena (2014) somewhat hastily concludes that institutional factors are more important than market forces, but she only "controls" for the latter by including additional demographic factors in her model. Finally, Jovicic

9. These results are not shown, but are available from the authors upon request.

(2015) presents a few simple correlations between labor market institutions and measures of wage inequality (all of which are significant and have the "right" sign), and concludes from this that "institutions have more power" in explaining international differences in wage inequality than skills do.

We will return to the importance of market forces in explaining higher inequality in the next section of this chapter. Before we do so, the final three columns in table 7.2 show the combined effect of skills and skills prices in explaining higher wage inequality in the United States. Only in the cases of Korea and Japan do these explain a negative part of the difference in wage inequality with the United States. In the other countries, the joint contribution of skills and skills prices ranges from 11.4 percent in the case of England/Northern Ireland to 49 percent in the case of Italy (excluding Estonia, which is a clear outlier). These results are not surprising given that they combine the modest, negative effects of skills endowments with the larger, positive effects of skills prices.

7.4   The Role of Demand and Supply

One weakness of the wage simulation method used above (but which applies equally to the methods used by Devroye and Freeman [2001], Blau and Kahn [2005], Jovicic [2015], Paccagnella [2015], and Pena [2014]) is that it analyzes the role of skills from a static perspective. However, as pointed out by Leuven, Oosterbeek, and van Ophem (2004), this is not realistic and the price of skills should be seen as reflecting at least in part the outcome of the dynamic interaction between demand and supply: if the supply of skills increases relative to demand, then one would expect both the price of skills and inequality to fall.

The idea that the returns to skill (and therefore inequality) depend on demand and supply factors was first introduced by Tinbergen (1975), who famously described inequality as a "race between education and technology." Technological change was argued to be skills biased—that is, it increases the demand for more skilled workers and therefore their wage premium in the labor market. To keep inequality in check, the supply of skills needs to increase to meet that demand. It is now widely accepted that the increase in inequality in the United States over the past few decades can be partly blamed on the fact that the supply of educated workers has not kept pace with the rise in demand for them (Juhn, Murphy, and Pierce 1993; Juhn 1999; Goldin and Katz 2008; Autor 2014). While more recent theories of routine-biased technological change have refined this argument somewhat, they still maintain a central role for skills in explaining rising wage inequality in the United States (Autor, Levy, and Murnane 2003; Autor, Katz, and Kearney 2006, 2008; Autor and Dorn 2013; Autor 2015).

The findings from the previous section, and the results obtained by Blau and Kahn (2005) and Paccagnella (2015), among others, therefore appear at odds with the story that rising wage inequality in the United States was to a large extent related to changes in the demand for, and the supply of, skills. One possible explanation for this inconsistency is that the decomposition methods used in the literature fail to account for the dynamic interaction between the demand and supply of skills. To gain a better understanding of how the supply of skills interacts with the demand for skills and what effect this may have on wage inequality (through its effect on the price of skills), this section applies a different methodology developed by Katz and Murphy (1992) and used by a number of researchers since to investigate the relationship between the net supply of cognitive skills and wage differentials between skill groups (Blau and Kahn 1996; Leuven, Oosterbeek, and van Ophem 2004). The only difference is that, instead of looking at wage differentials between skill groups, the analysis that follows focuses on standard, interdecile measures of wage inequality.

To implement the Katz and Murphy (1992) methodology, we follow an approach similar to both Blau and Kahn (1996) and Leuven, Oosterbeek, and van Ophem (2004). In a first step, the workforce of the average PIAAC country is divided into three skills groups of equal size corresponding to the low, medium, and high skilled, respectively. The thresholds defined by these groups (in numeracy points) are then applied to each of the twenty-two countries included in the sample to classify workers as either low, medium, or high skilled. Because the distribution of skills varies from country to country, applying these PIAAC average thresholds will result in different-sized groups of low-, medium-, and high-skilled workers in each one of these countries. For example, table 7.3 shows that in Japan, 47.4 percent of the working-age population is high skilled according to this definition, but that in both Italy and Spain more than 50 percent is low skilled. Equally, the workforce in the United States is relatively low skilled, with 45.8 percent low-skilled workers and only 24.6 percent high-skilled workers.

The next step is to construct indices that measure how the demand and supply for each skill group in the United States compare to those in the other PIAAC countries. We start by building a supply index $\text{Supply}_{s,x}$, which intends to measure the relative supply of skills group s in the United States compared to country x:

$$\text{Supply}_{s,x} = \ln \frac{\varepsilon_{s,\text{US}}}{\varepsilon_{s,x}}$$

where $\varepsilon_{s,x}$ and $\varepsilon_{s,\text{US}}$ are the shares of the labor force accounted for by skill group s in country x and the United States, respectively (as reported in table 7.3). Intuitively, the supply index compares the relative importance of each skill group in the United States labor force with country x's shares used as the norm.

Table 7.3   Proportion of high-, medium-, and low-skilled individuals in the labor force by country (percent)

Country	Low	Medium	High
Australia	34.3	32.2	33.4
Austria	28.0	35.2	36.9
Canada	36.5	31.4	32.0
Czech Republic	26.7	38.1	35.2
Denmark	26.8	32.7	40.5
England/N. Ireland (UK)	39.7	31.1	29.2
Estonia	28.8	37.9	33.3
Finland	24.7	31.7	43.6
Flanders (B)	25.7	32.1	42.2
France	43.8	31.0	25.2
Germany	31.9	31.5	36.6
Ireland	42.5	34.1	23.4
Italy	50.9	31.4	17.7
Japan	18.7	33.9	47.4
Korea	35.5	38.7	25.8
Netherlands	24.6	32.1	43.2
Norway	26.3	32.1	41.7
Poland	40.0	34.5	25.4
Slovak Republic	26.1	36.3	37.6
Spain	50.1	33.1	16.8
Sweden	25.6	32.3	42.1
United States	45.8	29.6	24.6

We then build a demand index $\text{Demand}_{s,x}$, which measures the degree to which the industry-occupation structure[10] in the United States favors skill group s in comparison to country x:

$$\text{Demand}_{s,x} = \ln\left(1 + \sum_o \frac{\theta_{s,o,x}}{\varepsilon_{s,x}}(\theta_{o,\text{US}} - \theta_{o,x})\right)$$

where $\theta_{s,o,x}$ is skill group s's share of employment in industry-occupation cell o in country x; $\theta_{o,x}$ and $\theta_{o,\text{US}}$ are the total shares of employment in cell o in country x and the United States, respectively, and $\varepsilon_{s,x}$ is the share of skill group s in the total workforce of country x. The demand index therefore represents the average difference in the employment shares of each industry/occupation between the United States and the comparator country—weighted by the skill intensity of each industry/occupation relative to the overall skill intensity in the comparator country.[11] If employment in the United States were strongly concentrated in industry/occupation cells that employ a large share of skilled workers compared to country x, the demand

10. Industry-occupation cells are defined in the same way as in Blau and Kahn (1996) and Leuven, Oosterbeek, and van Ophem (2004).
11. Country x is chosen to calculate these weights. This is an arbitrary choice, with no effect on the results.

Table 7.4   Difference in the demand for high-, medium-, and low-skilled workers
                   between the United States and other PIAAC countries

	Low	Medium	High
PIAAC	−0.026	−0.012	0.037
Australia	−0.013	−0.013	0.026
Austria	−0.031	−0.003	0.024
Canada	0.044	−0.012	−0.042
Czech Republic	−0.133	−0.019	0.109
Denmark	0.002	−0.003	0.001
England	−0.048	−0.008	0.071
Estonia	−0.052	0.000	0.042
Finland	−0.027	−0.017	0.026
Flanders	0.022	−0.008	−0.007
France	−0.072	0.010	0.103
Germany	−0.048	−0.030	0.063
Ireland	−0.049	−0.004	0.094
Italy	−0.092	0.023	0.205
Japan	−0.086	−0.015	0.039
Korea	−0.138	0.009	0.155
Netherlands	0.058	−0.003	−0.031
Norway	0.019	−0.013	−0.001
Poland	−0.102	−0.010	0.156
Slovakia	−0.055	0.014	0.025
Spain	−0.094	0.024	0.213
Sweden	−0.007	−0.009	0.011

index would be high (and vice versa).[12] Table 7.4 shows the difference between the demand index for the United States and every other country, and for each of the three skills groups. It shows clearly that the demand for high-skilled workers in the United States is higher than in most other countries, while the demand for low-skilled workers is lower. To some extent, this is driven by the industry-occupation structure of employment in the United States. Indeed, when we look at employment shares by industry-occupation in the United States compared to those of the other countries included in PIAAC (appendix table 7A.1), we notice that demand in the United States is relatively high in some high-skill industry/occupation combinations (e.g., managers and professionals in government and in finance, insurance, real estate, and services). By contrast, the employment share of craft workers, operatives, labor and service workers is relatively low in the United States.

In the final step, because market forces reflect the interaction between supply and demand, a "net supply" index is calculated by subtracting the demand index from the supply index:

12. This demand index implicitly assumes that the demand for labor is a derived demand reflecting the composition of output by industry and occupation. It therefore treats output as an intermediate product.

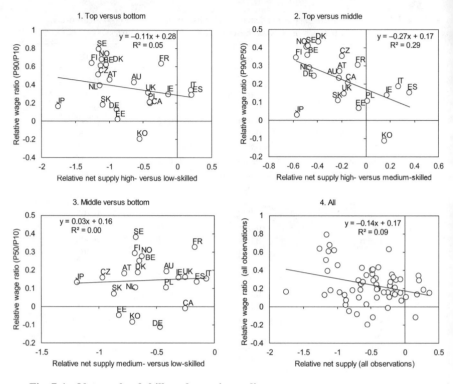

Fig. 7.4   Net supply of skills and wage inequality

$$\overline{\text{Supply}}_{s,x} = \text{Supply}_{s,x} - \text{Demand}_{s,x}.$$

The hypothesis we then want to test is whether differences across countries in the relative net supply of skills ($\overline{\text{Supply}}_{s,x} - \overline{\text{Supply}}_{s,x}$) can explain cross-country differences in wage inequality (as measured by interdecile wage ratios). Intuitively, the larger the supply of skill group s relative to demand in the United States compared to country x, the worse off we expect skill group s to be in the United States compared to country x. For example, if the net supply of high- relative to low-skilled workers is lower in the United States than it is in Sweden, then we would expect to see higher wage inequality in the United States than in Sweden. Indeed, juxtaposing the information from tables 7.3 and 7.4, we see that the United States combines a low supply of high-skilled workers with a high demand for such workers, while in Sweden the high demand for high-skilled workers is matched by a high supply—which would help explain why inequality is higher in the United States. While there are other countries with a low supply of high-skilled workers (e.g., Italy and Spain), these countries also have a low demand for high-skilled workers and, therefore, lower wage inequality than the United States.

The relationship between the relative net supply of skills and wage inequality is shown in graphical form in figure 7.4. The first graph plots the relationship between the relative net supply of high- versus low-skilled workers, on the one hand, and the P90/P10 wage ratio on the other. Each observation shows the extent to which the United States differs with respect to that particular country. Taking Sweden as an example again, the graph confirms that the United States has a much lower relative net supply of high- versus low-skilled workers, as well as a significantly higher P90/P10 wage ratio. While the relationship is negative overall, it is not particularly strong: only 5 percent of higher wage inequality in the United States can be explained by the higher net supply of skilled workers in other countries.

The second graph in figure 7.4 shows that the relationship is much stronger at the top of the wage distribution: the higher relative net supply of high- versus medium-skilled workers in other countries accounts for 29 percent of the higher P90/P50 ratio in the United States. The effect size is also quite large: a 1 percent increase in the relative net supply of high-skilled workers in the United States would reduce the top-half wage inequality by 0.27 percent. By contrast, the third graph shows that the net supply of skills explains nothing of the higher wage inequality at the bottom of the wage distribution (P50/P10).[13] Finally, the fourth graph combines all the observations of the previous three graphs and shows that, overall, differences in the relative net supply of skills can explain 9 percent of differences in wage inequality between the United States and other countries.[14]

13. Leuven, Oosterbeek, and van Ophem (2004) found that differences in the relative net supply of skills could account for 58 percent of the cross-country variance in skills premiums between medium- and low-skilled workers, and 44 percent in the case of high- versus low-skilled workers. There are some important differences between our analysis and that of Leuven, Oosterbeek, and van Ophem (2014). The first of these is that we focus on wage inequality while they look at relative skills premiums. The second difference lies in the fact that we define our skills groups using "absolute" thresholds based on the PIAAC average, while they define them relative to one specific country. Because their approach means that the results are sensitive to the choice of reference country, they repeat the analysis as many times as there are countries in their sample. This boosts their sample size which, in turn, increases their *R*-squared. When we repeat our analysis to replicate exactly the methodology used by Leuven, Oosterbeek, and van Ophem (2004), we find that the relative net supply of skills can explain 19 percent of the cross-country variance between medium- and low-skilled workers and 22 percent in the case of high- versus low-skilled workers. These estimates are considerably lower than those found by Leuven, Oosterbeek, and van Ophem (2014). It is difficult to say whether the difference represents a real change over time in the relationship between net skills supplies and relative wages of skills groups, or whether it can be explained by the difference in samples. Countries included in their sample but not in ours are Chile, Hungary, Slovenia, and Switzerland. Conversely, countries included in our sample, but not in theirs, are Australia, Austria, England/Northern Ireland, Estonia, Flanders, France, Ireland, Japan, Korea, the Slovak Republic, and Spain.

14. Blau and Kahn (1996) also carry out a demand and supply analysis to quantify the extent to which higher wage inequality in the United States could be explained by differences in the relative supply of, and demand for, educated workers—but they conclude that market forces appear to have little explanatory power. However, Blau and Kahn (1996) derive workers' skill levels simply from the number of years of schooling and work experience, and Leuven, Oosterbeek, and van Ophem (2004) show that the Blau and Kahn (1996) results change substantially once more direct measures of skills are used.

7.5   Controlling for Institutional Characteristics

The previous analysis demonstrated that the demand and supply of skills appear to be correlated with wage inequality. However, one may argue that this correlation is, in fact, driven by differences in labor market institutions that happen to be correlated with differences in skills demand and supply. To test for the robustness of the findings obtained in the previous section, we therefore run a series of regressions identical to those reported in figure 7.4, but add controls for labor market institutions, policies, and practices as well. The results from this analysis are reported in table 7.5. The first column of each panel simply reproduces the regressions from figure 7.4, which shows that a significant portion of the difference in top-half wage inequality between the United States and other countries can be explained by differences in the net supply of high- versus medium-level skills, but that skills do not appear to explain the higher inequality in the United States in the bottom half of the wage distribution.

In subsequent columns, we include a series of controls for labor market institutions, policies, and practices:[15] the level at which statutory minimum wages are set (with a dummy to control for countries that do not have a statutory minimum wage), the strictness of employment protection legislation, the bargaining coverage rate, the size of the public sector, and the generosity of unemployment benefits. In the final column, all controls are added simultaneously.

All the aforementioned institutions could be argued to reduce wage inequality, either directly or indirectly. The impact of statutory minimum wages is perhaps the most obvious one, as they directly boost the wages of workers at the bottom of the distribution.[16] Even in countries with no statutory minimum wage, a large part of the workforce is covered by wage floors specified in sector- and/or occupation-level collective agreements which, in combination with high collective bargaining coverage, are a functional equivalent of a binding minimum wage (Garnero, Kampelmann, and Rycx 2015). Wage inequality could therefore be expected to be lower in countries with higher bargaining coverage.[17] Strict employment protection legislation might have a more indirect effect by reducing employment overall, and of low-skilled, low-wage workers in particular. Because wages paid to low-skilled workers in the public sector may be higher than those that would be dictated by the market, the size of the public sector may also be inversely related with wage inequality. Finally, generous unemployment benefits may raise the reservation wages of the unemployed to the extent that low-skilled

15. These institutional controls are added one at the time to avoid issues of collinearity.
16. See DiNardo, Fortin, and Lemieux (1996), Lee (1999), and Autor, Manning, and Smith (2016) for evidence of the link between minimum wages and inequality in the United States.
17. See Blau and Kahn (1996), DiNardo, Fortin, and Lemieux (1996), and Firpo, Fortin, and Lemieux (2011) for the impact of falling union coverage on wage inequality in the United States.

Table 7.5   Net supply of skills, wage inequality, and labor market institutions

	(1)	(2)	(3)	(4)	(5)	(6)	(7)
	Panel A Dependent variable: P90/P10 (in logs, relative to the United States)						
Net supply of skills (high vs. low)	-0.111	-0.143*	-0.104	-0.121**	-0.110**	-0.131*	-0.138***
	(0.093)	(0.075)	(0.084)	(0.044)	(0.050)	(0.071)	(0.036)
Statutory minimum wage (MW dummy)[a]		-0.384**					0.003
		(0.146)					(0.149)
Level of minimum wage[b] × MW dummy		-0.987**					-0.198
		(0.456)					(0.405)
Employment protection legislation[c]			-0.377*				0.038
			(0.206)				(0.195)
Bargaining coverage				-0.306***			-0.242***
				(0.035)			(0.069)
Size of public sector[d]					-0.415***		-0.209
					(0.074)		(0.128)
Generosity of unemployment benefits[e]						-0.482***	0.041
						(0.135)	(0.114)
Constant	0.284***	0.395***	0.026	-0.134**	0.264***	0.097	-0.057
	(0.075)	(0.096)	(0.150)	(0.057)	(0.051)	(0.084)	(0.149)
N	21	21	21	21	21	21	21
R^2	0.053	0.427	0.133	0.712	0.571	0.289	0.818
Adjusted R^2	0.003	0.325	0.037	0.68	0.523	0.209	0.721

(continued)

Table 7.5 (continued)

	(1)	(2)	(3)	(4)	(5)	(6)	(7)
Panel B. Dependent variable: P90/P50 (in logs, relative to the United States)							
Net supply of skills (high vs. medium)	-0.270**	-0.198*	-0.263**	-0.179***	-0.187***	-0.250**	-0.163***
	(0.11)	(0.095)	(0.101)	(0.042)	(0.063)	(0.100)	(0.031)
Statutory minimum wage (MW dummy)[a]		-0.178*					0.039
		(0.088)					(0.075)
Level of minimum wage[b] × MW dummy		-0.325					0.136
		(0.285)					(0.226)
Employment protection legislation[c]			-0.237**				-0.029
			(0.110)				(0.109)
Bargaining coverage				-0.161***			-0.123***
				(0.019)			(0.029)
Size of public sector[d]					-0.207***		-0.105**
					(0.045)		(0.048)
Generosity of unemployment benefits[e]						-0.283***	-0.054
						(0.080)	(0.047)
Constant	0.170***	0.263***	0.006	-0.027	0.176***	0.073*	-0.028
	(0.032)	(0.039)	(0.087)	(0.027)	(0.024)	(0.036)	(0.088)
N	21	21	21	21	21	21	21
R²	0.288	0.507	0.385	0.812	0.657	0.539	0.889
Adjusted R²	0.25	0.42	0.317	0.792	0.619	0.488	0.829
Panel C. Dependent variable: P50/P10 (in logs, relative to the United States)							
Net supply of skills (medium vs. low)	0.027	-0.105	0.03	-0.053	-0.038	-0.003	-0.14
	(0.070)	(0.084)	(0.066)	(0.066)	(0.086)	(0.079)	(0.080)
Statutory minimum wage (MW dummy)[a]		-0.185**					-0.033
		(0.081)					(0.096)
Level of minimum wage[b] × MW dummy		-0.605**					-0.337
		(0.227)					(0.213)

	(1)	(2)	(3)	(4)	(5)	(6)
Employment protection legislation[c]	0.160***					0.067
	(0.048)					(0.185)
Bargaining coverage		-0.135	-0.131***			-0.119*
		(0.157)	(0.029)			(0.062)
Size of public sector[d]				-0.182***		-0.101
				(0.058)		(0.087)
Generosity of unemployment benefits[e]					-0.153	0.091
					(0.099)	(0.097)
Constant	0.135	0.067	-0.061	0.114**	0.089	-0.038
	(0.079)	(0.097)	(0.057)	(0.050)	(0.061)	(0.120)
N	21	21	21	21	21	21
R^2	0.29	0.042	0.411	0.347	0.087	0.58
Adjusted R^2	0.165	-0.065	0.346	0.274	-0.015	0.354

Source: OECD Statistics for EPL and minimum wage (2012); ICTWSS version 4 for bargaining coverage (latest available); PIAAC for share permanent and part time (2012); OECD Government at a Glance, 2013 (2011, 2010 for Germany, Ireland, Norway, Sweden, and the United Kingdom); OECD Tax-Benefit Model for unemployment benefits (2012).

Notes: All variables are relative to the United States (and in logs). Robust SE in parentheses.

[a] Dummy variable indicating countries that have a minimum wage. Countries that do not have a minimum wage are Finland, Sweden, Norway, Denmark, Germany, Austria, and Italy.

[b] Minimum wage relative to median wage of full-time workers.

[c] Strictness of employment protection legislation —individual and collective dismissal (regular contracts).

[d] Employment in general government as a percentage of the labor force.

[e] Net replacement ratio (NRR), which is defined as the average of the net unemployment benefit (including SA and cash housing assistance) replacement rates for two earnings levels, three family situations, and sixty months of unemployment.

***Significant at the 1 percent level.

**Significant at the 5 percent level.

*Significant at the 10 percent level.

workers decide not to work for low wages, indirectly compressing the wage distribution. Further details about the construction of the variables can be found in the notes to table 7.5.

The results show that the relative net supply of high- versus medium-level skills (panel B) always remains significant in explaining higher wage inequality in the United States, regardless of which institutional control is included in the regression. By contrast, the relative net supply of medium-versus low-skilled workers is never statistically significant (panel C). In panel A, which reports the results for the P90/P10 wage ratio, the coefficient of the skills variable is insignificant in the regression without institutional controls, but it turns statistically significant in most of the regressions with institutional controls. This suggests that differences in the net supply of skills can explain differences in the 90–10 gap within countries with similar institutional setups.

Overall, this robustness check corroborates the previous conclusion that the supply of skills seems to matter for wage inequality, particularly at the top of the wage distribution. All the institutional controls also have the expected, negative impact on inequality. However, it is worth repeating that, based on the analysis presented here, these relationships cannot necessarily be interpreted as causal. As mentioned above, there is a high degree of collinearity between the institutional variables. Indeed, institutions within a country do not evolve in isolation, and one would therefore expect a high degree of interdependence between them. Also, the analysis treats policies as exogenous factors affecting inequality, but there may be reason to be concerned by endogeneity: institutions may be introduced or adjusted in response to changes in inequality. Given that data are only available for one point in time, we cannot include country fixed effects and country-level institutions at the same time in the regression model. The results from these regressions should therefore not be interpreted as causal links, but rather as interesting statistical correlations.

7.6   Wage Compression and Employment Effects

So far, we have shown that wage inequality is significantly higher in the United States than it is in most other OECD countries. We have also argued that differences in skills are likely to play some role in explaining this higher wage inequality. However, skills could only explain part of the gap and, as seen in section 7.5, labor market policies and institutions also have a compressing effect on the wage distribution. One key mechanism through which they achieve this is by artificially raising the wages of those at the bottom of the distribution, possibly above the level that would arise under free market conditions. By looking at wages alone, we may therefore be ignoring another important aspect of inequality, which is inequality in employment outcomes. Indeed, in countries with stronger labor market institutions wage

Table 7.6   Employment and unemployment rates by skill group and country (percent)

	Employment rate			Unemployment rate		
	Low skilled	Medium skilled	High skilled	Low skilled	Medium skilled	High skilled
Australia	61.8	76.8	81.9	8.0	4.8	5.0
Austria	64.0	72.7	81.2	5.9	4.8	3.3
Canada	66.3	78.4	84.3	8.4	4.8	3.5
Czech Republic	56.1	65.0	73.4	10.8	7.2	3.7
Denmark	57.0	73.9	83.8	9.7	7.8	3.7
England/N. Ireland (UK)	59.9	74.4	81.7	13.6	6.8	3.6
Estonia	60.9	71.8	81.6	12.4	8.5	3.8
Finland	54.1	70.8	78.5	10.1	5.9	4.4
Flanders (B)	56.4	70.0	78.2	4.0	2.8	2.4
France	57.0	65.6	73.9	11.8	9.2	5.6
Germany	63.2	77.6	84.1	9.6	4.8	2.6
Ireland	51.6	64.5	74.2	17.6	12.0	7.9
Italy	48.7	59.2	73.6	17.5	12.9	7.2
Japan	65.7	70.1	76.4	1.9	3.6	2.4
Korea	66.7	68.1	67.2	4.4	3.7	4.2
Netherlands	60.7	75.9	84.8	8.7	5.2	3.1
Norway	65.0	77.7	88.1	7.1	4.3	2.3
Poland	53.1	63.3	72.0	13.2	9.3	6.8
Slovak Republic	42.1	62.8	71.6	23.0	9.0	6.4
Spain	48.4	64.9	76.5	25.7	15.4	10.1
Sweden	57.9	73.9	83.0	12.4	7.0	3.4
United States	63.5	78.4	85.7	14.5	8.3	4.0
PIAAC average	58.2	70.7	78.9	11.4	7.2	4.5

Note: PIAAC average is the unweighted average of the country employment and unemployment rates.

inequality might be lower, but so might the employment rates of the least skilled. If unemployment and other out-of-work benefits are lower than what individuals would earn in the labor market, more compressed wage distributions could result in more unequal earnings distributions if a large portion of low-skilled workers are forced out of a job.

In this section, we explore to what extent higher wage inequality in the United States might be compensated for by higher employment rates among the low skilled. To shine light on this issue, we once again split the workforce of each country into high-, medium-, and low-skilled groups using the same skill group definitions derived in section 7.4. Table 7.6 shows the employment and unemployment rates of each of these skills groups by country. Employment rates are generally higher in the United States than they are in other countries. However, the differences in employment rates between the various skill groups in the United States are comparable to those observed on average across the PIAAC countries. In the United States, the low skilled (medium skilled) are 26 percent (9 percent) less likely to be employed than the high skilled, while the equivalent PIAAC averages are 26 percent and 10 percent, respectively. The least skilled in the United States are therefore not more likely to be in employment relative to the more skilled—which contradicts the wage compression hypothesis. Overall, there is a slight negative relationship between wage inequality (as measured by the P90/P10) and the percentage difference in employment rates between high- and low-skilled groups (although this is significant only at the 10 percent level). Countries like Japan and Korea have relatively high wage inequality, but small differences in the employment rates of different skills groups, while Scandinavian countries tend to have low wage inequality, but relatively large differences in the unemployment rates of different skills groups.

Turning to unemployment rates, there is even less support for the wage compression hypothesis in the United States: the low skilled (medium skilled) are 3.6 (2.1) times more likely to be unemployed than the high skilled. The equivalent PIAAC average ratios are 2.5 and 1.6, respectively. Again, there is very little evidence of a relationship between wage inequality and the relative unemployment rates of skills groups across countries. Some countries with much lower wage inequality than the United States have similar unemployment ratios between skills groups (e.g., Sweden), while others have much higher unemployment gaps (e.g., Flanders). Overall, these results do not suggest that higher wage inequality in the United States results in better relative employment outcomes for the low skilled—which is consistent with earlier findings from Nickell and Bell (1996), Freeman and Schettkat (2001), and Howell and Huebler (2005), as well as with more recent analysis by Jovicic (2015).

An alternative way of assessing the employment effects of wage compression is to look at whether the skills of the unemployed differ from the skills of the employed. If wage compression were pushing the least skilled into unemployment, one would expect the unemployed to be significantly less skilled than the employed. Table 7.7 reports the average numeracy scores for the unemployed and employed by country. While the average skill level of the unemployed is (nearly) always lower than that of the employed, the employed-to-unemployed average skills ratio ranges from 1 in Korea to 1.14 in England/Northern Ireland. In the United States, this ratio (1.10) tends to be quite high as well (i.e., the unemployed are relatively less skilled compared to the employed than they are in other countries). Once again this is inconsistent with the idea that higher wage inequality might be the price paid for higher employment rates among the low skilled.

While table 7.7 looked at the average skills of the employed and unemployed in each country, figure 7.5 sheds some light on how these skills are distributed. It shows the proportion of the employed and unemployed who are low, medium, and high skilled, respectively. Compared to the PIAAC average, the unemployed in the United States are disproportionately low

Table 7.7   Average skills by employment status and country (points)

	Employed	Unemployed	P-value
Australia	275	262	0.002
Austria	280	265	0.001
Canada	272	249	0.000
Czech Republic	281	259	0.000
Denmark	286	265	0.000
England/N. Ireland (UK)	270	237	0.000
Estonia	278	258	0.000
Finland	290	271	0.000
Flanders (B)	287	278	0.036
France	261	245	0.000
Germany	278	248	0.000
Ireland	264	247	0.000
Italy	255	236	0.000
Japan	291	286	0.286
Korea	264	264	0.925
Netherlands	287	265	0.000
Norway	285	257	0.000
Poland	267	251	0.000
Slovak Republic	285	258	0.000
Spain	256	235	0.000
Sweden	287	255	0.000
United States	260	236	0.000

Notes: PIAAC average is the unweighted average of the country skill levels. The P-values reported are from a test of the equality of mean skill levels between the employed and unemployed.

skilled, but this will partly reflect the fact that skills are generally lower in the United States. More important, the proportion of unemployed among the low skilled is 1.63 times the proportion of employed among the low skilled, while this ratio is 1.54 across PIAAC countries on average.

7.7   Conclusion

The collection and publication of new data from internationally comparable assessments of cognitive skills has sparked renewed interest in the relationship between skills and wage inequality (e.g., Jovicic 2015; Paccagnella 2015; Pena 2014). While the earlier literature on this topic was divisive and did not come to any definite conclusions about the role of skills, the more recent literature has tended to ignore an entire side of the earlier argument and claims that skills matter very little to explaining international differences in wage inequality. This assertion seems counterintuitive, however, given (a) that skills play an important role at the individual level in terms of determining wages (Hanushek et al. 2015), and (b) that skills-/routine-biased technological change has played a crucial role in labor market polarization

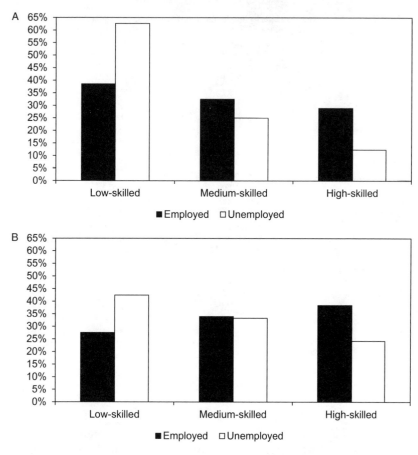

Fig. 7.5   Distribution of skill levels among employed and unemployed. *A*, **United States;** *B*, **PIAAC average.**

Note: PIAAC average is the unweighted average of the country shares.

and rising inequality (Juhn 1999; Goldin and Katz 2008; Autor and Dorn 2013; Autor, Katz, and Kearney 2006). The primary purpose of this chapter was therefore to fully revive the earlier literature on cognitive skills and wage inequality and to show that, despite the availability of new data, this earlier polemic remains unsettled. Indeed, as the results in this chapter have shown, there does appear to be a role for skills in explaining international differences in wage inequality, which operates primarily through the relative balance between supply and demand. What has been missing to date, however, is the methodology to make comparable assessments of the importance of skills and labor market institutions in determining wage inequality. This would require a unified framework for analysis, and should be a priority for future research.

ppendix

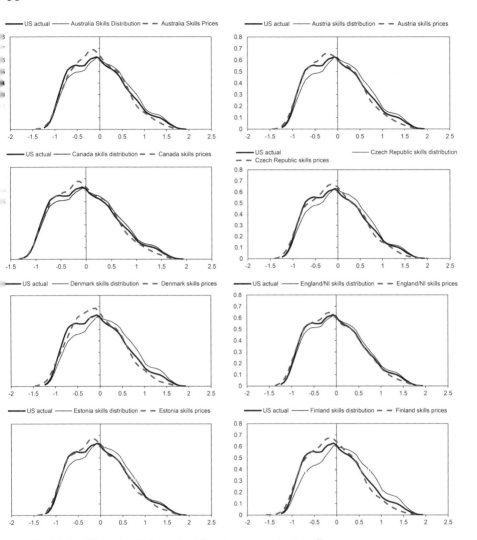

Fig. 7A.1   Wage simulations of skill endowment and price effects

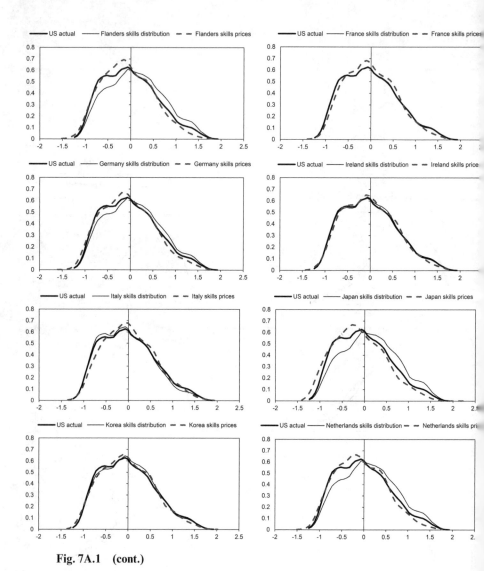

Fig. 7A.1   (cont.)

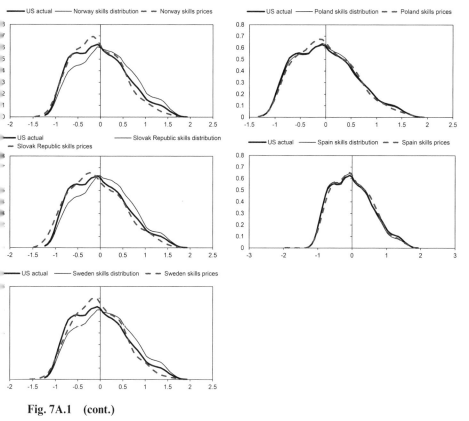

Fig. 7A.1   (cont.)

Table 7A.1   Industry-occupation employment shares by country (percent)

Industry	Agriculture	Mining, manufacturing, and construction	Transportation, communication, and public utilities	Trade	Finance, insurance, real estate, and services	Government
PIAAC	2.45	14.41	3.82	2.91	3.69	2.42
Australia	2.00	13.16	3.82	4.25	3.79	1.78
Austria	3.91	13.27	3.38	3.03	3.51	2.94
Canada	1.70	10.61	3.16	3.75	2.95	1.40
Czech Republic	1.81	24.96	5.14	3.04	2.15	2.54
Denmark	1.79	12.80	3.09	3.43	3.80	2.71
England/N. Ireland (UK)	0.67	11.39	5.48	3.17	4.29	1.55
Estonia	3.37	19.80	4.52	2.43	3.37	2.27
Finland	3.10	13.58	4.60	2.87	3.13	2.79
Flanders (B)	1.31	12.01	3.65	2.32	3.65	3.60
France	2.69	11.71	3.44	3.34	4.59	7.41
Germany	1.42	16.82	3.69	2.76	4.29	2.35
Ireland	4.40	11.11	3.99	2.84	3.99	2.74
Italy	4.14	21.59	4.37	2.83	4.59	2.52
Japan	2.19	15.63	3.68	2.09	2.44	1.63
Korea	3.12	16.83	4.19	3.46	4.03	1.79
Netherlands	0.64	8.70	2.31	2.84	2.98	1.94
Norway	1.69	9.44	2.96	1.64	2.98	1.85
Poland	7.36	21.19	3.75	2.58	2.63	2.41
Slovak Republic	1.84	21.27	4.38	2.29	2.33	2.61
Spain	3.60	11.53	3.40	4.13	7.15	2.42
Sweden	1.95	12.62	3.85	1.70	3.66	1.39
United States	0.69	9.85	3.63	3.32	5.52	1.40
Average numeracy score	251.9	258.2	259.0	255.0	242.9	250.3

Occupation: Craft workers, operatives, laborers, and service workers

Occupation: Clerical and sales workers

Industry	Agriculture	Mining, manufacturing, and construction	Transportation, communication, and public utilities	Trade	Finance, insurance, real estate, and services	Government
PIAAC	0.13	2.32	1.98	7.35	7.57	8.36
Australia	0.18	2.28	1.62	7.60	7.12	7.48
Austria	0.07	2.42	2.06	8.29	8.70	5.92
Canada	0.07	1.27	1.41	6.99	7.88	5.87
Czech Republic	0.27	3.21	3.11	6.02	5.20	5.36
Denmark	0.10	1.54	1.56	6.80	5.77	9.12
England/N. Ireland (UK)	0.07	1.42	1.62	7.47	8.95	14.15
Estonia	0.09	1.11	1.72	5.75	4.83	4.83
Finland	0.09	1.43	2.03	6.51	7.63	9.38
Flanders (B)	0.03	3.05	2.73	5.35	5.32	8.06
France	0.06	1.28	1.89	5.77	6.27	9.70
Germany	0.19	4.03	2.95	7.97	7.60	7.06
Ireland	0.20	1.61	1.47	8.15	9.71	10.9
Italy	0.08	2.19	1.99	7.89	10.63	4.55
Japan	0.04	3.72	2.39	10.06	10.89	9.84
Korea	0.01	6.70	1.98	11.14	12.18	5.76
Netherlands	0.10	2.72	2.14	6.68	6.35	9.85
Norway	0.18	1.27	1.51	10.61	4.82	13.88
Poland	0.03	2.17	1.65	8.12	5.26	4.07
Slovak Republic	0.33	1.60	2.00	6.90	5.60	4.86
Spain	0.59	3.59	2.41	6.76	11.9	7.55
Sweden	0.09	1.11	1.80	5.11	5.29	14.28
United States	0.11	1.37	1.55	6.04	9.69	9.89
Average numeracy score	**257.6**	**276.7**	**275.6**	**266.0**	**263.2**	**262.1**

(*continued*)

Table 7A.1   (continued)

Industry	Agriculture	Mining, manufacturing, and construction	Occupation: Managers and professionals Transportation, communication, and public utilities	Trade	Finance, insurance, real estate, and services	Government
PIAAC	0.25	7.06	4.44	3.62	9.58	17.62
Australia	0.13	5.20	4.78	5.52	11.33	17.98
Austria	0.37	8.03	3.81	3.91	9.05	17.35
Canada	0.18	8.20	5.06	5.02	14.05	20.43
Czech Republic	0.18	9.03	4.85	3.53	7.81	11.79
Denmark	0.35	6.13	6.06	3.40	9.10	22.46
England/N. Ireland (UK)	0.14	5.31	4.61	2.60	11.36	15.75
Estonia	0.60	8.12	5.15	5.62	8.60	17.83
Finland	0.29	7.22	4.37	3.23	9.38	18.35
Flanders (B)	0.16	8.93	4.48	3.85	9.43	22.06
France	0.23	9.26	4.49	3.57	7.71	16.59
Germany	0.16	7.18	3.50	2.09	8.81	17.14
Ireland	0.25	4.76	3.96	2.61	10.64	16.68
Italy	0.33	5.86	2.90	2.34	8.49	12.69
Japan	0.05	8.89	4.88	2.88	5.96	12.73
Korea	0.01	5.00	2.12	2.02	8.28	11.37
Netherlands	0.17	7.33	4.73	4.56	11.35	24.60
Norway	0.06	6.47	4.99	4.58	8.95	22.12
Poland	0.45	7.18	3.85	3.31	6.57	17.43
Slovak Republic	0.73	9.99	5.85	4.25	8.62	14.54
Spain	0.39	4.21	2.95	2.73	8.59	16.13
Sweden	0.21	5.64	5.22	4.76	12.55	18.77
United States	0.23	7.06	4.33	2.47	12.71	20.13
Average numeracy score	**290.1**	**296.2**	**304.0**	**288.4**	**295.4**	**289.1**

References

Autor, D. H. 2014. "Skills, Education, and the Rise of Earnings Inequality among the Other 99 Percent." *Science* 344 (6186): 843–51.

———. 2015. "Why Are There Still So Many Jobs? The History and Future of Workplace Automation." *Journal of Economic Perspectives* 29 (3): 3–30.

Autor, D. H., and D. Dorn. 2013. "The Growth of Low-Skill Service Jobs and the Polarization of the US Labor Market." *American Economic Review* 103 (5): 1553–97.

Autor, D. H., L. F. Katz, and M. S. Kearney. 2006. "The Polarization of the U.S. Labor Market." *American Economic Review* 96 (2): 189–94.

———. 2008. "Trends in U.S. Wage Inequality: Revising the Revisionists." *Review of Economics and Statistics* 90 (2): 300–323.

Autor, D. H., F. Levy, and R. J. Murnane. 2003. "The Skill Content of Recent Technological Change: An Empirical Exploration." *Quarterly Journal of Economics* 118 (4): 1279–333.

Autor, D. H., A. Manning, and C. L. Smith. 2016. "The Contribution of the Minimum Wage to U.S. Wage Inequality over Three Decades: A Reassessment." *American Economic Journal: Applied Economics* 8 (1): 58–99.

Blau, F. D., and L. M. Kahn. 1996. "International Differences in Male Wage Inequality: Institution versus Market Forces." *Journal of Political Economy* 104 (4): 791–837.

———. 2005. "Do Cognitive Test Scores Explain Higher U.S. Wage Inequality?" *Review of Economics and Statistics* 87 (1): 184–93.

Cingano, F. 2014. "Trends in Income Inequality and Its Impact on Economic Growth." OECD Social, Employment and Migration Working Papers no. 163, Organisation for Economic Co-operation and Development.

Devroye, D., and R. Freeman. 2001. "Does Inequality in Skills Explain Inequality of Earnings across Advanced Countries?" NBER Working Paper no. 8140, Cambridge, MA.

DiNardo, J., N. M. Fortin, and T. Lemieux. 1996. "Labor Market Institutions and the Distribution of Wages, 1973–1992: A Semiparametric Approach." *Econometrica* 64 (5): 1001–44.

Firpo, S., N. M. Fortin, and T. Lemieux. 2011. "Occupational Tasks and Changes in the Wage Structure." IZA Discussion Paper no. 5542, Institute for the Study of Labor.

Fortin, N. M., T. Lemieux, and S. Firpo. 2010. "Decomposition Methods in Economics." NBER Working Paper no. 16045, Cambridge, MA.

Freeman, R., and R. Schettkat. 2001. "Skill Compression, Wage Differentials, and Employment: Germany vs. the US." *Oxford Economic Papers* 3:582–603.

Garnero, A., S. Kampelmann, and F. Rycx. 2015. "Sharp Teeth or Empty Mouths? European Institutional Diversity and the Sector-Level Minimum Wage Bite." *British Journal of Industrial Relations* 53 (4): 760–88.

Goldin, C. D., and L. F. Katz. 2008. *The Race between Education and Technology.* Cambridge, MA: Harvard University Press.

Hanushek, E. A., G. Schwerdt, S. Wiederhold, and L. Woessmann. 2015. "Returns to Skills around the World: Evidence from PIAAC." *European Economic Review* 73 (C): 103–30.

Howell, D. R., and F. Huebler. 2005. "Wage Compression and the Unemployment Crisis: Labour Market Institutions, Skills, and Inequality-Unemployment Trade-offs." In *Fighting Unemployment: The Limits of Free Market Orthodoxy*, edited by D. R. Howell. Oxford: Oxford University Press.

Jovicic, S. 2015. "Wage Inequality, Skill Inequality, and Employment: Evidence from PIAAC." Schumpeter Discussion Paper no. 2015-007, University of Wuppertal.

Juhn, C. 1999. "Wage Inequality and Demand for Skill: Evidence from Five Decades." *Industrial and Labor Relations Review* 52 (3): 424–43.

Juhn, C., K. M. Murphy, and B. Pierce. 1993. "Wage Inequality and the Rise in the Returns to Skill." *Journal of Political Economy* 101 (3): 410–42.

Katz, L. F., and K. M. Murphy. 1992. "Changes in Relative Wages, 1963–1987: Supply and Demand Factors." *Quarterly Journal of Economics* 107 (1): 35–78.

Krueger, A. 2012. "The Rise and Consequences of Inequality." Remarks delivered at the Center for American Progress, Washington, DC, Jan. 12.

Lee, D. S. 1999. "Wage Inequality in the United States during the 1980s: Rising Dispersion or Falling Minimum Wage?" *Quarterly Journal of Economics* 114 (3): 977–1023.

Lemieux, T. 2002. "Decomposing Changes in Wage Distributions: A Unified Approach." *Canadian Journal of Economics* 35 (4): 646–88.

———. 2010. "What Do We Really Know about Changes in Wage Inequality?" In *Labor in the New Economy*, edited by K. G. Abraham, J. R. Spletzer, and M. Harper. Chicago: University of Chicago Press.

Leuven, E., H. Oosterbeek, and H. van Ophem. 2004. "Explaining International Differences in Male Skill Wage Differentials by Differences in Demand and Supply of Skills." *Economic Journal* 114 (495): 466–86.

Nickell, S., and B. Bell. 1996. "Changes in the Distribution of Wages and Unemployment in OECD Countries." *American Economic Review* 86 (2): 302–8.

Organisation for Economic Co-operation and Development (OECD). 2013. *OECD Skills Outlook 2013: First Results from the Survey of Adult Skills*. Paris: OECD Publishing.

———. 2014. *OECD Employment Outlook 2014*. Paris: OECD Publishing.

Ostry, J. D., A. Berg, and C. G. Tsangarides. 2014. "Redistribution, Inequality, and Growth." IMF Staff Discussion Notes no. 14/02, International Monetary Fund.

Paccagnella, M. 2015. "Skills and Wage Inequality: Evidence from PIAAC." OECD Education Working Papers no. 114, Organisation for Economic Co-operation and Development.

Pena, A. A. 2014. "Revisiting the Effects of Skills on Economic Inequality: Within- and Cross-Country Comparisons Using PIAAC." Presentation at Taking the Next Step with PIAAC: A Research-to-Action Conference, Arlington, VA, Dec. 11–12.

Pickett, K., and R. Wilkinson. 2011. *The Spirit Level: Why Greater Equality Makes Societies Stronger*. New York: Bloomsbury Press.

Salverda, W., and D. Checchi. 2014. "Labour-Market Institutions and the Dispersion of Wage Earnings." IZA Discussion Paper no. 8820, Institute for the Study of Labor.

Stiglitz, J. 2012. *The Price of Inequality: How Today's Divided Society Endangers Our Future*. New York: W. W. Norton & Company.

Suen, W. 1997. "Decomposing Wage Residuals: Unmeasured Skill or Statistical Artifact?" *Journal of Labor Economics* 15 (3): 555–66.

Tinbergen, Jan. 1975. *Income Distribution: Analysis and Policies*. Amsterdam: Elsevier.

Yun, M. 2009. "Wage Differentials, Discrimination and Inequality: A Cautionary Note on the Juhn, Murphy and Pierce Decomposition Method." *Scottish Journal of Political Economy* 56 (1): 114–22.

Comment   Frank Levy

The authors have written an interesting chapter addressing an important question: To what extent does a nation's earnings inequality reflect market forces versus weak labor market institutions? In the United States, the question is quite timely. Recent discussion of labor market institutions includes potential increases in the minimum wage, the number of workers covered by overtime pay, and whether California Uber drivers are Uber employees or independent contractors—an issue that will eventually extend to other parts of the "gig" economy (Office of the President, n.d.; Memoli 2016; Isaac and Singer 2015).

As the authors note, their chapter is the latest in a substantial body of research on the market/institution question. In this literature, a central methodology involves using decomposition to assess whether earnings inequality is better explained by a nation's wage dispersion or its distribution of skills. Consider, for example, inequality in the US earnings distribution compared to earnings inequality in each of two counterfactual distributions:

- The earnings distribution created by valuing the US distribution of workers at different educational (skill) levels with, say, German wages rates for workers at those educational levels.
- The earnings distribution created by valuing the German distribution of workers at different educational (skill) levels with US wage rates for workers at those educational levels.

These comparisons suggest the dispersion of US wage rates (skill prices), rather than the US skills distribution, is the main source of US earnings inequality. Many authors interpret this wage rate dispersion as reflecting relatively weak US labor market institutions (e.g., Paccagnella 2015). Leuven, Oosterbeek, and van Ophem (2004), however, challenged this interpretation, arguing that a large dispersion of relative wages may arise from not only weak labor market institutions but from a shortage of skilled workers relative to the country's demand.

In this chapter, Broecke, Quintini, and Vandeweyer address the institutions/market question using numeracy scores from the internationally administered PIAAC tests, a potentially better measure of adult skills than the standard years of schooling measure. The authors are not the first to use the PIAAC data in this way (Paccagnella 2015; Pena 2014), but they have access to scores from a larger sample of countries than previous studies and they are the first to analyze the PIAAC data that adjusts for the Leuven,

Frank Levy is the Daniel Rose Professor (emeritus) at the Massachusetts Institute of Technology and a research associate at the Department of Health Care Policy at Harvard Medical School.

For acknowledgments, sources of research support, and disclosure of the author's material financial relationships, if any, please see http://www.nber.org/chapters/c13704.ack.

Oosterbeek, and van Ophem critique. Their analysis establishes three main points:

- A country's net supply of numeracy skills (i.e., supply minus projected demand at different skill levels) has modest power in explaining cross-country differences in the 90–10 earnings gap.
- The modest explanatory power is the average of significant power in explaining the 90–50 gap and no power in explaining the 50–10 gap.
- Even in skills regressions explaining the 90–50 gap, adding variables that describe labor market institutions significantly increases explanatory power.

The authors' arguments are convincing and my comments focus on how their work might be extended.

The first line of inquiry involves cross-country differences in industrial structure. By constructing the net-supply numbers, the authors' estimates implicitly capture cross-country variations in the demand for labor skills. It would be useful to explore the demand side further by examining differences in industrial structure.

Stijn Broecke was good enough to send me tabulations of the industry of employment for persons in the 45th–55th percentiles of the earnings distribution and the 90th percentile and above. Figure 7C.1 shows the industry composition for the 90th percentile and above in the United States versus all other countries in the chapter's sample—that is, the comparative industry sources of high earnings.

Compared to the average of other PIAAC countries, the top earnings decile of workers in the United States shows significantly smaller shares of workers in manufacturing and education industries with relatively equal pay. Conversely, the top US earnings decile shows relatively large shares of workers in finance, professional-scientific-technical activities (presumably including lawyers)—industries with significant earnings inequality—and two other industries, one of which is health and social work activities, which includes physicians. A next step would involve exploring whether industries with relatively high levels of pay in the United States also have relatively high levels of pay in other countries. If they do, this suggests that one source of inequality may be an industrial structure that emphasizes industries that themselves pay wages that are relatively high or low.

A second line of inquiry involves utilizing the one-digit PIAAC occupational data. As part of their analysis, the authors attempt to use net supplies of medium- and low-skilled workers to explain cross-country variation in the 50–10 ratio. Here, however, they can find no relationship.

A possible explanation for the lack of a relationship is the hollowing out of the occupational structure of the kind proposed by Autor, Levy, and Murnane (2003) and Goos, Manning, and Salomons (2014), among others. In this story, some combination of computer-based technical change

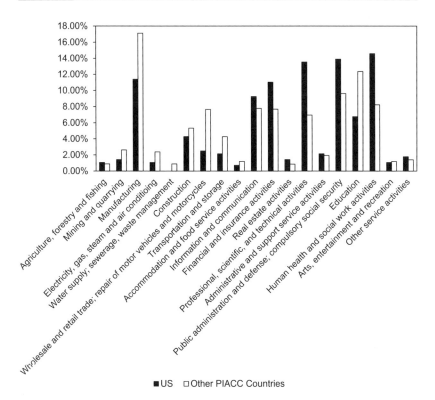

■ US   □ Other PIACC Countries

Fig. 7C.1   Industrial composition of 90th–100th earnings percentiles: United States, other PIAAC countries
Source: Tabulations of OECD Survey of Adult Skills (PIAAC 2012).

and offshoring eliminate highly structured jobs that largely occur in the middle of the earnings distribution. The immediate result is the displacement of medium-skilled workers. To the extent these workers lack the skills to move up in the earnings distribution, they move down where they compete with less skilled workers for available jobs. This pattern of displacement could account for the chapter's finding of people with both low and middle numeracy skills occupying similar low-paying jobs. It may be that comparing occupational distributions at the 10th and 50th earnings percentiles can shed some light on the relevance of this explanation.

Beyond the exploration of demand, the chapter could usefully remind the reader of the difficulty in distinguishing market factors from institutional factors. The current chapter improves on the standard wage/skill decompositions described above by starting with a regression that uses only a country's net supplies of high- and low-skilled workers to explain the 90–10 earnings difference. The authors then examine how this regression changes when institutional variables are added. The results suggest that institutional variables are important—net skill supplies are only statistically significant when institutional variables are included in the regression. But there is significant multicollinearity among the institutional variables suggesting that specific labor market institutions may be the endogenous results of culture as much as strictly exogenous policies.

As another example of the difficulty in attribution, the authors show that an individual's skill attainment can explain much of the cross-country earnings gap between individuals whose mother had tertiary education and individuals whose mother had lower secondary education. In proximate terms this is a skills story, but as the authors acknowledge, it might be in part a genetic story and it could be an institutional story. In particular, the OECD Skills Outlook for 2013 points out this relationship:

> Social background has a strong impact on skills in some countries. . . . In England/Northern Ireland (UK), Germany, Italy, Poland and the United States, social background has a major impact on literacy skills. In these countries more so than in others, the children of parents with low levels of education have significantly lower proficiency than those whose parents have higher levels of education, even after taking other factors into account. (OECD 2013, 30)

The quote underlines the obvious: an adult's skills may reflect the education to which he (she) had access—that is, their country's institutions.

Finally, it would be interesting to see the authors speculate a little on how the relationships they examine might change in the future. The current chapter makes the standard assumption that industrial economies will continue to experience stable or increasing demands for skill. There is, however, some evidence suggesting the demand for skills may be weakening. Beaudry, Green, and Sand (2013) discuss a declining demand for cognitive skills after 2000. David Autor and Brendan Price, applying a task framework, show a declining intensity of analytical tasks after 2000 (personal communication). My work with Alan Benson and Krishna Esteva shows lower rates of return to college in 2010 than in 2000 (Benson, Esteva, and Levy 2013).

This slowdown has many potential explanations, but a possibility worth considering is the slowing rate of population growth and, in particular, labor force growth (figure 7C.2).

For the last half century, demographic discussions in labor economics largely focused on the baby boom cohorts. Because of the baby boom, adequate population and growth—enough to simulate investment in new capital equipment—was taken for granted.

That may be changing. The slow recovery from the 2008 recession involved weak macroeconomic policy, but it also raised the possibility that slow population growth in the United States and other countries was creating a significant policy headwind. Larry Summers noted this possibility in

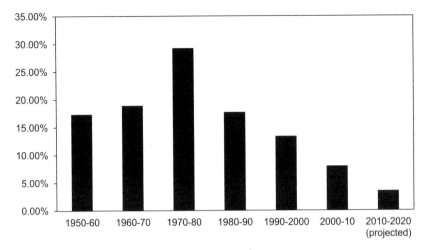

35.00%
30.00%
25.00%
20.00%
15.00%
10.00%
5.00%
0.00%

1950-60   1960-70   1970-80   1980-90   1990-2000   2000-10   2010-2020
(projected)

Fig. 7C.2   Labor force growth per decade, 1950–2020

talking about secular stagnation, something that had not been much thought about since World War II.

In the labor force per se, slow population growth has combined with the baby boomers' retirement. As a result, labor force growth between 2000 and 2010 was the lowest in the last sixty years, and growth for 2010–2014 projected over a decade is significantly slower still.

I appreciate that many other factors are involved in determining labor demand numbers, but it is worth exploring whether the combination of slow force growth and an aging population exert systematic effects on the demand for labor that help to shape what may be a slowdown in the demand for bachelor's degrees versus other levels of education.

In conclusion, Broecke, Quintini, and Vandeweyer have made good use of the PIAAC data to advance the discussion of skills versus institutions in explaining cross-country earnings inequality. My hope is that they will further develop this work to give us a better understanding of what remains a central economic issue.

References

Autor, David, Frank Levy, and Richard J. Murnane. 2003. "The Skill Content of Recent Technological Change: An Empirical Exploration." *Quarterly Journal of Economics* 118 (4): 1279–334.
Beaudry, Paul, David A. Green, and Benjamin M. Sand. 2013. "The Great Reversal in the Demand for Skill and Cognitive Tasks." NBER Working Paper no. 18901, Cambridge, MA.
Benson, Alan, Raimundo Esteva, and Frank S. Levy. 2013. "Dropouts, Taxes and

Risk: The Economic Return to College under Realistic Assumptions." Working paper, Carlson School of Management, University of Minnesota. September. https://papers.ssrn.com/sol3/papers.cfm?abstract_id=2325657.

Goos, Maarten, Alan Manning, and Anna Salomons. 2014. "Explaining Job Polarization: Routine-Biased Technological Change and Offshoring." *American Economic Review* 104 (8): 2509–26.

Isaac, Mike, and Natasha Singer. 2015. "California Says Uber Driver Is Employee, Not a Contractor." *New York Times*, June 17. Accessed Oct. 18, 2016. http://www .nytimes.com/2015/06/18/business/uber-contests-california-labor-ruling-that-says -drivers-should-be-employees.html.

Leuven, E., H. Oosterbeek, and H. van Ophem. 2004. "Explaining International Differences in Male Skill Wage Differentials by Differences in Demand and Supply of Skills." *Economic Journal* 114 (495): 466–86.

Memoli, Michael A. 2016. "Obama Administration Announces Final Overtime Rule, Boosting Pay for Millions." *Los Angeles Times*, May 17. Accessed on Oct. 18, 2016. http://www.latimes.com/politics/la-na-obama-overtime-pay-20160517-snap -story.html.

Office of the President. n.d. "Raise the Wage." Washington, DC. Accessed Oct. 15, 2016. https://www.whitehouse.gov/raise-the-wage.

Organisation for Economic Co-operation and Development (OECD). 2013. *OECD Skills Outlook 2013: First Results from the Survey of Adult Skills*. Paris: OECD Publishing. http://dx.doi.org/10.1787/9789264204256-en.

Paccagnella, M. 2015. "Skills and Wage Inequality: Evidence from PIAAC." OECD Education Working Paper no. 114, Organisation for Economic Co-operation and Development.

Pena, A. A. 2014. "Revisiting the Effects of Skills on Economic Inequality: Within- and Cross-Country Comparisons Using PIAAC." Presentation at the Taking the Next Step with PIAAC: A Research-to-Action Conference, Alexandria, VA, Dec. 11–12.

8

Education and the Growth-Equity Trade-Off

Eric A. Hanushek

Considerable discussion surrounds the interrelationship of economic growth and the distribution of income. A common consideration generally underlying discussions about both growth and the character of the income distribution is the human capital of the population. But it has been unclear how human capital and, particularly, policies designed to improve human capital might affect growth-equity outcomes. The discussion here builds on recent analyses that focus on the interplay of cognitive skills with long-run growth and with individual earnings. This new focus provides a different perspective on how human capital development fits into the aggregate picture and suggests that the impact of various human capital policies is likely to be heterogeneous with some policies leading to growth-equity trade-offs and others to growth-equity complementarities.

Much of the growth-equity discussion has been motivated by examination of the Kuznets curve, which relates income levels of a country to an inverted U-shaped curve of income inequality (Kuznets 1955). Recent work in this heavily traveled area has gone in a variety of directions. Much of the related work has stayed at the aggregate level, focusing on variations across countries or the impact of various redistribution policies (e.g., Ostry, Berg, and Tsangarides 2014; Brueckner, Dabla Norris, and Gradstein 2015). Other work has gone into detail on various subparts such as the relationship between human capital and income inequality (e.g., Castelló-Climent and

Eric A. Hanushek is the Paul and Jean Hanna Senior Fellow at the Hoover Institution at Stanford University and a research associate of the National Bureau of Economic Research.

For acknowledgments, sources of research support, and disclosure of the author's material financial relationships, if any, please see http://www.nber.org/chapters/c13938.ack.

Doménech 2008, 2014).[1] Most, however, has only indirectly provided guidance on specific policy choices.

This chapter does not attempt to reconcile these different perspectives. Instead, the discussion focuses on recent research that links growth and individual incomes through the importance of cognitive skills. Historically, this linkage has not been the focus, even though both growth and income determination have been closely linked to ideas of human capital. The ubiquitous measurement of human capital by school attainment provides a biased view of the role of skills and leads to policy conclusions that are not suggested by a skills formulation and measurement focus. Not only are growth and individual incomes closely related to differential cognitive skills as measured by standardized achievement tests, but also this direct measure of human capital is closely aligned with many current policy discussions.

The next section describes basic results of empirical growth models. This is followed by conclusions about individual earnings determination. These discussions, which differ from many of the common developments, form the basis for considering the relationship between education policies and growth and equity objectives. With this background, it is possible to present illustrative schooling policies that produce growth-equity complementarities and that produce growth-equity trade-offs.

8.1   Long-Run Growth

Modern growth theory has investigated a variety of explanations for what fundamentally determines economic growth (Hanushek and Woessmann 2008). The focus has been different underlying models of how resources and institutions affect growth. And, in the empirical analysis there has been a broad attempt to discover how various factors from politics to geography enter into growth differences across countries. Important for the purposes of this discussion, virtually all developments—both theoretical and empirical—maintain a key role for the skills of workers—that is, for human capital.

In the late 1980s and early 1990s, macroeconomists launched extensive efforts to explain differences in growth rates around the world. A variety of different issues have consumed much of the theoretical growth analysis that developed with the resurgence of growth analysis. At the top of the list is whether growth should be modeled in terms of the level of income or in terms of growth rates of income. The former is typically thought of as neoclassical growth models (e.g., Mankiw, Romer, and Weil 1992), while the latter is generally identified as endogenous growth models (e.g., Lucas 1988; Romer 1990).

The two different perspectives have significantly different implications for

---

1. For a broad review of the theoretical history and modeling of growth, human capital, and income inequality, see Galor (2011).

the long-run growth and income of an economy. In terms of human capital, the focus of this discussion, an increase in human capital would raise the level of income but would not change the steady-state rate of growth in the neoclassical model. But, increased human capital in the endogenous growth model will lead to increases in the long-run growth rate. The theoretical distinctions have received a substantial amount of theoretical attention, although relatively little empirical work has attempted to provide evidence on the specific form (see Benhabib and Spiegel 1994; Hanushek and Woessmann 2008; Holmes 2013).

Both views can be considered in a stylized form of an empirical growth model:

(1)   growth = $\alpha_1$human capital + $\alpha_2$other factors + ε.

A country's growth rate is described as a function of workers' skills along with other systemic factors including economic institutions, initial levels of income, and technology. As noted, there have been distinct differences in how skills are seen as affecting the economy, but little of the broad theoretical work has focused on the measurement of relevant skills. Measurement issues are crucial to any empirical considerations of human capital and growth, yet surprisingly, human capital measurement has also received relatively little attention in the associated empirical analysis.

Owing to the ready availability of data (and to the standard labor economics perspective below), the quantity of schooling became virtually synonymous with human capital, so much so that the choice in empirical work is seldom explicitly considered. Thus, when growth modeling required a measure of human capital, measures of school attainment were seldom questioned. The early data construction of Barro and Lee (1993) provided the necessary data on school attainment supporting international growth work.[2] Thus, equation (1) could be estimated by substituting school attainment, S, for human capital and estimating the growth relationship directly.

While using school attainment to measure human capital generally arouses little attention, this presents huge difficulties in an international setting. In comparing human capital across countries, it is impossible to believe that schools in Singapore yield, on average, the same learning per year as those in Brazil.

This formulation of the growth model also presumes schooling is the only source of human capital and skills. Yet, the very large literature on education production functions (Hanushek 2002) focuses both on differences in school quality and on other inputs including families, health, and abilities of a general form such as

---

2. There were some concerns about accuracy of the data series, leading to alternative developments (Cohen and Soto 2007) and to further refinements by Barro and Lee (2010), but the availability of this as a suitable measure of human capital has seemed clear over the past two decades.

(2)   human capital $= \beta_1$schools $+ \beta_2$families $+ \beta_3$ability $+ \beta_4$health.

$+ \beta_5$other factors $+ \upsilon$

Unless families, health, and school quality are unrelated to school attainment, empirical growth modeling that simply substitutes school attainment for human capital in equation (1) will yield biased estimates of how human capital affects growth. Indeed, this observation is consistent with the early findings about the sensitivity of empirical growth models to model specification and the range of alternative factors considered (Levine and Renelt 1992).

An alternative approach is to measure human capital directly. Consistent with the educational production function literature and with the educational accountability movement, one can use standardized achievement tests of students as a direct measure of the relevant skills of individuals. This proves to be a very productive way to proceed in empirical growth models.

Cross-country skill differences can be constructed from international assessments of math and science (see the description in Hanushek and Woessmann [2011a]). These assessments, conducted over the past half century, provide a common metric for measuring cognitive skill differences across countries.[3] This aggregate measure of a country's skills, labeled the *knowledge capital* in order to distinguish it from school attainment, provides for testing directly the fundamental role of human capital in growth, as found in equation (1). This approach to modeling growth as a function of international assessments of skill differences was introduced in Hanushek and Kimko (2000) and has been extended in Hanushek and Woessmann (2007, 2015a).

The fundamental idea is that skills as measured by achievement, A, can be used as a direct indicator of the knowledge capital of a country in equation (1) and, as described in equation (2), can be thought of as combining the skills of individuals from different sources in different countries.[4]

The impact of both school attainment and knowledge capital can be seen in the basic long-run growth models displayed in table 8.1. The table presents simple models of long-run growth over the period 1960–2000 for the set of fifty countries with required data on growth, school attainment, and achievement. Growth is measured by increases in real gross domestic product (GDP) per capita. The inclusion of initial income levels for countries is

3. Note that the various assessments over the past half century have not been designed to provide longitudinal information. It is possible to construct a longitudinal measure, however, by linking all international tests to US performance, which is independently measured over time with the National Assessment of Educational Progress (NAEP). See Hanushek and Woessmann (2015a).
4. Note, however, that the test scores at a given age or point in time are interpreted as an index of the skills of individuals. It is not the specifically tested information that is important, but instead the indication of relative learning levels that can be applied across the schooling spectrum.

Table 8.1   Alternative estimates of long-run growth models with knowledge capital

	(1)	(2)	(3)
Cognitive skills (A)		2.015	1.980
		(10.68)	(9.12)
Years of schooling 1960 (S)	0.369		0.026
	(3.23)		(0.34)
GDP per capita 1960	−0.379	−0.287	−0.302
	(4.24)	(9.15)	(5.54)
No. of countries	50	50	50
R^2 (adj.)	0.252	0.733	0.728

Source: Hanushek and Woessmann (2015a).
Notes: Dependent variable: average annual growth rate in GDP per capita, 1960–2000. Regressions include a constant; *t*-statistics in parentheses.

quite standard in this literature, permitting the convergence of incomes. In simplest terms, it reflects the fact that countries starting behind can grow rapidly simply by copying the existing technologies in other countries while more advanced countries must develop new technologies (see Hanushek and Woessmann 2012).

The estimates in column (1), which mirror the most common historical approach, rely just on years of schooling to measure human capital and show a significant relationship between school attainment and growth. It explains one-quarter of the international variation in growth rates. Much of the existing empirical growth analysis was designed to go beyond this and to explain a portion of the remaining variation in growth, generally by adding additional measures of country differences including institutions, international trade, political stability, and the like.

The second column substitutes knowledge capital, the direct measure of skills derived from international math and science tests for school attainment, for years of schooling. Not only is there a significant relationship of knowledge capital with growth but also this simple model now explains three-quarters of the variance in growth rates. The final column includes both measures of human capital, that is, knowledge capital and school attainment. Importantly, once direct assessments of skills are included, years of school is not significantly related to growth, and the coefficient on school attainment is very close to zero.

These models, of course, do not say that schooling is worthless. They do say, however, that it is the portion of schooling directly related to skills that has a significant and consistent impact on cross-country differences in growth. The importance of skills and conversely the unimportance of just extending schooling that does not produce higher levels of skills has a direct bearing on human capital policies for both developed and developing countries.

Two aspects of these estimates are relevant for policy consideration. First, it is the case that countries with higher skill levels also invest more in years of schooling. This holds for both developed and developing countries. Second, and very important for thinking about these results, education is a cumulative process, and later learning always builds on earlier learning. James Heckman and his colleagues describe it as dynamic complementarities, such that "skill begets skill" (Cunha et al. 2006; Cunha and Heckman 2007). The idea is very simple—schools not only build upon early learning, but the path of output follows a multiplicative function.

The estimated growth impacts of knowledge capital, scaled in standard deviations of achievement in table 8.1, are very large. The estimates imply that a one standard deviation difference in performance equates to 2 percent per year in average annual growth of GDP per capita.

Finally, estimating models in this form with a convergence term permits some assessment of the differences between the endogenous and neoclassical growth models, although full discussion is beyond this chapter. In the neoclassical model, the cumulative increases in GDP that emanate from increased human capital are approximately one-third less over a seventy-five-year period than those from the endogenous growth model, but they are still very substantial (see Hanushek and Woessmann 2011b). It remains difficult, however, to distinguish between the two models with existing data because insufficient data about changes in knowledge capital over time are available and because the impacts on growth are seen only in the distant future (see Holmes 2013).

A major concern with empirical growth modeling is that the estimated relationships do not measure causal influences but instead reflect reverse causation, omitted variables, cultural differences, and the like. This concern has been central to the interpretation of much of the prior work in empirical growth analysis, and indeed some have rejected the entire body of work on the basis of concerns about causation. Fully considering these issues goes beyond what can be presented here (see Hanushek and Woessmann 2012, 2015a), but it is possible to give some sense of the issues and their resolution.

An obvious issue is that countries that grow faster have added resources that can be invested in schools, implying that growth could cause higher scores. However, the lack of relationship across countries in the amount spent on schools and the observed test scores that has been generally found provides evidence against this (Hanushek and Woessmann 2011a). Moreover, a variety of sensitivity analyses show the stability of these results when the estimated models come from varying country and time samples, varying specific measures of cognitive skills, and alternative other factors that might affect growth (Hanushek and Woessmann 2012).

It is possible to address the main causation concerns with a series of alternative analyses, even if none of the tests is completely conclusive. To rule out simple reverse causation, Hanushek and Woessmann 2012 estimate the effect of scores on tests conducted until the early 1980s on economic growth in 1980–2000, finding an even larger effect of knowledge capital in the later period. Additional analysis considers the earnings of immigrants to the United States and cognitive skills in order to address the idea that cognitive skills are unimportant and that is just correlated with other causal factors. This analysis finds that the international test scores for their home country significantly explain US earnings, but only for those educated in their home country and not for those educated in the United States. This finding addresses simple issues of cultural differences because immigrants from the same country (but educated differently) are directly compared. By observing impacts within a single labor market, it also addresses possible concerns that countries with well-functioning economies also have good schools without the good schools driving growth.

Another analysis shows that changes in test scores over time are systematically related to changes in growth rates over time. In other words, it implicitly holds the country constant while looking at whether changing scores have the impact on changing growth rates that is predicted in table 8.1.

Finally, it is possible to exploit institutional features of school systems as instrumental variables for test performance. By employing only the variation in test outcomes emanating from country differences because of the use of central exams, decentralized decision-making, and privately operated schools, this instrumental variable approach both supports a causal interpretation and suggests that schooling can be a policy instrument contributing to economic outcomes.

While concerns about issues of causation still remain, the tests that have been done provide a prima facie case that improving cognitive skills and the knowledge capital of a country can be expected to improve economic growth. Each of the causation tests points to the plausibility of a causal interpretation of the basic models. But, even if the true causal impact of cognitive skills is less than suggested in table 8.1, the overall finding of the importance of such skills is unlikely to be overturned.

With this foundation of the relationship between knowledge capital and growth, it is possible to turn to issues affecting the distribution of income.

8.2   Individual Earnings

The overall distribution of income depends on a variety of factors including labor force participation, taxes, subsidies, international competition, firm ownership, and the like. Nonetheless, individual earnings will have a substantial influence on the ultimate distribution of income.

Importantly, cognitive skills of individuals have a clear and strong relationship to individual earnings and incomes. There has been a long history of investigating the determination of incomes and the role of human capital. While the relationship of skills to productivity of individuals dates back to

Sir William Petty (Petty [1676] 1899) and Adam Smith (Smith [1776] 2010), the modern consideration of earnings determination is dominated by Jacob Mincer (Mincer 1970, 1974).

With a simple investment model, Mincer related school attainment (years of schooling) to individual earnings. Perhaps no other empirical relationship has had more influence than the Mincer earnings function.[5] Over time this structure has been almost universally applied, and virtually any analysis considering individual variations in human capital measures skill differences primarily by years of schooling. Not only is there the conceptual support for this from Mincer's work and from subsequent developments, but also it was expedient because measures of years of schooling are ubiquitous in census and survey data.[6]

Unfortunately, characterizing the human capital of individuals simply by years of schooling ignores other elements of human capital determination and also eliminates most of the relevant policy deliberations about investments in human capital. As noted, there is extensive evidence from the educational production function literature that highlights the central role of families, peers, and neighborhoods—in addition to schools—on the achievement and skills of individuals (Hanushek 2002). As with growth modeling, this suggests that the typical estimates of the impact of human capital on earnings from a Mincer earnings function is actually the combined effect of added schooling and of the correlated influence of these other factors. Additionally, when any policy discussion turns to the influence of schools, the interest is more focused on issues of school quality than school quantity. While there is some discussion about school completion and about college access, most of the policy concerns are focused on aspects of school quality, something that is generally neglected in the analysis of individual earnings.

An alternative formulation that acknowledges these shortcomings in standard analyses of earnings determination is again to focus on individual measures of cognitive achievement as a direct measure of human capital. Such analysis has not been very common because of the general lack of measures of achievement or skills in surveys that have information about earnings and labor market activities. Recent data, however, are particularly apropos to understanding how skills relate to individual earnings.

The Programme for the International Assessment of Adult Competencies (PIAAC) provides labor market information and assessments of cognitive skills for a random sample of the population age sixteen to sixty-five in thirty-two separate countries (Organisation for Economic Co-operation and

5. The standard Mincer earnings function has log earnings as a linear function of years of schooling and a quadratic in potential experience (i.e., in years since completing schooling). It may then also include other specific factors influencing earnings.
6. To give an international view, Psacharopoulos and Patrinos (2004) estimated Mincer earnings functions for ninety-eight countries.

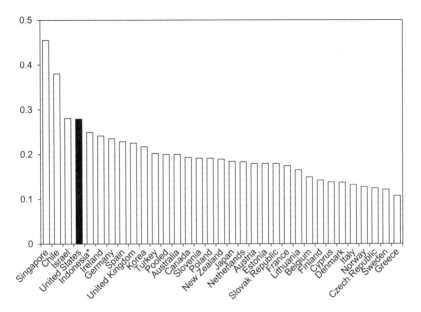

Fig. 8.1   Returns to numeracy skills
Source: Hanushek, Schwerdt, Wiederhold, and Woessmann (2017).
Note: Coefficient estimates on numeracy score (standardized to std. dev. 1 within each country) in a regression of log gross hourly wage on numeracy, gender, and a quadratic polynomial in age, sample of full-time employees age thirty-five to fifty-four.
*Jakarta only. *Data source:* PIAAC 2016.

Development [OECD] 2016). This survey of individuals collected demographic background along with labor market history. Sampled individuals also took tests in numeracy, literacy, and problem solving in technology-rich environments. These data are particularly useful for understanding the returns to skills. First, they provide information on earnings during mid- and later life-cycle periods, when the value of skills becomes most observable.[7] Second, by observing variations in returns across countries, it is possible to get suggestive insights into underlying causes of skill differences (Hanushek, Schwerdt, Wiederhold, and Woessmann 2017).

Estimates of the earnings-skills gradient, shown in figure 8.1, indicate that the United States has close to the highest return to skills across the thirty-two countries. These estimates for numeracy skills indicate that a person one standard deviation above the mean numeracy score will, on average,

---

7. Most of the available evidence on returns to skills comes from US panel survey information where, unfortunately, the observations occur early in the work life. But early career returns provide underestimates of the full value of skills (Hanushek et al. 2015), perhaps because employers are still learning about individual skills (Altonji and Pierret 2001).

earn 28 percent more per year throughout the working life. But this estimate also shows that low achievement is harshly dealt with by the labor market—because somebody at the 16th percentile of the achievement distribution (one standard deviation below the mean) will earn 28 percent below the average achieving worker.

These calculations underscore a basic fact: upgrading the skills of workers makes them more productive, which in turn raises their own incomes and improves overall growth of GDP. Pulling low achievers toward the mean implies lowering the variance in earnings while increasing the rate of growth of the economy—a point highlighted below.

There is, however, one additional aspect of the returns to individual skills that is relevant for consideration of growth-equity choices. Consistent with the arguments of Nelson and Phelps (1966), Welch (1970), and Schultz (1975), returns to skills appear to be higher when there is more economic change. Specifically, Hanushek, Schwerdt, Wiederhold, and Woessmann (2017) show that differences in returns to skills across countries are correlated with economic growth rates. In other words, growth and skills are complementary—higher skills imply greater growth that in turn implies greater returns to those higher skills.

8.3   Illustrative Human Capital Policies and Growth-Equity Outcomes

Most discussions of the human capital impact on both growth and distribution have looked exclusively at school attainment, and this has led to distortions of the policy discussions. Essentially discussions of alternative public policies have been inappropriately separated from the discussions of possible growth-equity trade-offs and growth-equity complementarities.

The evidence on growth and on individual earnings suggests that policies that improve learning while reducing the variance in achievement and skills will promote higher and more equitable incomes. Perhaps the most obvious program in this category involves preschool programs, although other ideas surrounding lifelong learning are also relevant.

8.3.1   Early Childhood Education

There is a broad consensus that the United States should expand its current preschool programs, particularly for disadvantaged students. From the demand side, there is little question that there are significant variations in the preparation of children for schooling and that these variations are systematically related to families' socioeconomic status. On the supply side, we have credible evidence that quality preschool can significantly improve achievement and life outcomes of disadvantaged students.

Evidence from a wide variety of sources indicates that disadvantaged students have less education in the home before entry into school. The Coleman Report, the massive governmental report mandated by the 1964 Civil Rights

Act, first documented early achievement differences by family background (Coleman et al. 1966). These differences, documented in 1965, focused on racial differences. Another important investigation looked at the vocabulary of children and found dramatic differences by parents' socioeconomic status (Hart and Risley 1995). Both the amount and quality of parent-child interactions differed significantly, leading to large differences in vocabularies that directly reflected parental background. More recently, data from the Early Childhood Longitudinal Study documents the continuing early achievement deficits that accompany family background. Fryer and Levitt (2004) identify gaps in scores by socioeconomic status, while Reardon (2008) suggests that these gaps may have widened over many years.

How important are these initial gaps? Considerably so: while there is some disagreement about whether they shrink, expand, or hold constant over time in school, there is no evidence that they actually disappear.[8]

The final demand-side element for preschool is the significant impact on individuals' future incomes. The most direct relationship between early test performance and earnings is found in Chetty et al. (2011), which traces kindergarten performance directly to college completion and early career earnings. While recent public and media focus has largely concentrated on the top 1 percent of earners, such results point to the enormous implications of skill gaps *within* the remaining 99 percent of earners.[9]

The importance of early childhood learning in the overall growth-equity discussions is clear: the evidence suggests that high-quality programs tend to enhance the achievement of disadvantaged students—lifting the mean of the achievement distribution while lowering the variance. Thus, if effective, such programs both promote higher growth and more equity.

On the supply side, the existing evaluation literature generally suggests that preschool programs can be effective in raising achievement and other outcomes. Well-publicized studies with strong research designs, based on random assignment of students to programs, suggest high efficacy: the Perry Preschool Project, the Carolina Abecedarian Project, and the Early Training Project provide important evidence in favor of early childhood education (Schweinhart et al. 2005; Witte 2007).[10] The experimental evidence has been supplemented by observational studies. Chicago's Child-Parent Center program (Reynolds et al. 2002), studies on preschool outcomes in Tulsa, Oklahoma (Gormley et al. 2005), and Georgia's universal pre-K program (Cascio and Schanzenbach 2013), generally indicate a positive impact for disadvantaged children (but no impact for more advantaged kids). Offsetting these results to some extent is the federal Head Start program, which

---

8. See, for example, the projections of racial gaps in achievement starting with those found in the Coleman Report (Hanushek 2016a).

9. See also the discussion in Autor (2014).

10. A comprehensive review of different pre-K programs and their evaluations can be found in Besharov et al. (2011).

has been extensively evaluated and shows little success.[11] One recent high-quality evaluation, for instance, found that any achievement gains produced by Head Start disappear by third grade. Puma et al. (2012), the first random-assignment evaluation of Head Start, assessed a variety of child outcomes with none showing significant impact by third grade.

The caveat to this discussion is that little is currently known about the characteristics of effective preschool programs. The considerable discussion of various input requirements suggested for preschool programs has not been matched with evidence about the impact of different inputs.[12] Understanding how to structure effective preschool programs and how to price and provide access to them are remaining questions that are central to developing actual policies.

8.3.2 Lifelong Learning

Changing the skill level of youth, while effective in improving both long-run growth and equity according to existing research, does take a long time to have its economic impact (Hanushek and Woessmann 2015a, 2015b). This suggests short- to middle-range economic effects that might be different. Specifically Autor (2014), in summarizing a number of studies, shows how the income distribution has widened in the United States in recent decades and relates this to differential skills. Specifically, more educated workers have been able to adjust to changed demands and have seen their earnings diverge from those of less educated. While the central focus is on differences in school attainment, it is almost certainly true for cognitive skills.

One aspect of this adaptation to change has been the ability of the more skilled to train for different job demands. As noted above, the ability to continually train and adapt is reflected by the higher returns to skills that accompany faster growth (Hanushek, Schwerdt, Wiederhold, and Woessmann 2017). This adjustment to change has led to continual calls for enhancing lifelong learning, particularly by those in jobs subject to more intense competition and by those currently receiving less continual training and upgrading.

If effective, enhanced lifelong learning would tend to make growth and equity more complementary because it is the lower skilled that generally receive less training throughout their career. Moreover, as discussed below, the need for improved career training has been emphasized for workers with vocational training, who generally have more specific skills that are more subject to lessened demand with changes in job demands. For this reason,

---

11. In practice, Head Start is not a unified program but rather a funding stream with loose regulations on the character of actual programs. As such, Head Start programs display considerable heterogeneity.

12. For discussion from the policy perspective, see Hanushek (2015).

regular calls for support of lifelong learning are more common in the European Union with its more plentiful use of vocational training. The problem from a policy viewpoint is that ideas about the appropriate policies to support lifelong learning generally fall short of the appeals for expansion. Little empirical knowledge exists about appropriate incentives to individuals or firms that would effectively expand lifelong learning.[13]

8.3.3 Vocational Education

Of course, not all education policies produce a long-run felicitous growth-equity outcome.[14] For example, there are many examples of ineffective policies that fail to yield improved student outcomes. This fact is easiest to see in both cross-country and within-country analyses of the inconsistent relationship between resources to schools and outcomes (see, e.g., Hanushek and Woessmann 2011a; Hanushek 2003).[15] Little evidence suggests that just spending more on schools within the current institutional arrangements is likely to lead to much improvement.

More interestingly, there are educational programs that meet their declared goals but that might simultaneously suggest trade-offs between growth and equity. This includes intensive vocational education and programs that skew education toward the elite.

One major educational policy decision countries face is how much to emphasize vocational education, that is, education that is designed to produce more job-related skills, rather than the standard general education program. Vocational education programs, popular in both Europe and many developing countries, aim to ease the school-to-work transition of youth by directly providing skills that industries demand. Attention was particularly focused on these programs following the 2008 recession, in part due to the success of the German economy that is built on its apprenticeship program and intensive vocational education. And, while the United States has largely dismantled its vocational education program, there has been more recent attention to the possibility of reinstating at least part of the system (e.g., Lerman 2009).

Most of the attention has focused on the school-to-work transition. The evidence on the impact of vocational education on labor market entry is somewhat ambiguous because of the selectivity of choice across school

13. In closely related work, governments often have training programs for unemployed adults. These programs are sometimes effective, but it is hard to describe precisely when they are successful or what are the characteristics of successful programs (McCall, Smith, and Wunsch 2016).

14. Much discussion surrounding short-run growth and employment focuses on such things as labor and product market regulations, taxes, and subsidies. In the long run, however, these do not show any relationship with growth (Hanushek and Woessmann 2015a).

15. Strictly speaking, poor education programs may lower growth and widen the income distribution.

types, but there is a general sense that vocational education does in fact make career entry easier (Ryan 2001). However, the impact of vocational education on the growth-equity relationship proves to be more complicated.

In the short run, expanding vocational education programs would, if they get youth into the labor market more quickly, tend to lead to expansion of the economy and to higher incomes at the lower end of the income distribution—a case of an education program that moves toward more growth and more equity.

The long run may, however, be different. Krueger and Kumar (2004) suggest that a significant contributor to the overall lower growth rates in Europe as opposed to the United States may be the reliance on vocational education, particularly in the face of labor market regulations that lead to market distortions. The idea is that firms choose lower-skill technologies when workers have more skill-based training as opposed to general training. The advantage of the United States is that broad general education and limited labor market regulation allows firms to seek better technologies.

From a different perspective, Hanushek, Schwerdt, Woessmann, and Zhang (2017) look at the life-cycle impacts of vocational versus general education. They test the simple hypothesis that individuals with vocational education are less able to adapt to changed technologies and thus their employment opportunities later in life are diminished. For countries with the most intensive vocational education—apprenticeship countries—there is a clear lessening of employment later in the life cycle when compared to those with general education. The lowered employment later in the life cycle is also found in other countries with less intensive vocational education programs, but the decline in employment is not as sharp.

Others have subsequently looked at the same hypothesis with somewhat varying results. Hampf and Woessmann (2017) confirm the major findings using more recent data across a larger number of countries. Forster, Bol, and van de Werfhorst (2016) find the same overall life-cycle pattern across countries but do not find the strong differences by intensity of the vocational system in different countries; they find the pattern to be consistent across a wide range of countries. For Britain, Brunello and Rocco (2017) find employment declines for those with vocational education, but the later declines do not appear to be large enough to offset the initial employment gains.

Taken together, the evidence suggests that movement toward expanded vocational education is unlikely to lead to more rapid long-run growth. Moreover, it does not appear to lead to more equitable outcomes, even if it has a short-run impact of improving the school-to-work transition. These issues are especially important in developing countries where the main focus is on increased growth. To the extent that a country experiences more rapid growth, the economy is going through larger changes—and this is just where individuals with vocational education tend to be at a larger disadvantage over time.

The existing evidence does not argue against all vocational education. The analysis in Hanushek, Schwerdt, Woessmann, and Zhang (2017) indicates that one of the elements of the improved life-cycle employment of those with general education is that they tend to get more ongoing education through their careers. Thus, programs that ensured continued education for those with vocational—that is, lifelong learning—could ameliorate the later life disadvantage of vocational education. But, as noted previously, the potential desirability of lifelong learning has not been matched by programs or institutions that have been very effective in its provision.

8.3.4 Higher Education and Elite Programs

Perhaps the most common educational policy initiative today is a call for expansion of college and university training. The growth models in table 8.1, however, indicate that once direct assessments of skills are included, school attainment is not significantly related to growth, and the coefficient on school attainment is very close to zero. These results hold even if the amount of tertiary education is separately considered (Hanushek 2016b).

These models, of course, do not say that schooling is worthless. They do say, however, that it is the portion of schooling directly related to skills that has a significant and consistent impact on cross-country differences in growth. The importance of skills and conversely the unimportance of just extending schooling that does not produce higher levels of skills has a direct bearing on human capital policies for both developed and developing countries.[16]

Of course, there are no scientists and engineers without higher education, so the insignificance for growth of having more college education appears strange. But, this can be interpreted as just a special case of the dynamic complementarities discussed previously (Cunha et al. 2006; Cunha and Heckman 2007). The idea is very simple—schools not only build upon early learning, but the path of output follows a multiplicative function. Students who enter college better prepared can be expected to learn more and be more productive on graduation, and this skill differential over less prepared students dominates any productivity effect of adding a greater number of less prepared graduates.

The one potential anomaly about tertiary education is that the growth models appear slightly different for just OECD countries. In the presence of knowledge capital, years of tertiary schooling has a positive effect (significant at the 10 percent level) for the twenty-four OECD countries in the sample (Hanushek 2016b). But this effect is entirely driven by the United States. If the United States is dropped, the estimated impact of higher education falls and is statistically insignificant.

---

16. Holmes (2013) also shows that neither the level nor the change in tertiary schooling for a larger group of countries is positively related to growth, even in the absence of knowledge capital measures.

How should this apparent impact in the United States be considered? It turns out that the United States has grown faster than would be predicted by the basic growth models with knowledge capital (i.e., the United States has a positive residual in the regression models of table 8.1). The United States is generally regarded as having the best universities, and this quality may make the difference. But, perhaps more importantly, the United States has been able to attract highly skilled immigrants. The latter argument is quite consistent with the previous growth results, because the measure of achievement of US students would not capture the skills of the immigrants. Hanson and Slaughter (chapter 12, this volume) find that 55 percent of PhD workers in the United States in science, technology, engineering, and mathematics (STEM) fields were foreign born. In other words, the United States is able to bring in highly skilled individuals who frequently get PhDs at US universities and then remain to work in the United States.[17] In short, it is difficult to attribute the faster-than-expected growth in the United States just to the impact of higher education for US students.

Even though expanding higher education may not have any clear impact on growth rates, it would be expected to add to income inequality. With increases in the labor market returns to higher education, past expansion of college education has led to increased income inequality, and, while not certain in the future, might be expected have similar impacts in the future (Autor 2014).

A slightly different perspective focuses on whether the education system favors providing basic skills or developing high performers. The previous growth models uniformly considered just country-average skills. Yet, particularly in developing countries, there is often a large variance in performance with some very high performers and many very low performers (see Hanushek and Woessmann 2008). These choices can, however, also be seen in developed countries, as with the US accountability system that has emphasized bringing all students up to a minimum achievement level.

In terms of modeling growth, it is possible to separate the impacts of the proportion of high performers and the proportion with basic literacy as assessed by the cognitive skills tests. Importantly, both broad basic skills ("education for all" in terms of achievement) and high achievers have a separate and statistically significant impact on long-term growth (Hanushek and Woessmann 2015a). These estimates, while suggestive, do not answer the overall policy question about where to invest resources. To address that question, it is necessary to know more about the relative costs of producing more basic and more high performers. In fact, no analysis is available to describe the costs of producing varying amounts of skills.

17. The United States has also had generally the strongest economic institutions for growth— free and open labor and capital markets, limited government regulation, secure property rights, and openness to trade. These institutions could further add to the explanation of the faster-than-expected growth.

At the same time, in terms of the interplay between growth and equity, investing relatively more in the top-end skills would clearly lead to a wider income distribution compared with investing at the bottom end. Thus, understanding whether growth and equity move together or not depends on the magnitude of potential changes in the distribution of achievement.

8.4 Conclusions

This chapter considers how recent analyses of the role of skills in long-run economic growth and in individual earnings changes significant parts of the discussion of possible growth-equity trade-offs. The key driver for individual incomes and for economic growth from this work is the cognitive skills of the individual—skills that are developed not only in schools, but also in the family and in neighborhoods. This perspective changes conclusions about policies considerably.

Human capital is always mentioned as part of both aggregate growth and individual incomes. But, if human capital is thought of just as it is commonly measured—by school attainment—the policy discussions become very distorted. Moreover, in discussions of the potential growth-equity trade-offs that frequently occur, the message of improved human capital can be quite misunderstood.

The common policy discussion in education is largely around the quality of schools. That, in fact, is the correct focus because the skills that are important for growth and for individual incomes involve achievement and learning as opposed to just years spent in school.

While the distribution of income involves many factors, a key element is the distribution of earnings. In that regard, many policies that improve school quality will lead to growth-equity complementarities.

References

Altonji, Joseph G., and Charles R. Pierret. 2001. "Employer Learning and Statistical Discrimination." *Quarterly Journal of Economics* 116 (1): 313–50.

Autor, David H. 2014. "Skills, Education, and the Rise of Earnings Inequality among the 'Other 99 Percent.'" *Science* 344 (843): 843–51.

Barro, Robert J., and Jong-Wha Lee. 1993. "International Comparisons of Educational Attainment." *Journal of Monetary Economics* 32 (3): 363–94.

———. 2010. "A New Data Set of Educational Attainment in the World, 1950–2010." NBER Working Paper no. 15902, Cambridge, MA.

Benhabib, Jess, and Mark M. Spiegel. 1994. "The Role of Human Capital in Economic Development: Evidence from Aggregate Cross-Country Data." *Journal of Monetary Economics* 34 (2): 143–74.

Besharov, Douglas J., Peter Germanis, Caeli Higney, and Douglas M. Call. 2011. *Assessing the Evaluations of Twenty-Six Early Childhood Programs.* College Park,

MD: Welfare Reform Academy, University of Maryland, July. http://www.welfare academy.org/pubs/early_education/index.shtml.

Brueckner, Markus, Era Dabla Norris, and Mark Gradstein. 2015. "National Income and Its Distribution." *Journal of Economic Growth* 20 (2): 149–75.

Brunello, Giorgio, and Lorenzo Rocco. 2017. "The Labor Market Effects of Academic and Vocational Education over the Life Cycle: Evidence from Two British Cohorts." *Journal of Human Capital* 11 (1): 106–66.

Cascio, Elizabeth U., and Diane W. Schanzenbach. 2013. "The Impacts of Expanding Access to High-Quality Preschool Education." *Brookings Papers on Economic Activity* Fall: 127–78.

Castelló-Climent, Amparo, and Rafael Doménech. 2008. "Human Capital Inequality, Life Expectancy and Economic Growth." *Economic Journal* 118 (528): 653–77.

———. 2014. "Human Capital and Income Inequality: Some Facts and Some Puzzles." BBVA Research Working Paper no. 12/ 28, Madrid, Banco Bilbao Vizcaya Argentaria. March.

Chetty, Raj, John N. Friedman, Nathaniel Hilger, Emmanuel Saez, Diane Whitmore Schanzenbach, and Danny Yagan. 2011. "How Does Your Kindergarten Classroom Affect Your Earnings? Evidence from Project STAR." *Quarterly Journal of Economics* 126 (4): 1593–660.

Cohen, Daniel, and Marcelo Soto. 2007. "Growth and Human Capital: Good Data, Good Results." *Journal of Economic Growth* 12 (1): 51–76.

Coleman, James S., Ernest Q. Campbell, Carol J. Hobson, James McPartland, Alexander M. Mood, Frederic D. Weinfeld, and Robert L. York. 1966. *Equality of Educational Opportunity*. Washington, DC: US Government Printing Office.

Cunha, Flavio, and James J. Heckman. 2007. "The Technology of Skill Formation." *American Economic Review* 97 (2): 31–47.

Cunha, Flavio, James J. Heckman, Lance Lochner, and Dimitriy V. Masterov. 2006. "Interpreting the Evidence on Life Cycle Skill Formation." In *Handbook of the Economics of Education*, vol. 1, edited by Eric A. Hanushek and Finis Welch, 697–812. Amsterdam: North Holland.

Forster, Andrea G., Thijs Bol, and Herman G. van de Werfhorst. 2016. "Vocational Education and Employment over the Life Cycle." *Sociological Science* 3:473–94.

Fryer, Roland G., Jr., and Steven D. Levitt. 2004. "Understanding the Black-White Test Score Gap in the First Two Years of School." *Review of Economics and Statistics* 86 (2): 447–64.

Galor, Oded. 2011. "Inequality, Human Capital Formation, and the Process of Development." In *Handbook of the Economics of Education*, vol. 4, edited by Eric A. Hanushek, Stephen Machin, and Ludger Woessmann, 441–93. Amsterdam: North Holland.

Gormley, Jr., William T., Ted Gayer, Deborah Phillips, and Brittany Dawson. 2005. "The Effects of Universal Pre-K on Cognitive Development." *Developmental Psychology* 41 (6): 872–84.

Hampf, Franziska, and Ludger Woessmann. 2017. "Vocational vs. General Education and Employment over the Life-Cycle: New Evidence from PIAAC." *CESifo Economic Studies* 63 (3): 255–69.

Hanushek, Eric A. 2002. "Publicly Provided Education." In *Handbook of Public Economics*, vol. 4, edited by Alan J. Auerbach and Martin Feldstein, 2045–141. Amsterdam: North Holland.

———. 2003. "The Failure of Input-Based Schooling Policies." *Economic Journal* 113 (485): F64–98.

———. 2015. "The Preschool Debate: Translating Research into Policy." In *The Next*

Urban Renaissance: How Public-Policy Innovation and Evaluation Can Improve Life in America's Cities, edited by Ingrid Gould Ellen, Edward L. Glaeser, Eric A. Hanushek, Matthew E. Kahn, and Aaron M. Renn, 25–40. New York: Manhattan Institute for Policy Research.

———. 2016a. "What Matters for Achievement: Updating Coleman on the Influence of Families and Schools." *Education Next* 16 (2): 22–30.

———. 2016b. "Will More Higher Education Improve Economic Growth?" *Oxford Review of Economic Policy* 32 (4): 538–52.

Hanushek, Eric A., and Dennis D. Kimko. 2000. "Schooling, Labor Force Quality, and the Growth of Nations." *American Economic Review* 90 (5): 1184–208.

Hanushek, Eric A., Guido Schwerdt, Simon Wiederhold, and Ludger Woessmann. 2015. "Returns to Skills around the World: Evidence from PIAAC." *European Economic Review* 73:103–30.

———. 2017. "Coping with Change: International Differences in the Returns to Skills." *Economic Letters* 153 (April): 15–19.

Hanushek, Eric A., Guido Schwerdt, Ludger Woessmann, and Lei Zhang. 2017. "General Education, Vocational Education, and Labor-Market Outcomes over the Life-Cycle." *Journal of Human Resources* 52 (1): 48–87.

Hanushek, Eric A., and Ludger Woessmann. 2007. *Education Quality and Economic Growth*. Washington, DC: World Bank.

———. 2008. "The Role of Cognitive Skills in Economic Development." *Journal of Economic Literature* 46 (3): 607–68.

———. 2011a. "The Economics of International Differences in Educational Achievement." In *Handbook of the Economics of Education*, vol. 3, edited by Eric A. Hanushek, Stephen Machin, and Ludger Woessmann, 89–200. Amsterdam: North Holland.

———. 2011b. "How Much Do Educational Outcomes Matter in OECD Countries?" *Economic Policy* 26 (67): 427–91.

———. 2012. "Do Better Schools Lead to More Growth? Cognitive Skills, Economic Outcomes, and Causation." *Journal of Economic Growth* 17 (4): 267–321.

———. 2015a. *The Knowledge Capital of Nations: Education and the Economics of Growth*. Cambridge, MA: MIT Press.

———. 2015b. *Universal Basic Skills: What Countries Stand to Gain*. Paris: Organisation for Economic Co-operation and Development.

Hart, Betty, and Todd R. Risley. 1995. *Meaningful Differences in the Everyday Experience of Young American Children*. Baltimore: Paul H. Brookes Publishing Co.

Holmes, Craig. 2013. "Has the Expansion of Higher Education Led to Greater Economic Growth?" *National Institute Economic Review* 224 (1): R29–47.

Krueger, Dirk, and Krishna B. Kumar. 2004. "Skill-Specific Rather Than General Education: A Reason for US-Europe Growth Differences?" *Journal of Economic Growth* 9 (2): 167–207.

Kuznets, Simon. 1955. "Economic Growth and Income Inequality." *American Economic Review* 45 (1): 1–28.

Lerman, Robert I. 2009. *Training Tomorrow's Workforce: Community College and Apprenticeship as Collaborative Routes to Rewarding Careers*. Washington, DC: Center for American Progress, December.

Levine, Ross, and David Renelt. 1992. "A Sensitivity Analysis of Cross-Country Growth Regressions." *American Economic Review* 82 (4): 942–63.

Lucas, Robert E., Jr. 1988. "On the Mechanics of Economic Development." *Journal of Monetary Economics* 22 (1): 3–42.

Mankiw, N. Gregory, David Romer, and David Weil. 1992. "A Contribution to the Empirics of Economic Growth." *Quarterly Journal of Economics* 107 (2): 407–37.

McCall, Brian, Jeffrey Smith, and Conny Wunsch. 2016. "Government-Sponsored Vocational Education for Adults." In *Handbook of the Economics of Education*, edited by Eric A. Hanushek, Stephen Machin, and Ludger Woessmann, 479–652. Amsterdam: Elsevier.

Mincer, Jacob. 1970. "The Distribution of Labor Incomes: A Survey with Special Reference to the Human Capital Approach." *Journal of Economic Literature* 8 (1): 1–26.

———. 1974. *Schooling, Experience, and Earnings*. New York: National Bureau of Economic Research.

Nelson, Richard R., and Edmund Phelps. 1966. "Investment in Humans, Technology Diffusion and Economic Growth." *American Economic Review* 56 (2): 69–75.

Organisation for Economic Co-operation and Development (OECD). 2016. *Skills Matter: Further Results from the Survey of Adult Skills*. Paris: OECD.

Ostry, Jonathan D., Andrew Berg, and Charalambos G. Tsangarides. 2014. "Redistribution, Inequality, and Growth." IMF Staff Discussion Note no. SDN/14/02, Washington, DC, International Monetary Fund, February.

Petty, Sir William. (1676) 1899. "Political Arithmetic." In *The Economic Writings of Sir William Petty*, edited by Charles Henry Hull, 233–313. Cambridge: Cambridge University Press.

Psacharopoulos, George, and Harry A. Patrinos. 2004. "Returns to Investment in Education: A Further Update." *Education Economics* 12 (2): 111–34.

Puma, Michael, Stephen Bell, Ronna Cook, Camilla Heid, Pam Broene, Frank Jenkins, Andrew Mashburn, and Jason Downer. 2012. *Third Grade Follow-Up to the Head Start Impact Study Final Report*. Washington, DC: Office of Planning, Research and Evaluation, Administration for Children and Families, US Department of Health and Human Services.

Reardon, Sean F. 2008. "Differential Growth in the Black-White Achievement Gap during Elementary School among Initially High- and Low-Scoring Students." IREPP Working Paper no. 2008-07, Institute for Research on Education Policy and Practice, Stanford University, March.

Reynolds, Arthur J., Judy A. Temple, Dylan L. Robertson, and Emily A. Mann. 2002. "Age 21 Cost-Benefit Analysis of the Title I Chicago Child-Parent Centers." *Educational Evaluation and Policy Analysis* 24 (4): 267–303.

Romer, Paul. 1990. "Endogenous Technological Change." *Journal of Political Economy* 99 (5, pt. 2): S71–102.

Ryan, Paul. 2001. "The School-to-Work Transition: A Cross-National Perspective." *Journal of Economic Literature* 39 (1): 34–92.

Schultz, Theodore W. 1975. "The Value of the Ability to Deal with Disequilibria." *Journal of Economic Literature* 13 (3): 827–46.

Schweinhart, Lawrence J., Jeanne Montie, Zongping Xiang, W. Steven Barnett, Clive R. Belfield, and Milagros Nores. 2005. *Lifetime Effects: The High/Scope Perry Preschool Study through Age 40*. Ypsilanti, MI: High/Scope Press.

Smith, Adam. (1776) 2010. *The Wealth of Nations*. Hollywood, FL: Simon and Brown.

Welch, Finis. 1970. "Education in Production." *Journal of Political Economy* 78 (1): 35–59.

Witte, John F. 2007. "A Proposal for State, Income-Targeted, Preschool Vouchers." *Peabody Journal of Education* 82 (4): 617–44.

Recent Flattening in the Higher Education Wage Premium
Polarization, Skill Downgrading, or Both?

Robert G. Valletta

9.1 Introduction

Holding a four-year college degree confers a distinct advantage to workers in the US labor market. The wage gaps between college-educated working adults and those with a high school degree—higher education wage premiums—are large and have grown substantially over the past thirty-five years. These gaps may have been bolstered by technological advances in the workplace, notably the growing reliance on computers and related technologies, because the skills that are needed to master and apply these technologies are often acquired through or associated with higher education (Krueger 1993; Autor, Katz, and Krueger 1998; Autor, Levy, and Murnane 2003; Acemoglu and Autor 2011).

The expansion of the higher education wage premium has not been completely uniform over time, however, with rapid growth in the 1980s followed by progressively slower growth ("flattening"). During the years 2000 through 2010, the wage premium for college-educated workers rose by only a small amount. Most recently, from 2010 to 2015, the wage premium for those with

Robert G. Valletta is a vice president and economist at the Federal Reserve Bank of San Francisco.

The author thanks David Autor for his highly constructive and detailed discussant comments; the editors of this volume, Charles Hulten and Valerie Ramey, for their guidance with revisions; and participants in this conference and also the May 2016 Society of Labor Economists annual meetings for additional helpful comments. He also thanks Catherine van der List for outstanding research assistance. The views expressed in this chapter are solely those of the author and are not attributable to the Federal Reserve Bank of San Francisco or the Federal Reserve System. For acknowledgments, sources of research support, and disclosure of the author's material financial relationships, if any, please see http://www.nber.org/chapters/c13705.ack.

college and graduate degrees was largely unchanged, suggesting that the factors propelling its earlier rise have disappeared.

While the wage advantage associated with higher education remains large, the lack of growth in recent years represents a departure from the earlier pattern. This change may have important implications for the value of higher education as an individual and social investment, and consequences for economic growth as well. Despite the voluminous literature on returns to education, little attention has been paid to slower growth in the college wage premium and differences between these higher education groups (Lindley and Machin [2016] is an exception).

In this chapter, I assess and attempt to explain the stalling of the higher education wage premium and its variation across the college-only and graduate-degree groups. I focus on two primary, related explanations for changing returns to higher education.

The first potential explanation is labor market "polarization" (Acemoglu and Autor 2011). This theory emphasizes a shift away from medium-skill occupations driven largely by technological change. It provides a broad, cohesive explanation for changes in employment patterns in the United States and other advanced economies in recent decades. Polarization may account for the slowdown in the college wage premium through a shift in the occupational distribution of college graduates toward jobs that are being displaced by automation technologies and related factors (such as outsourcing and rising trade). At the same time, rising demand for the cognitive skills possessed by graduate-degree holders may help maintain and expand their wage advantage relative to those holding a four-year college degree only (Lindley and Machin 2016).

I will refer to the second broad potential explanation for the flattening of higher education wage premiums as "skill downgrading," based on the recent work of Beaudry, Green, and Sand (2016). They emphasize a general weakening since the year 2000 in the demand for cognitive tasks in the workplace, reflecting a maturation in the information technology (IT) revolution and consequent slowdown in workplace IT investments. Skill downgrading in their framework refers to the process by which weaker demand for advanced cognitive skills cascades down the skill distribution as highly skilled workers, such as those possessing advanced degrees, increasingly compete with and replace lower-skilled workers in occupations that rely less heavily on advanced cognitive skills.

I begin my empirical assessment in the next section by establishing the basic facts regarding changes in educational attainment and the higher education wage premiums, distinguishing between individuals with a four-year college degree and those with graduate degrees. The analyses throughout are based primarily on data from the Current Population Survey (CPS) monthly earnings files (monthly outgoing rotation groups, or MORG) spanning the period 1979–2015. I also conduct selected parallel analyses using the CPS Annual Social and Economic Supplement files (March CPS-ASEC), which at the time of this writing provide earnings data through 2014. Standard wage regressions that adjust for changing workforce composition highlight the flattening of the higher education wage premiums noted above.

To help interpret these empirical findings, I then discuss the polarization and skill-downgrading arguments in more detail. Observed occupational employment shifts indicate the potential importance of polarization for the flattening of the college wage premium. The Beaudry, Green, and Sand skill-downgrading narrative takes polarization as its starting point but emphasizes different dynamics over time, with weaker demand for cognitive skills arising as a consequence of a slowdown in technology investment.

To assess the effects of polarization and skill downgrading on higher education wage premiums, I examine changing premiums within and between the broad occupation categories that are used to identify the extent of polarization. The results of these analyses suggest that polarization and skill downgrading have both contributed to the flattening of the wage premium for individuals with a four-year college degree or postgraduate degree. Consistent with the polarization story, the flattening in the wage premium is partly explained by shifting employment and relative wages across broad occupation groups, mainly for those with a college degree but no graduate degree. However, a substantial contribution also comes from the slowdown in the wage premium within broad occupation categories, consistent with skill downgrading and heightened competition between educational groups for similar jobs. In the conclusion, I discuss the implications of these findings for future research on the returns to higher education and its role in economic growth.

9.2   Changes in the Higher Education Wage Premium

The wage premium earned by individuals with higher educational attainment is commonly attributed to the more extensive skills that they possess (Card 1999; Goldin and Katz 2008). To save space, I will not review the voluminous and well-known literature on estimating and interpreting the returns to education, but will instead turn directly to updated estimates of the returns to higher educational attainment (college degrees and above).

9.2.1   Data and Descriptive Statistics

Because the data and processing procedures I use are well known, I describe them only briefly here, with additional details relegated to appendix A. The primary data used are from the CPS MORG files, compiled by the National Bureau of Economic Research (NBER) and available for the years 1979–2015 when this chapter was written. These files contain data for the quarter sample of the monthly CPS that receives survey questions regarding earnings and related variables in currently held jobs. I also use the complete monthly CPS files for selected tabulations that do not involve wages.

The data handling and processing procedures largely follow those detailed in Lemieux (2006a, 2010). These include elimination of observations with imputed values of earnings or hours and adjustments for changing top-codes. I use hourly wages as my earnings measure, either reported directly by hourly workers or formed as usual weekly earnings divided by usual weekly hours worked for salaried workers. All wage and earnings variables are deflated by the annual average value of the gross domestic product (GDP) deflator for personal consumption expenditures (and expressed in 2015 terms for ease of interpretation). For all of the analyses in this chapter, the samples are restricted to wage and salary workers age twenty-five to sixty-four (with farming and resource occupations excluded).

The basis for the measurement of educational attainment in the CPS switched in 1992 from the highest grade attained and completed to the highest degree received. I formed educational categories that are largely consistent over time following the guidance of Jaeger (1997).[1] Individuals with a graduate degree, along with information about the type of degree, are directly identified beginning in 1992. Graduate-degree holders prior to 1992 are identified as those reporting at least eighteen years of completed education. I code individuals who report seventeen years of completed schooling in the pre-1992 period as possessing a four-year college degree, but not a graduate degree.[2]

For comparison purposes, I also use data from the March CPS files to estimate changes in the higher education earnings premium. Compared with the MORG data, which provides information on earnings in the current reference week, the March CPS data refer to earnings in the complete prior calendar year. Following standard practice, I restrict the March CPS sample to full-time, full-year workers and use weekly earnings (annual labor earnings divided by weeks worked) as the earnings measure, once again dropping observations with imputed earnings or hours and adjusting for changing top-codes (e.g., Autor, Katz, and Kearney 2008). These files are currently available through 2015. Since the data refer to the prior calendar year, the reference period for the March data ends one year earlier than the MORG data (2014 rather than 2015).

---

1. Relative to Jaeger (1997), in the 1992-forward data I include individuals who report twelve years of schooling but no diploma in the "no degree" group rather than the "high school degree" group, to be consistent with the emphasis on degree attainment beginning in 1992.

2. Lindley and Machin (2016) take a similar approach, which groups individuals who drop out of a graduate program after one year or complete a one-year master's degree program with those who complete a four-year college degree only. This approach generates a slight discontinuity in the relative college/graduate shares in 1992, but the discontinuity is larger if instead such individuals are treated as having a graduate degree.

Table 9.1   Educational attainment shares and real hourly wages

	1980 (1)	1990 (2)	1992 (3)	2000 (4)	2010 (5)	2015 (6)
Panel A. Employment share						
No degree (< 12 yrs. education)	0.197	0.130	0.115	0.099	0.082	0.077
High school degree	0.371	0.368	0.358	0.314	0.280	0.256
Some college	0.205	0.238	0.259	0.280	0.280	0.278
College only (4-year)	0.158	0.183	0.177	0.205	0.232	0.247
Graduate degree	0.069	0.081	0.090	0.103	0.126	0.143
Graduate degree by type						
Master's			0.068	0.075	0.094	0.107
Professional			0.012	0.014	0.016	0.016
Doctoral			0.010	0.013	0.016	0.019
Panel B. Real hourly wage (2015$) (averages by group)						
No degree (< 12 yrs. education)	14.19	12.84	12.47	13.03	13.22	13.56
High school degree	16.33	15.99	15.87	17.20	17.77	17.98
Some college	18.80	19.29	19.16	20.84	21.47	21.59
College only (4-year)	22.85	25.32	25.18	28.98	30.49	30.93
Graduate degree	27.27	31.43	31.66	36.40	39.70	39.48
Graduate degree by type						
Master's			29.94	33.99	36.85	36.83
Professional			38.32	45.01	50.75	50.51
Doctoral			35.83	41.44	46.43	45.70

Notes: Author's calculations from CPS monthly files (panel A) and MORG files (panel B); sample weights used. See table 9.2 note for MORG sample description and counts. Master's degrees include MBAs along with a wide set of other master's degrees; professional degrees are JD, MD, and related.

Table 9.1 displays descriptive statistics for employment shares (panel A) and average real wages (panel B) by educational attainment, calculated using the full monthly CPS files for the employment shares and the MORG files for the wage data. These are provided for ten-year intervals that largely span the sample frame. The table also lists statistics for selected other years, including the year that the education variables changed (1992) to bridge the gap in definitions, and a listing for the final data year (2015).

Panel A of table 9.1 illustrates the well-known, steady decline in the employment share of individuals whose educational attainment is a high school degree or less accompanied by a steady rise in the share of individuals possessing a four-year college degree or graduate degree. As of 2015, nearly 40 percent of employed individuals age twenty-five to sixty-four held at least a college degree, and one in seven held a graduate degree, accounting for slightly more than a third of employed college graduates. Master's degrees (which include MBAs) account for most of the level and change in the fraction holding graduate degrees, along with a large proportional increase for the small share of doctoral degrees.

Panel B of table 9.1 illustrates the large wage gaps between the educational attainment groups, with the spread in real wages between the graduate-degree group and those with less than a high school degree widening approximately from a factor of two to a factor of three over the sample frame. Average real wages changed little over the sample frame for those with a high school degree or less. For those with at least some college education, average real wages rose somewhat between 1980 and 2000, with larger increases evident for those with higher educational attainment. Between 2000 and 2010, only holders of graduate degrees saw any meaningful increase in real wages. Between 2010 and 2015, real wages were flat to down slightly for all groups. The gap in average real wages between individuals with a four-year college degree or graduate degree and high school graduates rose from 40 to 67 percent in 1980 to 72 to 120 percent as of 2015.

9.2.2   Composition-Adjusted Estimates of Wage Gaps

To assess the changing wage premium associated with higher educational attainment, I estimate standard log-wage equations of the following form (where i indexes individuals):

$$(1) \qquad \mathrm{Ln}(w_i) = X_i\beta + S_i\Gamma + \varepsilon_i,$$

where X_i represents a set of demographic controls and S_i represents educational attainment (measured in discrete categories). This equation is estimated separately for each year using the MORG and March CPS data as described above. The control variables in the vector X include dummy variables for seven age groups (e.g., thirty to thirty-four, etc., with twenty-five to twenty-nine omitted), three racial/ethnic groups, gender, marital status, gender * marital status, and geographic location (nine census divisions). These controls adjust for the changing composition of the estimation sample, so that the results for the education categories reflect the average wage premium associated with educational attainment for an individual with a fixed set of demographic characteristics (X).[3]

Our interest centers on the estimated vector of coefficients (Γ) on a set of dummy variables representing discrete categories of educational attainment (S). Table 9.2 lists the numerical results for selected years, while figure 9.1 displays the results for the complete sample period of 1979 through 2015 (2014 for the March CPS).[4] For both displays, panel A lists the results for the MORG data, while panel B lists the results for the March CPS. The results are expressed in natural log terms. These conditional wage gaps are displayed

3. The results reported below are very similar when this set of control variables is replaced by complete interactions between four decadal age categories, four race/ethnic categories, the two genders, and marital status (married spouse present or not), for a total of sixty-four demographic cells.
4. The estimated coefficients for college and postgraduate educational attainment are highly statistically significant in virtually all cases reported below, with the exception of a few group-specific estimates reported in table 9.3.

Table 9.2   Composition-adjusted wage/earnings differentials (log points, relative to high school graduates)

	1980 (1)	1990 (2)	1992 (3)	2000 (4)	2010 (5)	2015 (6)
		Panel A. CPS MORG data				
Full sample						
College degree or higher	0.304	0.449	0.464	0.518	0.566	0.566
	(.003)	(.003)	(.003)	(.004)	(.004)	(.005)
College only (4-year)	0.270	0.402	0.403	0.451	0.475	0.477
	(.004)	(.004)	(.004)	(.005)	(.005)	(.005)
Graduate degree	0.383	0.553	0.581	0.648	0.727	0.712
	(.005)	(.005)	(.005)	(.006)	(.006)	(.006)
Observations	121,001	123,111	119,014	83,314	85,397	76,789
College degree or higher sample						
Graduate degree	0.111	0.149	0.170	0.194	0.245	0.226
	(.006)	(.006)	(.006)	(.007)	(.006)	(.006)
Observations	27,042	33,334	32,684	26,789	32,305	31,572
		Panel B. CPS March data				
Full sample						
College degree or higher	0.293	0.449	0.477	0.538	0.579	0.576
	(.006)	(.006)	(.006)	(.006)	(.006)	(.007)
College only (4-year)	0.260	0.400	0.415	0.468	0.488	0.488
	(.007)	(.007)	(.007)	(.006)	(.007)	(.007)
Graduate degree	0.368	0.557	0.593	0.680	0.740	0.725
	(.009)	(.009)	(.009)	(.008)	(.008)	(.008)
Observations	34,258	38,123	37,143	52,489	45,575	43,435
College degree or higher sample						
Graduate degree	0.102	0.155	0.174	0.206	0.244	0.230
	(.011)	(.010)	(.010)	(.009)	(.008)	(.009)
Observations	8,184	10,630	10,709	16,350	17,608	17,540

Notes: Estimated coefficients from ln(wage or earnings) regressions for the years indicated in the column labels; horizontal lines identify coefficients obtained from separate regressions. Standard errors in parentheses. Samples are wage and salary workers age twenty-five to sixty-four for both data sources, restricted to full-time, year-round workers (annual hours ≥ 1,750) in the CPS March data. Dependent variable is ln(hourly earnings) for the MORG data and ln(weekly earnings) for the CPS March data, with allocated values dropped and top-code adjustments (see the text and appendix). Composition adjustment relies on the inclusion of the following control variables (all categorical): seven age, three race/ethnic, married, female, married × female, and eight geographic divisions.

for three educational groupings: the broad group of all workers with at least a four-year college degree, and the two subgroups consisting of those with a four-year degree only ("college only"), and those who hold a postgraduate degree as well. The results for the "college degree or higher" group are based on regressions that are estimated separately from the one used to estimate the returns for the two subgroups (as indicated by the horizontal lines in

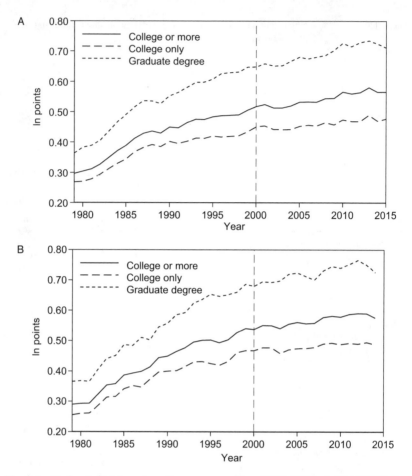

Fig. 9.1   Estimated higher education wage premium, 1979–2015. *A*, CPS MORG data (1979–2015); *B*, March CPS data (1979–2014).

Notes: Author's calculations using CPS MORG and March data (see table 9.2 note). Differentials expressed relative to high school graduates.

the table). The higher education wage premiums are first expressed relative to the wages of high school graduates. In addition, separate estimates are provided for those holding a graduate degree. These are based on the restricted sample of individuals who have at least a college degree, hence they represent the graduate wage premium relative to the wages of the college-only group.

The estimates in table 9.2 and figure 9.1 show that the wage premiums for higher education generally have been rising over time. However, both data sets show that the growth has slowed in recent decades, with the slowdown for the graduate group lagging behind that for the college-only group. The

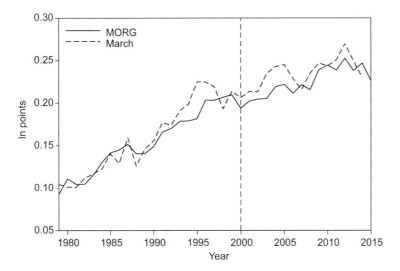

Fig. 9.2   Estimated graduate-degree wage premium
Notes: Author's calculations using CPS MORG and March data (see table 9.2 note). Differentials expressed relative to four-year college graduates.

rate of growth in the college-only wage premium was cut approximately in half between the 1980s and 1990s and then slowed virtually to a standstill after 2000. It rose about 2 to 2.5 log points through 2010 and then was unchanged between 2010 and 2015.

For the graduate-degree group, the slowdown over time is most evident based on the results for the college or higher sample. These are displayed at the bottom of both panels in table 9.2 and also in figure 9.2, where the results for the MORG and March data sets are directly compared. The estimated wage premiums are very similar in the two data sources, with somewhat greater annual volatility evident in the March data for the college-only sample in figure 9.2 due to its smaller sample. Relative to the college-only group, individuals with a graduate degree saw consistent wage premium gains of about 4 to 5 log points in each of the decades of the 1980s, 1990s, and first decade of the twenty-first century. During this time frame, their wage advantage over college-only workers grew steadily, reaching nearly 25 log points by 2010. However, since 2010, the graduate-degree premium is down slightly in both data sources (through 2015 in the MORG data and 2014 in the March data).

9.2.3   Robustness Checks and Disaggregation by Age and Gender

One potential concern with respect to these results is the possibility that they reflect underlying changes in employment conditions among narrow worker groups or industries. Such narrow changes may be independent of the broad occupational changes and shifting labor market competition related to polarization and skill downgrading (which are discussed and analyzed below, in sections 9.3 and 9.4). One such narrow group is teachers, who constitute a substantial but declining share of employed college graduates.[5] Excluding educator and librarian occupations from the regressions raises the estimated higher education wage premiums by 2–4 log points in general. However, the pattern over time is unchanged relative to the full sample results, with progressive flattening in the wage premiums and no change from 2010 forward.

It is also important to consider the potential influence of changing conditions in key industries that employ large numbers of college graduates. One such industry is the financial sector, for which the housing bust and financial crisis tied to the Great Recession of 2007–2009 destroyed a disproportionate number of jobs. Many finance-sector jobs are highly paid, and their disappearance may have affected the higher education wage premium. However, exclusion of workers employed in the financial, insurance, and real estate sectors from the regression analysis has virtually no impact on the estimated wage premiums and their pattern over time.[6] Similarly, Beaudry, Green, and Sand (2016) highlight the role of the business and management services industries for their findings, emphasizing substantial employment changes for young college graduates in this sector. Exclusion of individuals employed in these industries does not affect the estimated college-only wage premium. It does raise the level of the graduate school wage premium, suggesting a relatively low value for graduate degrees in this industry. Nonetheless, the pattern of the higher education wage premiums over time, as reflected in the results from table 9.2 and figure 9.1, is unaffected.

It is also instructive to examine the higher education wage premium decomposed by age group and gender. Analyses of employment and wage patterns for the college educated often highlight younger workers, who are likely to experience the most immediate effects of changing employment conditions across educational attainment groups (e.g., Beaudry, Green, and Sand 2014). Figure 9.3 parallels figure 9.1 (panel A, MORG), but displays wage premiums for the youngest decadal age group in my sample (age twenty-five to thirty-four) in panel A and an older group (age forty-five to fifty-four) in panel B.[7] For younger workers, movements in the wage premi-

---

5. Among workers with at least a four-year college degree in the MORG data, the fraction of educators and librarians declined by a third over my sample frame, from 24 percent in 1979 to about 16 percent in 2015.

6. Separate analyses by broad industry, also noted in section 9.4, show that the higher education wage premiums within the finance sector broadly track the patterns evident for the overall economy.

7. The underlying regressions used to produce the results in figure 9.3 are identical to those used for table 9.1 and figure 9.1, except the samples are restricted to the indicated age groups and the age category controls are adjusted accordingly. I use age forty-five to fifty-four rather than the oldest group in my sample, age fifty-five to sixty-four, to minimize the influence of partial retirement decisions on relative earnings over time.

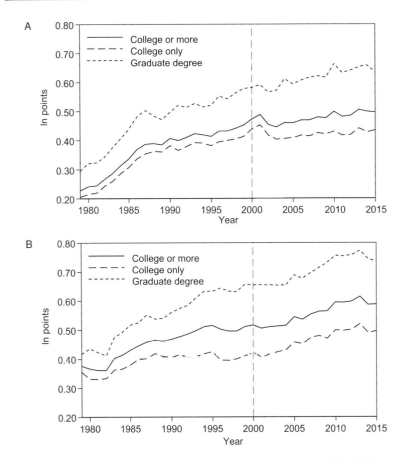

Fig. 9.3   Estimated higher education wage premium by age group, 1979–2015. *A*, age twenty-five to thirty-four; *B*, age forty-five to fifty-four.
Notes: Author's calculations using CPS MORG data (see table 9.2 note). Differentials expressed relative to high school graduates.

ums over time largely parallel those for the complete sample in figure 9.1, with large gains in the 1980s followed by slower gains in the 1990s and the first decade of the twenty-first century, and no change since 2010. By contrast, for older workers the college-only premium was largely flat in the 1990s, perhaps because this group did not readily adapt to the new information technologies introduced during that decade. The college-only premium for older workers picked up in the early twenty-first century, although like the graduate-degree premium, it has been flat since 2010. Comparison across the two panels in figure 9.3 also indicates that the higher education wage premiums are larger for older than for younger workers, by about 5 to 10 log points, on average. This likely arises due to important interaction or

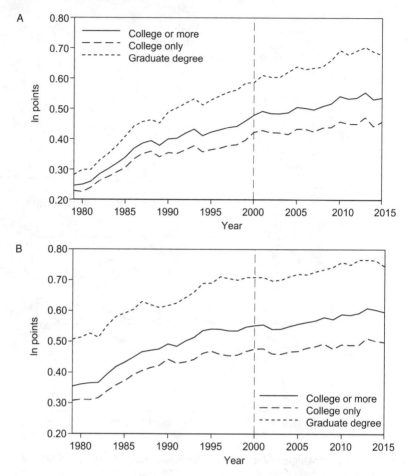

Fig. 9.4   Estimated higher education wage premium by gender, 1979–2015. *A*, men; *B*, women.
Notes: Author's calculations using CPS MORG data (see table 9.2 note). Differentials expressed relative to high school graduates.

reinforcing effects between higher education and the subsequent acquisition of on-the-job skills that raise wages as workers age.

Given the well-known increase in the attainment of higher education for women relative to men, it is also informative to examine the wage premiums by gender. These are displayed in figure 9.4 (panel A for men, panel B for women). The series represent the composition-adjusted higher education wage premiums by gender; as such, they reflect relative wages within gender group and hence should not be interpreted as capturing wage differences between men and women. The higher education wage premiums are larger for women than for men, although the gap has closed over time, especially for graduate degrees. The pattern over time for both genders is similar to that for the overall sample in figure 9.1 (panel A), with a flattening of the wage premiums over time and essentially no change since 2010.

9.2.4   Summing Up: Higher Education Wage Premiums over Time

The results presented in this section indicate general flattening in the wage premiums associated with four-year college and graduate degrees. The sharp increases observed in the 1980s have been followed by much slower gains. Since the year 2000, the wage premium associated with a four-year college degree has changed little. By contrast, from 2000 to 2010, the wage premium for holders of graduate degrees relative to those with four-year college degrees continued to grow at its previous pace, contributing to increasing "convexification" in the returns to higher education (Lemieux 2006b; Lindley and Machin 2016). Since 2010, however, wage premiums for both groups have sputtered. They remain large but were essentially unchanged for the college-only group and down slightly for holders of graduate degrees. These patterns indicate that the factors propelling earlier increases in the returns to higher education have dissipated.

Because of the significant time required for individual investments in higher education—four years or more—the flatter wage premiums may reflect a delayed response of the supply of college-educated individuals to earlier increases in demand (Acemoglu and Autor 2011; Autor 2014). However, given the relatively consistent increase over time in the college-educated employment share listed in table 9.1 (panel A), factors on the demand side that affect relative productivity and employers' preference for workers with higher education merit further consideration.

9.3   Potential Explanations: Polarization and Skill Downgrading

The slower growth and eventual flattening in the wage premium for higher education documented in the preceding section raises the possibility that the factors propelling rising wage premiums for highly skilled workers have dissipated. Past accounts of rising wage premiums for skilled workers generally revolved around the skill-biased technological change (SBTC) explanation of labor market developments. Under SBTC, rising reliance on sophisticated workplace technologies boosts the employment and wages of workers, mainly the highly educated, whose skills enable them to apply those technologies (e.g., Bound and Johnson 1992; Autor, Katz, and Krueger 1998). Recent research has pointed to factors that may alter or offset this process. I focus on two broad explanations: labor market polarization and skill downgrading.

9.3.1   Polarization and Skill Downgrading: The Basics

The "polarization" hypothesis is a leading explanation for recent employment developments in the United States and other advanced countries (Goos and Manning 2007; Acemoglu and Autor 2011; Autor 2015; Goos, Manning, and Salomons 2014). This is a refinement of the SBTC story that accounts for excess employment growth in the top and bottom portions of the wage distribution, with erosion in the middle.

In the polarization framework, evolving workplace technologies undermine demand for "routine" jobs, in which workers and the tasks they perform are readily substituted by computer-intensive capital equipment and processes. They include white-collar office jobs (e.g., bookkeeping and clerical work), termed "routine cognitive" jobs, and blue-collar occupations that involve repetitive production or monitoring activities, termed "routine manual" jobs. These routine jobs are concentrated toward the middle of the wage and skill distribution. By contrast, workers in high-wage "nonroutine cognitive" (or "abstract") jobs tend to have skills that are complementary with computer-based technologies, while low-wage service workers in "nonroutine manual" jobs are neither substitutes nor complements with computer-based technologies. Polarization arising from changes in domestic production technologies may be reinforced by related changes in overseas production technologies through the impact of offshoring and import competition (see, e.g., Autor, Dorn, and Hanson 2013).

Beaudry, Green, and Sand (2016) provide a related but alternative framework for understanding changing occupational employment patterns over the past few decades. They rely on a basic variant of the polarization hypothesis as their starting point, but they emphasize a slowdown in IT investments that has undermined the demand for cognitive skills since the year 2000. In their narrative, weaker demand for cognitive skills and the consequent impact on highly skilled workers has cascaded down the skill distribution, undermining the demand for lesser skilled workers as well. They refer to this process as "skill downgrading," which contrasts with the opposite pattern of "skill upgrading" that occurs during the initial period of accelerating IT investments.

The similarities and contrasts between the polarization and Beaudry, Green, and Sand skill-downgrading scenarios can be readily summarized with reference to the production functions and associated objective functions that underlie the two models. In a basic model of polarization, firms rely on cognitive and routine task inputs supplied by workers for production, combined with inputs of computer capital (see, e.g., Autor, Levy, and Murnane 2003).[8] The firm aims to maximize profits π by choosing appropriate input combinations given its production function F:

$$(2) \qquad \max_{\Omega, L_c, L_r} = p * F(\Omega, L_c, L_r, \theta) - r\Omega - w_c L_c - w_r L_r$$

where p is the price of the firm's output, Ω is a form of technological (computer) capital with per-unit rental rate r, L_c and L_r are inputs of cognitive

---

8. This representation is adapted from Autor, Levy, and Murnane (2003), modified to be broadly consistent with the notation and framework in Beaudry, Green, and Sand (2016). Nonroutine manual jobs are largely ignored here for simplicity and because they have limited relevance for college-educated workers.

and routine labor with wage rates w_c and w_r, and θ is a technology parameter that shifts the level of output for a given set of inputs (assumed constant in this basic version of the model, but allowed to change in the Beaudry, Green, and Sand variant below). The production function $F(\cdot)$ is assumed to reflect constant returns to scale and hence diminishing marginal productivity for individual inputs.

Production efficiency requires hiring labor inputs up to the point where each input's marginal product equals its market wage or rental rate. Importantly, Autor, Levy, and Murnane (2003) assume that computer capital is perfectly substitutable with routine labor inputs, implying complementarity between computers and cognitive (nonroutine) labor in their setting. In this framework, as the price of computer capital falls, production techniques shift toward greater reliance on cognitive labor inputs and less on routine labor inputs, with corresponding reductions in the relative wage paid for routine labor inputs. Because routine tasks are common among many jobs toward the middle of the wage distribution, polarization will tend to erode or "hollow out" middle-class jobs and wages.

Beaudry, Green, and Sand extend the basic polarization model by incorporating the key feature that cognitive labor inputs create a stock of organizational capital for firms, which enables them to develop and utilize new technologies. This is captured in the following modification of equation (2), which is a discrete-time version of the objective function from equation (1) in Beaudry, Green, and Sand (2016):

(3)
$$\max_{L_c, L_r} \pi = p * F(\Omega, L_r, \theta) - w_c L_c - w_r L_r$$

$$s.t.\ \Delta\Omega = L_c - \delta\Omega_{-1}.$$

Relative to the production function in equation (2), Ω in equation (3) represents intangible "organizational capital" rather than tangible computer capital. In this modified framework, cognitive labor inputs do not directly affect current production but instead contribute to output through the accumulation of organizational capital (which depreciates at the rate δ). The first-order conditions for production efficiency are similar to the basic polarization model from equation (2).

This modified model is distinguished by its dynamic properties in response to a technological shift, or change in θ. Beaudry, Green, and Sand assume that an increase or improvement in the technology factor θ raises the productivity of the organizational capital accumulated through the use of cognitive labor inputs but has no direct effect on the productivity of routine labor inputs. These model features generate a "boom-bust cycle" in the demand for cognitive tasks and overall labor demand in response to technological improvement. In particular, the dynamics of the model predict that the stock of cognitive tasks/skills grows during the boom, as the economy adjusts to the need for additional organizational capital to manage the new technology.

Once the level of organizational capital becomes sufficiently large for appropriate use of the new technology, the demand for cognitive tasks declines as their use is shifted from expanding organizational capital to maintaining it by offsetting depreciation (similar to the pattern in existing models of technology diffusion and capital investment).

The Beaudry, Green, and Sand model can predict the strong growth in demand and wages for workers in jobs that rely heavily on cognitive tasks/skills up to the year 2000—the boom phase—followed by a decline thereafter—the bust phase. The demand reversal during the bust phase causes high-skilled workers to move down the occupational ladder and replace lower-skilled workers, pushing the latter group further down the occupational ladder ("skill downgrading") and perhaps out of the labor market entirely.[9]

9.3.2 Descriptive Evidence

Broad empirical evidence suggests that polarization and skill downgrading are both contributing to changing employment patterns and hence may be affecting higher education wage premiums.

Patterns of occupational job growth in recent decades confirm the relevance of the polarization narrative. Labor demand and job growth have been relatively rapid in the high-wage nonroutine cognitive and low-wage nonroutine manual categories, with the middle-wage routine jobs experiencing downward pressure. This pattern can be seen in figure 9.5, which displays annual rates of job growth for the four broad polarization categories over four subperiods (classified using the broad occupational scheme from Acemoglu and Autor [2011]; see appendix B for the correspondence).[10] The figure shows substantial growth in the 1980s, followed by a slowdown in the 1990s for all groups (reflecting in part the impact of the early 1990s recession).[11] Polarization is evident in the 1990s, reflected in a sharper slowdown for the routine versus the nonroutine categories. This process appeared to accelerate after the year 2000, with substantial gains for nonroutine jobs and substantial net losses for routine jobs, particularly during the Great Recession of 2007–2009 and the subsequent recovery.[12]

9. As Beaudry, Green, and Sand note in their introduction: "In this maturity stage, having a college degree is only partly about obtaining access to high-paying managerial and technology jobs—it is also about beating out less educated workers for barista and clerical-type jobs."

10. Autor (2015) relabeled the nonroutine categories and collapsed the two routine categories into a single one. I maintain the original four-group categorization based on the cognitive/manual and routine/nonroutine distinctions due to the preponderance of college graduates in each of the cognitive categories.

11. The start year of 1983 was dictated by the availability of official BLS occupational employment data beginning in that year, and the change between 1999 and 2000 is omitted to eliminate the influence of a discontinuity in occupation category definitions.

12. The differential growth rates across the broad occupation categories have generated significant changes in their employment shares over time. Nonroutine cognitive jobs are the largest category: their share rose from about 30 percent to slightly over 40 percent of all jobs during the sample frame. The share of routine jobs declined from nearly 60 percent to about 45 percent. The share of nonroutine manual jobs rose from about 12 to about 15 percent, mostly since the year 2000.

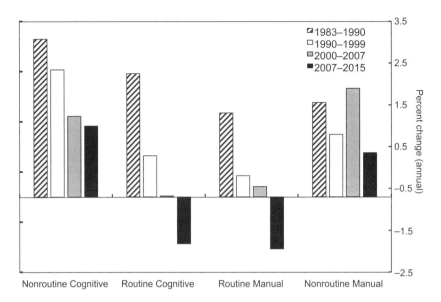

Fig. 9.5   Employment growth by broad occupation category, subperiods from 1983 to 2015

Notes: Author's calculations from Bureau of Labor Statistics data. See text and appendix table 9B.1 for occupation category definitions.

Polarization will differentially affect highly educated and less educated groups due to their very different occupational distributions. Figure 9.6 shows the shares of the college-only and graduate-degree groups in the nonroutine cognitive (panel A) and routine cognitive (panel B) categories. Workers with at least a college degree account for a large and rising share of nonroutine cognitive jobs, reaching nearly 70 percent by 2014 (panel A). Underlying this pattern is a significant rise in the share of nonroutine cognitive jobs held by individuals possessing a graduate degree, with little change in the share from the college-only group. This pattern is consistent with rising demand for the most highly educated individuals in jobs that require extensive nonroutine cognitive skills. The college-only group share also has grown in the routine cognitive category (panel B), commensurate with their rising share of the overall workforce.

Figure 9.7 reverses the figure 9.6 calculations by displaying the share of nonroutine cognitive jobs within the college-only and graduate-degree groups. Among the college-only group, the fraction employed in nonroutine cognitive jobs declined between 2000 and 2015, from about 68 to 64 percent. By contrast, the share of graduate-degree holders employed in nonroutine cognitive jobs has been largely stable at about 90 percent in recent years, while their overall workforce share has grown.

These tabulations suggest that polarization may be an important factor underlying the rising relative return to postgraduate education. As discussed

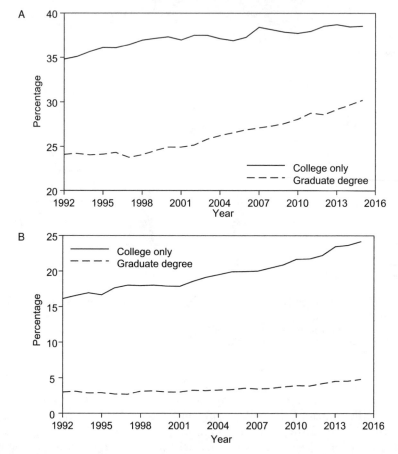

Fig. 9.6   Higher educational attainment shares by occupation category (selected), 1992–2015. *A*, nonroutine cognitive; *B*, routine cognitive.

Notes: Author's calculations using monthly CPS files. See text and appendix table 9B.1 for occupation category definitions. Series are shares of educational attainment groups in the broad occupation categories.

by Autor (2015), the wage impacts of polarization depend not only on skill/ technology complementarity, but also on (a) the demand elasticity for products and services that rely heavily on the different skill/task groups, and (b) labor supply elasticities for the different skill/task groups. In regard to nonroutine cognitive jobs, both factors imply that workers in these jobs are likely to see their wages rise in response to rising reliance on computer and automation technologies (assuming that their skills are complementary with computers). Demand for their output is relatively elastic, and an inelastic supply response due to the time required for acquiring additional education implies that the supply of such workers does not respond quickly to rising

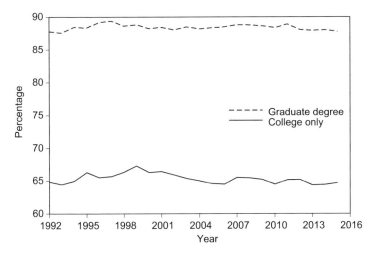

Fig. 9.7   Share of nonroutine cognitive employment by educational attainment, 1992–2015

Notes: Author's calculations using monthly CPS files. See text and appendix table 9B.1 for occupation category definitions. Series are nonroutine cognitive jobs as a share of employment within each educational attainment group.

demand. As such, ongoing polarization should put upward pressures on the relative wages of individuals employed in nonroutine cognitive jobs, most of whom have college or graduate degrees.[13]

As discussed above, the Beaudry, Green, and Sand "skill downgrading" alternative takes polarization as its starting point, but emphasizes a more general decline in demand for cognitive skills, which may affect all educational attainment groups. Beaudry, Green, and Sand present evidence to support the claim that the demand for cognitive and technological skills in the US labor market has weakened since the year 2000. They focus on broad patterns in employment across occupational and educational attainment groups, distinguishing between jobs that are intensive in cognitive versus routine or manual skills. The patterns in employment growth that they document are consistent with a reversal in the demand for cognitive skills, notably a slowdown in the relative rate of employment growth for occupations that are toward the top end of the wage distribution. They also use a more detailed identification scheme for cognitive-task-intensive jobs and confirm the shift out of such jobs by college graduates implied by my figure 9.7 (see their figure 10).

---

13. Based on these considerations, Autor (2014) notes that while polarization is likely to lower wages of workers in routine skill/task occupations, wages for workers in nonroutine manual jobs are likely to be relatively unaffected by polarization, despite the favorable polarization effects on employment for that group.

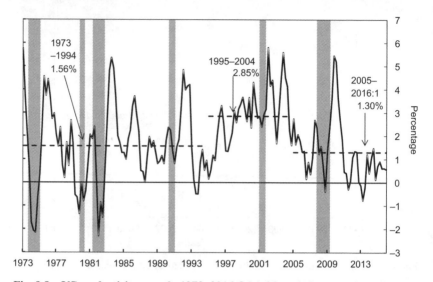

Fig. 9.8   US productivity growth, 1973–2016:Q1 (with period averages)
Notes: Author's calculations from US Bureau of Labor Statistics data. Series displayed is labor productivity in the nonfarm business sector, percentage change from four quarters earlier. Gray areas denote NBER recession dates.

One key element of the Beaudry, Green, and Sand framework and predictions is a pickup followed by a decline in technological advance, which generates the boom-bust cycle for cognitive employment and eventual skill downgrading in their model. This assumption is supported by patterns in US productivity growth in recent decades, depicted in figure 9.8.[14] The growth in output per worker measured by productivity gains generally reflects improvements in production technologies. Figure 9.8 shows a sharp productivity acceleration from 1995 to 2004, which corresponds roughly to the period of diffusion for new IT technologies that motivated the Beaudry, Green, and Sand model.[15] This was followed by an even more pronounced downshift in productivity growth. Productivity gains were especially slow from 2010 forward, the period during which the wage premium for higher education was flat or down (as discussed earlier in section 9.2). This correspondence suggests that the Beaudry, Green, and Sand narrative of a technology slowdown is relevant for understanding the recent pattern in the higher education wage premium.

Beaudry, Green, and Sand note that their model has limited implications for relative wages across skill groups. They also note, however, that a simple

14. I thank my colleague John Fernald for his advice with this display; see also Fernald (2015).
15. Beaudry, Green, and Sand focus on the year 2000 as a dividing line for the slowdown in demand for cognitive skills. However, it is likely that firms' ability to utilize new organizational capital associated with the IT revolution, and hence increase measured productivity, continued for a time after investment in that capital and corresponding rapid expansion of cognitive jobs largely came to an end.

parametrization of their model generates the slowdown in wage growth across the skill spectrum observed during the "bust" phase, consistent with the observed slowdown in real wages beginning in the year 2000 (see my table 9.1, panel A).[16] The skill-downgrading narrative also can explain the recent elevated level of "underemployment" of young college graduates, defined as the tendency for them to work in jobs that do not strictly require a college degree (see Abel and Deitz, chapter 4, this volume).

9.4   Wage Effects of Polarization and Skill Downgrading

The confluence of polarization and skill-downgrading influences on the labor market in recent years has been noted by others. Autor (2015), Lindley and Machin (2016), and Beaudry, Green, and Sand (2016) all provide a balanced, informed discussion and interpretation of labor market developments from 2000 forward and acknowledge the possibility that polarization and skill downgrading may both be playing a role. Each may have contributed to the flattening of the higher education wage premiums documented in section 9.2.

No sharp dividing lines between the two explanations are readily apparent. However, some insight can be gleaned by examining the wage premium patterns within and across the four broad occupation groups used in the polarization typology. The descriptive evidence presented in section 9.3.2 showed complex changes in the employment and wage patterns of highly educated individuals across the broad polarization occupation grouping in recent years. A within-between analysis is a relatively straightforward means for combining these changes into a single set of summary results.

This analysis begins with the same wage regressions as reported in section 9.2, but with separate regressions run for each of the four broad occupation groups from the polarization typology. Let γ represent a higher education wage premium (college or more, college only, or graduate degree) estimated for a specific year based on equation (1) and reported in table 9.2. The overall premium estimate can be decomposed as follows:

$$(4) \qquad \gamma = (\text{within effect}) + (\text{between effect})$$

$$= \sum_{j=1}^{4} w_j * \gamma_j + (\text{between effect})$$

where j subscripts the four broad occupation groups in the polarization typology, the γ_j's are occupation-specific estimates of the higher education premium, and the weights w_j are set equal to the share of each occupation group in total employment.

The within component in equation (4) is defined as the employment-

---

16. The slowdown in real wage growth for all educational groups displayed in table 9.1 is maintained when the data are adjusted for the same individual characteristics as used for the regression analyses in tables 9.2 and 9.3 (using a reweighting methodology).

weighted sum of the estimated occupation-specific wage premiums. It represents the higher education wage premium conditional on occupational skill/task group. It can be interpreted as the competitive advantage enjoyed by individuals with higher educational attainment when competing directly with less educated individuals for similar jobs (within the broad polarization occupation groupings). As such, a decline in the within component likely reflects a Beaudry, Green, and Sand skill-downgrading effect, which causes enhanced competition across educational groups for similar jobs. The between component is obtained as the difference between the total estimate and the within component.[17] It does not have a precise interpretation in the context of the polarization and skill-downgrading narratives: a relative increase in the shares of college-educated workers in routine jobs could reflect ongoing polarization in the distribution of jobs or the process of skill downgrading. However, it is informative nonetheless to assess whether the changes in the wage premium are associated with shifts in the occupational distribution of employment by education group.

I conduct this analysis by first estimating higher education wage premiums within each of the four broad polarization occupation groups, which provide the inputs into equation (4) above. The regressions are otherwise identical to those reported in table 9.2. Table 9.3 lists the regression results, focusing on the college-only premium (measured relative to high school graduates) and the graduate-degree premium (measured relative to the college-only group), with results for the same set of years as table 9.2 listed. The panel immediately below the regressions lists the decomposition of the "total" effect into "within" and "between" components.

The regression results in table 9.3 indicate that the higher education wage premiums are widely dispersed and their changes over time have been relatively consistent across the occupation groups. The exception is routine manual jobs, in which the higher education wage premium is relatively small for both education groups: the college-only premium is about half its size relative to the estimates for the other three groups, and the graduate premium is not statistically different from zero.

These patterns imply that increases in the total effect over the complete sample frame have been primarily driven by changes in the within component, with limited movement in the between component. This is confirmed by the decomposition results listed in table 9.3 for selected years and displayed in figure 9.9 for the complete sample frame. The within component, representing a competitive advantage to higher education within broad occupation groups, accounts for virtually all of the increase in the higher education wage premiums over time. However, the between component con-

---

17. Note that the total effect corresponds to the full-sample estimates from table 9.2. For example, the first total effect listed in column (1) of table 9.3, 0.270, corresponds to the college-only estimate from column (1) of panel A in table 9.2.

ıble 9.3   **Within-between analysis of higher education wage premiums (CPS MORG data, regressions by broad occupation groups)**

	1980 (1)	1990 (2)	1992 (3)	2000 (4)	2010 (5)	2015 (6)
ollege only versus high school degree (full sample) egressions						
Nonroutine cognitive	0.215	0.303	0.305	0.350	0.378	0.392
	(.007)	(.007)	(.007)	(.009)	(.009)	(.011)
Routine cognitive	0.134	0.255	0.265	0.309	0.346	0.327
	(.007)	(.007)	(.007)	(.009)	(.009)	(.010)
Routine manual	0.056	0.142	0.134	0.131	0.163	0.160
	(.010)	(.011)	(.011)	(.014)	(.014)	(.014)
Nonroutine manual	0.166	0.246	0.256	0.286	0.297	0.325
	(.014)	(.014)	(.014)	(.017)	(.013)	(.014)
ecomposition						
Within component share of total	0.135	0.237	0.241	0.278	0.314	0.322
	0.501	0.590	0.598	0.616	0.662	0.674
Between component share of total	0.135	0.165	0.162	0.173	0.160	0.155
	0.499	0.410	0.402	0.384	0.338	0.326
Total	0.270	0.402	0.403	0.451	0.475	0.477
raduate degree (college degree or higher sample) egressions						
Nonroutine cognitive	0.068	0.096	0.109	0.128	0.170	0.154
	(.006)	(.006)	(.006)	(.007)	(.007)	(.007)
Routine cognitive	0.101	0.085	0.112	0.157	0.147	0.135
	(.021)	(.018)	(.018)	(.025)	(.022)	(.022)
Routine manual	0.034	0.032	0.033	0.072	0.041	0.017
	(.031)	(.033)	(.037)	(.040)	(.042)	(.042)
Nonroutine manual	−0.032	−0.008	0.015	0.151	0.194	0.130
	(.048)	(.046)	(.044)	(.050)	(.035)	(.036)
ecomposition						
Within component share of total	0.068	0.087	0.103	0.131	0.163	0.144
	0.612	0.587	0.605	0.674	0.667	0.637
Between component share of total	0.043	0.061	0.067	0.063	0.081	0.082
	0.388	0.413	0.395	0.326	0.333	0.363
Total	0.111	0.149	0.170	0.194	0.245	0.226

ɔtes: See note to table 9.2 for basic data and specifications. Coefficients listed with standard errors in pa-
ıntheses; regressions run separately for each of the broad occupation groups listed, by year. See the text for
ʝescription of the decomposition.

tributed to a slight increase in the wage premiums for both higher education
groups up to the year 2000, indicating an ongoing shift toward higher-paid
cognitive jobs for college-educated workers.

Our primary goal is to understand and interpret the changes in the
within and between components since the year 2000. For the college-only
group, the within component continued to grow after the year 2000 at

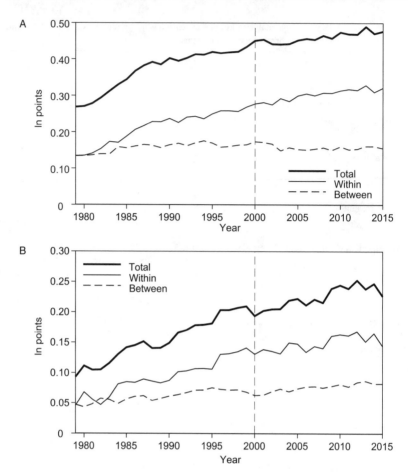

Fig. 9.9   Total and within/between wage premiums, 1979–2015. *A*, college only (relative to high school graduates); *B*, graduate degree (relative to college graduates).

Notes: See text and table 9.3 for data and methods. Based on broad polarization occupation grouping (nonroutine cognitive, routine cognitive, routine manual, nonroutine manual).

nearly the same rate as in the 1990s, rising by about 3.5 log points through 2010 and another 1.0 points through 2015. However, this was offset by a decline in the between component of about 2 log points from 2000 through 2015. This suggests that despite the increase in the college-only group's competitive advantage over lesser educated individuals within broad occupation groups, their overall wage advantage has been eroded slightly by a shift toward routine jobs. This is consistent with polarization or skill downgrading.

For the graduate-degree group, the between effect did not decline after 2000, indicating that their occupational distribution has not shifted away from highly paid cognitive jobs. However, between 2010 and 2015, the wage gap between graduate-degree holders and the college-only group within the same broad occupations fell by about 2 log points, suggesting that the direct competitive edge afforded by graduate training may be eroding. This erosion of the within effect for the graduate group suggests that skill downgrading may be playing an increasingly important role at the top of the skill distribution.

These results are robust to alternative definitions for the four broad occupations groups, including reorganization of the routine and nonroutine manual categories and separate treatment of selected services industries.[18] I also investigated alternative decompositions based on industry rather than occupation categories. These analyses indicated that the level and changes in higher education wage premiums are almost entirely determined within industries, with virtually no contribution coming from differences or changes over time in the wage premium and higher education shares across industries.

Overall, the results suggest rising competition between education groups for increasingly scarce well-paid jobs. Some of this is reflected in the movement of individuals holding only a college degree into routine jobs, consistent with polarization or skill downgrading, and some is reflected in the wage advantage of those with graduate degrees over the college-only group within broad occupations, suggesting skill downgrading.

9.5   Discussion and Conclusions

I have documented a flattening in the US higher education wage premium over the last few decades. In particular, after rising substantially in the 1980s, growth in the wage gap between individuals with a four-year college or graduate degree and those with a high school degree slowed progressively. The gaps have changed little since the year 2000, and they were flat to down during the period 2010 2015. These patterns suggest that the previously growing complementarity between highly educated labor and new production technologies, especially those that rely on computers and related organizational capital, may be leveling off.

I investigated these patterns with reference to two related explanations for changing US employment patterns: (a) a shift away from medium-skill occupations driven largely by technological change ("polarization"; e.g., Acemoglu and Autor 2011), and (b) a general weakening in the demand for advanced cognitive skills that cascades down the skill distribution ("skill downgrading"; Beaudry, Green, and Sand 2016). Descriptive evidence and comparison of the higher education wage premiums within and between

---

18. I thank my discussant David Autor for emphasizing the importance of investigating alternative occupational groupings, based on his recent research (e.g., Autor and Dorn 2013).

broad occupation groups suggests that both factors have played a role in the flattening of the overall premiums. Occupational employment shifts have held down the college-only premium somewhat since the year 2000, suggesting that college-educated workers are increasingly sliding down into routine jobs. This is consistent with polarization or skill downgrading. More recently, since 2010, the wage gap between graduate-degree holders and the college-only group within the same broad occupations has declined somewhat, suggesting that graduate training may be providing less of a competitive edge than it has in the past. This suggests that skill downgrading may be playing an increasingly important role at the top of the skill distribution. Overall, the results suggest rising competition between education groups for increasingly scarce well-paid jobs.

These findings should not be interpreted as suggesting that college and graduate training are no longer sound financial investments, from an individual or social perspective. Recent analyses indicate that relative to financing costs, higher education yields positive net returns for most individuals who complete college (Abel and Deitz 2014; Autor 2014; Daly and Cao 2015). On the other hand, it is important to note that the wage premiums to higher education are likely to vary substantially across individuals. Although higher education may be financially advantageous on average, the flattening of returns as costs have continued to rise suggests that college may be an unfavorable financial investment for rising numbers of individuals. In these circumstances, individual variation in returns looms as an increasingly important issue for future research.

I have focused on demand-side factors propelling the relative wages of college graduates, but supply-side factors may be important as well (Acemoglu and Autor 2011; Autor 2014). Sorting out the relative contributions of a demand slowdown and supply speed-up may be a worthwhile endeavor for future research. Related to overall supply trends, the composition of the college-educated workforce may have shifted in important ways. Enrollment at for-profit colleges has expanded rapidly since the year 2000, and subsequent wage increases appear limited, even for those who complete four-year undergraduate and graduate degrees at for-profit institutions (Cellini and Turner 2016). This may be holding down the returns to college estimated from the population of employed college graduates.

With these caveats in mind, I will conclude by noting that my findings raise the possibility of an eroding relationship between technological advance and the returns to investment in higher education. If this interpretation proves to be correct and durable, it has potentially important implications for this volume's primary themes. Human capital has been a key engine of growth in developing and advanced economies alike. Slower growth in the returns to higher education suggest that this connection may be fraying, raising the possibility of continued slow growth ahead.

Appendix A
MORG and March CPS Data

The data handling and definitions for the CPS MORG and March data generally follow Lemieux (2006a, 2010) and Autor, Katz, and Kearney (2008; see also Buchmueller, DiNardo, and Valletta 2011). All analyses are limited to wage and salary workers age twenty-five to sixty-four (with farming and resource occupations excluded), and appropriate survey weights are used for all tabulated results.

MORG Data (Definitions, Top-Coding, and Imputation)

As noted in the text, I use hourly wages as my earnings measure, either reported directly by hourly workers or formed as usual weekly earnings divided by usual weekly hours worked for salaried workers. Wage levels are expressed in real terms using the GDP deflator for personal consumption expenditures.

Following Lemieux (2006a), the wage analyses are limited to individuals whose hourly wage is greater than $1 and less than $100 (in 1979 dollars); only a small number of observations are dropped due to this restriction. Recorded earnings are subject to maximum limits ("top-codes") in the public-use data files, which change over time. I multiplied the value of top-coded earnings observations by 1.4. This largely follows Lemieux (2006a), with the exception that for the sake of consistency over time, I did not rely on the higher top-code enabled by the use of unedited earnings values for the years 1989–1993.

As noted in past research, nonresponse to the earnings and hours questions in the CPS data and the consequent need to impute their values is substantial and has grown over time, potentially distorting analyses of wage differentials. Following common practice, I dropped observations with imputed values of earnings or hours worked from all wage analyses. I followed the procedures outlined in Lemieux (2006a) for identifying imputed earnings observations. This includes the comparison of unedited and edited earnings values during the years 1989–1993, when the earnings imputation flags are incorrect. Imputation flags are missing for 1994 and most of 1995, which precludes dropping observations with imputed values during this period.

CPS March Annual Demographic Supplement Data
(Definitions, Top-Coding, and Imputation)

I supplement the CPS MORG analyses using data from the CPS March Annual Demographic Supplement files. These files are currently available through 2015. Income data from the annual March supplement refer to the prior calendar year, so the reference period for the March data that I use ends one year earlier than the MORG data (2014 rather than 2015).

As noted in the text, following standard practice, I restrict the March CPS sample to full-time, full-year workers and use weekly earnings (annual labor earnings divided by weeks worked) as the earnings measure. The sample restriction with respect to real hourly wages (in 1979 dollars), the treatment of top-coded values, and the elimination of imputed earnings values are the same as described for the MORG data.

Appendix B
Polarization Occupational Coding (Excluding Agriculture/Resources)

Table 9B.1   Polarization occupational coding

Broad polarization category	Standard Occupational Classification (SOC) major occupation groupings
Nonroutine cognitive	Management, business, and financial operations (SOC 11, 13) Professional/technical (SOC 15–29)
Routine cognitive	Sales and related (SOC 41) Office and administrative support (SOC 34)
Routine manual	Construction and extraction (SOC 47) Installation/maintenance/repair (SOC 49) Production (SOC 51) Transportation and material moving (SOC 53)
Nonroutine manual	Health care support (SOC 31) Protective services (SOC 33) Food preparation and serving (SOC 35) Building and grounds (SOC 37) Personal care and service (SOC 39)

Note: 2010 SOC codes listed; earlier period codes harmonized.

References

Abel, Jaison R., and Richard Deitz. 2014. "Do the Benefits of College Still Outweigh the Costs?" Federal Reserve Bank of New York *Current Issues in Economics and Finance* 20 (3). https://www.newyorkfed.org/medialibrary/media/research/current _issues/ci20-3.pdf.
Acemoglu, Daron, and David Autor. 2011. "Skills, Tasks and Technologies: Implications for Employment and Earnings." In *Handbook of Labor Economics*, vol. 4b,

edited by Orley Ashenfelter and David Card, 1043–171. Amsterdam: Elsevier-North Holland.

Autor, David H. 2014. "Skills, Education, and the Rise of Earnings Inequality among the 'Other 99 Percent.'" *Science* 344 (6186): 843–51.

———. 2015. "Polanyi's Paradox and the Shape of Employment Growth." Federal Reserve Bank of St. Louis: Economic Policy Proceedings, *Reevaluating Labor Market Dynamics*, 129–77. http://economics.mit.edu/files/9835.

Autor, David H., and David Dorn. 2013. "The Growth of Low-Skill Service Jobs and the Polarization of the US Labor Market." *American Economic Review* 103 (5): 1553–97.

Autor, David H., David Dorn, and Gordon H. Hanson. 2013. "The China Syndrome: Local Labor Market Effects of Import Competition in the United States." *American Economic Review* 103 (6): 2121–68.

Autor, David H., Lawrence F. Katz, and Melissa S. Kearney. 2008. "Trends in U.S. Wage Inequality: Revising the Revisionists." *Review of Economics and Statistics* 90 (2): 300–323.

Autor, David H., Lawrence F. Katz, and Alan B. Krueger. 1998. "Computing Inequality: Have Computers Changed the Labor Market?" *Quarterly Journal of Economics* 113 (4): 1169–214.

Autor, David H., Frank Levy, and Richard J. Murnane. 2003. "The Skill Content of Recent Technological Change: An Empirical Exploration." *Quarterly Journal of Economics* 118 (4): 1279–333.

Beaudry, Paul, David A. Green, and Benjamin M. Sand. 2014. "The Declining Fortunes of the Young Since 2000." *American Economic Review* 104 (5): 381–86.

———. 2016. "The Great Reversal in the Demand for Skill and Cognitive Tasks." *Journal of Labor Economics* 34 (S1, pt. 2): S199–247.

Bound, John, and George Johnson. 1992. "Changes in the Structure of Wages in the 1980s: An Evaluation of Alternative Explanations." *American Economic Review* 83:371–92.

Buchmueller, Thomas C., John DiNardo, and Robert G. Valletta. 2011. "The Effect of an Employer Health Insurance Mandate on Health Insurance Coverage and the Demand for Labor: Evidence from Hawaii." *American Economic Journal: Economic Policy* 3:25–51.

Card, David. 1999. "The Causal Effect of Education on Earnings." In *Handbook of Labor Economics*, vol. 3, edited by Orley Ashenfelter and David Card. Amsterdam: Elsevier.

Cellini, Stephanie Riegg, and Nicholas Turner. 2016. "Gainfully Employed? Assessing the Employment and Earnings of For-Profit College Students Using Administrative Data." NBER Working Paper no. 22287, Cambridge, MA.

Daly, Mary C., and Yifan Cao. 2015. "Does College Pay?" In *Does College Matter?* Federal Reserve Bank of San Francisco 2014 Annual Report. http://sffed-education.org/annualreport2014/#college-essays/does-college-pay.

Fernald, John. 2015. "Productivity and Potential Output before, during, and after the Great Recession." *NBER Macroeconomics Annual 2014*, vol. 29, edited by Jonathan Parker and Michael Woodford, 1–51. Chicago: University of Chicago Press.

Goldin, Claudia, and Lawrence F. Katz. 2008. *The Race between Education and Technology*. Cambridge, MA: Harvard University Press.

Goos, Maarten, and Alan Manning. 2007. "Lousy and Lovely Jobs: The Rising Polarization of Work in Britain." *Review of Economics and Statistics* 89 (1): 118–33.

Goos, Maarten, Alan Manning, and Anna Salomons. 2014. "Explaining Job Polarization: Routine-Biased Technological Change and Offshoring." *American Economic Review* 104 (8): 2509–26.

Jaeger, David. 1997. "Reconciling the Old and New Census Bureau Education Questions: Recommendations for Researchers." *Journal of Business and Economic Statistics* 15 (3): 300–309.

Krueger, Alan. 1993. "How Computers Have Changed the Wage Structure: Evidence from Microdata, 1984–1989." *Quarterly Journal of Economics* 108 (1): 33–60.

Lemieux, Thomas. 2006a. "Increasing Residual Wage Inequality: Composition Effects, Noisy Data, or Rising Demand for Skill?" *American Economic Review* 96 (3): 461–98.

———. 2006b. "Postsecondary Education and Increasing Wage Inequality." *American Economic Review* 96 (2): 195–99.

———. 2010. "What Do We Really Know About Changes in Wage Inequality?" In *Labor in the New Economy*, edited by Katharine G. Abraham, James R. Spletzer, and Michael Harper. Chicago: University of Chicago Press.

Lindley, Joanne, and Stephen Machin. 2016. "The Rising Postgraduate Wage Premium." *Economica* 83:281–306.

Comment   David Autor

Robert Valletta's chapter illuminates one of the leading puzzles for contemporary US labor economics: the unexpected "flattening" of the premium to higher education in the United States in the first decade of the twenty-first century. This single metric—the college/high school wage premium—has been the North Star guiding neoclassical analysis of the evolution of wage inequality during a period of rapidly shifting wage structures. Two impactful papers by Beaudry, Green, and Sand (2014, 2016,) argue that since approximately the year 2000, this North Star has become an increasingly dubious point of navigation. Specifically, Beaudry, Green, and Sand highlight the failure of the college premium to rise in the first decade of the twenty-first century following two decades of steep increases. They interpret this deceleration as reflecting the maturation of the information technology revolution, which in turn has spurred a slackening in the pace of workplace IT investments and a consequent slowdown in the trend of rising demand for highly educated labor. A key piece of evidence favoring Beaudry, Green, and Sand's narrative is the precipitous fall in US investment in information-processing equipment and software in the United States after 1999 (figure 9C.1), which seems to have precisely the right timing to explain a falloff in IT augmentation of skilled labor demand.

Valletta's careful analysis extends and probes the Beaudry, Green, and Sand findings, verifies their robustness, and considers their interpretation

David Autor is the Ford Professor of Economics and associate head of the Department of Economics at the Massachusetts Institute of Technology and a research associate and codirector of the Labor Studies Program at the National Bureau of Economic Research.

For acknowledgments, sources of research support, and disclosure of the author's material financial relationships, if any, please see http://www.nber.org/chapters/c13706.ack.

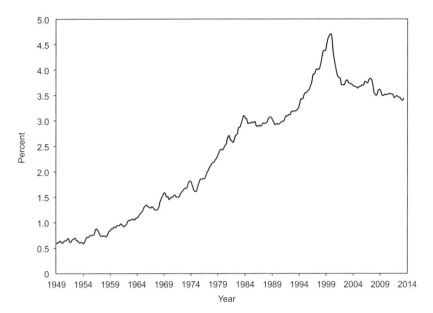

Fig. 9C.1   Private fixed investment in information-processing equipment and software as a percentage of gross domestic product, 1949–2014
Source: FRED, Federal Bank of St. Louis (http://research.stlouisfed.org/fred2/graph/?g =GXc; accessed 8/3/2014). This graphic originally appeared in Autor (2015).

in the light of both their conceptual framework and an alternative framing offered by Acemoglu and Autor (2011). There are many things to admire about Valletta's chapter: it is empirically rigorous, intellectually ecumenical, and commendably ambitious in synthesizing and adjudicating between two conceptual models that are not, to a first approximation, speaking the same language. My remarks focus exclusively on one question that is core to both Valletta's and Beaudry, Green, and Sand's work: When did rising demand for college-educated labor decelerate? I argue below that (a) the recent flattening of the skill premium in the first decade of the twenty-first century is *not* surprising in light of the canonical supply-demand model, and (b) what *is* surprising is that the underlying demand for college labor decelerated sharply and (to date) inexplicably almost a decade beforehand. These observations render the phenomenon that Valletta tackles no less consequential, but they may suggest a different set of explanations for the slowdown than those focusing on discontinuous changes in economic trends in the first decade of the twenty-first century.

Modeling School

Following an extraordinarily influential series of papers that includes Goldin and Margo (1992), Katz and Murphy (1992), Murphy and Welch

(1992), Card and Lemieux (2001), and Goldin and Katz's magisterial 2008 volume *The Race between Education and Technology*, labor economists have applied a remarkably simple and surprisingly powerful calibrated supply-demand model (the "canonical model") to rationalize the fluctuations over time in the skill premium and the accompanying evolution of wage inequality. This so-called canonical model takes its inspiration from the observation by Nobel Laureate Jan Tinbergen in 1974 that there appears to be an ongoing "race" between technology and schooling, with techno-logical advancements progressively raising the demand for educated labor and the school system simultaneously secularly raising its supply. When technological advancement surges faster than educational production, the relative scarcity of educated labor rises, and the skill premium rises with it—that is, technology pulls ahead of education in this two-person race. Conversely, when educational production surges ahead of technologically induced demand shifts, the skill premium falls.

While many elements of this description seem far too simple (e.g., history provides many examples of technologies that replace rather than comple-ment skills), this framework provides a surprisingly good high-level descrip-tion of what we see in the data. The canonical model provides a benchmark for interpreting the evolution of the skill premium. I apply this model here to address the question of whether we should be surprised—and if so, how much—by the slowdown in the skill premium after 2000. Before applying the model, I review its rudiments, and I refer readers to Acemoglu and Autor (2011) for a fuller development.

The canonical model posits two skill groups, high and low. It draws no distinction between skills and occupations (tasks), so that high-skilled work-ers effectively work in separate occupations (perform different tasks) from low-skilled workers. In most empirical applications of the canonical model, it is natural to identify high-skilled workers with college graduates (or in dif-ferent eras, with other high-education groups), and low-skilled workers with high school graduates (or in different eras, those with less than high school). Critical to the two-factor model is that high- and low-skilled workers are imperfect substitutes in production. The elasticity of substitution between these two skill types is central to understanding how changes in relative sup-plies affect skill premiums.

Suppose that the total supply of low-skilled labor is L and the total supply of high-skilled labor is H. Naturally not all low- (or high-) skilled workers are alike in terms of their marketable skills. As a simple way of introducing this into the canonical model, suppose that each worker is endowed with either high or low skill, but there is a distribution across workers in terms of efficiency units of these skill types. In particular, let $\mathcal{L}$ denote the set of low-skilled workers and $\mathcal{H}$ denote the set of high-skilled workers. Each low-skilled worker $i \in \mathcal{L}$ has l_i efficiency units of low-skilled labor and each high-skilled worker $i \in \mathcal{H}$ has h_i units of high-skilled labor. All workers supply their efficiency units inelastically. Thus the total supply of high-skilled and low-skilled labor in the economy can be written as

$$L = \int_{i \in \mathcal{L}} l_i di \text{ and } H = \int_{i \in \mathcal{H}} h_i di.$$

The production function for the aggregate economy takes the following constant elasticity of substitution form

(1) $$Y = [(A_L L)^{(\sigma-1)/\sigma} + (A_H H)^{(\sigma-1)/\sigma}]^{\sigma/(\sigma-1)},$$

where ($\sigma \in 0, \infty$) is the elasticity of substitution between high-skilled and low-skilled labor, and A_L and A_H are factor-augmenting technology terms.[1] The elasticity of substitution between high- and low-skilled workers plays a pivotal role in interpreting the effects of different types of technological changes in this canonical model. We refer to high- and low-skilled workers as *gross substitutes* when the elasticity of substitution $\sigma > 1$, and gross complements when $\sigma < 1$.

In this framework, technologies are *factor augmenting*, meaning that technological change serves to increase the productivity of either high- or low-skilled workers (or both). This implies that there are no explicitly skill-replacing technologies. Depending on the value of the elasticity of substitution, however, an increase in A_H or A_L can act either to complement or (effectively) substitute for high- or low-skilled workers (see below).

Assuming that the labor market is competitive, the low-skill unit wage is simply given by the value of the marginal product of low-skilled labor, which is obtained by differentiating equation (1) as

(2) $$w_L = \frac{\partial Y}{\partial L} = A_L^{(\sigma-1)/\sigma}[A_L^{(\sigma-1)/\sigma} + A_H^{(\sigma-1)/\sigma}(H/L)^{(\sigma-1)/\sigma}]^{1/(\sigma-1)}.$$

Similarly, the high-skill unit wage is

(3) $$w_H = \frac{\partial Y}{\partial H} = A_H^{(\sigma-1)/\sigma}[A_L^{(\sigma-1)/\sigma}(H/L)^{-[(\sigma-1)/\sigma]} + A_H^{(\sigma-1)/\sigma}]^{1/(\sigma-1)}.$$

Combining equations (2) and (3), the skill premium—the high-skill unit wage divided by the low-skill unit wage—is

(4) $$\omega = \frac{w_H}{w_L} = \left(\frac{A_H}{A_L}\right)^{(\sigma-1)/\sigma} \left(\frac{H}{L}\right)^{-(1/\sigma)}.$$

Equation (4) can be rewritten in a more convenient form by taking logs

(5) $$\ln \omega = \frac{\sigma-1}{\sigma} \ln\left(\frac{A_H}{A_L}\right) - \frac{1}{\sigma} \ln\left(\frac{H}{L}\right).$$

---

1. This production function is typically written as $Y = [\gamma(A_L L)^{(\sigma-1)/\sigma} + (1-\gamma)(A_H H)^{(\sigma-1)/\sigma}]^{\sigma/(\sigma-1)}$, where A_L and A_H are factor-augmenting technology terms and γ is the distribution parameter. I suppress γ (i.e., set it equal to 1/2) to simplify notation.

The log skill premium, ln ω, has been a central object of study in the empirical literature on the changes in the earnings distribution. Equation (5) shows that there is a simple log-linear relationship between the skill premium and the relative supply of skills as measured by H/L. Equivalently, equation (5) implies

$$(6) \qquad \frac{\partial \ln \omega}{\partial \ln H/L} = -\frac{1}{\sigma} < 0.$$

This relationship corresponds to the second of the two forces in Tinbergen's race (the first being technology, the second being the supply of skills): for a *given skill bias of technology*, captured here by A_H/A_L, an increase in the relative supply of skills reduces the skill premium with an elasticity of $1/\sigma$. Intuitively, when high- and low-skilled workers are producing the same good but performing different functions, an increase in the number of high-skilled workers will necessitate a substitution of high-skilled workers for the functions previously performed by low-skilled workers.[2] The downward-sloping relationship between relative supply and the skill premium implies that if technology, in particular A_H/A_L, had remained roughly constant over recent decades, the remarkable increase in the supply of skills (seen, e.g., in table 9.1 of Valletta's chapter) would have led to a significant decline in the skill premium. The lack of such a decline is a key reason why economists believe that the first force in Tinbergen's race—changes in technology increasing the demand for skills—must have also been important throughout the twentieth century (cf. Goldin and Katz 2008).

More formally, differentiating equation (5) with respect to A_H/A_L yields

$$(7) \qquad \frac{\partial \ln \omega}{\partial \ln(A_H/A_L)} = \frac{\sigma - 1}{\sigma}.$$

Equation (7) implies that if $\sigma > 1$, then relative improvements in the high-skill-augmenting technology (i.e., in A_H/A_L) increase the skill premium. This can be seen as a shift out of the relative demand curve for skills. The converse is obtained when $\sigma < 1$: that is, when $\sigma < 1$, an improvement in the productivity of high-skilled workers, A_H, relative to the productivity of low-skilled workers, A_L, shifts the relative demand curve inward and reduces the skill premium. Nevertheless, the conventional wisdom is that the skill premium increases when high-skilled workers become relatively more—not relatively less—productive, which is consistent with $\sigma > 1$. Most estimates put σ in this

2. In this interpretation, we can think of some of the "tasks" previously performed by high-skilled workers now being performed by low-skilled workers. Nevertheless, this is simply an interpretation, since in this model there are no tasks and no endogenous assignment of tasks to workers. One could alternatively say that the H and L tasks are imperfect substitutes, and hence an increase in the relative supply of H labor means that the H task is used more intensively but less productively at the margin.

context to be somewhere between 1.4 and 2 (Johnson 1970; Freeman 1986; Heckman, Lochner, and Taber 1998).

The key equation of the canonical model links the skill premium to the relative supply of skills, H/L, and to the relative technology term, A_H/A_L. This last term is not directly observed. Nevertheless, the literature has made considerable empirical progress by taking a specific form of Tinbergen's hypothesis, and assuming that there is a log-linear increase in the demand for skills over time coming from technology, captured in the following equation:

$$(8) \qquad \ln\left(\frac{A_{H,t}}{A_{L,t}}\right) = \gamma_0 + \gamma_1 t,$$

where t is calendar time and variables written with t subscripts refer to these variables at time t. Substituting this equation into equation (8), we obtain

$$(9) \qquad \ln \omega_t = \frac{\sigma - 1}{\sigma}\gamma_0 + \frac{\sigma - 1}{\sigma}\gamma_1 t - \frac{1}{\sigma}\ln\left(\frac{H_t}{L_t}\right).$$

Equation (9) implies that "technological developments" take place at a constant rate, while the supply of skilled workers may grow at varying rates at different points in times. Therefore, changes in the skill premium will occur when the growth rate of the supply of skills differs from the pace of technological progress. In particular, when H/L grows faster than the rate of skill-biased technical change, $(\sigma - 1)\gamma_1$, the skill premium will fall. And when the supply growth falls short of this rate, the skill premium will increase. Surprisingly, this simple equation provides considerable explanatory power for the evolution of the skill premium—though its limitations are also immediately evident.

Doing the Katz-Murphy

Using data from Autor (2014), I fit this simple model to fifty years of US data for 1963–2012. Figure 9C.1 provides the key input into this estimation: the observed log relative supply of US college versus noncollege labor for years 1963–2012, measured in efficiency units and normalized to zero in the base year.[3] Figure 9C.2 highlights the steep rise in production of college-educated labor in the United States in the postwar period—specifically until the late 1970s—followed by a sharp deceleration after 1980. This deceleration is frequently interpreted as the key driver of the rapid rise in the skill premium after 1980 (Katz and Murphy 1992). Notably, there is also a steep *acceleration* of supply after 2004. All else equal, one would except this supply acceleration to depress the skill premium absent any slowdown of the secular

---

3. Extensive details on the calculation of these series are provided in Acemoglu and Autor (2011).

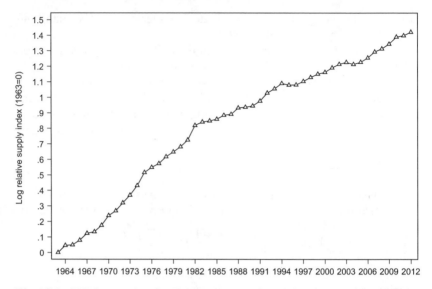

Fig. 9C.2   Efficiency units of college versus noncollege labor supply, 1963–2012
Source: Data from Autor (2014).

trend rise in relative demand after 2004. This observation highlights that the evolution of the skill premium is *not* a sufficient statistic for fluctuations in demand for skilled labor; one must also account for supply.

Using the data series in figure 9C.2, I fit equation (9) to obtain the following estimate:

$$
(10) \qquad \ln \omega_t = \text{constant} + \underset{(0.0013)}{.0151} \times t - \underset{(0.0429)}{0.302} \cdot \ln\!\left(\frac{H_t}{L_t}\right).
$$

This simple ordinary least squares (OLS) model implies that (a) the relative demand curve for college versus noncollege labor is shifting outward by approximately 1.5 log points per year, and (b) that increases in the relative supply of skilled labor buffer the impact of shifting demand on wage inequality. Specifically, the point estimate of −0.30 on the relative supply term implies an elasticity of substitution of $\hat{\sigma} = 1/3.31$. While the explanatory power of this time-series model is high ($R^2 = 0.94$), the point estimate for the elasticity of substitution is more than *twice* as high as Katz-Murphy's 1992 estimate of 1.41. This implies that either the elasticity of substitution is changing over time or that the linear time trend is not doing an adequate job of capturing trends in relative demand.

Figure 9C.3 explores these possibilities. The series plotted with circular markers corresponds to the measured (i.e., observed) skill premium in each year. This series depicts the now familiar rise in the skill premium from the

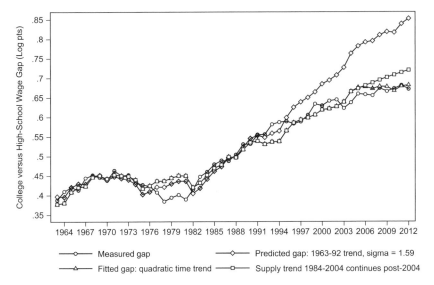

Fig. 9C.3   Observed, predicted, and fitted evolution of the log college/noncollege hourly earnings gap, 1963–2012
Source: Data from Autor (2014).

early 1960s (start of the series) to the early 1970s, the sharp fall between 1971 and 1981, the steep and continuous rise from 1982 to 1999, and then the much shallower rise from 2000 to 2012 (end of the series). The series with diamond markers performs a within-series extrapolation by reestimating equation (10) using only data from 1962 to 1992 (the period of best fit), and recovering estimates of the time trend and the elasticity of substitution ($\hat{\gamma} =$ 0.028, $\hat{\sigma} = 1/-0.631 = -1.59$). The plotted series then projects this estimate forward to 2012 using the estimated parameters from the 1962 to 1992 fit in combination with the *observed* evolution of aggregate skill supplies (ln H_t/L_t). Notably, the time trend and elasticity recovered from this procedure are extremely similar to those obtained by Katz-Murphy's in 1992, and using data for 1963 through 1987. The similarity of the current estimates implies that Katz-Murphy's within-sample point estimates continue to closely track the observed data for an additional five years *out of sample*.

As the figure reveals, however, this projection badly misses the mark after 1992. Adjusting for the evolution of aggregate skill supplies, the growth in the skill premium is far more modest after 1992 than the extrapolation projects. Between 1992 and 2012, the observed college/noncollege log earnings gap rises by 11.6 log points. But the projection based on data to 1992—applying the observed evolution of skill supplies to 2012—predicts an increase of 30.4 log points, nearly three times as large as what occurred. A summary judgment is that the evolution of the skill premium has been *surprising* since 1992.

The Element of Surprise

Economic literature noted this surprise some time ago. Card and DiNardo (2002) first pointed out this discrepancy in their broad critique of the skill-biased technical change literature. Autor, Katz, and Kearney (2008) proposed an ad hoc workaround, which was to allow for a trend deceleration in the evolution of skill demands after 1992. Goldin and Katz (2008) and Autor (2014) pursue a related approach by applying a quadratic time trend in the time-series model, thereby allowing a smooth deceleration of the trend demand shift. The series in figure 9C.3 labeled "Fitted gap: quadratic trend" (triangular marker) shows just how well this works. Conditional on the quadratic trend the fit is impressively close. But of course, this is simple reverse engineering. This flexibility was added to the model because the data demanded it, not because the theory suggested it.

These various exercises raise an urgent question: After accounting for fluctuations in the supply of skilled labor, when did the "flattening" of demand for skill commence? Here, I draw a distinction between flattening in the *skill premium* and flattening (or deceleration) in the movement of the underlying demand schedule. As noted above, it would be entirely possible for the skill premium to decline even as demand was accelerating—if skill supplies rose fast enough. Figure 9C.1 makes clear that skill supplies *accelerated* after 2004. Was this *supply-side* change an important contributor to the observed flattening of the skill premium?

The series plotted in square markers in figure 9C.3 addresses this question. The log relative supply of college workers (figure 9C.2) rose at an annual rate of 4.31 log points between 1963 and 1982, by 1.79 log points between 1982 and 2004, and by 2.61 log points between 2004 and 2012 (i.e., a 45 percent *increase* after 2004). The series with square markers in figure 9C.3 (labeled "Supply trend 1984–2004 continues post-2004") replaces the observed values of $\ln(H_t/L_t)$ with a counterfactual series in which log relative supply rises at the 1963–1982 of 1.79 log points per annum. Surprisingly (at least to me), this substitution makes a *substantial* difference for inference. The estimated college premium rose by only 1.65 log points between 2004 and 2012. This exercise implies that had the relative supply of college-educated labor *not* accelerated after 2004, the skill premium would have risen by 5.47 log points rather than a measly 1.65 log points. I submit based on this evidence that had there been no *supply acceleration* after 2004, Beaudry, Green, and Sand would have had a more difficult time making the argument that there was a *demand deceleration* in the first decade of the twenty-first century.

How Long Has This Been Going On?

The evidence in figure 9C.3 in no way obviates the claim that demand for college workers flattened according to the canonical model. It instead under-

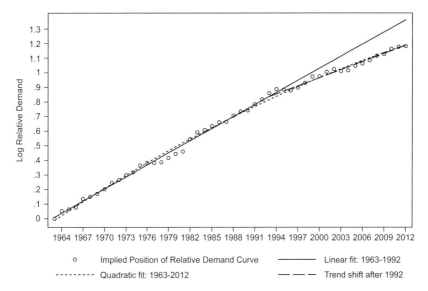

Fig. 9C.4   **Implied evolution of the demand for college versus noncollege labor using σ = 1.59, 1963–2012**
Source: Data from Autor (2014).

scores that the raw skill premium, not purged of the impact of supply forces, could generate misleading inferences about the trajectory of the demand for skilled labor.

To address this shortcoming, figure 9C.4 plots the implied log relative demand shift favoring college versus noncollege labor for 1962–2012, again using the estimated value of σ = 1.59 based on fitting equation (9) to data for 1962–1992. The plotted (scatter) points in figure 9C.4 are not regression estimates. They correspond instead to the calculated values of $\gamma_t = \omega_t - (1/\hat{\sigma})\ln(H_t/L_t)$ in each year, where we treat σ as known,[4] To guide interpretation of these data points, the figure also contains three regression lines. The solid line depicts a pure linear extrapolation, fitted and projected using data for 1962–1992. This corresponds to the implied path of relative demand from 1992 through 2012 had there been no deviation after 1992. The short-dashed series is the quadratic fit to this set of scatter points. The long-dashed series is a linear spline that allows for a discrete slope shift in 1992 (and otherwise fully overlays the initial trend from 1963 to 1992).

This plot highlights three key patterns. A first is that the trajectory of (implied) relative demand for educated labor is astonishingly linear for the initial thirty years of the series, 1963–1992. This linearity is in no sense

4. Equivalently, they are the time dummies from a saturated regression (no error term) of $\omega_t - (1/\hat{\sigma})\ln(H_t/L_t)$ on a full set of year indicators.

mechanical: the relative demand shift estimates plotted in figure 9C.4 are extracted from a college wage premium series that fluctuates dramatically over three decades, rising for the first ten years of the time interval, falling for the next nine, and then increasing with remarkable rapidity thereafter. The linearity of the (implied) underlying demand trend therefore reflects the uncanny success of the relative supply term ln H_t/L_t in explaining the fluctuations in the premium, leaving little behind but a smooth secular underlying demand shift favoring college-educated labor.

The second pattern immediately visible in figure 9C.4 is that the steady secular demand shift favoring college-educated labor *decelerates* after 1992, and does so abruptly. Estimates of equation (9), fit using a linear spline (long-dashed series), imply that the relative demand for college labor rose by 2.80 log points per year between 1963 and 1992 and then *decelerated* to 1.84 points thereafter (a fall of one-third). This pattern, while occasionally noted in the literature (cf. Acemoglu and Autor 2011), has not been rigorously explained by any formal model—though of course there are many informal explanations.

The third takeaway from figure 9C.4 is that it is hard to see any evidence of a discontinuous deceleration in the demand for educated labor in the first decade of the twenty-first century. Whether fit using the linear spline, allowing all the post-1992 points to cluster along one axis, or a quadratic trend, allowing the deceleration to cumulate over the full sample, there is almost nothing in this figure that suggests a trend break in demand in the first decade of the twenty-first century.[5] Rather, this evidence suggests that the trend movements in relative demand early in the twenty-first century were a continuation of those commencing circa 1992.

Conclusion: Timing Is Everything

These fact patterns lead me to draw a distinct inference from Beaudry, Green, and Sand: we *should not* be surprised by the evolution of the skill premium—or even the weaker job prospects of college-educated workers—in the early twenty-first century. These outcomes are consistent with steadily rising demand for college-educated labor and a surprising surge in new college entrants in the US labor market after 2003, which depressed the skill premium as it would be predicted to do. We *should* however be deeply puzzled by the sudden trend shift in demand after 1992, which ushered in (at least) twenty years of slower (though still nonnegligible) growth in demand for skilled labor.

5. If one squints, it is possible to see that some of the points immediately after 2000 fall slightly below the regression line, whereas those immediately before fall slightly above it—implying a possible further deceleration after 2000. But then the last three points in the series (2010–2012) again lie slightly above the regression line, suggesting a tiny reacceleration. This is pretty thin evidence.

This development is not altogether bad news, however. Had demand for skilled labor continued to rise after 1992 at its pre-1992 pace, the estimates in figure 9C.3 suggest that the United States would have seen *substantially more growth* of between-group inequality—specifically, a meteoric 30 log point rise in the college premium between 1992 and 2012, nearly three times as large as the economically significant rise of 11 log points that actually occurred. This "good news" is at best partial, however. In the canonical model, relative demand shifts intrinsically convey good economic news because they imply ongoing factor-augmenting technological progress.[6] Thus, this slowdown may be read to support Beaudry, Green, and Sand's view that as information technology has matured, the pace of accompanying labor augmentation has slackened. If so, however, we would want to caveat their conclusion to note that this slowdown started about ten years prior to the date that Beaudry, Green, and Sands pinpoint, and that it occurred during a period in which aggregate productivity growth was robust and US IT investment was rising extraordinarily rapidly (figure 9C.1).

References

Acemoglu, Daron, and David H. Autor. 2011. "Skills, Tasks and Technologies: Implications for Employment and Earnings." *Handbook of Labor Economics* 4:1043–171.

Autor, David H. 2014. "Skills, Education, and the Rise of Earnings Inequality among the 'Other 99 Percent.'" *Science* 344 (6186): 843–51.

———. 2015. "Why Are There Still So Many Jobs? The History and Future of Workplace Automation." *Journal of Economic Perspectives* 29 (3): 3–30.

Autor, David H., Lawrence F. Katz, and Melissa S. Kearney. 2008. "Trends in U.S. Wage Inequality: Revising the Revisionists." *Review of Economics and Statistics* 90 (2): 300–323.

Beaudry, Paul, David A. Green, and Benjamin M. Sand. 2014. "The Declining Fortunes of the Young Since 2000." *American Economic Review* 104 (5): 381–86.

———. 2016. "The Great Reversal in the Demand for Skill and Cognitive Tasks." *Journal of Labor Economics* 34 (S1, pt. 2): S199–247.

Card, David, and John E. DiNardo. 2002. "Skill-Biased Technological Change and Rising Wage Inequality: Some Problems and Puzzles." *Journal of Labor Economics* 20 (4): 733–83.

Card, David, and Thomas Lemieux. 2001. "Can Falling Supply Explain the Rising Return to College for Younger Men? A Cohort-Based Analysis." *Quarterly Journal of Economics* 116 (2): 705–46.

Freeman, Richard B. 1986. "Chapter 6 Demand for Education." *Handbook of Labor Economics* 1:357–86.

Goldin, Claudia, and Lawrence F. Katz. 2008. *The Race between Education and Technology.* Cambridge, MA: Harvard University Press.

Goldin, Claudia, and Robert A. Margo. 1992. "The Great Compression: The Wage

6. This is also true for technological progress that raises A_L or both A_H and A_L. Presuming as the model does that technological progress is always factor augmenting—there are no factor-retarding technological regresses—any movement of A_H or A_L must correspond to an *increase* of either or both terms and hence rising productivity.

Structure in the United States at Mid-Century." *Quarterly Journal of Economics* 107 (1): 1–34.

Heckman, James J., Lance Lochner, and Christopher Taber. 1998. "Explaining Rising Wage Inequality: Explorations with a Dynamic General Equilibrium Model of Labor Earnings with Heterogeneous Agents." *Review of Economic Dynamics* 1 (1): 1–58.

Johnson, George E. 1970. "The Demand for Labor by Educational Category." *Southern Economic Journal* 37 (2): 190–204.

Katz, Lawrence F., and Kevin M. Murphy. 1992. "Changes in Relative Wages, 1963–1987: Supply and Demand Factors." *Quarterly Journal of Economics* 107 (1): 35–78.

Murphy, Kevin M., and Finis Welch. 1992. "The Structure of Wages." *Quarterly Journal of Economics Structure* 107 (1): 285–326.

Tinbergen, Jan. 1974. "Substitution of Graduate Labor by Other." *Kyklos* 27 (2): 217–26.

IV

The Supply of Skills

Accounting for the Rise in College Tuition

Grey Gordon and Aaron Hedlund

10.1 Introduction

Over the past thirty years, the perceived necessity of a college degree and a growing college earnings premium have led to record enrollments and greater degree attainment in higher education. However, a dramatic escalation in tuition looms over the heads of prospective students and their parents and serves as a stark reminder to graduates saddled with large student loans. From 1987 to 2010, sticker price tuition and fees ballooned from $6,630 to $14,510 in 2010 dollars. After subtracting institutional aid, net tuition and fees still grew by 92 percent, from $5,720 to $11,000. To provide perspective, had net tuition risen at the rate of much maligned *health care costs*, tuition would have only risen 32 percent to $7,550 in 2010.[1]

In this chapter, we seek to account for the college tuition increase by quantitatively evaluating existing explanations using a structural model of higher education and the macroeconomy. We divide our hypotheses about driving forces into supply-side changes (Baumol's cost disease and exogenous

Grey Gordon is assistant professor of economics at Indiana University. Aaron Hedlund is assistant professor of economics at the University of Missouri.

We thank Kartik Athreya, Sue Dynarski, Gerhard Glomm, Bulent Guler, Kyle Herkenhoff, Jonathan Hershaff, Brent Hickman, Felicia Ionescu, John Jones, Michael Kaganovich, Oksana Leukhina, Lance Lochner, Amanda Michaud, Urvi Neelakantan, Chris Otrok, Fang Yang, Eric Young, and participants at Midwest Macro 2014 and the brown bags at Indiana University and the University of Missouri. We also thank the editors, Chuck Hulten and Valerie Ramey; our discussant, Sandy Baum; and conference participants. All errors are ours. The web appendix for this chapter is available at http://www.nber.org/data-appendix/c13711/appendix.pdf. For acknowledgments, sources of research support, and disclosure of the authors' material financial relationships, if any, please see http://www.nber.org/chapters/c13711.ack.

1. Calculations used the health care personal consumption expenditures price index deflated by the CPI.

changes to nontuition revenue), demand-side changes (notably, expansions in grant aid and loans), and macroeconomic forces (namely, skill-biased technical change resulting in a higher college earnings premium). Our quantitative model shows that the combined effect of these changes more than accounts for the tuition increase and provides key insights about the role of individual factors as well as their complementary effects.

Existing hypotheses of why college tuition is increasing largely fall into two camps: those that emphasize the unique virtues and pathologies of higher education and those that place rising higher education costs into a broader narrative of increasing prices in many service industries. Advocates of the latter approach look to cost disease and skill-biased technical progress as drivers of higher costs in service industries that employ highly skilled labor. Cost disease, which dates back to seminal papers by Baumol and Bowen (1966) and Baumol (1967), posits that economy-wide productivity growth pushes up wages and creates cost pressures on service industries that do not share in the productivity growth. To cope, these industries increase their relative price, passing their higher costs onto consumers.

By contrast, theories emphasizing the uniqueness of higher education take several forms. Falling within our notion of supply-side shocks, state and local funding for higher education fell from $8,200 per full-time equivalent (FTE) student in 1987 to $7,300 in 2010, all while underlying costs and expenditures were rising. Several studies, including a notable study commissioned by Congress in the 1998 reauthorization of the Higher Education Act, attribute a sizable fraction of the increase in public university tuition to these state funding cuts. We take a somewhat broader view in this chapter by looking at how exogenous changes to *all* sources of nontuition revenue impact the path of tuition.

On the demand side, several expansions in financial aid have occurred over the past several decades. During our period of analysis, annual and aggregate subsidized Stafford loan limits were increased in 1987 and five years later in 1992. The Higher Education Amendments of 1992 also established a program of supplementary unsubsidized Stafford loans and increased the annual PLUS loan limit to the cost of attendance minus aid, thereby eliminating aggregate PLUS loan limits. Interest rates on student loans also fell considerably during the first decade of the twenty-first century. In a famous 1987 *New York Times* op-ed titled "Our Greedy Colleges," then-secretary of education William Bennett asserted that "increases in financial aid in recent years have enabled colleges and universities blithely to raise their tuitions" (Bennett 1987). We evaluate this claim through the lens of our model, and we also cast light on the tuition impact of the 53 percent rise in *nontuition* costs (such as those arising from the greater provision of student amenities), which has the effect of increasing subsidized loan eligibility.

Last, we quantify the impact of macroeconomic forces—specifically, rising labor market returns to college—on tuition changes. Autor, Katz, and

Kearney (2008) find that, from the mid-1980s to 2005, the overall earnings premium for having a college degree increased from 58 percent to over 93 percent. Ceteris paribus, such an increase in the return to college has assuredly driven up demand for a college degree. We use our model to quantify how much this increase in demand translates to higher tuition and how much it contributes to higher enrollments.

Our quantitative findings can be summarized as follows:

1. The combined effect of the aforementioned shocks generates a 102 percent increase in equilibrium tuition. This result compares to a 92 percent increase in the data.

2. The rise in the college earnings premium alone causes tuition to increase by 21 percent. With all other shocks present *except* the college premium hike, tuition increases by 81 percent.

3. The demand-side shocks by themselves cause tuition to jump by 91 percent. With all other changes *except* the demand-side shocks, tuition only increases by 14 percent.

4. The supply-side shocks by themselves cause tuition to *decline* by 8 percent. With all other changes *except* the supply-side shocks, tuition increases by 116 percent.

The model we construct to arrive at these conclusions embeds a rich higher education framework based off of Epple, Romano, and Sieg (2006) and Epple et al. (2013) into a life-cycle environment with heterogeneous agents, incomplete markets, and student loan default. Imperfectly competitive colleges in the model set differential tuition and admissions policies to maximize quality, which, as a proxy for reputation, depends on investment per student and the average academic ability of the heterogeneous student body. In this chapter, we restrict attention to the case of a representative nonprofit institution that has limited market power because of unobservable student preference shocks. Even with these shocks, the representative college assumption still abstracts from important heterogeneity and strategic interactions in the higher education market. For this reason, the findings in this chapter should be used to guide further research rather than viewed authoritatively. To further simplify matters, we treat all nontuition revenue as exogenous (e.g., endowment income and state funding), which implies that the college faces a balanced budget constraint each period that equates total revenue with total spending on investment and non-quality-enhancing custodial costs. On the household side, we include several important features: heterogeneity in ability and parental income dimensions, college financing decisions, college dropout risk, and student loan repayment decisions.

Our assumption that colleges maximize quality—in line with what Clotfelter (1996) calls the "pursuit of excellence"—implicitly incorporates another prominent hypothesis for rising tuition, namely, Bowen's (1980) "Revenue Theory of Costs." Ehrenberg (2002, 11) states it best:

The objective of selective academic institutions is to be the best they can in every aspect of their activities. They aggressively seek out all possible resources and put them to use funding things they think will make them better. To look better than their competitors, the institutions wind up in an arms race of spending.

To make matters concrete, quality in our setting depends on investment per student and the average ability of the student body. As a result, students act both as customers and as inputs to the production of quality via peer effects, as described by Winston (1999). This unique feature of higher education gives colleges an additional motive to engage in price discrimination beyond the usual monetary rent extraction—namely, to attract high-ability students by offering generous institutional aid.

To discipline the model, we use a combination of calibration and estimation. Rather than ex ante assume cost disease or a particular production structure (e.g., number of faculty, administrators, etc., needed to run a college), we directly estimate a reduced-form custodial cost function and track its changes over the period 1987–2010. Similarly, we compute average nontuition revenue per FTE student using Delta Cost Project data and feed it into the model. On the household side, we use earnings premium estimates by Autor, Katz, and Kearney (2008) and construct time series for federal student loan program (FSLP) variables.

As mentioned previously, we find that the combined effects of the supply-side changes, demand-side changes, and increases in the college earnings premium can fully account for the mean net tuition increase. Looking at individual factors, we find that expansions in borrowing limits drive 54 percent of the tuition jump and represent the single most important factor.[2] To grasp the magnitude of the change in borrowing capacity, first note that real aggregate borrowing limits increased by 56 percent between 1987 and 2010, from $26,200 to $40,800 in 2010 dollars.[3] Second, the reauthorization of the Higher Education Act in 1992 introduced a major change along the extensive margin by establishing an unsubsidized loan program alongside the subsidized loans. We also find that increased grant aid contributes 18 percent to the rise in tuition, which mirrors the 21 percent impact of the higher college earnings premium. These results give credence to the Bennett (1987) hypothesis.

Last, our results, while preliminary and subject to the caveat mentioned above regarding the representative college assumption, paint a more nuanced picture of cost disease as a driver of higher tuition. Although our estimated cost function shifts upward from 1987 to 2010, this isolated effect *reduces*

2. For this calculation, we take one minus the tuition increase without the borrowing limit expansion relative to the increase with the expansion, that is, $1 - (\$9,066 - \$6,146)/(\$12,428 - \$6,146)$. Adding the percentage contribution from each exogenous driving force need not yield 100 percent because of interaction effects.
3. We use the limits in place from 1981 to 1986 as our figure for 1987.

average tuition (a contribution of −16 percent). Importantly, our estimates suggest that the upward shift in the cost function between 1987 and 2010 comes largely in the form of higher fixed costs rather than higher marginal costs, which has important implications for how colleges respond. Intuitively, colleges face a trade-off between raising tuition and retaining high-ability students when they experience a balance sheet deterioration. If they increase tuition, fewer high-ability students may enroll, which drives down quality. Alternatively, a decision to not raise tuition forces colleges to cut back on quality-enhancing investment expenditures. We find that colleges take this latter route to the tune of almost $2,800 in cuts per student as a response to higher custodial costs. This result comports with the behavior we observe among many public universities across the country of replacing tenured faculty with less expensive non-tenure-track positions. Additionally, changes in nontuition revenue have almost no impact on tuition (a contribution of 2 percent).

We do not claim that Baumol's cost disease or changes in state support have no importance for tuition increases. Rather, we suspect that these factors affect some colleges more than others. For instance, if private research universities experience cost disease, they may increase their tuition. However, higher tuition may induce substitution of students into lower-cost universities. Given the absence of competition and college heterogeneity in our model, our estimation implicitly incorporates substitution of households across college types and any corresponding composition effects.

10.1.1   Relationship to the Literature

This chapter relates to two broad strands of the literature. First, the chapter relates to a large empirical literature that estimates the effects of macroeconomic factors and policy interventions on tuition and enrollment. Second, this chapter relates to a growing body of literature employing structural models of higher education. With a few exceptions, these models focus on student demand and abstract from many distinguishing features of the supply side.

Empirical Literature

In discussing related work, we map our categorization of supply-side shocks, demand-side shocks, and macroeconomic forces into the existing empirical literature. For supply-side shocks, we analyze the impact of upward shifts in custodial (non-quality-enhancing) costs as well as changes in nontuition revenues. The literature on Baumol's cost disease most closely relates to the former, while the literature analyzing the effect of the decline in state appropriations for higher education addresses the latter.

Supply Shocks: Cost Disease. The origins of cost disease emerge from seminal works by Baumol and Bowen (1966) and Baumol (1967). They lay out a clear mechanism: productivity increases in the economy at large drive up wages everywhere, which service sectors that lack productivity growth pass along by increasing their relative prices. Recently, Archibald and Feldman (2008) use cross-sectional industry data to forcefully advance the idea that cost and price increases in higher education closely mirror trends for other service industries that utilize highly educated labor. In short, they "reject the hypothesis that higher education costs follow an idiosyncratic path."

We find that the form of the cost increase matters. In particular, our estimates uncover a large increase in the fixed cost of operating a college from $12 billion to $30 billion in 2010 dollars. To pay for the higher fixed cost, the college in our model lowers per-student investment and increases enrollment, which lowers average tuition by a composition effect.

Supply Shocks: Cuts in State Appropriations. Heller (1999) suggests a negative relationship between state appropriations for higher education and tuition, asserting that "the higher the support provided by the state, the lower generally is the tuition paid by all students." Recent empirical work by Chakrabarty, Mabutas, and Zafar (2012), Koshal and Koshal (2000), and Titus, Simone, and Gupta (2010) support this hypothesis, but notably, Titus, Simone, and Gupta (2010) show that this relationship only holds up in the short run. Last, in a large study commissioned by Congress in the 1998 reauthorization of the Higher Education Act of 1965, Cunningham et al. (2001) conclude that "Decreasing revenue from government appropriations was the most important factor associated with tuition increases at public four-year institutions."

While our model fails to confirm this idea in the aggregate—that is, lumping public and private colleges together—cuts in appropriations could potentially play a role in driving up public school tuition. Extending our model to incorporate heterogeneous colleges with detailed, disaggregated funding data will shed further light on this issue.

Demand Shocks: The Bennett Hypothesis. For demand-side shocks, we focus on the effects of increased financial aid. We address the extent to which changes in loan limits and interest rates under the FSLP as well as expansions in state and federal grants to students drive up tuition—famously known as the Bennett hypothesis. A long line of empirical research has studied this hypothesis with mixed results.

Broadly speaking, we can divide the literature into those papers that find at least *some* support for this hypothesis and those that are highly skeptical. In the first group, McPherson and Shapiro (1991) use institutional data from 1978 to 1985 and find a positive relationship between aid and tuition at public universities, but not at private universities. Singell and Stone (2007), using panel data from 1983 to 1996, find evidence for the Bennett hypothesis among top-ranked private institutions but not among public and lower-ranked private universities. They also found evidence in favor of the Bennett hypothesis for public *out-of-state* tuition. Rizzo and Ehrenberg (2004,

339) come to the mirror opposite conclusion: "We find substantial evidence that increases in the generosity of the federal Pell Grant program, access to subsidized loans, and state need-based grant aid awards lead to increases in in-state tuition levels. However, we find no evidence that nonresident tuition is increased as a result of these programs." Turner (2012) shows that tax-based aid crowds out institutional aid almost one-for-one. Turner (2014) also finds that institutions capture some of the benefits of financial aid, but at a more modest 12 percent pass-through rate. Long (2004a, 2004b) uncovers evidence that institutions respond to greater aid by increasing charges, in some cases by up to 30 percent of the aid. Cellini and Goldin (2014) compare for-profit institutions that participate in federal student aid programs to those that do not participate. Institutions in the former group charge tuition that is about 78 percent higher than those in the latter group. Most recently, Lucca, Nadauld, and Shen (2015) find a 65 percent pass-through effect for changes in federal subsidized loans and positive but smaller pass-through effects for changes in Pell grants and unsubsidized loans.

In contrast to the previous literature, several papers reject or find little evidence for the Bennett hypothesis. For example, in their commissioned report for the 1998 reauthorization of the Higher Education Act, Cunningham et al. (2001, x) conclude that "the models found no associations between most of the aid variables and changes in tuition in either the public or private not-for-profit sectors." These sentiments are echoed by Long (2006). Last, Frederick, Schmidt, and Davis (2012) study the response of community colleges to changes in federal aid and find little evidence of capture.

Our model likely exaggerates the impact of the Bennett hypothesis. As we discuss in section 10.4, the representative college engages in an implausibly high degree of rent extraction despite the presence of preference shocks. We suspect that more competition in our model of the higher education market would temper the magnitude of the tuition increase attributable to the Bennett hypothesis.

Macroeconomic Forces: Rising College Earnings Premiums. According to data from Autor, Katz, and Kearney (2008), the college earnings premium increased from 58 percent in the mid-1980s to 93 percent in 2005. While we remain agnostic about the cause of the increasing premium, several papers, including Autor, Katz, and Kearney (2008), Katz and Murphy (1992), Goldin and Katz (2007), and Card and Lemieux (2001), ascribe it to skill-biased technological change combined with a fall in the relative supply of college graduates.

In recent work, Andrews, Li, and Lovenheim (2012) study the *distribution* of college earnings premiums and find substantial heterogeneity attributable to variation in college quality. Hoekstra (2009) looks at earnings of white males ten to fifteen years after high school graduation and finds a premium of 20 percent for students who attended the most selective state university relative to those who barely missed the admissions cutoff and went elsewhere. Incorporating this heterogeneity in college earnings premiums may help explain why tuition increases at selective schools (such as public and private research universities) have outpaced those at less selective schools.

Quantitative Models of Higher Education

Our chapter also fits into a growing body of papers that employ structural models of higher education such as Abbott et al. (2013), Athreya and Eberly (2013), Ionescu and Simpson (2016), Ionescu (2011), Garriga and Keightley (2010), Lochner and Monge-Naranjo (2011), Belley and Lochner (2007), and Keane and Wolpin (2001). In the interest of space, we discuss only the most closely related papers.

Recent work by Jones and Yang (2016) closely mirrors the objectives of this chapter. They explore the role of skill-biased technical change in explaining the rise in college costs from 1961 to 2009. Their paper differs from our chapter in several ways. First, whereas they explore the effect of cost disease on higher college costs, we quantify the role of supply-side as well as demand-side shocks. Second, Jones and Yang (2016) analyze college costs—which increased by 35 percent in real terms between 1987 and 2010—whereas we address the increase in net tuition, which went up by 92 percent. Also, whereas they use a competitive framework, we employ a model with peer effects, imperfect competition with price discrimination, and student loan borrowing with default. Fillmore (2014) also analyzes a model of price discriminating colleges, but he treats peer effects in a reduced-form way. Fu (2014) considers a rich game-theoretic framework of college admissions and enrollment but does not allow for price discrimination.

10.2 The Model

The model embeds a college sector into a discrete-time open economy. A fixed measure of heterogeneous households enter the economy upon graduating high school, make college enrollment decisions, and then progress through their working life and into retirement. A monopolistic college with the ability to price discriminate transforms students into college graduates (with dropout risk), and the government levies taxes to finance student loans.

10.2.1 Households

We describe sequentially the environment faced by youths, college students, and finally, workers and retirees. We immediately follow this discussion with a description of colleges in the model. Section 10.2.4 gives the decision problems for all agents in the economy.

Youths

Youths enter the economy at $j = 1$ (corresponding to high school graduation at age eighteen), at which point they draw a two-dimensional vector of characteristics $s_Y = (x, y_p)$ consisting of academic ability x and parental income y_p from a distribution G. Youths make a once-and-for-all choice to either enroll in college or enter the workforce. In addition to the explicit pecuniary and nonpecuniary benefits of college that we will describe momentarily, youths receive a preference shock $(1/\alpha)\varepsilon$ of attending college, where $\alpha > 0$ and ε comes from a type 1 extreme distribution. Colleges cannot condition tuition on the preference shock.

College Students

Newly enrolled students enter college with their vector of characteristics s_Y and a zero initial student loan balance, $l = 0$. Colleges charge type-specific net tuition $T(s_Y)$—equal to sticker price $\bar{T}$ minus institutional aid—which they hold fixed for the duration of enrollment.

Students also face nontuition expenses ϕ that act as perfect substitutes for consumption c. Direct government grants $\zeta(T + \phi, EFC(s_Y))$ offset some of the cost of attendance, where $EFC(s_Y)$ represents the expected family contribution—a formula used by the government to determine eligibility for need-based grants and loans. After taking into account both forms of aid, the net cost of attendance comes out to $NCOA(s_Y) = T(s_Y) + \phi - \zeta(T(s_Y) + \phi, EFC(s_Y))$.

While enrolled, college students receive additively separable flow utility $v(q)$, which increases in college quality q.[4] In order to graduate, students must complete J_Y years of college. Students in class j return to college each year with probability $\pi_{j+1} \equiv \mathbf{1}_{[j+1 \leq J_Y]}$; otherwise, they either drop out or graduate.[5]

Students can borrow through the FSLP. Of primary interest, the FSLP features subsidized loans that do not accrue interest while the student is in college, where eligibility depends on financial need (NCOA less EFC). Since 1993, students can borrow additional funds up to the net cost of attendance using unsubsidized loans. Students face annual and aggregate limits for subsidized and combined borrowing.

Denote the annual and aggregate combined limits by $\bar{b}_j$ and $\bar{l}$, respectively.[6] Because students can borrow only up to the net cost of attendance, their annual combined subsidized borrowing b_s and unsubsidized borrowing b_u must satisfy

(1) $$b_s + b_u \leq \min\{\bar{b}_j, NCOA(s_Y)\}.$$

---

4. To improve tractability while computing the transition path, we assume students receive $v(q)$ each year based on the college's quality q at the time of *initial* enrollment. In the computation, we make the isomorphic assumption that students receive the net present value of $v(q)$ at the time of enrollment.

5. We do not allow endogenous dropout for reasons of tractability.

6. The aggregate limit caps maximum loan balances the period after borrowing, inclusive of interest.

Similarly, define $\bar{b}_j^s$ as the statutory annual subsidized limit and $\bar{I}_j^s$ as the statutory aggregate subsidized limit. The actual amount $\tilde{b}_j^s(s_Y)$ that students can borrow in subsidized loans depends on their net cost of attendance and the expected family contribution, both of which vary with student type. Last, define $\tilde{I}_j^s(s_Y)$ as the maximum amount of subsidized loans that students can accumulate by year j in college. Mathematically,

(2) $\tilde{b}_j^s(s_Y) = \min\{\bar{b}_j^s, \max\{0, \text{NCOA}(s_Y) - \text{EFC}(s_Y)\}\}$

$\tilde{I}_j^s(s_Y) = \min\{\bar{I}^s, \Sigma_{i=1}^j \tilde{b}_i^s(s_Y)\}.$

Given the superior financial terms of subsidized loans, we assume that students always exhaust their subsidized borrowing capacity before taking out any unsubsidized loans. Furthermore, to increase tractability, we assume that borrowers can carry over unused *subsidized* borrowing capacity into subsequent years. These two assumptions reduce the state space and significantly simplify the student's debt portfolio choice problem.

Apart from loans, students have two other means of paying for college. First, they have earnings e_Y, which we treat as an endowment.[7] Second, they receive a parental transfer $\xi \text{EFC}(s_Y)$, where $0 \leq \xi \leq 1$ is a parameter.

Workers/Retirees

Working and retired households receive earnings e that depend on a vector of characteristics s that includes their level of education, age/retirement status, and a stochastic component. Each period, households face a proportional earnings tax τ.

These households value consumption according to a period utility function $u(c)$ and discount the future at rate β. Workers with student loans face a loan interest rate of i and amortization payments of $p(l, t) = l\{[i(1 + i)^{t-1}]/[(1 + i)^t - 1]\}$ where l represents the loan balance and t the remaining duration. All households can use a discount bond to save at the risk-free rate r^* and borrow up to the natural borrowing limit $\bar{a}$ at rate $r^* + \iota$, where ι is the interest premium on borrowing. The price of the bond is denoted $(1 + r(a'))^{-1}$.

10.2.2   Colleges

There is one representative college. Following Epple, Romano, and Sieg (2006), the college seeks to maximize its quality (or prestige), q, which depends on the average academic ability θ of the student body and on investment expenditures per student, I. The college's other expenses include non-quality-enhancing custodial costs $F + C(\{N_j\}_{j=1}^{J_Y})$, where F represents a fixed cost and C is an increasing, twice-differentiable, convex function of enrollment $\{N_j\}_{j=1}^{J_Y}$.

---

7. We abstract from labor supply choice and the trade-off between increased earnings and studying.

The college finances its expenditures with two sources of revenue. First, the college has exogenous nontuition revenue per student E, which includes endowment income, government appropriations, and revenues from auxiliary enterprises. Second, the college has endogenous tuition revenue, a function of enrollment decisions and type-specific net tuition $T(s_y)$. The college is a nonprofit and, given our assumption of an exogenous endowment stream, runs a balanced budget period-by-period.[8]

In order to avoid dealing with issues such as the college's discount factor—not to mention other difficulties associated with the transition path computation—we make the college problem static through four assumptions. First, we assume that college quality $q(\theta, I)$ depends on the academic ability of *freshmen* and investment expenditures per *freshman* student.[9] Second, we assume that colleges face a quadratic cost function for each *class* given by

$$(3) \qquad F + C(\{N_j\}_{j=1}^{J_Y}) = F + \sum_{j=1}^{J_Y} c(n_j)$$

where N_j is the population measure in class j ($j = 1$ for freshmen, $j = 2$ for sophomores, etc.) and $n_j \equiv N_j / (1/J)$ is the measure relative to the age-eighteen population (for scaling purposes in the estimation). Third, we assume the college has no access to credit markets. Last, we isolate the effect of current tuition and spending decisions on *future* budget conditions. Specifically, we assume that each year the college exchanges the rights to all future budget flows generated by contemporaneous tuition and expenditure decisions in exchange for an immediate net present value payment from the government. This last assumption implicitly rules out any "quality smoothing" on the part of the college and captures the fact that administrators typically have short tenures that may make borrowing against expected future flows challenging.[10]

10.2.3   Legal Environment and Government Policy

Consistent with US law, workers in the model cannot liquidate their student loan debt through bankruptcy. However, they *can* skip payments and become delinquent. Upon initial default, workers enter delinquency status and face a proportional loan penalty of η that accrues to their existing balance. In subsequent periods, delinquent workers face a proportional wage garnishment of γ until they rehabilitate their loan by making a payment. Upon rehabilitation, the loan duration resets to the statutory value t_{max} and the amortization schedule adjusts accordingly.

The government operates the student loan program and finances itself

8. Technically, the nonprofit status of the college only implies that it cannot distribute dividends. However, we abstract from strategic decisions regarding endowment accumulation.
9. We assume the college commits to a level of I for the duration of each incoming cohort's enrollment.
10. The average tenure of a dean is five years (Wolverton et al. 2001).

with a combination of taxation on labor earnings, funds from loan repayments and wage garnishments, and the revenue flows generated by colleges discussed above. We assume that the government sets the tax rate τ to balance its budget period-by-period.

10.2.4   Decision Problems

Now we work backward through the life cycle to describe the household-decision problem. Afterward, we describe the college's optimization problem.

Workers/Retirees

Households start each period with asset position a, student loan balance l and duration t, characteristics s, and delinquency status $f \in \{0,1\}$, where $f = 0$ indicates good standing. Households in good standing on their student loans choose consumption, savings, and whether to make their scheduled loan payment. These households have the value function

$$(4) \qquad V(a,l,t,s,f = 0) = \max\{V^R(a,l,t,s), V^D(a,l(1 + \eta),s)\}$$

where V^R is the utility of repayment and V^D is the utility of delinquency. Note that η increases the stock of outstanding debt in the case of a default.

Households in bad standing face the decision of whether to rehabilitate their loan or remain delinquent. Their value function is

$$(5) \qquad V(a,l,s,f = 1) = \max\{V^R(a,l,t_{\max},s), V^D(a,l,s)\}.$$

Household utility conditional on repayment or rehabilitation is given by

$$(6) \qquad V^R(a,l,t,s) = \max_{c \geq 0, a' \geq \underline{a}} u(c) + \beta \mathbb{E}_{s'|s} V(a',l',t',s',f' = 0)$$

subject to

$$c + a'/(1 + r(a')) + p(l,t) \leq e(s)(1 - \tau) + a$$
$$l' = (l - p(l,t))(1 + i), \quad t' = \max\{t - 1, 0\}.$$

The value of defaulting (if $f = 0$) or not rehabilitating a loan (if $f = 1$) is[11]

$$(7) \qquad V^D(a,l,s) = \max_{c \geq 0, a' \geq \underline{a}} u(c) + \beta \mathbb{E}_{s'|s} V(a',l',s',f' = 1)$$

subject to

$$c + a'/(1 + r(a')) \leq e(s)(1 - \tau)(1 - \gamma) + a$$
$$l' = \max\{0, (l - e(s)(1 - \tau)\gamma)(1 + i)\}.$$

In the last period of life, households have no continuation utility and no ability to borrow or save. We allow households to die with student loan debt.

---

11. In the case of a default, note that η has already been applied to the loan balance in equation (4).

College Students

College students with characteristics $s_Y = (x, y_p)$ and debt l choose consumption and additional loans, $l' \geq l$ (to speed up computation, we assume that students do not pay back their loans while in college). We also introduce an annual *unsubsidized* borrowing limit $\bar{b}_j^u$ that equals either the combined limit or zero (the latter case captures the pre-1993 environment).

Taking college quality q and the net tuition function $T(\cdot)$ as given, students solve

$$(8) \quad Y_j(l, s_Y; T, q) = \max_{c \geq 0, l' \geq l} u(c + \phi) + v(q) + \beta \left[\begin{array}{l} \pi_{j+1} Y_{j+1}(l', s_Y; T) + (1 - \pi_{j+1}) \\ \times \mathbb{E}_{s'|j, s_Y} V(a' = 0, l', t_{max}, s', 0) \end{array} \right]$$

subject to

$$c + \text{NCOA}(s_Y) \leq e_Y + \xi \text{EFC}(s_Y) + b_s + b_u$$

$$(l'_s, l'_u) = \begin{cases} (l', 0) & \text{if } l' \leq \tilde{l}_j^s(s_Y) \\ (\tilde{l}_j^s(s_Y), l' - \tilde{l}_j^s(s_Y)) & \text{otherwise} \end{cases}$$

$$(l_s, l_u) = \begin{cases} (l, 0) & \text{if } l \leq \tilde{l}_{j-1}^s(s_Y) \\ (\tilde{l}_{j-1}^s(s_Y), l - \tilde{l}_{j-1}^s(s_Y)) & \text{otherwise} \end{cases}$$

$$b_s = l'_s - l_s$$

$$b_u = \frac{l'_u}{1 + i} = l_u$$

$$l' + \frac{l'_u}{1 + i} \leq \bar{l}$$

$$b_u \leq \min\{\bar{b}_j^u, \text{NCOA}(s_Y)\}$$

$$b_s + b_u \leq \min\{\bar{b}_j, \text{NCOA}(s_Y)\}.$$

Note from these equations that our setup allows us to easily decompose student debt into its subsidized and unsubsidized components. We deflate l'_u by $1 + i$ in the aggregate borrowing constraint because the loan limit is inclusive of interest accrued by unsubsidized loans.

Youth

Youth making their college enrollment decisions have value function

$$(9) \quad \max \left\{ \underbrace{\mathbb{E}_{s|s_Y} V(a = 0, l = 0, t = 0, s)}_{\text{enter the labor force}}, \underbrace{Y_1(l = 0, s_Y; T, q) + \frac{1}{\alpha} \varepsilon}_{\text{attend college}} \right\}$$

where ε denotes the college preference shock and s is the initial worker characteristics draw.

Colleges

The college problem can be written as

$$(10) \qquad \max_{I \geq 0, T(\cdot)} q(\theta, I)$$

subject to

$$\mathcal{E} + \mathcal{T} = F + C(N_1) + \mathcal{J}$$

$$N_1 = \int \mathbb{P}(\text{enroll}|s_Y; T(\cdot), q) d\mu_0(s_Y)$$

$$\theta N_1 = \int x(s_Y) \mathbb{P}(\text{enroll}|s_Y; T(\cdot), q) d\mu_0(s_Y)$$

$$\mathcal{T} = \sum_{j=1}^{J_Y} \frac{\pi^{j-1} \int T(s_Y) \mathbb{P}(\text{enroll}|s_Y; T(\cdot), q) d\mu_0(s_Y)}{(1 + r^*)^{j-1}}$$

$$\mathcal{E} = E \sum_{j=1}^{J_Y} \frac{\pi^{j-1} N_1}{(1 + r^*)^{j-1}}$$

$$C(N_1) = \sum_{j=1}^{J_Y} \frac{c\{\pi^{j-1}[N_1/(1/J)]\}}{(1 + r^*)^{j-1}}$$

$$\mathcal{J} = I \sum_{j=1}^{J_Y} \frac{\pi^{j-1} N_1}{(1 + r^*)^{j-1}}$$

where $\mu_0(s_Y) \equiv G(s_Y)/J$ is the distribution of characteristics across the age-eighteen population.

The first constraint reflects the college balanced budget requirement, while the remaining constraints establish the definitions of enrollment, average freshman ability, tuition revenues, nontuition revenues, custodial costs, and investment expenditures, respectively.

10.2.5   Steady-State Equilibrium

A steady-state equilibrium consists of household value and policy functions, a tax rate, college policies and quality, and a distribution of households such that

1. The household value and policy functions satisfy (4–9).
2. The college policies and quality satisfy (10).
3. The government budget balances.
4. The distribution is invariant.

10.3   Data and Estimation

We calibrate the model to replicate key features of the US economy and higher education sector in 1987. These initial conditions set the stage for the results section, which feeds in the observed changes between 1987 and 2010 described in the introduction to assess their impact on equilibrium tuition. We proceed through our description of the calibration and estimation in the same order as we described the model.

10.3.1   Households

Youths

We determine the distribution G of youth characteristics $s_Y = (x, y_p)$ using data from the National Longitudinal Survey of Youth 1997 (NLSY97). The ability measure comes from percentiles on the Armed Services Vocational *Aptitude* Battery (ASVAB) test. For parental income, we use the household income measure from 1997 in those cases where the data correspond to the parents rather than the youth (98.0 percent of cases).

Conditional on our ability measure, parental income resembles a truncated normal distribution. This can be seen in figure 1 of web appendix A (http://www.nber.org/data-appendix/c13711/appendix.pdf). To handle truncation from above due to top-coding and truncation from below, we estimate a Tobit model where parental income depends on ability. Specifically, we estimate

(11) $y_i^* = \beta_0 + \beta_1 x_i + \varepsilon_i$

                      $y_i = \min\{\max\{0, y_i^*\}, \bar{y}\}$

where y_i is the observed parental income, y_i^* is the "true" parental income, and $\varepsilon_i \sim N(0, \sigma^2)$.[12] The parameter $\bar{y}$ corresponds to the 2 percent top-coded level implemented in the NLSY97 (we find $\bar{y} = \$226,546$ in 2010 dollars). In 2010 dollars, we find $\beta_0 = \$40,006$, $\beta_1 = \$614.6$, and $\sigma = \$48,012$, with standard errors of \$1,529, \$25.95, and \$543.4, respectively. By the construction of x in NLSY97, $x \sim U[0,100]$. Hence, our estimation implies that, all else equal, parents of children at the top of the ability distribution earn \$152,900 more on average than parents of children at the bottom of the ability distribution. We assume the joint distribution is time invariant.

Table 10.1 reports the correlation between ability, observed parental income, and enrollment. All the correlations are significant at more than a 99.9 percent confidence level. We use the correlation between ability and

---

12. The NLSY97 top-codes at the 2 percent level by replacing the true value with the conditional mean of the top 2 percent. In this estimation, we bound the observed value at the 2 percent threshold value.

Table 10.1   Correlations between ability, parental income, and enrollment

	Ability	Parental income	Enrollment
Ability	1.0000		
Parental income	0.3164	1.0000	
Enrollment	0.5216	0.2952	1.0000

enrollment as a calibration target and the correlation between enrollment and parental income as an untargeted prediction of the model.

College Students

For our specification of the expected family contribution function $EFC(s_Y)$, we use an approximation from Epple et al. (2013) to the true statutory formula. Specifically, we assume a mapping between raw and adjusted gross parental income of $\tilde{y}(y_p) = y(1 + .07 \cdot \mathbf{1}[y \geq \$50000])$ and an EFC formula given by $EFC(y_p) = \max\{\tilde{y}(y_p)/5.5 - \$5,000, \tilde{y}(y_p)/3.2 - \$16,000, 0\}$ in 2009 dollars.

We assume that the government grants $\zeta(T + \phi, EFC(s_Y))$ are given by

$$(12) \quad \zeta(T(s_Y) + \phi, EFC(s_Y)) = \begin{cases} \zeta^F \bar{\zeta} & \text{if } \zeta^F \bar{\zeta} \leq T(s_Y) + \phi - EFC(s_Y) \\ 0 & \text{otherwise} \end{cases},$$

which reflects their progressive nature. First, we estimate the average value of government grants $\bar{\zeta}$ from the college-level Integrated Postsecondary Education Data (IPEDS) published by the National Center for Education Statistics (NCES). Then, we calibrate $\zeta^F \geq 1$ to match average grants per student, $\bar{\zeta}$, in the initial steady state. Over the transition path we keep ζ^F constant but vary $\bar{\zeta}$.

The utility function $u(c) = c^{1-\sigma}/(1-\sigma)$ for students as well as workers and retirees features constant relative risk aversion. We use the standard parametrization of $\sigma = 2$ and $\beta = 0.96$. We assume utility from college quality is linear, $v(q) = q$ (and so all curvature comes from the production function $q(\theta, I)$).

To determine student earnings e_Y while in college, we again turn to the NLSY97. For our sample, students enrolled in a four-year college earn on average \$7,128 (in 2010 dollars).[13] We convert this to model units and set e^Y equal to it. The mapping from dollars into model units is discussed in the web appendix, section B.1.

Recall that the annual retention rate satisfies $\pi_{j+1} = \pi\mathbf{1}[j + 1 \leq J_Y]$, which implies constant progression probabilities for students in years $1, \cdots, J_Y - 1$. Students in their last year, which we set to $J_Y = 5$, successfully graduate and

---

13. Students work an average of 824 hours a year in the NLSY97. Using different data, Ionescu (2011) reports similar results of 46 percent of full-time students working with mean worker earnings of \$20,431 in 2007 dollars.

earn a diploma with this same probability. We set $\pi = 0.556^{1/J_Y}$ to match the aggregate completion rate of 55.6 percent reported by Ionescu and Simpson (2016).

Last, we allow the nontuition cost of attending college, ϕ, which plays a significant part in determining eligibility for subsidized loans, to vary over the transition path. We measure ϕ using room-and-board estimates from the NCES (NCES 2015c).

Workers/Retirees

The earnings process for working households follows

(13) $\log e_{ijt} = \lambda_t h_i / J_Y + \mu_j + z_{ij} + \nu$

$$z_{i,j+1} = \rho z_{ij} + \eta_{i,j+1}$$

$$\eta_{i,j+1} \sim N(0, \sigma_z^2)$$

where h_i is the number of completed years of college, i is an individual identifier, j is age, and t is time. Households who begin working at age j draw z_{ij} from an unconditional distribution with mean zero and variance $\sigma_z^2(1 + \ldots + \rho^{2(j-1)})$. For the persistent shock, we use Storesletten, Telmer, and Yaron's (2004) estimates in setting $(\rho, \sigma_z) = (0.952, 0.168)$.[14] The deterministic earnings profile μ_j is a cubic function of age with coefficients also taken from Storesletten, Telmer, and Yaron (2004).[15]

In the model, λ_t represents the earnings premium for college graduates relative to high school graduates. We compute λ_t using the estimates from Autor, Katz, and Kearney (2008), which range from roughly 0.43 in the 1960s and 1970s to 0.65 in the early twenty-first century. To deal with the fact that Autor, Katz, and Kearney (2008) estimate values only up until 2005, we fit a quadratic polynomial over 1988–2005 and extrapolate for 2006–2010.[16] We use the fitted values (both in-sample and out-of-sample) for λ_t, and they are presented in web appendix A (web appendix B gives a comparison of the raw and fitted values).

Retired households ($j > J_R = 48$) have constant earnings given by $\log e_{ijt} = \log(0.5) + \lambda_t h_i / J_Y + \mu_{J_R} + \nu$, which yields an average replacement rate of roughly 50 percent.

10.3.2   Legal Environment and Government Policy

We set the duration of loan repayment to its value in the federal student loan program, $t_{\max} = 10$. Two parameters—the loan balance penalty η and

---

14. Storesletten, Telmer, and Yaron (2004) let σ vary with the business cycle and estimate $\sigma = .211$ for recessions and $\sigma = .125$ for expansions. We average these.

15. In principle, one could include a cohort-specific term that allows for average log earnings in the economy to grow over time. However, we found that such a term is negligible in the data as we show in web appendix B.1.

16. The "1987" college premium corresponds to the average from 1981 to 1987.

garnishment rate γ—control the cost of student loan delinquency. Various changes in student loan default laws between 1987 and 2010 render obtaining values for these parameters less than straightforward.[17] Our approach sets $\eta = 0.05$ (which is half the value in Ionescu [2011], and only a fifth of the current statutory maximum) and then pins down γ in the joint calibration to match the 17.6 percent student loan default rate in 1987.

10.3.3   Colleges

We need to parametrize and provide estimates for the per-student endowment E, the quality production function $q(\theta, I)$, and custodial costs $F + C(\{N_j\}_{j=1}^{J_y})$.

Institution-Level Data

Our primary source for college revenue and expenditures is institution-level data from the Delta Cost Project (DCP), which is drawn from the National Center for Education Statistics Integrated Postsecondary Education Data System (IPEDS). One important distinction between our DCP-based average tuition measures and those reported by the nces330p10y15 (in table 330.10) is that, for public colleges, the NCES only uses in-state tuition.[18] Consequently, the gross tuition and fees in our data are larger than those reported by the NCES. However, despite this discrepancy in levels, figure 10.1 shows that the trend growth in gross tuition and fees between the two measures is nearly identical.

For sample selection, we restrict attention to four-year, nonprofit, non-specialty institutions (according to their Carnegie classification) that have nonmissing enrollment and tuition data in every year of the DCP data from 1987 to 2010.[19] Additionally, we drop institutions with fewer than 100 FTE students or net tuition per FTE outside of the 1st–99th percentile range.

The college budget constraint in the model features custodial costs, endowment income, quality-enhancing investment, and tuition. The corresponding data measures are as follows:

- Endowment: total nontuition revenue, which is the sum of (non-Pell) grants at the federal, state, and local levels plus all auxiliary revenue.
- Investment: total education and general expenditures including sponsored research but excluding auxiliary enterprises.
- Tuition: net tuition and fees revenue.
- Custodial costs: a residual computed as the endowment plus tuition less investment.

Web appendix A provides more details on our use of the DCP data.

---

17. See Ionescu (2011) for changes in student loan default laws.
18. This difference in methodologies accounts for the mismatch in reported tuition numbers brought up by our discussant, Sandy Baum.
19. The DCP data is released at a multiyear lag, and all indications are that changes in college tuition continue to outpace inflation.

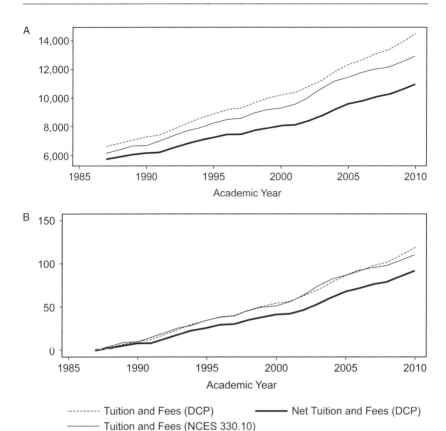

Fig. 10.1   College tuition trends: DCP versus NCES. *A*, real tuition per FTE; *B*, real tuition per FTE, percentage change since 1987.

Notes: 2010 dollars per FTE. The DCP series are authors' calculations using Delta Cost Project data. NCES 330.10 from https://nces.ed.gov/programs/digest/d13/tables/dt13_330.10.asp, retrieved 3/28/16. NCES 330.10 conversion to 2010 dollars is authors' calculation. NCES 330.10 assumes only in-state tuition is charged.

Calibrated Parameters

We set the per-student endowment E equal to nontuition revenues per FTE student in the 1987 IPEDS data, and then we vary E along the transition path. Figure 10.2 plots the time series for E and other key aggregates. For college quality, we follow Epple et al. (2013) and choose a Cobb-Douglas functional form, $q(\theta, I) = \chi_q \theta^{\chi_\theta} I^{\chi_I}$, where $\chi_I = 1 - \chi_\theta$.[20]

The local first-order conditions of the college problem provide some

---

20. In principle, $q(\theta, I)$ need not satisfy constant returns to scale. With one college, it is difficult to pin down—using only steady state information—what the returns should be. With multiple colleges, dispersion in θ and I translates into dispersion in q that is controlled by returns to scale.

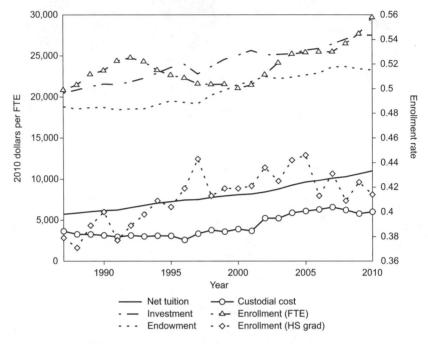

Fig. 10.2   College cost, expenditure, and enrollment trends

insight into calibrating χ_θ and χ_q. The key tuition-pricing condition comes out to

$$(14) \quad T(s_Y) + \frac{\mathbb{P}(\text{enroll}\,|\,s_Y;T(\cdot),q)}{\partial\mathbb{P}(\text{enroll}\,|\,s_Y;T(\cdot),q)\,/\,\partial T} = C'(N) + I + \frac{q_\theta}{q_I}(\theta - x(s_Y))$$

where $\mathbb{P}(\text{enroll}\,|\,s_Y;Ts_Y,q)$ comes from the decision rule of youths for whether to attend college, taking into account the idiosyncratic preference shock ε. Epple et al. (2013) label the collected right-hand-side terms the "effective marginal cost" (EMC) of a type-s_Y student, which captures the fact that students act both as customers *and* as inputs to the production of quality (an argument put forth by Winston [1999] and others). The above equation states that colleges admit any student to whom they can charge at least EMC(s_Y).

With our Cobb-Douglas specification, $q\theta/q_I = (\chi_\theta/\chi_I)(I/\theta) = [\chi_\theta/(1 - \chi_\theta)](I/\theta)$. The degree to which EMC(s_Y), and therefore tuition $T(s_Y)$, varies by student type depends on χ_θ. This price discrimination generates cross-sectional enrollment patterns that we use to target χ_θ and χ_q. Specifically, we target overall enrollment and the correlation between parental income and enrollment.

Cost Function Estimation

Like in Epple, Romano, and Sieg (2006), we estimate the college's custodial cost function directly. In particular, we assume that the custodial costs by class, $c(n)$, have the functional form $C^1 n + C^2 n^2$. When we explicitly allow for time-varying coefficients, custodial costs satisfy

$$(15) \qquad F_t + C_t(\{N_{jt}\}_{j=1}^{J_Y}) = F_t + C_t^1 \sum_{j=1}^{J_Y} n_{jt} + C_t^2 \sum_{j=1}^{J_Y} n_{jt}^2$$

where $n_{jt} \equiv N_{jt}/(1/J)$ is class j enrollment in year t relative to the age-eighteen population.

To identify F_t, C_t^1, and C_t^2, we estimate cost functions for individual colleges using IPEDS data and then aggregate them. Let college i's cost function at time t be given by

$$(16) \qquad c_{it} = \alpha_i + c_t^0 + c_t^1 \sum_{j=1}^{J_Y} n_{ijt} + c_t^2 \sum_{j=1}^{J_Y} n_{ijt}^2 + \varepsilon_{it}.$$

Here, α_i is a fixed effect and both α_i and ε_{it} are i.d.d. normally distributed with mean zero.

The IPEDS data contains enrollment information but not its composition by class. To deal with this problem and to create consistency with the model, we assume a constant retention rate π and a five-year college term, $J_Y = 5$. Given π, J_Y, and total FTE enrollment data by school relative to the age-eighteen population, we calculate implied class j enrollment as $n_{ijt} = \pi^{j-1} \text{FTE}_{it} / \sum_{\iota=1}^{J_Y} \pi^{\iota-1}$. Thus, the two summation terms in the cost function come out to $\sum_{j=1}^{J_Y} n_{ijt} = \text{FTE}_{it}$ and $\sum_{j=1}^{J_Y} n_{ijt}^2 = \text{FTE}_{it}^2 \sum_{j=1}^{J_Y} \pi^{2(j-1)} / (\sum_{j=1}^{J_Y} \pi^{j-1})^2$. As a result,

$$(17) \qquad c_{it} = \alpha_i + c_t^0 + c_t^1 \text{FTE}_{it} + c_t^2 \text{FTE}_{it}^2 \frac{\sum_{j=1}^{J_Y} \pi^{2(j-1)}}{\left(\sum_{j=1}^{J_Y} \pi^{j-1}\right)^2} + \varepsilon_{it}.$$

As in Epple, Romano, and Sieg (2006), we measure custodial costs as a residual in the college budget constraint, which gives us

$$(18) \qquad c_{it} \equiv e_{it} + t_{it} - i_{it}.$$

The first term, e_{it}, represents total nontuition revenue in IPEDS (which consists mostly of endowment revenue and government appropriations), while t_{it} and i_{it} equal net tuition revenues and total education and general (E&G) expenditures, respectively. Intuitively, our cost measure reflects the fact that, holding investment i_{it} constant, higher costs must accompany any observed increase in revenues in order to maintain a balanced budget. Using these definitions, we run the fixed effects panel regression above to obtain $\{(c_t^0, c_t^1, c_t^2)\}_{t=1987}^{2010}$.

To translate the individual cost function estimates into the aggregate cost function, we sum costs over colleges. In particular, to calculate the total cost

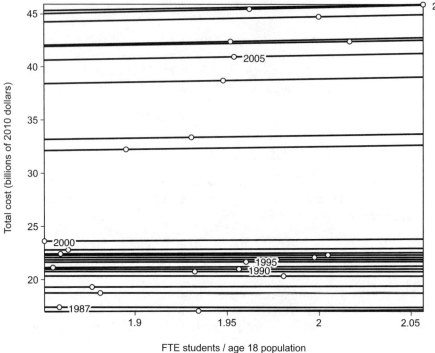

Fig. 10.3   Estimated aggregate cost function by year

of educating $\{N_{jt}\}_{j=1}^{J_Y}$ students, we assume students sort across colleges $i = 1$, ..., K in proportion to the observed share in the data.[21] Define $s_{ijt} \equiv N_{ijt}/N_{jt} = n_{ijt}/n_{jt}$ as the share of students in class j at time t who attend college i. From our assumption of geometric retention probabilities, this share does not vary with j, that is, $s_{ijt} = s_{it}$. Thus, $N_{ijt} = s_{it}N_{jt}$ and $n_{ijt} = s_{it}n_{jt}$ for all j, which gives us[22]

$$(19) \qquad F_t + C_t(\{N_{jt}\}_{j=1}^{J_Y}) = K\mathfrak{c}_t^0 + \mathfrak{c}_t^1 \sum_{j=1}^{J_Y} n_{jt} + \left(\mathfrak{c}_t^2 \sum_{i=1}^{K} s_{it}^2\right) \sum_{j=1}^{J_Y} n_{jt}^2.$$

This mapping between individual colleges and the representative college yields $F_t = K\mathfrak{c}_t^0$, $C_t^1 = \mathfrak{c}_t^1$, and $C_t^2 = \mathfrak{c}_t^2 \sum_i s_{it}^2$.

The web appendix presents the estimates. We found it necessary to impose $\mathfrak{c}_t^1 = 0$ to ensure an increasing aggregate cost function over the relevant range

---

21. We allow K to vary over time in the estimation (it is the number of colleges in the sample) but treat it as fixed here to simplify the exposition.

22. We assume that $\sum_i \alpha_i = 0$ and $\sum_i \varepsilon_{it} = 0$, where the first assumption is required for identification in the fixed effects regression.

of N. Figure 10.3 plots the aggregate cost function over time and circles the realized values from each year.

10.3.4   Joint Calibration

We determine the remaining parameters (ν, ξ, γ, χ_θ, χ_q, ζ^F, α) jointly such that the initial steady state matches the following moments in 1987: average earnings, average net tuition, the two-year cohort default rate, the correlation between parental income and enrollment, the enrollment rate, the average grant size, and the percent of students with loans.[23]

Table 10.2 summarizes the calibration. Note that, while the table associates each parameter in the joint calibration with an individual moment, the calibration identifies the parameters simultaneously, rather than separately. We discuss model fit next.

10.3.5   Model Fit

Table 10.3 presents key higher education statistics from the model and the data. The calibration of the initial steady state directly targets the first set of statistics from 1987, while the remaining statistics act as an informal test of the model. Note that, while the calibration matches mean earnings, net tuition, and the two-year default rate from 1987 quite well, the model generates too little enrollment and too many students with loans.

We pinpoint two sources for these shortcomings. First, the presence of only one college in the model generates too much market power, which results in a small calibrated value for the parental transfers parameter ξ in order to still match average net tuition. Thus, students rely more on borrowing. Second, by omitting ability terms in the postcollege-earnings process, we implicitly attribute the entire college premium to the sheepskin effect of a diploma (as opposed to selection effects). This exaggerated sheepskin effect generates a larger surplus from attending college, which the college partially captures through higher tuition.

Despite the presence of too many student borrowers, the model actually generates smaller average loans than in the data—$4,600 versus $7,100. Last, the model nearly matches investment per student of $20,300 in 1987 and the ratio of assets to income of about three. The matching of the asset-to-income ratio reflects the fact that our model of households is, at its core, a standard incomplete markets life-cycle model.

10.4   Results

Now we present the main results. First, we compare the model's initial and terminal steady states to the data from 1987 and 2010. Next, we evaluate the

---

23. The correlation between parental income and enrollment is from NLSY97 (and so is not a 1987 moment).

Table 10.2   **Model calibration**

Description	Parameter	Value	Data	Model	Target/reason
Calibration: Independent parameters					
Discount factor	β	0.96			Standard
Risk aversion	σ	2			Standard
Savings interest rate	r^*	0.02			Standard
Borrowing premium	ι	0.107			12.7 percent rate on borrowing
Earnings in college	e_Y	$\$7{,}128_{2010}$			NLSY97
Loan balance penalty	η	0.05			Ionescu (2011)
Loan duration	t_{max}	10			Statutory
Retention probability	π	$0.554^{1/5}$			55.4 percent completion rate
Earnings shocks	(ρ, σ_z)	(0.952, 0.168)			Storesletten, Telmer, and Yaron (2004)
Age-earnings profile	μ_j	Cubic			Storesletten, Telmer, and Yaron (2004)
College premium	$\{\lambda\}$	Web appendix A			Autor, Katz, and Kearney (2008)
Nontuition costs	$\{\phi\}$	Web appendix A			IPEDS
Student loan rate	$\{i\}$	Web appendix A			Statutory
Annual loan limits	$\{\bar{b}^s, \bar{b}^u, \bar{b}\}$	Web appendix A			Statutory
Aggregate loan limits	$\{\bar{T}^s, \bar{T}^u, \bar{T}\}$	Web appendix A			Statutory
Custodial costs	$\{F, C^2\}$	Web appendix A			IPEDS regression
Endowment flow	$\{E\}$	Web appendix A			IPEDS
Grant aid	$\{\bar{\zeta}\}$	Web appendix A			IPEDS
Calibration: Jointly determined parameters					
Earnings normalization	ν	−1.25	31,385	31,686	Mean earnings
Parental transfers	ξ	0.208	5,723	6,146	Mean net tuition
Garnishment rate	γ	0.158	0.176	0.165	Two-year default rate
Ability input to quality	χ_θ	0.252	0.295	0.244	Corr. (p. income, enroll)
College quality loading	χ_q	2.68	0.379	0.358	Enrollment rate
Grant progressivity	ζ^F	1.85	0.027	0.025	Average grant size
Preference shock size	α	290	0.357	0.488	Percent with loans

Note: $\{x\}$ means x has a transition path given in table 2 in web appendix A; $\$x_{yyyy}$ means $\$x$, measured nominally in $_{yyyy}$ dollars, converted to model units.

Table 10.3   **Steady-state statistics**

	Model 1987	Data 1987	Model Final SS	Data 2010
Statistics targeted in 1987				
Mean earnings^z ($)	31,686	31,385*	37,301	36,200
Mean net tuition^z ($)	6,146	5,723*	12,428	10,999
Two-year default rate^a	0.165	0.176*	0.167	0.091
Enrollment rate^b	0.358	0.379*	0.560	0.414
Graduation rate^c	0.554	0.554*	0.554	0.594
Attainment rate (grad × enroll)^z	0.198	0.210*	0.310	0.246
Percent taking out loans^{e,f}	48.8	35.7*	100.0	52.9
Corr. (parental income, enrollment)	0.244	—	0.301	0.295*
Untargeted statistics				
Investment per student^z ($)	21,921	20,475	30,701	27,534
Average EFC^{d,e,f,z} ($)	18,288	16,270	16,514	13,042
Average annual loan size for recipients^{d,e,f,z} ($)	4,589	7,144	6,873	8,414
Total assets/total income^{d,g,z}	3.05	2.94	3.07	3.06
Student loan volume/total income^{d,h,z}	0.012	—	0.053	0.050
Newly defaulted/non-defaulted loans^{h,z}	0.045	—	0.054	0.019
Newly defaulted/good standing borrowers^{h,z}	0.029	—	0.046	0.032
Pop. with loans/age 18 + pop^{h,i,z}	0.040	—	0.140	0.146
Ability of college graduates^z	0.728	—	0.701	0.716
Corr. (ability, enrollment)	0.588	—	0.782	0.522
Nongarnishment payments/total income	0.002	—	0.006	—
Garnishments/total income	0.000	—	0.001	—

Note: Dashes indicate unknown values.
^{a}US Department of Education (2015b)
^{b}NCES (2015a)
^{c}NCES (2015b)
^{d}FRED (2015)
^{e}Tables 2 and 7 in Wei et al. (2004)
^{f}Tables 2.1-C and 3.3 Bersudskaya and Wei (2011)
^{g}BEA (2015)
^{h}US Department of Education (2015a)
^{i}Howden and Meyer (2011)
^{z}authors' calculations
*Targeted.

transition path of the model in light of the time-series data. Last, we undertake a number of counterfactual experiments to quantify the explanatory power of each tuition inflation theory.

10.4.1   Steady-State Comparisons

Tuition

Of central importance, the model generates a 102 percent increase in average net tuition—from approximately $6,100 to $12,400—between the initial

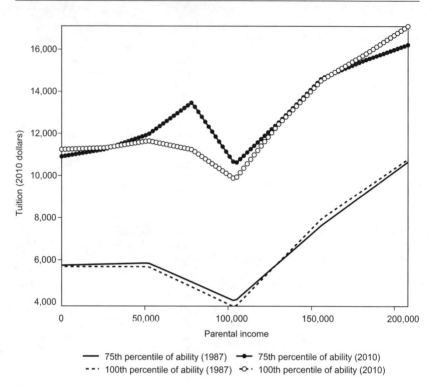

Fig. 10.4   Slices of the tuition function (equilibrium tuition for select ability levels)

and terminal steady states. This jump compares to a 92 percent increase in the data. To illustrate how tuition changes, figure 10.4 plots slices of the tuition function (web appendix C gives the entire function).

In both steady states, tuition does not move monotonically with income. Instead, tuition in the initial steady state first increases with parental income before it starts to decline at income levels between $50,000 and $100,000 as financial aid eligibility tightens and grants decline. After $100,000, tuition resumes its ascent as student ability to pay increases. The tuition curves shift up noticeably between the two steady states, though not in a parallel fashion. In particular, the region of declining tuition compresses to the range between $75,000 and $100,000, which is largely due to the expansion in aid between 1987 and 2010.

The college engages in less price discrimination by academic ability than by parental income.[24] Inspection of the 100th percentile and 75th curves in

24. In fact, theoretically, tuition should be monotonically decreasing in ability. However, due to computational cost, we have parametrized the tuition function more flexibly in the income dimension to account for more variation there. See web appendix C for computation details.

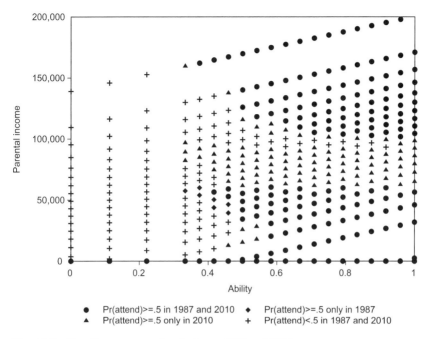

Fig. 10.5   **Enrollment comparison between 1987 and 2010**

1987 reveals that tuition never differs by more than $700 between moderate-and high-ability students. By 2000, the largest tuition difference between the 75th and 100th percentiles of the ability distribution rises to $2,000.

When weighing whether to offer tuition discounts to high-ability students, colleges face the trade-off between a higher-ability student body and the need for resources to fund quality-enhancing investment expenditures. In our calibration, the latter effect dominates. The data provides supporting evidence. For instance, table 10.3, which presents selected statistics from the data and the initial and terminal steady states, shows that investment in the model increases by 40 percent between the two steady states. This increase approximates well the untargeted 34 percent rise in the data. While we lack data on student ability in 1987, the model's mean college graduate ability of 0.701 in 2010 closely matches the untargeted 0.716 from the data.

Enrollment

Figure 10.5 reveals how the enrollment patterns change between the steady states. Recall that the calibration targets the correlation between parental income and enrollment, and observe that average student ability aligns closely with the data in table 10.3. However, figure 10.5 unveils a striking polarization of enrollment by income in the initial steady state.

Specifically, middle-income students find themselves priced out of college, enrolling at a rate of less than 50 percent.

As shown in equation (14), colleges set tuition by charging each student their type-specific effective marginal cost $EMC(s_Y)$ plus a markup that reflects the student's willingness to pay. Given that effective marginal cost only depends on the ability component $x(s_Y)$ of each student's type, all tuition variation within ability types derives from the impact of parental income and access to financial aid on student willingness to pay.[25] Furthermore, in the absence of preference shocks (the limiting case as $\alpha \to \infty$), colleges first only admit students that have a willingness to pay that exceeds their effective marginal cost, and then they proceed to charge tuition that extracts the entire surplus.

High-income students have a high willingness to pay because of parental transfers, while low-income students, despite lacking parental resources, have a high willingness to pay because of access to financial aid. Middle-income students find both of these avenues closed, in large part because each $1 increase in parental income reduces access to subsidized borrowing by $1 but only delivers $\xi \approx .21$ dollars of additional resources to the student. Consequently, these students cannot afford to pay the full net tuition directly and also lack eligibility for subsidized loan borrowing, which represents the only form of student loans accessible in 1987. The college responds to the higher demand elasticity of these students by reducing their tuition, but the decrease does not prove sufficient to prevent low enrollment of middle-income students in the initial steady state.

By 2010, the introduction of unsubsidized loans and repeated expansions in grants and subsidized borrowing induces middle-income students to flood into higher education. These innovations partly explain the increase in enrollment from 36 percent to 56 percent across steady states, as reported in table 10.3. The data show a more subdued rise from 38 percent to 41 percent.

Borrowing and Default

As we just explained, the enrollment surge between the initial and terminal steady states comes primarily from high-ability, middle-income youths who benefit from the introduction of unsubsidized loans and expansion of subsidized aid. In fact, in the terminal steady state, every single college student participates at least minimally in student borrowing (recall that $\beta = 0.96$ and the loan interest rate in 2010 is 3 percent, which makes student loans an attractive form of borrowing). Empirically, the percentage of students with loans increases more moderately from 35.7 percent to 52.9 percent. That

25. Replicated here:

$$T(s_Y) + \underbrace{\frac{\mathbb{P}(\text{enroll} \mid s_Y; T(\cdot), q)}{\partial \mathbb{P}(\text{enroll} \mid s_Y; T(\cdot), q) / \partial T}}_{(\partial \log \mathbb{P} / \partial T)^{-1}} = \underbrace{C'(N) + I + \frac{q_\theta}{q_I}(\theta - x)}_{EMC(s_Y)}$$

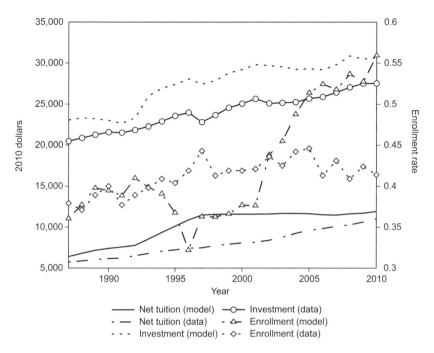

Fig. 10.6   Comparison of model and data over the transition

said, although the model greatly overestimates participation in the student loan program, it generates an average loan size of only $6,900 compared to $8,400 in the 2010 data.

The model delivers almost no change in the 17 percent student loan default rate across steady states. The data, by contrast, show a significant fall from 17.6 percent to 9.1 percent. This discrepancy largely comes from the fact that legal changes between 1987 and 2010 increased the cost of student loan default, whereas we abstract from such changes in the model.

10.4.2   Transition Path Dynamics

Given that we have constructed a rich time series of borrowing limits, the college premium, college endowments, and measured custodial costs, we can gain further insights by analyzing the entire transition path of the model. Figure 10.6 plots the path of net tuition, enrollment, and investment expenditures in both the model and the data.

While investment per student in the model lines up well with the data, equilibrium net tuition follows a different trajectory than net tuition in the data. In particular, equilibrium net tuition in the model rises by a similar amount to the data, but whereas model net tuition rises rapidly between 1993 and 1997 before stagnating, empirical net tuition increases gradually during the entire time period. As the next section will make clear, equilibrium net tuition in the model reacts strongly to the expansion in financial aid (especially the introduction of unsubsidized loans) following the reauthorization of the Higher Education Act in 1992. Although the college premium increased from 0.46 to 0.58 log points between 1987 and 1993, many middle-income households lacked the resources or borrowing capacity to take advantage by enrolling in college.

We can only speculate as to why net tuition in the data does not accelerate in 1993. To the extent that political concerns partially govern the setting of tuition, colleges may prefer to spread out tuition increases over longer time horizons rather than announce rapid escalations. Alternatively, students may not have accurately forecasted the persistent rise in the college premium, whereas our solution method assumes perfect foresight. Last, colleges may engage in some form of tacit collusion that takes time to implement, which our model does not capture because of the representative college assumption.

The overly rapid tuition increases in model may also explain the divergent pattern in enrollments between 1993 and 1998. In particular, the data enrollments increase steadily whereas model enrollments fall substantially. Had the college in the model "smoothed" tuition over this period, enrollments might not have fallen so sharply.

10.4.3   Assessing the Theories of Tuition Inflation

Our model successfully replicates the rapid increase in net tuition, and hence it is useful to now ask our main question of *why* net tuition has almost doubled since 1987. We quantify the role of the following factors in this tuition rise: (a) changes in custodial costs and nontuition sources of revenue, such as endowments and state support (supply shocks); (b) changes in student loan borrowing limits, interest rates, grant aid, and nontuition costs, such as room and board (demand shocks); and (c) macroeconomic forces, namely, the rise in the college wage premium.

We undertake the tuition decomposition from two different angles. First, we progressively solve the model by implementing only *one* of the broad categories of shocks at a time, which answers the question "How much would tuition have gone up if *only* X had occurred?" Then we sequentially shut down the supply shocks, demand shocks, and the college wage premium one at a time. This approach allows us to answer the question "How much would tuition have gone up if X had *not* occurred?" Last, we break down the effect of the individual factors that constitute our categorizations. In all the experiments, we solve for the tax rate that ensures a balanced budget for the government.

Demand Shocks: The Bennett Hypothesis

Table 10.4 summarizes the decomposition through some key statistics. With all factors present, net tuition increases from $6,100 to $12,400. As column (4) demonstrates, the demand shocks—which consist mostly of changes in financial aid—account for the lion's share of the higher tuition.

Table 10.4   **Experiments**

Statistic	1987	Experiment						2010
College costs					*	*	*	*
College endowment					*	*	*	*
Borrowing limits		*		*			*	*
Interest rates		*		*			*	*
Nontuition cost		*		*			*	*
Grants		*		*			*	*
College premium	*				*	*		*
Mean net tuition ($)	6,146	7,412	11,733	5,681	13,274	7,020	11,131	12,428
Std. net tuition ($)	1,263	1,328	1,347	1,558	1,270	1,138	1,405	1,320
Enrollment rate	0.36	0.37	0.38	0.53	0.35	0.54	0.52	0.56
Two-year default rate	0.17	0.15	0.32	0.17	0.17	0.15	0.32	0.17
Mean loan (recipients) ($)	4,589	4,690	6,876	4,692	6,872	4,676	6,877	6,873
Pct. taking out loans	48.8	54.1	100.00	49.6	100.00	58.6	100.00	100.00
Mean earnings ($)	31,686	34,179	31,870	33,445	33,884	37,001	33,306	37,301
Corr. (p. income, enroll)	0.24	0.20	0.36	0.20	0.33	0.10	0.32	0.30
Corr. (ability, enroll)	0.59	0.63	0.73	0.51	0.68	0.51	0.78	0.78
Ability of graduates	0.73	0.74	0.77	0.64	0.77	0.64	0.72	0.70
Investment ($)	21,921	23,304	27,653	23,684	29,019	25,140	29,007	30,701
Average EFC ($)	18,288	17,140	18,892	16,509	18,487	14,833	16,992	16,514
Ex ante utility	−40.98	−40.92	−40.84	−40.61	−40.72	−40.49	−40.51	−40.19

*The value changed over the transition.

Specifically, with demand shocks alone, equilibrium tuition rises by 91 percent, almost fully matching the 102 percent from the benchmark. By contrast, with all factors present *except* the demand shocks (column [7]), net tuition only rises by 14 percent.

These results accord strongly with the Bennett hypothesis, which asserts that colleges respond to expansions of financial aid by increasing tuition. In fact, the net tuition response to the demand shocks in isolation restrains enrollment to only grow from 36 percent to 38 percent. Furthermore, the students who *do* enroll take out $6,900 in loans compared to $4,600 in the initial steady state. The college, in turn, uses these funds to finance an increase of investment expenditures from $21,900 to $27,700 and to enhance the quality of the student body. In particular, the average ability of graduates increases by 4 percentage points. Last, the model predicts that demand shocks in isolation generate a surge in the default rate from 17 percent to 32 percent. Essentially, demand shocks lead to higher costs of attendance and more debt, and in the absence of higher labor market returns, more loan default inevitably occurs.

Importantly, we view this effect as an upper bound for the Bennett hypothesis. Given our representative college assumption, only the unobservable preference shocks prevent the college from extracting the entire surplus from its student body. Table 10.4 illustrates this market power in the small variation in ex ante utility across the decompositions (for any experiment, the consumption equivalent variation is less than 2 percent relative to 1987).

Greater competition would restrict rent extraction and give rise to different pricing patterns.

Macroeconomic Forces: The Rising College Wage Premium

The rise in the college wage premium also contributes to higher tuition, albeit more modestly. If only the college wage premium had changed between 1987 and 2010, the model predicts that net tuition would have gone up by 21 percent. In its absence, but with all other shocks present, tuition would have gone up by 81 percent. Interestingly, the rise in the college wage premium generates barely any increase in enrollment. Instead, average student body ability rises by 1 percentage point, and the correlation between ability and enrollment increases from 0.59 to 0.63, while the correlation between parental income and enrollment falls from 0.24 to 0.2. Limitations in borrowing capacity for (mostly middle-income) students in 1987 act as a binding constraint that prevents enrollments from responding strongly to labor market changes.

Supply Shocks: Cost Disease and Changing Nontuition Revenue

Last, our results paint a nuanced picture of how cost disease and movements in nontuition revenue (e.g., state support) affect tuition. In the model, tuition actually *falls* in response to the supply shocks alone. Specifically, when we feed in the empirical time-series estimates for custodial costs and college endowments (which summarize all nontuition revenue) but leave all other parameters at their initial 1987 levels, equilibrium tuition decreases from $6,100 to $5,700. Enrollment, by contrast, surges from 36 percent to 53 percent.

Table 10.5 decomposes the impact of each supply shock. As shown in column (2) of the experiments, omitting the change in college endowments has no impact on average net tuition relative to the 2010 equilibrium, which incorporates the endowment change. Note, however, that by aggregating all sources of nontuition revenue and lumping together public and private institutions, this analysis does not directly address the issue of stagnant state support raised by our discussant, Sandy Baum. In fact, according to figure 10.2, total nontuition revenue actually *increases* by approximately $4,500 between 1987 and 2010. Even restricting attention to public institutions, figure 10.7 shows that the growth in auxiliary revenues dominates the initially stagnant and subsequently declining trend in state support. In future work, we plan to directly address the impact of declining state support in a disaggregated framework that explicitly distinguishes between public and private institutions.[26]

---

26. The negative relationship between tuition/fees and state funding per FTE mentioned by Sandy Baum—which can also be found in figure 12A of Ma et al. (2017)—has multiple possible interpretations. One way is to view state-funding reductions as a causal mechanism for tuition hikes. Alternatively, legislative delays that cause state appropriations to be adjusted with a lag may explain the correlation. In this scenario, if demand increases, students are willing to pay higher tuition while state funding per FTE falls mechanically because of higher enrollment. The countercyclicality of enrollment (established by Betts and McFarland [1995] and Dellas and Koubi [2003]) and procyclicality of public appropriations lend some credibility to this argument, but more research is needed to weigh the merits of each interpretation.

Table 10.5   Experiments

Statistic	Experiment							2010
College costs		*	*	*	*	*	*	*
College endowment	*		*	*	*	*	*	*
Borrowing limits	*	*		*	*	*	*	*
Interest rates	*	*	*		*	*	*	*
Nontuition cost	*	*	*	*		*	*	*
Grants	*	*	*	*	*		*	*
College premium	*	*	*	*	*	*		*
Mean net tuition ($)	13,424	12,432	9,066	12,397	12,289	11,319	11,131	12,428
Std. net tuition ($)	1,182	1,265	1,958	1,312	1,463	1,916	1,405	1,320
Enrollment rate	0.34	0.56	0.53	0.52	0.52	0.56	0.52	0.56
Two-year default rate	0.17	0.17	0.07	0.19	0.17	0.17	0.32	0.17
Mean loan (recipients) ($)	6,873	6,873	4,746	6,856	6,872	6,871	6,877	6,873
Pct. taking out loans	100.00	100.00	73.8	100.00	100.00	100.00	100.00	100.00
Mean earnings ($)	33,605	37,256	36,767	36,681	36,594	37,217	33,306	37,301
Corr. (p. income, enroll)	0.27	0.26	−0.07	0.28	0.28	0.48	0.32	0.30
Corr. (ability, enroll)	0.67	0.77	0.52	0.77	0.76	0.79	0.78	0.78
Ability of graduates	0.77	0.70	0.64	0.71	0.71	0.71	0.72	0.70
Investment ($)	33,467	26,230	27,060	30,344	30,186	29,550	29,007	30,701
Average EFC ($)	17,620	16,041	12,256	16,412	16,640	18,331	16,992	16,514
Ex ante utility	−40.76	−40.35	−40.30	−40.37	−40.38	−40.36	−40.51	−40.19

*The value changed over the transition.

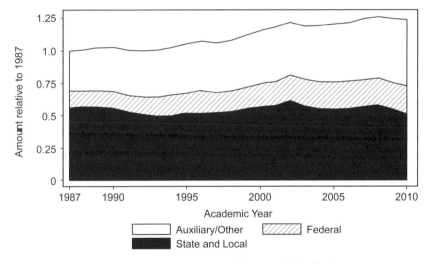

Fig. 10.7   Growth in nontuition revenue per FTE at public institutions
Note: Constant dollars, public institutions.

Table 10.5 also addresses the isolated impact of custodial costs. Perhaps surprisingly, upward shifts in the custodial cost function between 1987 and 2010 actually *reduce* tuition inflation by approximately $1,000, as seen by comparing the first experiment with the 2010 column. Rather than raise tuition, the college responds to higher custodial costs by cutting quality through reduced investment and expanded enrollment of lower-ability students. Two factors account for this divergence from the familiar cost disease narrative: the quality-maximizing objective function of the college and the role of fixed costs.

For intuition, consider a simplified framework with homogeneous students who each have ability x and some fixed parental income. Further, assume there are no preference shocks. In this context, the college sets tuition T to extract the entire student surplus, independent of the custodial cost function. Thus, given T (which is common across students due to their homogeneity), the college simply chooses the number of students to admit:

$$\max_{I,N} q(x,I) \text{ s.t. } IN + C(N) = TN + EN \Leftrightarrow \max_{N} q\left(x, T + E - \frac{C(N)}{N}\right).$$

With x constant, quality is effectively only a function of investment I, which the college maximizes by minimizing average costs $C(N)/N$. In the case of a quadratic cost function, $C(N) = c_0 + c_1 N + c_2 N^2$, average costs are minimized at $N = \sqrt{c_0/c_2}$, which is *increasing* in the fixed cost term and *does not depend* on the marginal cost term c_1. Consequently, in this simple model, higher fixed costs lead to increased enrollment, unchanged tuition, and reduced investment. By contrast, if the college were to maximize *total* investment IN, enrollment would satisfy $T + E = C'(N)$, which more closely resembles the familiar optimality condition of a profit-maximizing firm where changes in fixed costs have no effect on the optimal quantity (here, enrollment) choice.

Our regression estimates show that rising fixed costs between 1987 and 2010 are the dominant cost trend, and the simple model provides some intuition as to why the college responds by increasing enrollment. With student heterogeneity, the increased enrollment results in admission of lower-ability students and/or students with lower willingness to pay. The result is lower expenditures I (as in the simple example) and lower average ability 0.

Several factors caution us from boldly claiming that Baumol's cost disease is unimportant for tuition increases. First, the current model abstracts from the possibility of a rising relative price of college investment (i.e., pI instead of I). Second, we assume that colleges can freely reoptimize each period without regard for their previous investment and hiring choices. In reality, the need to pay the salaries of tenured faculty and cover maintenance on existing buildings may alter a college's response to shifting costs. Last, even if Baumol's cost disease were to cause higher tuition at an individual college, aggregate tuition may be unaffected if students substitute into lower-cost colleges. Our representative college framework does not allow us to explore the heterogeneous response of tuition across different college types. Even with these caveats, however, our finding that the *form* of cost increases (i.e., fixed vs. marginal) matters for tuition is an important and novel finding.

10.5   Conclusion

Existing demand-side and supply-side theories can explain the full increase in net tuition between 1987 and 2010. However, our model suggests that demand-side theories—namely, the role of financial aid expansions and the rise in the college premium—generate the strongest effects. However, given the limitation of our representative college assumption, the results likely exaggerate the quantitative sensitivity of tuition to changes in students' willingness to pay. Interestingly, upward shifts in the cost structure consistent with Baumol's cost disease have different effects on tuition depending on whether marginal costs or fixed costs move by more. We plan on addressing issues related to college heterogeneity in future work.

References

Abbott, B. G. Gallipoli, C. Meghir, and G. Violante. 2013. "Education Policy and Intergenerational Transfers in Equilibrium." NBER Working Paper no. 18782, Cambridge, MA.

Andrews, R. J., J. Li, and M. F. Lovenheim. 2012. "Quantile Treatment Effects of College Quality on Earnings: Evidence from Administrative Data in Texas." NBER Working Paper no. 18068, Cambridge, MA.

Archibald, R. B., and D. H. Feldman. 2008. "Explaining Increases in Higher Education Costs." *Journal of Higher Education* 79 (3): 268–95.

Athreya, K., and J. Eberly. 2013. "The Supply of College-Educated Workers: The Roles of College Premia, College Costs, and Risk." Working paper no. 13-02, Federal Reserve Bank of Richmond.

Autor, D. H., L. F. Katz, and M. S. Kearney. 2008. "Trends in U.S. Wage Inequality: Revising the Revisionists." *Review of Economics and Statistics* 90 (2): 300–323.

Baumol, W. J. 1967. "Macroeconomics of Unbalanced Growth: The Anatomy of Urban Crisis." *American Economic Review* 57 (3): 415–26.

Baumol, W. J., and W. G. Bowen. 1966. *Performing Arts: The Economic Dilemma; A Study of Problems Common to Theater, Opera, Music, and Dance.* Cambridge, MA: Twentieth Century Fund.

Belley, P., and L. Lochner. 2007. "The Changing Role of Family Income and Ability in Determining Educational Achievement." *Journal of Human Capital* 1 (1): 37–89.

Bennett, W. J. 1987. "Our Greedy Colleges." *New York Times*, Feb. 18. https://www.nytimes.com/1987/02/18/opinion/our-greedy-colleges.html.

Bersudskaya, V., and C. C. Wei. 2011. "Trends in Student Financing of Undergraduate Education: Selected Years 1995–96 to 2008–08." Report NCES 2011-218, National Center for Education Statistics.

Betts, J., and L. McFarland. 1995. "Safe Port in a Storm: The Impact of Labor Market Conditions on Community College Enrollments." *Journal of Human Resources* 30:741–65.

Bowen, H. R. 1980. *The Costs of Higher Education: How Much Do Colleges and Universities Spend per Student and How Much Should They Spend?* San Francisco, CA: Jossey-Bass Publishers.

Bureau of Economic Analysis (BEA). 2015. "Table 2-1, Current-Cost Net Stock of Private Fixed Assets, Equipment, Structures, and Intellectual Property Products by Type." Washington, DC, Bureau of Economic Analysis. Accessed June 18, 2015. http://www.bea.gov.

Card, D., and T. Lemieux. 2001. "Can Falling Supply Explain the Rising Return to College for Younger Men?" *Quarterly Journal of Economics* 116:705–46.

Cellini, S. R., and C. Goldin. 2014. "Does Federal Aid Raise Tuition? New Evidence on For-Profit Colleges." *American Economic Journal: Economic Policy* 6 (4):174–206.

Chakrabarty, R., M. Mabutas, and B. Zafar. 2012. "Soaring Tuitions: Are Public Funding Cuts to Blame?" *Liberty Street Economics*, Federal Reserve Bank of New York. http://libertystreeteconomics.newyorkfed.org/2012/09/soaring-tuitions-are -public-funding-cuts-to-blame.html. Accessed Aug. 28, 2015.

Clotfelter, C. T. 1996. *Buying the Best: Cost Escalation in Elite Higher Education.* Princeton, NJ: Princeton University Press.

Cunningham, A. F., J. V. Wellman, M. E. Clinedinst, J. P. Merisotis, and C. D. Carroll. 2001. *Study of College Costs and Prices, 1988–89 to 1997–98*, vol. 1. Report NCES 2002-157, National Center for Education Statistics.

Dellas, H., and V. Koubi. 2003. "Business Cycle and Schooling." *European Journal of Political Economy* 19:843–59.

Ehrenberg, R. G. 2002. *Tuition Rising: Why College Costs So Much.* Cambridge, MA: Harvard University Press.

Epple, D., R. Romano, S. Sarpca, and H. Sieg. 2013. "The U.S. Market for Higher Education: A General Equilibrium Analysis of State and Private Colleges and Public Funding Policies." NBER Working Paper no. 19298, Cambridge, MA.

Epple, D., R. Romano, and H. Sieg. 2006. "Admission, Tuition, and Financial Aid Policies in the Market for Higher Education." *Econometrica* 74 (4): 885–92.

Federal Reserve Economic Database (FRED). 2015. FRED Data Series (CPI-AUCSL, GDP, LFWA64TTUSA647N, DHLCRG3Q086SBEA). Federal Reserve Economic Database. Accessed June 18, 2015. https://research.stlouisfed.org /fred2/.

Fillmore, I. 2014. "Price Discrimination and Public Policy in the U.S. College Market." Working paper, Washington University in St. Louis.

Frederick, A. B., S. J. Schmidt, and L. S. Davis. 2012. "Federal Policies, State Responses, and Community College Outcomes: Testing an Augmented Bennett Hypothesis." *Economics of Education Review* 31 (6): 908–17.

Fu, C. 2014. "Equilibrium Tuition, Applications, Admissions, and Enrollment in the College Market." *Journal of Political Economy* 122 (2): 225–81.

Garriga, C., and M. P. Keightley. 2010. "A General Equilibrium Theory of College with Education Subsidies, In-School Labor Supply, and Borrowing Constraints." Working paper, Federal Reserve Bank of St. Louis and Congressional Research Office.

Goldin, C., and L. F. Katz. 2007. "The Race between Education and Technology: The Evolution of U.S. Educational Wage Differentials, 1890 to 2005." NBER Working Paper no. 12984, Cambridge, MA.

Heller, D. E. 1999. "The Effects of Tuition and State Financial Aid on Public College Enrollment." *Review of Higher Education* 23 (1): 65–89.

Hoekstra, M. 2009. "The Effect of Attending the Flagship State University on Earnings." *Review of Economics and Statistics* 91 (4): 717–24.

Howden, L. M., and J. A. Meyer. 2011. "Age and Sex Composition: 2010, Table 1." United States Census Bureau. Accessed June 18, 2015. http://www.census.gov /prod/cen2010/briefs/c2010br-03.pdf.

Ionescu, F. 2011. "Risky Human Capital and Alternative Bankruptcy Regimes for Student Loans." *Journal of Human Capital* 5 (2): 153–206.

Ionescu, F., and N. Simpson. 2016. "Default Risk and Private Student Loans: Implications for Higher Education Policies." *Journal of Economic Dynamics and Control* 64:119–47.

Jones, J. B., and F. Yang. 2016. "Skill-Biased Technological Change and the Cost of Higher Education." *Journal of Labor Economics* 34 (3): 621–62.

Katz, L. F., and K. M. Murphy. 1992. "Changes in Relative Wages, 1963–87: Supply and Demand Factors." *Quarterly Journal of Economics* 107:35–78.

Keane, M. P., and K. I. Wolpin. 2001. "The Effect of Parental Transfers and Borrowing Constraints on Educational Attainment." *International Economic Review* 42 (4): 1051–103.

Koshal, R. K., and M. Koshal. 2000. "State Appropriation and Higher Education Tuition: What Is the Relationship?" *Education Economics* 8 (1): 81–89.

Lochner, L. J., and A. Monge-Naranjo. 2011. "The Nature of Credit Constraints and Human Capital." *American Economic Review* 101 (6): 2487–529.

Long, B. T. 2004a. "How Do Financial Aid Policies Affect Colleges? The Institutional Impact of the Georgia Hope Scholarship." *Journal of Human Resources* 39 (4): 1045–66.

———. 2004b. "The Impact of Federal Tax Credits for Higher Education Expenses." In *College Choices: The Economics of Where to Go, When to Go, and How to Pay for It*, edited by C. M. Hoxby, 101–68. Chicago: University of Chicago Press.

———. 2006. "College Tuition Pricing and Federal Financial Aid: Is There a Connection?" Technical report, Testimony before the US Senate Committee on Finance.

Lucca, D. O., T. Nadauld, and K. Shen. 2015. "Credit Supply and the Rise in College Tuition: Evidence from Expansion in Federal Student Aid Programs." Staff Report no. 733, Federal Reserve Bank of New York.

Ma, J., S. Baum, M. Pender, and M. Welch. 2017. *Trends in College Pricing*. New York: The College Board.

McPherson, M. S., and M. O. Shapiro. 1991. *Keeping College Affordable: Government and Educational Opportunity*. Washington, DC: Brookings Institution Press.

National Center for Education Statistics (NCES). 2015a. "Table 302.10. Recent High School Completers and Their Enrollment in 2-Year and 4-Year Colleges, by Sex: 1960 through 2013." National Center for Education Statistics. Accessed June 18, 2015. http://nces.ed.gov/programs/digest/d14/tables/dt14_302.10.asp.

———. 2015b. "Table 326.10. Graduation Rate from First Institution Attended for First-Time, Full-Time Bachelor's Degree-Seeking Students at 4-Year Postsecondary Institutions, by Race/Ethnicity, Time to Completion, Sex, Control of Institution, and Acceptance Rate: Selected Cohort Entry Years, 1996 through 2007." National Center for Education Statistics. Accessed June 18, 2015. http://nces.ed .gov/programs/digest/d14/tables/dt14_326.10.asp.

———. 2015c. "Table 330.10. Average Undergraduate Tuition and Fees and Room and Board Rates Charged for Full-Time Students in Degree-Granting Postsecondary Institutions, by Level and Control of Institution: 1963–64 through 2013–14." National Center for Education Statistics. Accessed June 20, 2015. http://nces.ed .gov/programs/digest/d14/tables/dt14_330.10.asp.

Rizzo, M. J., and R. G. Ehrenberg. 2004. "Resident and Nonresident Tuition and Enrollment at Flagship State Universities." In *College Choices: The Economics of Where to Go, When to Go, and How to Pay for It*, edited by C. M. Hoxby, 303–53. Chicago: University of Chicago Press.
Singell, Jr., L. D., and J. A. Stone. 2007. "For Whom the Pell Tolls: The Response of University Tuition to Federal Grants-in-Aid." *Economics of Education Review* 26:285–95.
Storesletten, K., C. Telmer, and A. Yaron. 2004. "Cyclical Dynamics in Idiosyncratic Labor Market Risk." *Journal of Political Economy* 112 (3): 695–717.
Titus, M. A., S. Simone, and A. Gupta. 2010. "Investigating State Appropriations and Net Tuition Revenue for Public Higher Education: A Vector Error Correction Modeling Approach." Working paper, Institute for Higher Education Law and Governance Institute Monograph Series.
Turner, L. J. 2014. "The Road to Pell Is Paved with Good Intentions: The Economic Incidence of Federal Student Grant Aid." Working paper, University of Maryland.
Turner, N. 2012. "Who Benefits from Student Aid: The Economic Incidence of Tax-Based Aid." *Economics of Education Review* 31 (4): 463–81.
US Department of Education. 2015a. "Federal Student Loan Portfolio." Accessed June 18, 2015. https://studentaid.ed.gov/sa/about/data-center/student/portfolio.
———. 2015b. "National Student Loan Two-Year Default Rates." Federal Student Aid. Accessed June 18, 2015. https://www2.ed.gov/offices/OSFAP/defaultmanagement/defaultrates.html.
Wei, C. C., X. Li, L. Berkner, and C. D. Carroll. 2004. "A Decade of Undergraduate Student Aid: 1989–90 to 1999–2000." Report NCES 2004-158, National Center for Education Statistics.
Winston, G. C. 1999. "Subsidies, Hierarchy, and Peers: The Awkward Economics of Higher Education." *Journal of Economic Perspectives* 13 (1): 13–36.
Wolverton, M., W. H. Gmeich, J. Montez, and C. T. Nies. 2001. *The Changing Nature of the Academic Deanship: ASHE-ERIC Higher Education Research Report.* San Francisco, CA: Jossey-Bass Publishers.

Comment   Sandy Baum

Gordon and Hedlund have developed a detailed model to shed light on the important question of why college prices rose so rapidly between 1987 and 2010. They appropriately focus on net tuition revenues of institutions, rather than on the sticker prices they charge. They consider both the demand and supply sides of the market.

The authors take many historical trends into account, including prices, student aid, the college earnings premium, and nontuition revenue sources. But as the authors acknowledge, the model makes many assumptions that

Sandy Baum is a senior fellow at the Urban Institute.
For acknowledgments, sources of research support, and disclosure of the author's material financial relationships, if any, please see http://wwwdev.nber.org/chapters/c13712.ack.

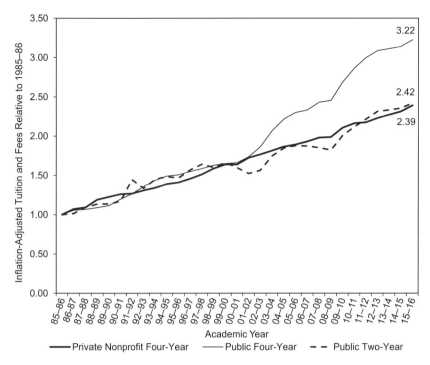

Fig. 10C.1   Inflation-adjusted published tuition and fees relative to 1985–1986, 1985–1986 to 2015–2016 (1985–1986 = 1.0)

Source: Ma et al. (2015).

are not consistent with how higher education institutions are structured and operate in the real world and with how students make decisions.

There is one representative institution, combining characteristics of the public and private nonprofit four-year sectors. There is no competition for the institution, which maximizes quality and prestige.

In reality, public and private institutions operate in very different worlds. They have very different funding sources and, as figure 10C.1 illustrates, the paths of tuition prices in these institutions have been quite different. The graph displays the path of sticker prices over time, illustrating the fact that prices in public four-year institutions rose much more rapidly than prices in the private nonprofit sector during the period of time covered by Gordon and Hedlund's work.

The model focuses on net tuition revenues, not sticker prices. But the discount rates at private institutions are higher and have increased more over time than those in the public sector, magnifying the divergence in prices. It is not at all clear that combining public and private price increases can generate an accurate estimate of the forces driving those price increases.

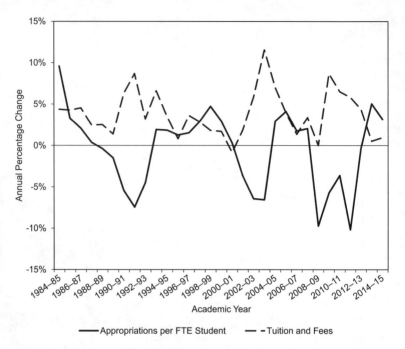

Fig. 10C.2   Annual percentage change in inflation-adjusted per-student state funding for higher education and in tuition and fees at public institutions, 1984–1985 to 2014–2015
Source: Ma et al. (2015).

The main conclusions emerging from the model in this chapter are both counterintuitive and inconsistent with existing evidence. In particular, the authors find that declines in nontuition revenues (including both state appropriations and endowments) are associated with price *reductions*, as are increases in institutional costs. These "reverse" effects leave increases in federal loan limits with the dominant positive impact on increasing tuition.

Looking at the actual patterns of changes in state funding for higher education and public college tuition levels, reported in figure 10C.2, raises serious questions about the conclusions emerging from the model. If the authors really want to argue that public colleges are not raising tuition to fill in the gaps left by declines in state per-student funding, they should provide strong logic and empirical evidence, not just the numbers that emerge from a model of a hypothetical institution.

Several other assumptions in the model deserve attention. The college only admits students who have a willingness to pay that exceeds marginal cost. When "custodial costs"—basically expenditures on student amenities—increase, colleges lower expenditures on instruction for fear that

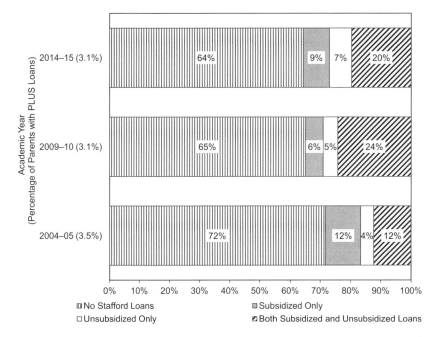

Fig. 10C.3   Percentage of undergraduate students borrowing federal subsidized and unsubsidized loans, 2004–2005, 2009–2010, and 2014–2015

Source: Baum et al. (2015).

they will lose high-quality students if they raise tuition. So costs do not drive tuition—in fact, the reverse is true.

In fact, the goal of maximizing quality and prestige that underlies the model actually applies only to a subset of four-year institutions—almost 50 percent of public and private nonprofit four-year colleges accept at least 75 percent of applicants (Ma et al. 2015). And selective institutions face considerable competition for students—a phenomenon not incorporated into the model.

A key question is which assumptions really matter for making the model a good representation of reality. Simplification is obviously necessary, but the model overestimates tuition increases and the number of students with loans. It underestimates enrollment. It predicts a 17 percent default rate on student loans over the entire time period. The authors acknowledge that this bears little relationship to reality, but nonetheless appear to have confidence about their analysis of the role of loans in driving net tuition.

According to the model, all undergraduates take federal students loans in 2010. But in fact, in any given year, the majority of undergraduates do not borrow (figure 10C.3). Just over two-thirds of bachelor's degree recipients graduate with debt. Many students borrow, but not every year.

Table 10C.1   Federal and nonfederal education borrowing, 2006–2005 to 2010–2011 ($)

	Federal loans	Nonfederal loans	Total borrowing
2005–06	70.5	20.8	91.3
2006–07	71.7	23.7	95.3
2007–08	78.5	25.6	104.1
2008–09	93.6	12.5	106.1
2009–10	110.7	9.0	119.6
2010–11	116.1	7.9	124.0

Source: Baum et al. (2015).

A critical question about the impact of the availability of federal student loans relates to alternative financing mechanisms, such as private loans. Nonfederal loans constituted 25 percent of education borrowing in 2007–2008, but declined sharply to 12 percent the following year, as financial markets collapsed and the federal government increased its borrowing limits. The increase in federal loan limits had almost no impact on total borrowing. As table 10C.1 shows, it was just associated with a substitution of federal borrowing for nonfederal borrowing. Is this reality consistent with such a large impact of federal loan limits on tuition prices? Information like this has to be incorporated into the logic and the conclusions of the model.

It is certainly useful to develop a stronger theoretical foundation for analyzing changes in college prices. But starting with a model that does not distinguish between public and private colleges or between endowments and state appropriations and that assumes that all colleges are selective—and presenting the results emerging from that model as reliable—has the potential to do real damage to the higher education financing system.

This chapter has already generated headlines including "Economists Confirm Financial Aid is Inflating Student Loan Bubble" (ShiffGold 2015). An article in *Forbes* titled "Cause of High Tuition? It's the Government, Stupid" reports that "Gordon and Hedlund attribute the big rise in tuition charges almost entirely to the federal student financial assistance programs. Bill Bennett is, by and large, right. Student loan programs do not help students, they help the permanent citizens of college campuses—the administrators, the faculty, the research assistants, and so forth" (Solis 2016).

The authors do not clearly distinguish between their measure of the net tuition revenue institutions receive and the net prices students pay. Even if federal aid does increase net tuition revenues of institutions, it can lower the net prices students pay. This is the case as long as the increase in net tuition per student is lower than the aid per student—a point the *Forbes* discussion misses. There is no measure of the distribution of those net tuition prices across students from different income categories, making it even more difficult to consider the impact on college access.

The question underlying this chapter is of critical importance in the real world. Of immediate concern, how should the student aid system be structured to meet goals of access, success, and attainment? How can any potential impact on increasing the net price of college for students be diminished? The main conclusion of the chapter is that increases in the availability of federal student loans more than account for the full increase in net tuition prices over the years in question. As the authors note, a number of empirical analyses by prominent higher education economists have generated results contradicting this conclusion. So the evidence behind this assertion should be strong.

If the availability of federal student loans is as significant a driver of college prices as this chapter suggests (despite much evidence contrary to this finding in existing literature)—is it time to abolish or dramatically reduce this stream of funding for students?

The authors acknowledge that their model "likely exaggerates the impact of the Bennett hypothesis. . . . The findings in this chapter should be viewed as an initial exploration to guide further research, rather than being authoritative or definitive." The exploration should continue before conclusions from this work become arguments for policy changes not really supported by evidence.

References

Baum, Sandy, Jennifer Ma, Matea Pender, and D'Wayne Bell. 2015. *Trends in Student Aid*. Report, Trends in Higher Education Series, The College Board. https:// files.eric.ed.gov/fulltext/ED572541.pdf.

Ma, Jennifer, Sandy Baum, Matea Pender, and D'Wayne Bell. 2015. *Trends in College Pricing 2015*. Report, Trends in Higher Education Series, The College Board. https://trends.collegeboard.org/sites/default/files/2015-trends-college-pricing -final-508.pdf.

ShiffGold. 2015. "Economists Confirm Financial Aid Is Inflating Student Loan Bubble." *ShiffGold*, Dec. 22. http://schiffgold.com/key-gold-news/economists -confirm-financial-aid-is-inflating-student-loan-bubble/.

Solis, Brian. 2016. "Cause of High Tuition? It's the Government, Stupid." *Forbes*, Feb. 10. http://www.forbes.com/sites/ccap/2016/02/10/cause-of-high-tuition-its -the-government-stupid/#21373b9155e8.

11

Online Postsecondary Education and Labor Productivity

Caroline M. Hoxby

11.1 The Promise and Possible Perils of Online Postsecondary Education

Could the availability of online postsecondary education substantially raise human capital and labor productivity in the United States and around the world? Online educational platforms potentially make postsecondary education available to people who, owing to their locations or time constraints, might otherwise lack access. Because the cost structure of online education differs from that of in-person education (online education is thought to have low marginal costs), the productivity (causal improvement in outcomes per dollar spent) of online schools could be high even if they did not improve students' outcomes more than in-person schools. Also, online platforms lend themselves to certain types of education, such as computer programming and technical design, where interacting with a computer is naturally an important part of the learning process. This suggests that online

Caroline M. Hoxby is the Scott and Donya Bommer Professor in Economics at Stanford University and a research associate and director of the Economics of Education Program at the National Bureau of Economic Research.

The opinions expressed in this chapter are those of the author alone and do not necessarily represent the views of the US Internal Revenue Service or the US Department of the Treasury. This work is a component of a larger project examining the effects of federal tax expenditures and on-budget expenditures related to higher education. The computations in this chapter were produced incidentally in the process of the Hoxby (2018) study, which relies on selected, deidentified data accessed through contract TIR-NO-12-P-00378 and TIR-NO-15-P-00059 with the Statistics of Income Division at the US Internal Revenue Service. The author gratefully acknowledges the help of Barry W. Johnson, Michael Weber, and Brian G. Raub of the Statistics of Income Division, Internal Revenue Service. The author is grateful for suggestions and very useful comments from her discussant, Nora Gordon, and from Katherine Abraham, John Bound, David Deming, Charles Hulten, Jennifer Hunt, Valerie Ramey, and conference participants. For acknowledgments, sources of research support, and disclosure of the author's material financial relationships, if any, please see http://www.nber.org/chapters/c13709.ack.

platforms might disproportionately expand the availability of education that trains people for technical, rapidly growing industries that routinely complain that they are unable to find a sufficient number of workers with the skills they require. Such hopeful views of online education are reflected in quotations like the following: "For those who believe that higher education should be personalized, inexpensive, as accessible to working mothers as it is to third-generation Yalies, and geared toward helping students acquire skills that employers actually desire, *utopia is on the horizon.*"[1]

On the other hand, the flexibility and paucity of face-to-face contact inherent in online education may mean that only highly self-disciplined students learn well on such platforms. These may not be the people who tend to enroll in online education. Indeed, online education is controversial among policymakers, especially federal ones, because the sector's students generate a disproportionate share of defaults on and repayment issues with student loans.[2] They also account for a disproportionate share of tax expenditures on tuition and fees (see below).[3] Moreover, in federal undercover investigations and audits, online postsecondary institutions have been disproportionately found engaging in deceptive marketing, fraud, academic dishonesty, low course-grading standards, and other violations of education regulations.[4]

In short, online postsecondary education may be a windfall for taxpayers and the economy more broadly: an inexpensive way for people to acquire the cutting-edge skills they need to be productive. Online students may earn returns disproportionate to their opportunity costs and direct schooling costs. Alternatively, online postsecondary education may be a liability for taxpayers and the economy: it may be a sector that takes funds from the federal taxpayers and students but that generates insufficient skills to repay those takings.

The first step in understanding whether online postsecondary education is a windfall or a liability is determining its return on investment (ROI) based on earnings. This is the primary goal of this study. Because proponents of online education also argue, however, that it enables people to reallocate themselves from slow-growing, obsolescent industries to fast-growing industries with rising labor productivity, this study also investigates direct evidence for that argument. Such reallocation could benefit all workers through general equilibrium effects. Thus, we are justified in looking for evidence

---

1. Beato (2014); emphasis added. His article emphasizes online courses that give students computer-related skills such as "Building a Search Engine," "Programming a Robotic Car," and "HTML5 Game Development." Clayton Christensen has made something of a career of arguing that online education will have low costs, generate instructional innovation, engage students with technology, and disproportionately fulfill employers' needs for cutting-edge skills. See, for instance, Christensen and Eyring (2011) and Christensen and Horn (2013). See also Waldrop (2013).

2. See Looney and Yannelis (2015).

3. Author's calculations. See below.

4. United States General Accountability Office (2010, 2011).

of reallocation, not merely evidence of increases in online students' own postenrollment earnings.

To achieve these goals, this study analyzes longitudinal data on nearly every person who engaged substantially in online postsecondary education between 1999 and 2011. These are ideal data for estimating ROIs and studying labor reallocation. As a result, this study is a good complement to (though not a good substitute for) previous studies of online education, which have often focused on a small number of online courses or a single provider of online education.[5] Such studies help us understand what happens in an online class, whereas this study should help us test broad theories about online education and help us evaluate its contribution to the economy.

The remainder proceeds as follows. In section 11.2, I define online postsecondary education and describe its explosive growth since 2005. The data are described in section 11.3. Section 11.4 describes who enrolls in online education, how long they engage in coursework, and how much they and taxpayers pay for it. In section 11.5, I use figures to show how earnings evolve before and after individuals' episodes of online enrollment. Although this section does not contain calculations of ROI, the figures contain so much information that readers will be able to anticipate ROIs. In section 11.6, I lay out my empirical strategy for estimating ROIs. The primary challenge is that some self-selection into online education may be driven by events that negatively affect earnings. This phenomenon, known as "Ashenfelter's Dip," was first identified as a problem in efforts to estimate the effects of job training.[6] When negative earnings events induce people to engage in training or online education, we have difficulty projecting what their earnings would have been in the absence of the training or education. In particular, we do not know whether their earnings would have bounced back on their own. Section 11.7 contains the ROI results. Section 11.8 investigates whether online students reallocate themselves toward industries that are associated with higher labor productivity, fast growth, or high technology. Section 11.9 makes calculations that show whether online education is a windfall or liability for taxpayers. Conclusions occupy section 11.10.

11.2 The Recent, Explosive Growth in Online Postsecondary Enrollment

Online postsecondary enrollment has grown very rapidly in recent years. Figures 11.1 and 11.2 show the number of students enrolled in coursework

---

5. For studies along these lines, see Bettinger et al. (2014), Bowen et al. (2014), Figlio, Rush, and Yin (2013), Xu and Jaggars (2013), Hart, Friedmann, and Hill (2014), and Streich (2014). Economic research regarding online education is still fairly limited. A brief survey might include Cowen and Tabarrok (2014), Deming, Goldin, and Katz (2012), Deming et al. (2015), Deming, Lovenheim, and Patterson (forthcoming), Ho et al. (2014), Hoxby (2014), and McPherson and Bacow (2015). None of the aforementioned studies have sufficient longitudinal data on earnings and costs to estimate ROIs.

6. The seminal paper is Ashenfelter and Card (1985).

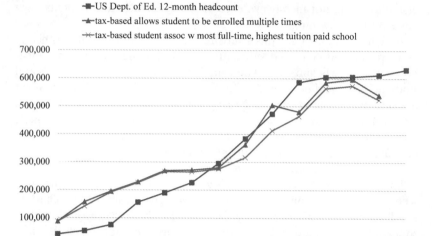

Fig. 11.1 Enrollment in postsecondary programs that are exclusively online (total: undergraduate and graduate, full- and part-time)

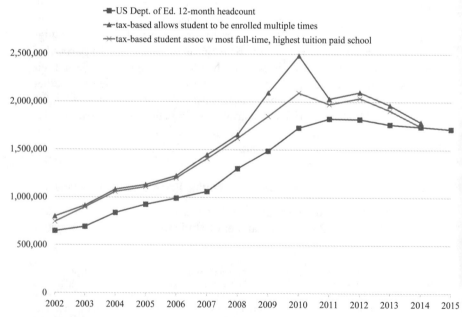

Fig. 11.2 Enrollment in postsecondary programs that are mainly online (total: undergraduate and graduate, full- and part-time)

that is, respectively, exclusively and mainly online. (The exact definition of "mainly online" is given below, but think of it as more than half online.) Both figures show that enrollment grew dramatically after 2005. This is not an accident or an effect of broadband access. Rather, 2005 corresponds to the year in which the US Department of Education eliminated the "50 percent rule" that required an institution's enrollment to be at least 50 percent in-person for its students to qualify for federal tax credits, tax deductions, grants, loans, and other financial aid.[7] This rule constrained the growth of online education because an institution had to recruit and have a campus (or campuses) to support one in-person student for each online student.

One line on each figure shows the US Department of Education's twelve-month head count, including all undergraduate and graduate students who enroll in a school year, regardless of whether they are full- or part-time. This number should correspond closely to enrollment figures based on tax data if we allow each student to count multiple times if he is enrolled at multiple institutions. This is the next line on each chart. The final line on each chart shows tax-data-based enrollment in which each student is counted only once and is associated with the institution where he is at least half-time and, if this leaves ambiguity, to which the highest tuition is paid on his behalf.[8]

Figure 11.1 shows that, up through 2002, fewer than 100,000 students enrolled each year in education that was exclusively online. By 2013, the number of students enrolled in exclusively online education was about 600,000, more than six times the number a decade previously. Walden University, Aspen University, and Argosy University are examples of exclusively online institutions. They truly have *no* campus or classrooms—only an office with staff who manage finances, keep records, and coordinate web-based instruction. They offer a variety of undergraduate and graduate degree programs but, typically, their programs are nonselective. That is, they enroll any student who has completed the previous level of education—a high school diploma or General Education Development (GED) certificate in the case of undergraduates.

Figure 11.2 shows that enrollment in mainly online education approximately tripled over the same period. It was about 700,000 in 2002, but it was about 1,700,000 in 2012–2015. However, this growth in overall enrollment understates the growth in *online* enrollment. Once they were released from the constraint of the 50 percent rule, mainly online institutions actually reduced the size and number of their brick-and-mortar campuses and

7. One might wonder how there could be any exclusively only students prior to 2005 given the 50 percent rule. First, some schools had experimental waivers from the rule. Second, the requirement applied to an institution as a whole, not program-by-program. Thus, the graduate students in a school could be exclusively only even if the undergraduates were not, and vice versa. See the next section for the question by which programs are classified.

8. See Hoxby (2018) for a detailed discussion and comparison of online enrollment data from the US Department of Education and from tax data.

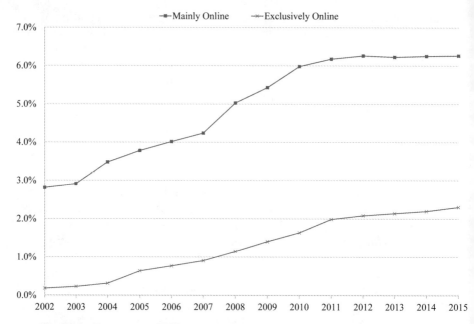

Fig. 11.3   Percentage of US postsecondary enrollment in exclusively and mainly online programs

shifted toward instruction that was increasingly online.[9] Examples of mainly online institutions are The University of Phoenix, Kaplan University, DeVry University, and Liberty University.

Although the growth of online education is striking and shows no signs of abating, it is important to keep in mind that it is still far from the norm. Figure 11.3 shows that, even in 2015, it accounted for only 8.3 percent of total enrollment. Thus, students who self-select into online education are unusual as a statistical matter. This is a fact to keep in mind because it affects the empirical strategy I adopt.

11.3   Data

This study employs deidentified data from an IRS database for the years 1999 to 2014. It includes all people who engaged in exclusively or mainly online postsecondary education. (See below for the definition of an enrollment episode.) From Form 1098-T, an information return that postsecondary institutions file, are derived tuition and fee payments, whether the student is enrolled at least half-time, whether the student is enrolled in graduate studies, and scholarships and grants received by the student. These variables

9. See Deming, Lovenheim, and Patterson (forthcoming).

are available regardless of whether the student actually files for tax credits or deductions for tuition and fees. From Forms 8917 and 8863, I derive the student's take-up of the tax credits and deductions for postsecondary tuition and fees. Wages and employment variables are derived from Form W-2, and these variables are available regardless of whether a person files an income tax return. From variants of Form 1040 are derived adjusted gross income and any postsecondary tax credits and deductions that are actually taken.[10]

For data on the share of an institution's courses that are taken online, I rely on the National Center for Education Statistics Integrated Postsecondary Education Data System (IPEDS). This is a data system to which nearly all postsecondary institutions must mandatorily report. IPEDS is also the source of numerous other institution-level, as opposed to student-level, variables: Pell grant revenue, total undergraduate student loans, total enrollment, and so on.

IPEDS asks postsecondary schools the following:

1. Are all programs at your institution offered exclusively via distance education?

2. How many degree/certificate-seeking undergraduates are (a) enrolled exclusively in distance education courses, (b) enrolled in some but not all distance education courses, or (c) not enrolled in any distance education course?

3. Repeat question (2) for non-degree/certificate-seeking undergraduates and for graduate students.

A student is classified as attending "exclusively online" if the answer to question (1) is yes or if the probability that he or she is enrolled in distance education is 100 percent based on the answers to questions (2) and (3). For instance, if a student were enrolled in graduate coursework, and all graduate students were enrolled exclusively in online courses (possibility 2[a]), then the student would be classified as exclusively online. Note that undergraduate and graduate students at the same institution could be classified differently.

A student's coursework is classified as "mainly online" if the probability that his or her courses are online is greater than 50 percent where the probability assigned to option (2)(a) is 100 percent, option (2)(b) is 50 percent, and option (2)(c) is 0 percent.[11] Unfortunately, it is not possible to classify mainly online experiences more precisely. Clearly, the mainly online category is imprecise and contains students with a variety of online experiences.

A student's coursework is classified as "hardly online" if the probability that his or her courses are online is 10 percent or less where the probabilities

---

10. Forms 8917 and 8863 ensure that tax credits and deductions are properly mapped from filer to student when they are not the same person.

11. I assign 50 percent to option (2)(b) because many institutions were tightly bound by the 50 percent rule up through 2005. In more recent years, the mainly online category has become, if anymore, more online.

assigned to options (2)(a), (2)(b), and (2)(c) are the same as given above. Finally, I classify a student as hardly online and at a nonselective institution if his or her school will enroll any student with a high school degree or GED in undergraduate coursework or enroll any student with a baccalaureate degree in graduate coursework. Because nearly all exclusively online and mainly online institutions are nonselective, this final category (hardly online and nonselective) is the best comparison for online schools. Indeed, recent evidence suggests that it is these institutions that are most likely to lose students to online postsecondary schools.[12] Put another way, students who attend nonselective institutions are more elastic between online and in-person settings than are students who attend selective ones.

The data include up to sixteen longitudinal observations for each person who enrolled between 1999 and 2014 in postsecondary education. However, given the explosive increase in online course-taking after 2005, the analysis of *online* students is strongly weighted toward the later years in the period. The descriptive statistics shown in the next section focus on students who were enrolled in 2013 so as to represent online education as its most current.[13] (Descriptive statistics based on earlier years are available from the author.)

The enrollment and other variables reported on Form 1098-T are for calendar years rather than school years. They are based on the calendar year in which the institution received payment for tuition and fees. In most cases, a school year is divided across two calendar years and the first calendar year is the lesser of the two years that make up the first school year. For instance, suppose a freshman enrolls for the 2012/13 school year. If she pays for the autumn semester in September 2012 and the spring semester in January 2013, she will have two years of 1098-T-based enrollment, even if she enrolls for only a single school year. She will appear in calendar-year data in 2012, even though at least half (and usually more) of the months in the school year are in 2013. Three calendar years usually correspond to two school years, four calendar years to three school years, five calendar years to four school years, and so on. This is not always true, however, because a student may pay for her spring term in December or may begin her enrollment episode in January. In such cases, a calendar year corresponds to a school year.

For the purpose of this chapter, I need to define postsecondary "episodes" over which to compute ROI. For instance, if a student were to take a single term off and then return to his degree program, it would make sense to treat his enrollment as a single episode. The interruption would be so short that his learning experience would be truly connected before and after the break. Moreover, it would be nearly impossible to assess his returns from only the

12. See Deming, Lovenheim, and Patterson (forthcoming).

13. The IPEDS data for 2014/15 are still preliminary, so I do not use 2014 for the descriptive statistics.

first part of his enrollment. There would be only a brief period for him to earn income without his work competing with his studying. But, if a break of a single term should not define the end of an enrollment episode, what length of break should? In the interests of estimating ROI at all well (see empirical strategy section below), I define an enrollment episode to begin when a person who was not enrolled in any of the three preceding calendar years enrolls. The episode ends when he discontinues his enrollment for three consecutive calendar years. (The results are not sensitive to switching the nonenrollment length to two years or four years.) A student may have multiple enrollment episodes but, as shown below, only a small share of people do.

Since the first year of wage data is 1999, the first calendar year in which an episode could begin is 2002. Since the last year of wage data is 2014, the last calendar year in an episode could end is 2011. Thus, the ROI calculations are for online students enrolled at some time between 2002 and 2011.

11.4 A Description of Online Education in the United States

This section attempts a rich description of online education in the United States, explaining who attends, the schools they attend, and how they pay for their coursework. Owing to the fact that the data are virtually population data, not sample data, *all* differences across the groups shown in the tables in this section are highly statistically significant. Therefore, I make no further mention of statistical significance.[14]

11.4.1 Who Enrolls in Online Education?

Table 11.1 shows us the characteristics of the students who enrolled in online postsecondary education in 2013. For comparison, it shows the same characteristics for 2013 students whose enrollment was (a) hardly online or (b) hardly online and nonselective. Table 11.2 shows the same characteristics broken down by undergraduate and graduate students.

The average age of online students is strikingly high: 36 for exclusively online students and 33.7 for mainly online students. Exclusively online and mainly online undergraduates average, respectively, 33.4 and 32.6 years of age. Exclusively online and mainly online graduate students average, respectively, 39.6 and 37. These ages are much higher than those of hardly online students (25.5) or hardly online nonselective students (27.1). Despite their relatively advanced ages, the exclusively online and mainly online students are more likely to be undergraduates than to be graduate students. Sixty percent of the exclusively online students are undergraduates and 77.2 percent of the mainly online students are undergraduates. While these percentages

---

14. The *p*-values on differences are always less than 0.0001, but there is, in any case, little consensus about how to interpret standard errors for population data.

Table 11.1   Characteristics of online and nononline students

| | 2013/14 students at postsecondary schools that are | | | |
	100% online	mainly online	hardly online	hardly online and nonselective
Average age	36.0	33.7	25.5	27.1
Probability of being male (%)	41.4	36.5	44.3	41.5
Probability enrolled at least half time (%)	93.2	87.2	90.9	85.3
Probability is an undergraduate (%)	60.0	77.2	89.0	99.0
Own wages while enrolled (includes zeros) ($)	33,195	24,641	14,335	12,058
Own household's income while enrolled (includes zeros and negative) ($)	49,051	40,006	20,836	17,920

Source: Data are from deidentified tax data combined with postsecondary institutions' classification data from IPEDS.

Notes: A postsecondary school is classified as "mainly online" if at least 50 percent of its courses are offered in an online or partially online way. A school is classified as "hardly online" if fewer than 10 percent of its courses are offered in an online or partially online way. Courses that serve "all students" are considered. A school is "nonselective" if any student with a high school diploma or GED may enroll in undergraduate classes or any student with a baccalaureate degree may enroll in graduate classes.

are lower than those for students who are hardly online (89 percent) or hardly online and nonselective (99 percent), we must conclude that many students who enroll online have been out of school for years or in school only sporadically since their teenage years.

The vast majority of exclusively online (93.2 percent) and mainly online (87.2 percent) students are enrolled at least half time. These percentages are fairly similar across undergraduate and graduate students. They are also fairly similar to those for hardly online and hardly online nonselective students. Thus, we can dismiss the idea that online education is dominated by students taking, say, a single course for professional development, as a hobby, or as an experiment. Most students appear to be attempting to complete coursework at a sufficient pace that they could potentially earn a degree or certificate.

In the same calendar year in which they are enrolled, exclusively online students earn average wages of $33,195: $27,118 for undergraduates and $42,039 for graduate students. These are not insubstantial amounts for students who are enrolled at least half time. Although the parallel average wage numbers for mainly online students are more modest—$24,641 for all, $21,640 for undergraduates, $34,780 for graduate students—they nevertheless suggest that those enrolled online are juggling school with a significant amount of work. Hardly online and hardly online nonselective students earn much less: $14,335 and $12,058, respectively. This is undoubt-

Table 11.2   Characteristics of online students, undergraduates versus graduate students

2013/14 school year	at schools that are 100% online in the courses that serve			at schools that are mainly online in the courses that serve		
	All students	Undergraduate students	Graduate students	All students	Undergraduate students	Graduate students
	all students	undergraduate students	graduate students	all students	undergraduate students	graduate students
Average age	36.0	33.4	39.6	33.7	32.6	37.0
Probability of being male (%)	41.4	47.9	30.5	36.5	38.0	33.7
Probability enrolled at least half time (%)	93.2	93.2	90.8	98.5	85.0	92.2
Probability is an undergraduate (%)	60.0	100.0	0.0	66.4	100.0	0.0
Own wages while enrolled (includes zeros) ($)	33,195	27,118	42,039	24,641	21,640	34,780
Own household's income while enrolled (includes zeros and negative) ($)	49,051	41,448	60,558	40,006	36,302	52,917

Source: Data are from deidentified tax data combined with postsecondary institutions' classification data from IPEDS.

Note: A postsecondary school is classified as "mainly online" if at least 50 percent of its courses are offered in an online or partially online way.

edly partly because they are younger and thus likely to earn less per hour. However, part of their lower earnings is likely due to their working fewer hours while enrolled.

In the calendar year they are enrolled, the households of exclusively online and mainly online students have moderate incomes: $49,051 and $40,006, respectively. This puts them around the 45th and 35th percentiles of the income distribution among households who file taxes. It is important to observe that the students are earning the majority of this household income themselves: they would probably not be well supported by another earner if they were to cease working altogether while they were enrolled.

The share of students who are male hovers around 40 percent for all student groups: exclusively online, mainly online, hardly online, and hardly online nonselective. This male share is typical of US postsecondary education. The only notable sex-related statistic is that only 30 percent of exclusively online graduate students are male. This may be because teachers and nurses, who receive wage boosts if they earn certain graduate certificates or degrees, make up a good share of exclusively online students.

11.4.2   Where Do Online Students Reside?

Well before 2013/14, the school year described in the tables, the internet was available in all parts of the United States and fixed-wire high-speed internet service was available in all areas defined as urban by the census. (Note that urban areas include towns and small cities.) About half of rural households had fixed-wire high-speed internet available and satellite dish-based high-speed internet was available to the remaining half.[15] Owing to online postsecondary education being potentially available almost everywhere, while brick-and-mortar schools were not, one might hypothesize that online students live disproportionately in small urban areas or sparsely populated areas. Table 11.3 (for all students) and table 11.4 (broken out for undergraduates and graduates) demonstrate that this hypothesis is correct only to a very slight extent. The tables are based on commuting zones (CZs), which combine counties into units that reflect common commutes between workers' homes and their job locations.[16] Because the typical student is in her midthirties and commutes to work, CZs are probably the geographic unit that best defines a student's brick-and-mortar postsecondary options.

Table 11.3 shows that 41.6 percent of students who attend exclusively online live in a CZ that has a population over the 90th percentile for CZs. Another 40.5 percent live in a CZ with a population between the 75th and 90th percentiles. Less than 6 percent live in a CZ with a population below

15. National Telecommunications and Information Administration, "US Broadband Availability: June 2010–June 2012." *A Broadband Brief*, published May 2013. https://www.ntia.doc.gov/files/ntia/publications/usbb_avail_report_05102013.pdf.

16. See United States Department of Agriculture, Economic Research Service (2016). See also Pew Charitable Trusts (2016).

Table 11.3   Location of online and nononline students (percent)

| | 2013/14 students at postsecondary schools that are | | | |
	100% online	mainly online	hardly online	hardly online and nonselective
CZ pop. ≤ 15,000 (≤ 10th percentile of CZs)	0.2	0.3	0.1	0.1
CZ pop. > 15,000 and ≤ 40,000 (10–25th percentile of CZs)	0.6	2.1	0.3	0.4
CZ pop. > 40,000 and ≤ 115,000 (25–50th percentile of CZs)	4.7	5.4	2.6	2.6
CZ pop. > 115,000 and ≤ 300,000 (50–75th percentile of CZs)	12.5	13.5	7.3	7.7
CZ pop. > 300,000 and ≤ 1,600,000 (75–90th percentile of CZs)	40.5	37.6	32.4	37.5
CZ pop. > 1,600,000 (> 90th percentile of CZs)	41.6	41.2	57.3	51.6
CZ density ≤ 7.75 (≤ 10th percentile of CZs)	1.0	1.4	0.3	0.5
CZ density > 7.75 and ≤ 24 (10–25th percentile of CZs)	3.4	5.0	1.7	1.7
CZ density > 24 and ≤ 63 (25–50th percentile of CZs)	10.1	10.4	5.4	6.4
CZ density > 63 and ≤ 143 (50–75th percentile of CZs)	20.7	18.8	13.3	14.7
CZ density > 143 and ≤ 320 (75–90th percentile of CZs)	26.7	27.6	25.1	21.2
CZ density > 320 (> 90th percentile of CZs)	38.2	36.9	54.2	55.5

Source: Data are from deidentified tax data combined with postsecondary institutions' classification data from IPEDS. The source of CZ data is United States Department of Agriculture, Economic Research Service (2016).

Notes: "CZ" means commuting zone. See text for a definition. A postsecondary school is classified as mainly online" if at least 50 percent of its courses are offered in an online or partially online way. A school is classified as "hardly online" if fewer than 10 percent of its courses are offered in an online or partially online way. Courses that serve "all students" are considered. A school is "nonselective" if any student with a high school diploma or GED may enroll in undergraduate classes or any student with a baccalaureate degree may enroll in graduate classes.

the 50th percentile for CZs. The numbers for students who attend mainly online are similar: 41.2 percent live in CZs with populations above the 90th percentile, 37.6 percent live in CZs with populations between the 75th and 90th percentiles, and less than 8 percent live in a CZ with a population below the 50th percentile. Thus, the notion that the typical online student lives in a small urban area is wrong.

Table 11.3 shows parallel statistics for students who attend schools that are hardly online or hardly online and nonselective. Interestingly, although such students are more likely to live in CZs with large populations, the differences are not stark. For instance, among students enrolled in schools that are hardly online and nonselective, 51.6 percent live in a CZ with a population above the 90th percentile, 37.5 percent live in a CZ with a population between the 75th and 90th percentiles, and 3.1 percent live in a CZ with a population below the 50th percentile.

The notion that the typical online student lives in a sparsely populated area is also wrong. Among students who attend exclusively online, only 4.4 percent live in CZs with a population density below the 25th percentile for CZs; 38.2 percent live in CZs with a population density above the 90th percentile. Similarly, among students who attend mainly online, only 6.4 percent live in CZs with a population density below the 25th percentile, and 36.9 percent live in CZs with a population density above the 90th percentile. Students who are enrolled at schools that are hardly online are somewhat more likely to live in densely populated CZs, but—again—the differences are not striking.

Table 11.4 shows that within a category (exclusively online, mainly online) undergraduates tend to be distributed across CZs in a manner that is very similar to how graduate students are distributed. Thus, both the typical online undergraduate and the typical online graduate student live in CZs with large, dense populations.

11.4.3   The Highest Degree and Control of Online Schools

US postsecondary institutions are often characterized by the highest degree they offer. This may be a certificate (a "less-than-two-year" school), an associate's degree (a "two-year" school), a baccalaureate degree (a "four-year" school), or some graduate degree (a "more-than-four-year" school). Most students who attend *nonselective* schools are at two-year or less-than-two-year institutions.[17]

In addition, each US postsecondary institution may be a public school (controlled by a government), a private nonprofit, or a private for-profit. Although the for-profit sector still accounts for a small share of total US enrollment, it has grown rapidly in recent years.[18] Much of this growth has occurred at schools that are exclusively or mainly online. Thus, it should be no surprise that online students disproportionately attend for-profit institutions. What may be more surprising, since exclusively online and mainly online institutions are nearly all nonselective, is that the vast majority of online students attend schools classified as four-year or more-than-four-year institutions.

Table 11.5, which contains results for all students, shows that 76.8 percent of students who are enrolled exclusively online attend for-profit schools that offer the baccalaureate or a higher degree. Another 21.0 percent attend private nonprofit schools that offer the baccalaureate or a higher degree. This leaves only tiny shares who attend public schools or who attend schools that do not offer at least a baccalaureate degree. Students who attend mainly online are more evenly split between schools that offer at least the baccalaureate degree and that are for-profit (37.7 percent) or nonprofit (44.2 percent). Nonnegligible shares attend public four-year institutions (10.2 percent) or public two-year institutions (6.7 percent).

All of this is in sharp contrast to the corresponding statistics for students

---

17. See table 11.5.
18. See Deming, Goldin, and Katz (2012).

Table 11.4   Location of online students, undergraduates versus graduate students (percent)

2013/14 school year	at schools that are 100% online in the courses that serve						at schools that are mainly online in the courses that serve					
	All students	Undergraduate students	Graduate students	all students	undergraduate students	graduate students	All students	Undergraduate students	Graduate students	all students	undergraduate students	graduate students
CZ pop. ≤ 15k (≤ 10th percentile of CZs)	0.2	0.2	0.2				0.3	0.4	0.2			
CZ pop. > 15k and ≤ 40k (10–25th percentile of CZs)	0.6	0.6	0.6				2.1	2.5	0.8			
CZ pop. > 40k and ≤ 115k (25–50th percentile of CZs)	4.7	5.1	4.2				5.4	5.7	4.5			
CZ pop. > 115k and ≤ 300k (50–75th percentile of CZs)	12.5	13.2	11.4				13.5	14.0	11.9			
CZ pop. > 300k and ≤ 1.6M (75–90th percentile of CZs)	40.5	42.1	38.1				37.6	38.0	36.5			
CZ pop. > 1.6M (> 90th percentile of CZs)	41.6	38.8	45.6				41.2	39.5	46.2			
CZ density ≤ 7.75 (≤ 10th percentile of CZs)	1.0	1.1	0.9				1.4	1.6	0.7			
CZ density > 7.75 and ≤ 24 (10–25th percentile of CZs)	3.4	3.6	3.1				5.0	5.6	3.1			
CZ density > 24 and ≤ 63 (25–50th percentile of CZs)	10.1	10.9	9.0				10.4	11.1	8.2			
CZ density > 63 and ≤ 143 (50–75th percentile of CZs)	20.7	22.7	17.6				18.8	19.1	18.0			
CZ density > 143 and ≤ 320 (75–90th percentile of CZs)	26.7	25.4	27.1				27.6	27.3	28.2			
CZ density > 320 (> 90th percentile of CZs)	38.2	35.4	42.3				36.9	35.3	41.9			

Sources: Data are from deidentified tax data combined with postsecondary institutions' classification data from IPEDS. The source of CZ data is United States Department of Agriculture, Economic Research Service (2016).

Notes: "CZ" means commuting zone. See text for a definition. A postsecondary school is classified as "mainly online" if at least 50 percent of its courses are offered in an online or partially online way.

Table 11.5   Sector of the schools attended by online and nononline students (percent)

	2013/14 students at postsecondary schools that are			
	100% online	mainly online	hardly online	hardly online and nonselective
Public, four-year or above	1.95	10.23	34.06	9.24
Private not-for-profit, four-year or above	21.03	44.15	33.17	4.23
Private for-profit, four-year or above	76.80	37.72	7.38	4.81
Public, two-year	0.02	6.67	13.74	45.53
Private not-for-profit, two-year	0.00	0.10	0.62	0.98
Private for-profit, two-year	0.09	0.84	5.99	16.87
Public, less than two-year	0.03	0.03	0.52	1.74
Private not-for-profit, less than two-year	0.00	0.00	0.14	0.40
Private for-profit, less than two-year	0.08	0.27	4.39	16.01

Source: Data are from deidentified tax data combined with postsecondary institutions' classification data from IPEDS.
Notes: A postsecondary school is classified as "mainly online" if at least 50 percent of its courses are offered in an online or partially online way. A school is classified as "hardly online" if fewer than 10 percent of its courses are offered in an online or partially online way. Courses that serve "all students" are considered. A school is "nonselective" if any student with a high school diploma or GED may enroll in undergraduate classes or any student with a baccalaureate degree may enroll in graduate classes.

who attend institutions that are hardly online and nonselective. Among these students, 45.5 percent attend public two-year schools. For-profit two-year and less-than-two-year schools account for, respectively, 16.9 percent and 16.0 percent of such students. Only 4.8 percent of students who attend a school that is hardly online and nonselective attend an institution that grants at least a baccalaureate degree.

The contrasting statistics are surprising because, as mentioned previously, online nonselective schools appear to be competing for the same students as brick-and-mortar nonselective schools (Deming, Lovenheim, and Patterson, forthcoming). Moreover, students who attend nonselective schools tend to be only marginally prepared for college and must often take remedial courses before beginning college-level work.[19] Thus, it is not obvious that a baccalaureate-granting institution is an appropriate fit for them. Also, table 11.1 gives us little reason to think that all online students are the "cream of the crop" of students who attend nonselective institutions. After all, most of them are still short of a baccalaureate degree (that is, still pursuing undergraduate education) even though they are in their midthirties. Furthermore, table 11.6 shows that it is not merely online graduate students who are almost exclusively in schools that grant at least the baccalaureate. Online undergraduate students are almost entirely in four-year-or-more schools, too.

19. See Long and Boatman (2013).

Table 11.6   Sector of the schools attended by online students, undergraduates versus graduate students (percent)

	at schools that are 100% online in the courses that serve			at schools that are mainly online in the courses that serve		
	All students	Undergraduate students	Graduate students	All students	Undergraduate students	Graduate students
	all students	undergraduate students	graduate students	all students	undergraduate students	graduate students
Public, four-year or above	1.95	2.40	2.89	10.23	10.54	19.74
Private not-for-profit, four-year or above	21.05	26.15	16.08	44.15	37.73	55.50
Private for-profit, four-year or above	76.79	71.10	80.05	37.72	41.09	23.25
Public, two-year	0.02	0.04	0.90	6.67	8.90	1.41
Private not-for-profit, two-year	0.00	0.00	0.01	0.10	0.14	0.01
Private for-profit, two-year	0.10	0.17	0.03	0.84	1.18	0.04
Public, less than two-year	0.00	0.00	0.01	0.03	0.04	0.01
Private not-for-profit, less than two-year	0.00	0.00	0.00	0.00	0.00	0.00
Private for-profit, less than two-year	0.09	0.14	0.03	0.27	0.38	0.03

Source: Data are from deidentified tax data combined with postsecondary institutions' classification data from IPEDS.

Note: A postsecondary school is classified as "mainly online" if at least 50 percent of its courses are offered in an online or partially online way.

Because the typical online student is in her midthirties, data on test scores and grades at the end of high school are often unavailable for her birth cohort. However, in other work (Hoxby 2015), I find little difference in end-of-high-school achievement between thirty-two-year-old students attending online nonselective and in-person nonselective schools in 2013. All this suggests that the same student is more likely to enroll in a four-year program if she attends online than if she attends in person. If online students complete their degree programs and learn a lot in them, it may be good that they attempted more ambitious degree programs. On the other hand, if they find themselves unable to learn the material, they might have been better off attempting a less ambitious degree program but completing it successfully. We cannot know without examining returns to education, so the contrast between online and in-person degree programs is an important reason to estimate ROIs.

11.4.4   How Long Are Online Students' Enrollment Episodes?

Sixty-one percent of online education is attributable to students with a single enrollment episode, 28 percent to students with two episodes, and 11 percent to students with three or more episodes. In order to avoid having an individual's experience counted more than once in what follows, I focus on the first enrollment episode. However, since multiple enrollment episodes are uncommon, the results are very similar if I choose one episode at random from each person's episodes, choose the most recent episode, or use all the episodes.

Figures 11.4 and 11.5 are histograms showing the length of the first enrollment episode for students who begin that enrollment episode at a school that is, respectively, exclusively online or mainly online. Note that if a student begins at a school that is exclusively online but later—in the same enrollment episode—switches to a school that is mainly online, the student is categorized as exclusively online for these histograms. The reverse is also true for switches from mainly to exclusively online.[20]

It is important to recall that an episode of length one (one calendar year) usually corresponds to less than one school year, an episode of length two usually corresponds to one school year, an episode of length three usually corresponds to two school years, and so on.

Figures 11.4 and 11.5 show that the lengths of exclusively online and mainly online enrollment episodes are distributed similarly. For both types of enrollment, the modal length of an episode is one calendar year or (probably) less than a single school year: these episodes represent 38 percent of episodes that are exclusively online and 50 percent of episodes that

20. Because the figures will show episode length to be similar across the two types of enrollment, the histograms would hardly change if I were to alter this method of categorization.

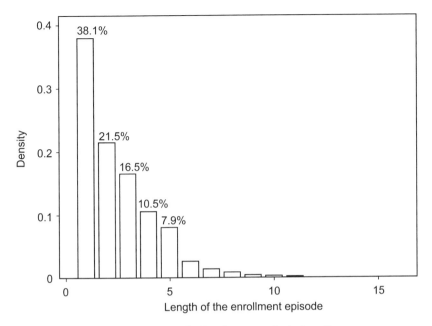

Fig. 11.4   Length of enrollment episodes that are exclusively online

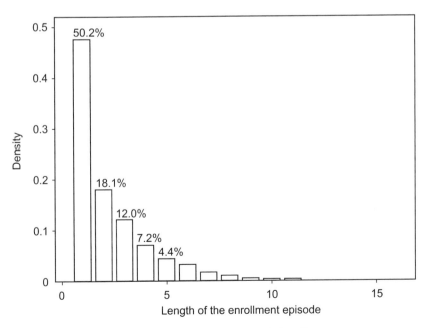

Fig. 11.5   Length of enrollment episodes that are mainly online

are mainly online. The next most common length is two calendar years or (probably) one school year: 22 percent of episodes that are exclusively online and 18 percent of episodes that are mainly online. For episodes that begin in exclusively online schools, the median length is 2 and the mean length is 2.5. The median is 1 and mean is 2.4 among episodes that begin in mainly online schools. Episodes of five or more calendar years, which are most likely to constitute a complete baccalaureate education, are rare: they constitute only 7.9 percent of episodes that begin at exclusively online schools and only 4.4 percent of episodes that begin at mainly online schools.

The preponderance of short enrollment episodes is striking because most online students are undergraduates and nearly all of them attend institutions that grant the baccalaureate degree at least. There are several possible interpretations. First, the vast majority of undergraduates could be attempting only to obtain a certificate or associate's degree even though their school is (more in theory than in practice) a baccalaureate degree-granting school. Second, students may be dropping out part of the way into a degree program (which may or may not be baccalaureate). Third, some of the students could have completed part of their postsecondary education prior to 1999. If so, they might only need a couple of years to complete their baccalaureate degree, especially if the online institutions are generous in allowing the transfer of credits from other institutions.

In the tax data, it is difficult to distinguish between these different explanations. However, the dropping-out explanation is indicated by IPEDS data on graduation rates. At exclusively and mainly online schools, only 22 percent of students complete the degree program in which they are enrolled within 150 percent of the normal time to degree completion. These statistics include students in all degree programs. Among these institutions' students who classify themselves as baccalaureate degree-seeking, only 2 percent complete a baccalaureate degree within six years. Another 14 percent complete some program such as a certificate or an associate's degree.[21]

---

21. Author's calculations based on IPEDS 2014 (the most recent) data. These contain graduation rates for students who commenced enrollment in 2008. Unfortunately, a good share of exclusively or mainly online postsecondary institutions did not report graduation rate data to IPEDS. In some cases, this is because the institution was so young in 2008. In other cases, the data are simply missing. The IPEDS degree completion rates are roughly consistent with the Beginning Postsecondary Students (BPS, 2004 and 2012 cohorts) longitudinal data. However, the BPS samples are even less representative than IPEDS. Given the size of BPS samples (approximately 16,700 students for the 2004 cohort, for instance), it is not possible to make the data, even with sampling weights, representative of a sector that accounts for a small share of enrollment, as the online postsecondary sector does. Moreover, the 2004 cohort only contains students who began their education before the 50 percent rule was dropped, so the data could not possibly describe the exclusively online sector. The most recent data available on the 2012 cohort is from 2014, when they could not be expected to have completed a baccalaureate degree. However, a good share of the online 2012 cohort has already experienced a gap in their enrollment.

Table 11.7   Costs of and payments for online and nononline education ($)

	2013/14 students at postsecondary schools that are			
	100% online	mainly online	hardly online	hardly online and nonselective
Instructional spending per FTE student	2,334	3,821	12,879	5,426
Academic support and student service spending per FTE student	2,469	3,318	6,491	2,740
Institution support per FTE student	3,522	3,241	4,686	2,981
Tuition *paid*	6,131	6,758	11,930	4,919
Total grants and scholarships received (includes zeros)	1,864	2,315	5,051	2,106
Pell grants received (includes zeros)	1,529	1,458	2,046	2,489
Other federal grants received (includes zeros)	69	89	198	117
State and local grants received (includes zeros)	73	138	294	188
Institutional grants received (includes zeros)	183	632	1,923	175
Amount of nonrefundable education credits taken (includes zeros)	1,369	1,407	1,443	1,247
Amount of refundable American Opportunity Tax Credit taken (includes zeros, 2008 onward only)	619	729	851	789
Amount of tuition and fees tax deduction taken (includes zeros)	24	16	7	3
Federal loans taken, undergraduates only (includes zeros)	4,228	5,075	4,424	4,259
Published undergraduate tuition and fees	9,548	14,193	18,841	6,483
Published graduate tuition and fees	9,730	10,890	17,354	11,542
Default rate in first fiscal year (%)	12.5	10.3	7.5	15.3

Source: Data are from deidentified tax data combined with postsecondary institutions' classification data from IPEDS.

Notes: A postsecondary school is classified as "mainly online" if at least 50 percent of its courses are offered in an online or partially online way. A school is classified as "hardly online" if fewer than 10 percent of its courses are offered in an online or partially online way. Courses that serve "all students" are considered. A school is "nonselective" if any student with a high school diploma or GED may enroll in undergraduate classes or any student with a baccalaureate degree may enroll in graduate classes.

11.4.5   How Much Does Online Postsecondary Education Cost and Who Pays for It?

In this subsection, I use IPEDS data to show how much instructional and other educational spending online students experience. I use both IPEDS and tax data to show who pays for this spending: the student himself, federal taxpayers, and so forth.

Table 11.7 shows statistics for all students, and table 11.8 shows them separately for undergraduate and graduate students. The data in the tables are for the 2013/14 *school year.*[22] Because the tax data are associated with

---

22. The tax deduction for tuition and fees is an exception (see below).

Table 11.8   Costs of and payments for online education, undergraduates versus graduate students ($)

	All students	Undergraduate students	Graduate students	All students	Undergraduate students	Graduate students
	at schools that are 100% online in the courses that serve			at schools that are mainly online in the courses that serve		
	all students	undergraduate students	graduate students	all students	undergraduate students	graduate students
Tuition paid	6,131	5,747	6,650	6,758	6,340	7,159
Grants received	1,864	2,795	475	2,315	2,690	876
Amount of nonrefundable education credits taken (includes zeros)	1,369	1,382	1,331	1,407	1,416	1,316
Amount of refundable American Opportunity Tax Credit taken (includes zeros, 2008 onward only)	619	845	286	729	860	289
Amount of tuition and fees tax deduction taken (includes zeros)	24	8	48	16	8	42

Source: Data are from deidentified tax data combined with postsecondary institutions' classification data from IPEDS.

Note: A postsecondary school is classified as "mainly online" if at least 50 percent of its courses are offered in an online or partially online way.

calendar years and because many enrollment episodes are so short that the calendar year represents less than a school year, I have adjusted the variables derived from tax data to make them as representative as possible of the 2013/14 school year.[23] For various reasons, readers should not expect payments, when totaled, to equal one of the two spending variables.[24]

IPEDS suggests that four types of spending are particularly relevant to students: an institution's spending on instruction, its spending on academic support,[25] its spending on student services,[26] and its spending on institutional support.[27] Together, these four categories make up "core" spending, which is intended to include costs associated with educating a student but to exclude spending on research, public service, maintenance and operations, construction, feeding students, and housing students. Since online schools organize the student experience differently than brick-and-mortar ones, it is useful to see student-related spending separately by instruction and the remainder of core spending.

Exclusively online schools spend $2,334 per full-time equivalent (FTE) student on instruction, but mainly online schools spend $3,821 or 64 percent more. Schools that are hardly online and nonselective spend much more: $5,426 (132 percent more than exclusively online schools and 42 percent more than mainly online schools). These numbers suggest that exclusively and mainly online schools are achieving substantial cost savings on instruction. However, these savings do not carry over to other per FTE core spend-

---

23. Specifically, I compute the ratio of (numerator) a student's school's published tuition and fees for the spring term of the 2012/13 school year and the fall term of the 2013/14 school year to (denominator) the sum of a person's payments and grants in 2013 calendar year. This ratio indicates the percentage of a school year that the tax-based variables likely represent. I multiply students' tax credits and deductions by this ratio to make those variables comparable to all the other variables in the tables, which are based on school years. For instance, if a student enrolled in an online school in the fall of 2013 but had no enrollment in the spring of 2013, his tax credits and deductions would reflect only *half* of a school year. To get a full school year's worth of credits and deductions, we would need to multiply by the ratio, which would be about 2 in his case.

24. First, neither instructional nor core spending are the total spending on a student's education. The ratio of core to total spending varies considerably by institution. Some institutions' spending is reported imprecisely because the institution must allocate overhead among its activities, which may include activities other than students' education. IPEDS does not force schools' spending, "saving," and other disbursements to equal their revenues. As a result, some institutions' spending is difficult to reconcile with their revenue. IPEDS (school year) and tax (calendar year) data are poorly aligned, even after the adjustments described in the previous footnote.

25. Academic support includes expenses that support instruction such as libraries, audiovisual services, academic administration, curriculum development, and so on.

26. Student support includes expenses for admissions, registrar activities, supplemental instruction, and student records. It also includes activities that contribute to students' development outside the formal instructional program. Examples of the latter would be student newspapers, curricular clubs (science club, French language society), and student government.

27. Institutional support includes expenses for the day-to-day operations including administrative services, central activities concerned with management and planning, legal, and fiscal operations, space management, human resources, records, purchasing, and so on.

ing, which is $5,991 at exclusively online schools, $6,559 at mainly online schools, and $5,721 at hardly online nonselective schools. Overall, the similar or somewhat greater spending on other core activities are not what one might expect if one thinks of online schools providing instruction but not other parts of a brick-and-mortar student experience: libraries, student newspapers, curricular clubs, in-person student advising, and the like. It must be that online schools provide instruction inexpensively but spend disproportionately (relative to instructional spending) on curriculum development, administrative services, and legal and fiscal operations.

Toward the bottom of table 11.7, there are rows that show schools' published tuition and fees for a full-time, full-year undergraduate or graduate student. It is useful to compare these numbers to core spending. For instance, exclusively online schools' core spending is $8,325, and their tuition and fees are $9,548 (undergraduates) and $9,730 (graduates). Given their ratio of undergraduate to graduate students (table 11.1), they have about $1,296 from tuition and fees to cover noncore costs and for profits. Mainly online schools' core spending is $10,480, and their tuition and fees are $14,193 (undergraduates) and $10,890 (graduates). They have an average of $2,960 dollars to cover noncore costs and for profits. Finally, the core spending of hardly online and nonselective schools is $11,147, while their tuition and fees are $6,483 (undergraduate) and $11,542 (graduate). In other words, many of these schools cannot meet their expenses with tuition revenue. The difference is made up by state and local government appropriations that effectively subsidize tuition. (Recall from table 11.5 that 56.5 percent of these students attend public schools.) For the perspective of current *students*, however, hardly online nonselective schools offer generous spending per dollar of tuition, relative to online schools.

Now consider who pays for the spending on students' education. Note that the following figures reflect what is actually paid, and not all students are full-time, full-year students. Thus, we should not expect these payments to reflect FTEs as the numbers in the previous paragraphs did.

The first noteworthy result in table 11.7 is that students *themselves* pay more for their education at online schools than do students who attend schools that are hardly online and nonselective. Students at exclusively online schools paid an average of $6,131 in tuition, and their counterparts at mainly online schools paid $6,758. In comparison, students at schools that are hardly online and nonselective paid an average of only $4,919.[28] These differences in tuition paid are mainly due to differences in published tuition and fees, not due to differences in grants (see below). In particular, the subsidized tuition at public hardly online nonselective schools plays the key role.

---

28. Tuition paid at schools that are hardly online but selective is of course much higher since such schools include the most resource-rich schools in the United States. They spend an order of magnitude more per student than do nonselective schools. See Hoxby and Avery (2013).

The second noteworthy result is that federal taxpayers would foot between 36 and 44 percent of the total cost of online education, even if students were to repay their federal loans fully. If they were to repay only 50 percent of their loans, federal taxpayers would fund 60 to 69 percent of the cost of online education. This heavy dependence on federal taxpayers arises because online students not only receive federal grants of around $1,600 per year, they also make disproportionate use of the federal tax credits and deductions for tuition and fees. The average student who is attending an exclusively online school takes a nonrefundable credit of $1,472 and a refundable credit of $867. The average student attending a mainly online school takes a nonrefundable credit of $1,492 and a refundable credit of $981. These are close to the maximum possible credits of $1,500 (nonrefundable) and $1,000 (refundable). These amounts are about 20 percent greater than those for the average student at hardly online nonselective schools. Online students also make disproportionate use of the tax *deduction* for tuition and fees. Compared to students at hardly online nonselective schools, exclusively online students take eight times the deduction and mainly online students take more than five times the deduction. All of the deduction amounts may seem small, but this is not because the deduction that students take, if they take it, is small. Rather, those who take it take close to the maximum possible deduction ($4,000), but in recent years the credits have been more generous than the deduction for most students. Thus, the apparently small amounts reflect students choosing a tax credit over the deduction. (A student cannot simultaneously take a credit and a deduction.)[29]

A third noteworthy finding is that other payments are fairly similar across schools that were exclusively online, mainly online, and hardly online and nonselective. For instance, grants and scholarships paid for an average of about $2,100 in both online and in-person schools.[30] The average student's federal loan was almost identical for students attending schools that were exclusively online ($4,228) and hardly online and nonselective ($4,259). Students at institutions that were mainly online had higher federal loans that averaged $5,075.

At this point, it is worthwhile taking a step back to assess online schools'

29. The tax deduction for tuition and fees is an "above-the-line" deduction, so a person need not itemize to take it. Its maximum possible value is $4,000 times the tax filer's tax rate—for instance, $1,200 for a taxpayer with a 30 percent rate. The Opportunity Tax Credit is a temporary credit with a maximum possible value of $2,500. The Hope Tax Credit and Tax Credit for Lifelong Learning are permanent credits with maximum possible values of $1,800 and $2,000, respectively. Because the tax deduction and credits have different eligibility criteria, some individuals maximize the tax expenditure on their education by taking the deduction, even if a credit would superficially appear to be more generous. If the Opportunity Tax Credit is not renewed or made permanent, the tax deduction will again be more used because it is more comparable in generosity to the Hope Tax Credit and Tax Credit for Lifelong Learning.

30. The small difference in the amount of Pell grants, in favor of hardly online students, is due to online students having incomes that are too high for eligibility.

costs and payments relative to what we might have expected based on the debate described in the introduction and based on the previous literature. Online schools do spend considerably less on instruction per FTE student, but they spend more on other core activities: academic support, student services, and institutional support. As a result, exclusively online schools are only 25 percent and mainly online schools are only 7 percent less costly than comparably selective schools in which the student experience is in person. These seem like modest cost savings relative to what was promised by supporters of online education—represented, for instance, by the quotation in the introduction. They are especially modest when one considers that exclusively online schools do not even attempt to replicate many dimensions of the in-person experience: libraries, laboratories, academic clubs, student music and drama, and so on. Furthermore, it is not obvious that online students would be glad to learn that all of the cost savings at their schools are achieved by spending less on instruction. Since instruction (and not the central office) is what they experience, one would presumably need to argue that the comparatively large amount spent on institutional support (administrative services, central management and planning, legal and fiscal operations, human resources, records, purchasing, etc.) is truly a modern form of instructional spending whereby central activities efficiently substitute for individual instructors.

Second, payments alone cannot explain why students are shifting to online enrollment and away from in-person enrollment at similarly nonselective schools. Online students are paying 25 to 37 percent more for an education that costs less to produce. Thus, we should consider other reasons why students may prefer online schools: the flexibility of course timing, the lack of a commute to and from campus, rationed classes at in-person schools, and so on. We should also consider some less positive reasons: lax academic standards, greater opportunities for cheating, and marketing that is more likely to promise exaggerated results (United States General Accountability Office 2010, 2011). It is difficult to separate these explanations using the data in this chapter. However, in Hoxby (2014), I found little or no evidence that students were engaging in online education because such schools offered advanced or exotic courses not available at brick-and-mortar, nonselective schools. Course-taking at online schools is highly concentrated in basic courses that are offered in nearly all nonselective postsecondary institutions: algebra, elementary accounting, data entry, reading comprehension, composition, and introductory courses in the social sciences.

11.5   Earnings before and after Online Enrollment

In this section, I use figures to illustrate how students' earnings evolve before and after an episode of enrollment at an institution that is exclusively

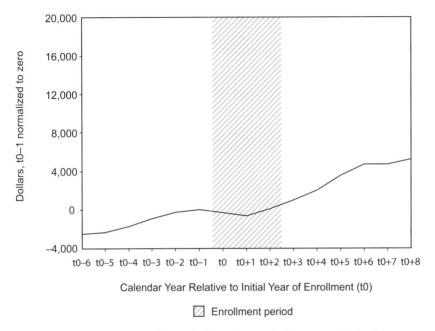

Calendar Year Relative to Initial Year of Enrollment (t0)

◫ Enrollment period

Fig. 11.6   Wages before, during, and after a three-calendar-year episode of post-secondary enrollment that is exclusively online

or mainly online. While these figures do not provide us with ROI estimates, they are designed to make transparent the data behind the estimates.

It is worthwhile describing the first of these figures, figure 11.6, carefully since all of the figures that follow have a similar basis. Figure 11.6 shows wage and salary earnings for all students who enrolled in exclusively online schools and whose enrollment episode lasted three calendar years (most likely two school years). I start with episodes that last three calendar years because they are fairly common (17 percent of exclusively online and 12 percent of mainly online) and because they are long enough for a person plausibly to earn an associate's degree, earn a master's degree, or complete a baccalaureate degree if the person already had a significant number of undergraduate credits when he enrolled. Rather than show raw wage and salary earnings for such students, I partial out calendar-year indicators and a quadratic polynomial in the person's age.

$$(1) \quad y_{it} = \alpha + \beta_{t_0-6} + \beta_{t_0-5} + \ldots + \beta_{t_0-2} + \beta_{t_0} + \beta_{t_0+1} + \beta_{t_0+2} + \ldots + \beta_{t_0+6}$$
$$+ \gamma_{\text{calendar year } t} + \delta_1 \text{age}_{it} + \delta_2 \text{age}_{it}^2 + \varepsilon_{it}.$$

The calendar-year fixed effects account for macroeconomic conditions, the price level, and changes in the online schools available each year. The quadratic in age accounts for smooth regularities in the relationship between age and earnings.[31] What the figure shows, therefore, are the estimates of β, the coefficients on indicators for the years leading into the enrollment ($t_0 - 6$, for instance), the year in which the episode begins (t_0), and the years following the commencement of the episode ($t_0 - 6$, for instance). Earnings are normalized to zero in year $t_0 - 1$.

Since the enrollment episode in figure 11.6 occurs over three calendar years, earnings in years t_0, $t_0 + 1$, and $t_0 + 2$ may be reduced directly because the student is spending his time studying instead of working. However, none of the other coefficients in β are directly affected by enrollment. Rather the preenrollment coefficients give us a sense of what triggered the episode, while the postenrollment coefficients give us a sense of postenrollment gains.

Figure 11.6 shows that students who will be enrolled for three calendar years at exclusively online schools have earnings that are growing at a modest rate of about \$504 per year prior to enrolling. There is some sign of Ashenfelter's Dip: earnings growth between $t_0 - 2$ and $t_0 - 1$ is only \$230. Earnings fall during the period of enrollment (t_0 through $t_0 + 2$), probably the direct effect of substituting study for work.[32] However, the decline in earnings is small: several hundred dollars, not several thousand. This suggests that people continue to work much as before when they enroll in exclusively online education. They are certainly not discontinuing work altogether or halving their work time. In the calendar years that strictly postdate the enrollment episode ($t_0 + 3$ through $t_0 + 8$), earnings grow at an average rate of \$853 a year, faster than preenrollment earnings.

Summing up, we see that three-calendar-year (probably two-school-year) episodes of exclusively online enrollment may lead to somewhat faster earnings growth. Countering this, society pays \$23,985 (\$7,995 for each of three years) for the education in the episode. Out of this, the student himself pays \$12,357 (\$4,119 for each of three years) if he eventually repays his loans in full. (The first year default rate is 12.5 percent.) In addition, the student earns slightly less while enrolled, but this change is so small that he can only be substituting out of work (and into study) to a very limited extent.

Figure 11.7 is exactly analogous to the previous figure except that it shows earnings for three-calendar-year episodes in schools that are mainly, rather than exclusively, online. Prior to enrollment, earnings are rising by an average of \$716 per year. There is a suggestion of Ashenfelter's Dip: earnings growth between $t_0 - 2$ and $t_0 - 1$ is only \$513. Earnings are lower during the period of enrollment, suggesting the substitution of studying for working.

---

31. Since age-earnings profiles can be well approximated by a quadratic, I obtain very similar figures if I use a cubic, quartic, or quintic in age. Such figures are available from the author. Moreover, the lack of additional explanatory power in polynomials beyond a quadratic motivates my use of the quadratic in formal ROI calculations (see below).

32. Notice that earnings are especially low in $t_0 + 1$, the only calendar year in which the person is probably enrolled for a full—as opposed to half—a school year.

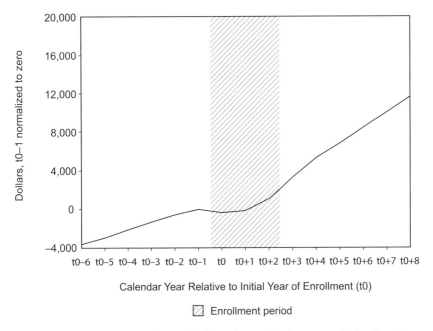

Fig. 11.7   **Wages before, during, and after a three-calendar-year episode of post-secondary enrollment that is mainly online**

Earnings grow by an average of $1,670 per year in the calendar years that strictly postdate the enrollment episode.

In short, the three-calendar-year (probably two-school-year) episodes of mainly online enrollment may be triggered by mild Ashenfelter's Dips. After the enrollment episode ends, earnings growth is higher than in the preenrollment period. For this apparent improvement, society pays $27,219 ($9,073 for each of three years). The student himself pays $13,818 ($4,606 for each of three years) if he eventually repays his loans in full. (The first year default rate is 10.3 percent.) In addition, the student faces lower earnings while enrolled. However, figure 11.7 suggests that these opportunity costs are small—in hundreds rather than thousands per year. Such small opportunity costs suggest that people continue to work without much change, even when they are enrolled at mainly online schools.

Figures 11.8, 11.9, 11.10, and 11.11 show earnings around exclusively online episodes with calendar-year lengths of one, two, four, and five years. Figures 11.12, 11.13, 11.14, and 11.15 do the same for mainly online episodes. They can be summarized briefly as follows. Occasionally, there is evidence for an Ashenfelter's Dip (actually, a mere slowdown in earnings growth) just prior to the enrollment episode, but some figures show no such evidence and there are only mild slowdowns in the figures that do show it. Earnings fall during the enrollment episode, but they never fall enough to

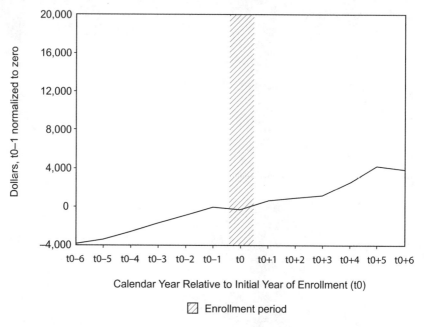

Fig. 11.8 Wages before, during, and after a one-calendar-year episode of post-secondary enrollment that is exclusively online

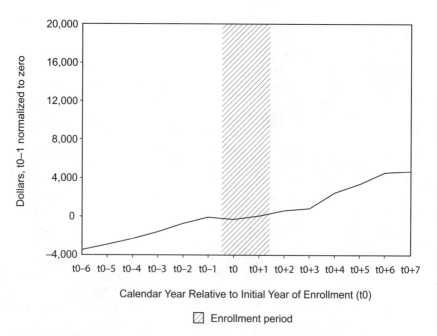

Fig. 11.9 Wages before, during, and after a two-calendar-year episode of post-secondary enrollment that is exclusively online

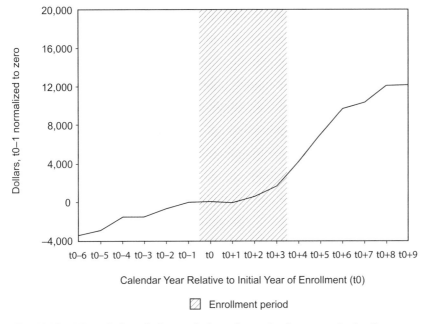

Fig. 11.10 Wages before, during, and after a four-calendar-year episode of post-secondary enrollment that is exclusively online

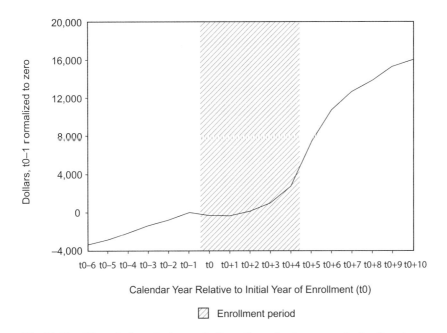

Fig. 11.11 Wages before, during, and after a five-calendar-year episode of post-secondary enrollment that is exclusively online

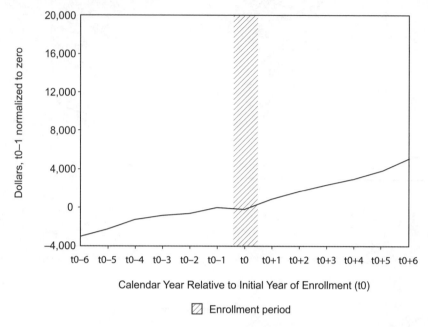

Fig. 11.12 Wages before, during, and after a one-calendar-year episode of post-secondary enrollment that is mainly online

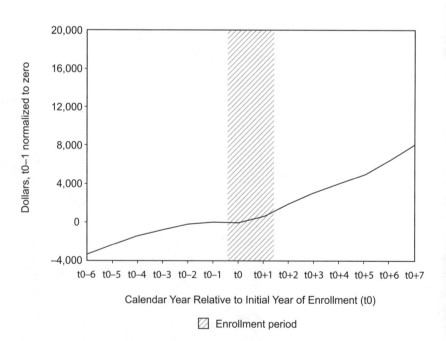

Fig. 11.13 Wages before, during, and after a two-calendar-year episode of post-secondary enrollment that is mainly online

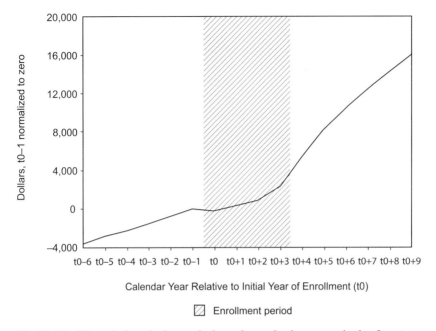

Fig. 11.14   Wages before, during, and after a four-calendar-year episode of post-secondary enrollment that is mainly online

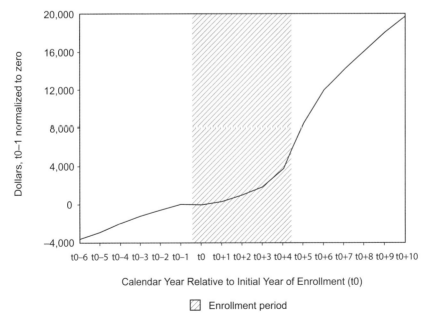

Fig. 11.15   Wages before, during, and after a five-calendar-year episode of post-secondary enrollment that is mainly online

be consistent with students leaving work entirely or even cutting their work hours substantially. Annual earnings growth is higher after the enrollment episode than before enrollment commences. The shorter the episode the smaller the before-versus-after increase in annual earnings growth. Indeed, for the short episodes of one and two calendar years the before-versus-after increase in growth is very modest, just a few hundred dollars per year. These episodes are important because they account for 60 percent of exclusively online enrollment and 68 percent of mainly online enrollment. The before-versus-after increase in earnings growth is most striking for episodes of five calendar years (probably four school years), but recall that such episodes are rare: 7.9 percent of exclusively online and 4.4 percent of mainly online enrollment.

One might consider several variants of the figures described above: figures for males only (since they have zero earnings less frequently than females), figures that include individual person fixed effects (to account for changing self-selection into online education over time), figures for graduate students only, and figures for undergraduate students only. Almost without exception, these variants generate figures (available from the author) that show patterns so similar to those already shown that it is hard to gain additional insights from them.

11.6   An Empirical Strategy for Estimating Return on Investment to Online Education

11.6.1   The Regularities Observed So Far

We have observed several regularities regarding earnings and online education. First, online education, though much more common than in the past, is still an uncommon way of pursuing postsecondary education. Thus, it is implausible that there is not selection into online education: it is unlikely to be a random choice. Second, most people who engage in online education are older than eighteen to twenty-four, the traditional age range for postsecondary enrollees. This, combined with the first observation, means that there is no obvious control group of people who are like the online students but who attend brick-and-mortar schools or who attend no postsecondary school at all. Third, in part because of their age, nearly all online students have earnings both before and after enrolling in online education. This is helpful empirically and is a feature that we usually lack when analyzing postsecondary students of traditional age. Fourth, although many subgroups of students show no sign of Ashenfelter's Dip prior to enrolling, at least some subgroups of students do appear to experience a mild Dip (really, just an earnings growth slowdown). The two most likely interpretations of such Dips are a deterioration in earnings opportunities at work (employer-driven events) or exogenous events (such as change in health) that cause earnings to slacken and also cause people to seek education. Regardless of which interpretation we adopt, an Ashenfelter's Dip makes it difficult to forecast what would have happened in the absence of online education because it suggests that at least some selection into online education is driven by changes in labor demand or labor supply. Sixth, during the years in which people are enrolled in online education, their earnings are slightly reduced, probably owing to their studying more. However, the reduction in earnings is so small that it is not consistent with students cutting their work hours more than slightly. Indeed, for some subgroups of students, opportunity costs appear to be negligible or even *negative*. We should not rule out negative opportunity costs as impossible because enrolling may give some students access to jobs that would otherwise be inaccessible to them. Finally, people pursue online education for varying periods, with everything from one to five calendar years being reasonably common. There is almost certainly considerable selection into these lengths: they are unlikely to be random.

Given these features of the data, the best empirical strategy to estimate returns would appear to be a within-person comparison of actual post-enrollment earnings with (counterfactual) projected earnings based on pre-enrollment earnings. This is not an infallible strategy, but it does take advantage of the availability of preenrollment earnings data for the vast majority of online students. It is also a strategy that can accommodate enrollment episodes of differing lengths since a person's direct costs can be measured over however many years he is enrolled. Conveniently, opportunity costs can be computed using counterfactual earnings based on preenrollment data. Put another way, the only serious challenge to this empirical strategy is Ashenfelter's Dip, which makes projecting earnings difficult. However, this challenge can be somewhat overcome by bounding the estimates (see below).

11.6.2   Alternative Empirical Strategies That Were Considered

Other empirical strategies would not merely use a person's own earlier self to generate counterfactual earnings. They would specify a control group and use that group's earnings to generate counterfactual earnings. Given the abovementioned features of the data, however, all of the likely control groups would be prone to introduce selection bias, often of an unknown sign and magnitude. Compared to such bias, the challenges posed by Ashenfelter's Dip seem manageable if for no other reason than that we can sign and bound its consequences.

For instance, one possible control group would consist of people who were like the online enrollees in terms of age, sex, location, industry, and earnings (in the enrollees' preenrollment years), but who did not engage in online education. This is an enormous set of people, however, and selection into online education is rare and likely triggered by events that the controls would not have experienced. With so many potential controls and so little capacity to match on triggering events, it would be disturbingly ad hoc to choose controls even with sophisticated methods like synthetic controls.[33] Moreover, the sign of the bias would be unclear: Would the controls be people whose experience was the same as the enrollees except that they lacked the initiative, motivation, and liquidity to pursue education? Or, would the controls be people who did not experience a triggering event or who had ways of responding to an event that were superior to online education? For instance, some controls might use a rich social network rather than online education to rebuild their earnings opportunities.

Another potential set of controls would be people similar to the online enrollees in terms of age, sex, location, industry, and earnings (in the enrollees' preenrollment years), but who enrolled in brick-and-mortar postsecondary education rather than online education. The difficulty here is that selection into online education is likely to be highly nonrandom. Since the timing of online classes is typically much more flexible, some part of the in-person versus online choice is probably due to factors such as whether a person is trying to simultaneously study and work or provide childcare. Some of the choice is probably due to the proximity of brick-and-mortar postsecondary institutions, but such proximity is distributed in a highly nonrandom way because brick-and-mortar institutions locate themselves near people who have a demand for education; the people who live close to brick-and-mortar institutions are considerably more prone to demand postsecondary education than those who live far.[34] Finally, some of the choice between online and brick-and-mortar education is probably due to the availability of high-speed internet, but the period of online education growth (from 2006 onward) was a period in which the only people without access to high-speed internet were truly rural, not merely people who lived in smaller or less densely populated labor markets.[35] Summing up, people who choose brick-and-mortar education are probably (a) less likely to need flexibility while enrolled, which may indicate lower opportunity costs of interfering with work but also indicate fewer conflicts with family-related demands; (b) located in areas where the general demand for postsecondary education is higher, suggesting that employers' demand for skilled labor is greater; and (c) located in urban areas that offer different job opportunities than arise in rural areas. These types of selection are highly problematic because we cannot even sign the resulting bias. Moreover, they represent only some of the forms that selection (between brick-and-mortar and online) might take.

In some research, the controls are those who opt into the same treatment

33. Synthetic control methods would seem to be best, under the circumstances, if this empirical strategy were pursued. Matching methods would be more arbitrary given the large number of people who would match to each online enrollee.

34. See Miller (2009).

35. See Federal Communications Commission (2009). The report can be downloaded from the Wireline Competition Bureau Statistical Reports internet site at https://www.fcc.gov /general/iatd-data-statistical-reports.

(online education) but at a different time. Thus, we might consider using people who will enroll in online education *in the future* as controls for people enroll in it now. The difficulty with doing this is that either we have to use controls who are in the same birth cohort as the treated people but who start online education at a later age or we have to use controls who are the same age as the treated people when they start online education but who start it in a different year. In the former case, the controls are problematic because the effect of education on earnings is likely to change with the age at which a person engages in education. In the latter case, the controls are problematic because online education (availability, curriculum, degree programs) has been changing fairly rapidly over recent years. Again, it is difficult to sign the biases created by using later online students as controls for earlier online students.

A final control group one might consider are students who enroll in online education but then drop it so quickly that it is implausible that it could affect their earnings much via the learning channel. The advantage of these potential controls is that their enrollment may have been triggered by an event, and is plausibly similar to the events that trigger other students' longer enrollment episodes. However, this group would be problematic because their lack of persistence in online education could be driven by their low effort or inability to master material. But, their lack of persistence could equally be caused by an improvement in their earnings opportunities. Thus, if we were to use them as controls, we would be unable even to sign the bias they would introduce.

11.6.3   Details on the Estimation Strategy Used

In short I dismiss, after careful consideration, strategies that generate counterfactual earnings for online enrollees based on other people. Thus, I focus on the narrower problem of using their own preenrollment earnings to project what their later earnings would have been had they not enrolled. We should be especially cautious of overreliance on the data in the years that immediately precede enrollment. If those data represent an inherently transitory event from which the person's earnings would have quickly rebounded in any case, we ought not to use data from that period to project future earnings. At the other extreme, those data could represent the "new normal": what the person's earnings trajectory would have been had he not enrolled in online education.

Given that economics and the data give us little indication of how to choose between these two extremes, it seems best to proceed with a bounding strategy. I therefore estimate earnings projections for each person that do and do not rely on data from the period just before enrollment (up to three years' worth).

A remaining issue is what specification to use in this estimation procedure. The (natural) log of earnings tends to have a steadier annual increase than the level of earnings. Also, earnings tend to be log-normally distributed rather than normally distributed. One would like to allow both the level and growth of a person's log earnings to shift after an episode of online enrollment. Most persons' log earnings evolve with age in a manner that resembles a quadratic relationship: log earnings grow swiftly in a person's early career, but growth slows and earnings eventually plateau or even decline close to retirement. Little would be added by estimating a higher-order polynomial in age (see footnote 30).

Thus, I estimate the following regression of log earnings

$$(2) \qquad \ln(\text{earnings}_{it}) = \alpha_i + \beta \text{age}_{it} + \gamma \text{age}_{it}^2 + \varepsilon_{it}$$

where α_i is a person fixed effect that is estimated three different ways: (a) using all preenrollment ($t \leq t_0$) observations, (b) using preenrollment observations except the two ($t \leq t_0 - 2$) that immediately precede enrollment, and (c) using preenrollment observations except the three ($t \leq t_0 - 3$) that immediately precede enrollment. Notice that none of these regressions makes use of earnings *during* enrollment since they are potentially directly affected by studying. Specifications (b) and (c) instantiate the theory that earnings immediately before enrollment reflect a transitory shock and that earnings would have bounced back to their previous path even in the absence of online education. Hereafter, I refer to specifications (a) through (c) as having "Ashenfelter discards" of zero, two, and three years.[36]

Once I have estimates from equation (2), I form ROIs using a discount rate of 3.5 percent because this corresponds approximately to annual wage inflation over the period I consider (Bureau of Labor Statistics 2017), but I obtain similar results using discount rates of 3 and 4 percent.[37] I compute ten-year ROIs (the ROI if we consider only the first ten years of earnings after the completion of the enrollment episode) because ten years is the standard length of a student loan. Thus, the ten-year ROI helps us understand whether a student could reasonably expect to pay off her loan if she borrowed to pay tuition. In addition, the ten-year ROI is appealing because it depends less on the discount rate and on the accuracy of the parameter estimates from equation (2).[38] Since ROIs may differ by type of school (exclu-

---

36. Another issue is online enrollees who have no reported wage or salary earnings in a pre-year. Such earnings are more likely to signal a period when the person was out of the labor force or when the person's reservation wage exceeded the wage he was offered. Because years with zero earnings are relatively rare, however, their exclusion has very little effect on the estimates. These alternative results are available from the author.

37. These results are available from the author.

38. In other words, if the parameter estimates are mistaken, the mistake may compound over the postenrollment years so that up-through-age-sixty-five estimates are more error prone than ten-year-out estimates. In particular, there are few data from people observed in their late fifties and sixties so predictions for those years are something of an extrapolation beyond the data. We would not want to focus on ROIs in which those predictions played much of a role.

sively or mainly online), by undergraduate versus graduate education, and by the length of enrollment, I show ROIs for all these relevant subgroups.

For instance, suppose that the estimated coefficients from the all-online-students regression that discards two preenrollment observations (but does not discard zero earnings) are designated by a circumflex (^). Then the numerator for ROI in this case is the estimated gain in lifetime earnings

$$(3) \quad \left[\sum_{\tau=0}^{\lambda-1} y_{i,t_0+\tau} + \sum_{\tau=\lambda}^{\lambda+9} \frac{\exp[(\hat{\alpha}_i + \hat{\mu}) + (\hat{\beta} + \hat{\nu})\text{age}_{i,t_0+\tau} + (\hat{\gamma} + \hat{\rho})\text{age}^2_{i,t_0+\tau}]}{(1 + \delta)^\tau} \right]$$

$$- \left[\sum_{\tau=0}^{\lambda+9} \frac{\exp(\hat{\alpha}_i + \hat{\beta}\text{age}_{i,t_0+\tau} + \hat{\gamma}\text{age}^2_{i,t_0+\tau})}{(1 + \delta)^\tau} \right]$$

where δ is the discount rate, λ is the length of the enrollment episode, and all amounts are discounted back to year t_0. Equation (3) is the difference in a person's lifetime earnings depending on whether she does or does not enroll. The first term is her actual earnings while enrolled and projected earnings assuming that she does enroll. The second term is projected earnings assuming (counterfactually) that she does not enroll.[39]

The denominator contains the schooling costs associated with generating the earnings gain

$$(4) \quad \sum_{\tau=0}^{\lambda-1} \frac{\text{Tuition Paid}_{it}}{(1 + \delta)^\tau}$$

or

$$(5) \quad \sum_{\tau=0}^{\lambda-1} \frac{\text{Social Cost}_{it}}{(1 + \delta)^\tau}$$

where expression (4) considers only the private costs paid by the student herself whereas expression (5) considers the full social cost of (equal to the core spending on) her education, regardless of who paid for it (the student herself or taxpayers).

If estimated ROI is greater than or equal to one, then the benefits generated by the enrollment episode apparently cover its direct costs. If estimated ROI is between zero and one, the episode generates benefits but they are insufficient to cover its direct costs. An ROI less than or equal to zero indicates that the episode generated no or negative benefits to earnings.

---

39. The parameters μ, ν, and ρ are from a regression that allows the level of earnings and its relationship with age to shift at the end of the enrollment episode. In the theoretical literature on returns to human capital investment, ROI is sometimes written as (postepisode earnings if enroll *minus* postepisode earnings if do not enroll) *divided by* (schooling cost *plus* opportunity cost). However, this formula generates oddities when opportunity costs are negative, as they sometimes are in practice. Thus, for ROI, I use (postenrollment earnings if enroll *minus* postenrollment earnings if do not enroll) *divided by* (schooling cost). Note the difference between "postepisode" and "postenrollment."

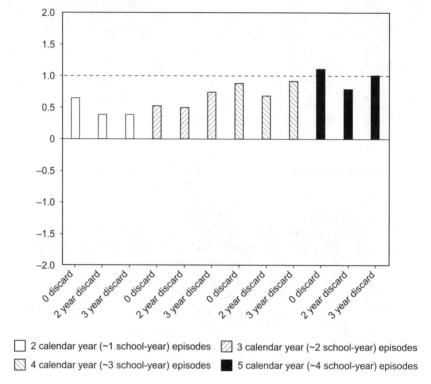

□ 2 calendar year (~1 school-year) episodes   ▨ 3 calendar year (~2 school-year) episodes

◩ 4 calendar year (~3 school-year) episodes   ■ 5 calendar year (~4 school-year) episodes

Fig. 11.16   ROIs of postsecondary episodes that were exclusively online

11.7   Findings: Returns on Investment to Online Education

The ROI estimates generated by the method just described are shown in figures 11.16–11.21, and appendix tables 11A.1–11A.3. The figures show only social ROIs and do not show ROIs for enrollment episodes that last only one calendar year (less than one school year) because they are somewhat unstable, owing to the small size of the denominator.[40] The tables, however, show both social and private ROIs and include the estimates for one-calendar-year episodes. Both figures and tables show the median ROI for each subcategory of enrollment and Ashenfelter discard. (The mean ROIs are similar but less stable.)

Figure 11.16, for example, shows estimated social ROIs for postsecondary episodes of two to five calendar years (one to four school years) that were exclusively online. Episodes are included regardless of whether they were undergraduate or graduate. There is a horizontal line at ROI = 1 as

40. In other words, very short enrollment episodes generate small direct costs, so modest changes in predicted earnings can generate large swings in ROIs.

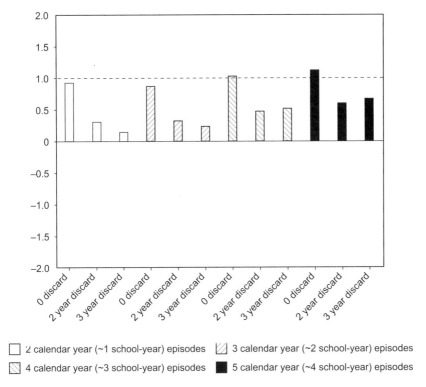

☐ 2 calendar year (~1 school-year) episodes   ▨ 3 calendar year (~2 school-year) episodes
▨ 4 calendar year (~3 school-year) episodes   ■ 5 calendar year (~4 school-year) episodes

Fig. 11.17   ROIs of postsecondary episodes that were mainly online

a reminder that, below this line, the estimated benefits of the enrollment episode do not cover its costs. The finding that is immediately striking is that, regardless of the Ashenfelter discard, estimated ROIs are nearly all less than one. That is, regardless of whether we treat immediate preenrollment earnings as predictive or not, the ten-year returns to most online episodes do not cover the direct costs to society. The exceptions to this statement are the comparatively rare episodes that last five calendar years. Depending on the Ashenfelter discard, such episodes generate ROIs that range from 0.79 to 1.11. Keep in mind that episode length is not random: a student who elects to study for five calendar years may differ from one whose episode lasts only two or three years. Thus, we cannot conclude that the students whose episodes are shorter (and whose ROIs are uniformly less than one) would have better ROIs if only they had continued their studies through a fifth calendar year.

Figure 11.17 is analogous to figure 11.16 except that it shows ROIs for episodes that are mainly, rather than exclusively, online. Again, most of the estimated ROIs are less than one. The zero discard ROIs for four-year and five-year episodes are slightly greater than one: 1.03 and 1.12, respectively.

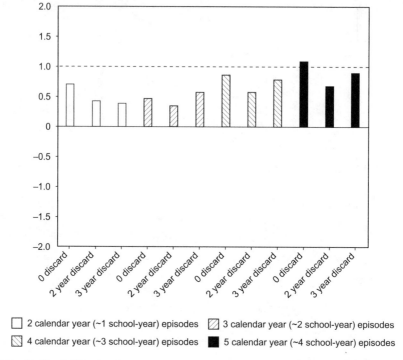

2 calendar year (~1 school-year) episodes   3 calendar year (~2 school-year) episodes

4 calendar year (~3 school-year) episodes   5 calendar year (~4 school-year) episodes

Fig. 11.18   ROIs of undergraduate postsecondary episodes that were exclusively online

However, even these modest results are not robust. The ROIs for these same episodes are substantially less than one with two-year or three-year Ashenfelter discards. Over all, the evidence suggests that the ten-year returns to mainly online episodes usually do not cover the direct costs to society.

Appendix table 11A.1 demonstrates that *private* ROIs are uniformly better than social ROIs. This is unsurprising, of course, because the private costs, which are in the denominator of the ROI equation, are uniformly smaller than social costs. (Recall the descriptive statistics in section 11.5.) The estimated private ROIs for two- and three-calendar-year episodes, which are fairly common, range from 0.61 to 1.17 for exclusively online education and range from 0.28 to 1.85 for mainly online education. However, two-thirds of the estimates for episodes of these lengths are less than one. Thus, such episodes *may* generate earnings that justify the tuition and fees that the person herself paid, but this conclusion depends on how one treats Ashenfelter's Dip. The estimated private ROIs for (comparatively rare) four- and five-calendar-year episodes are, with one exception, greater than one, suggesting that such episodes usually generate extra earnings that justify the tuition and fees paid by the student himself.

Figures 11.18 and 11.19 and appendix table 11A.2 show estimated ROIs

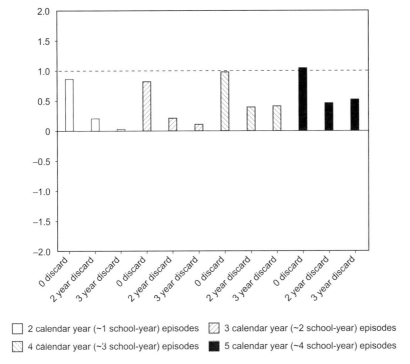

2 calendar year (~1 school-year) episodes   3 calendar year (~2 school-year) episodes
4 calendar year (~3 school-year) episodes   5 calendar year (~4 school-year) episodes

Fig. 11.19   ROIs of undergraduate postsecondary episodes that were mainly online

for *undergraduate* postsecondary episodes that are, respectively, exclusively and mainly online. The undergraduate ROIs are much like the overall ROIs shown in figures 11.16 and 11.17. Thus, the conclusions drawn in the paragraphs above also hold for undergraduate online education.

In contrast, figures 11.20 and 11.21 and appendix table 11A.3, which show estimated ROIs for *graduate* postsecondary episodes, suggest that the shorter (and much more common) episodes often generate negative ROIs. That is, when they enroll, people lose rather than gain earnings. The ROIs are especially poor for graduate education that is exclusively online. For two- and three-year episodes, they range from −0.91 to −1.73. Even four- and five-calendar-year episodes of exclusively online graduate education have ROIs far less than one. Mainly online graduate education has somewhat better ROIs. They are far below one (and sometimes negative) for two- and three-calendar-year episodes, hover around one for four-year episodes, and are always well above one for the comparatively rare five-year episodes.

Summing up, the evidence suggests that the vast majority of online postsecondary enrollment, which tends to be in short episodes, generates earnings benefits that never cover social costs and probably do not even cover students' private costs. Four- and five-calendar-year enrollment episodes,

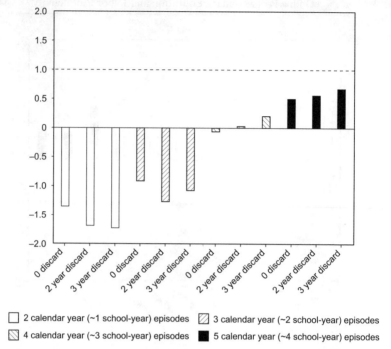

Fig. 11.20   ROIs of graduate postsecondary episodes that were exclusively online

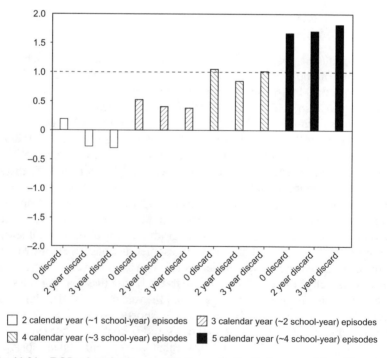

Fig. 11.21   ROIs of graduate postsecondary episodes that were mainly online

which are comparatively rare, usually generate benefits that do not cover social costs but that do cover private costs. The exception is five-calendar-year graduate episodes at mainly online schools: their benefits appear to cover social costs as well as private costs. However, we should remind ourselves that the better results for students with five-year episodes do not imply that other students could obtain similar results if only they were to persist longer in online education. Such a conclusion would be wholly unwarranted, especially given how rare it is for students to self-select into long enrollment episodes.

11.8   Do People Move into Higher Productivity Jobs after Online Enrollment?

So far, the evidence suggests that returns to online postsecondary education are modest. This seems at odds with the logic of commentators, such as Beato and Christiansen, who argue that online education will disproportionately educate people in cutting-edge skills they need for technical, often computer-related, jobs in rapidly growing industries. Their arguments would suggest that online education allows students to move into industries that are high technology, have unusually high predicted employment growth, use abstract rather than routine or manual skills, and that are not offshorable (Autor and Dorn 2013).[41] These are likely channels by which the online students' productivity might increase: improvements in their human capital that allow them to reallocate themselves to industries where productivity growth is higher.

In this section, I briefly examine evidence for these channels using figures much like those in section 11.6. Instead of showing earnings, however, the figures show (a) projected employment in a person's industry,[42] (b) the percentage of occupations that are high technology in a person's industry,[43] (c) the Autor-Dorn abstractness index for the skills required by a person's industry, (d) the analogous index for routine skills, (e) the analogous index for manual skills, and (f) the Autor-Dorn index of the offshorability of occupations in a person's industry. Figures 11.22–11.27 are for three-calendar-year enrollment episodes that are exclusively online. Figures 11.28–11.33 are for three-calendar-year enrollment episodes that are mainly online. I show only three-year episodes because they are common but nevertheless long enough for degree attainment. Also, the figures for episodes of other lengths

---

41. I rely throughout on Autor and Dorn's definitions of abstract, routine, and manual jobs. I also rely on their definition of offshorability.

42. I rely on the Employment Projects of the United States Bureau of Labor Statistics. Specifically, because its timing is appropriate for the students whose enrollment I examine, I use Figueroa and Woods (2007).

43. For data in which industries are high technology, I rely on Hecker (2005). However, the industries he defines as high technology in 2005 are nearly the same today. See Wolf and Terrell (2016).

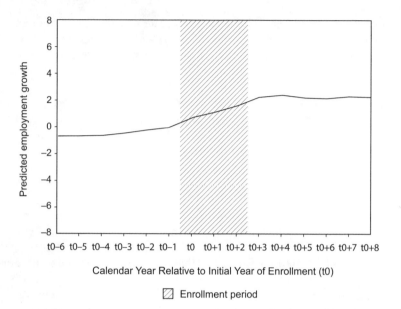

Calendar Year Relative to Initial Year of Enrollment (t0)

☐ Enrollment period

Fig. 11.22   Predicted employment growth before, during, and after a three-calendar-year episode of postsecondary enrollment that is exclusively online
Notes: Predicted employment growth in industry between 2012 and 2022. A standard deviation of this measure is 16 percentage points.

would not much affect the evidence.[44] All of the figures are designed so that the vertical axis represents one standard deviation in the measure of interest.

Consider figure 11.22, which shows the degree to which people move to a higher projected employment growth industry. There is certainly improvement on this measure: projected employment growth rises by 2.3 to 3 percentage points, which is 0.14 to 0.19 of a standard deviation. Interestingly, nearly all of the improvement occurs *while* the person is enrolled online.

Figure 11.23 shows, however, that the higher growth industries to which people are moving are not high-technology industries. In fact, the figure suggests that their employment in high technology is flat or slightly lower after their enrollment episode. However, the movement is very small. To see this, it is helpful to know how the Bureau of Labor Statistics flags determine the "high-tech-ness" of an industry. It flags certain occupations as indicators of high technology: engineers, life scientists, physical scientists, and computer systems managers. It then uses its industry-occupation matrix to assign high-tech-ness to an industry based on the percentage of its occupations that are high technology. If this percentage rises by 1 percentage point, this can be interpreted as one in four or five students taking a job in one of the forty-six

44. The author can make available figures for episodes of other lengths and figures that separate undergraduates from graduate students.

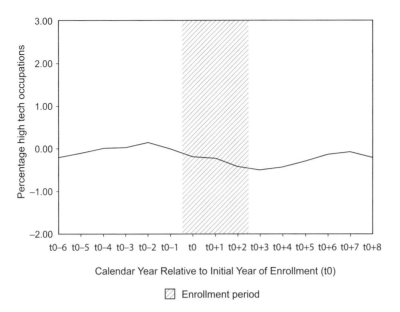

Fig. 11.23   Probability job is in high tech before, during, and after a three-calendar-year episode of postsecondary enrollment that is exclusively online
Notes: If percent of high-tech occupations rises by 1 percentage point, this can be interpreted as one in four or five students taking a job in one of the forty-six four-digit industries that are most high tech in the United States. This measure changes only when students change the industry of their jobs.

four-digit industries that are most high technology in the United States. A standard deviation in this measure is 5 percentage points.

Figures 11.24, 11.25, and 11.26 demonstrate that students are not moving into jobs that require greater abstract, routine, or manual skills after an enrollment episode that is exclusively online. A standard deviation in all these indices is 1 percentage point, and all of these indices fall by about 0.2 percentage points. The timing of the fall in abstractness hints that it is a causal effect of online enrollment. The declines in routineness and manualness appear simply to be trends that start before the episode and continue through it and after it. Together, this evidence suggests that students are moving (slightly) toward industries with occupations that could not be classified as abstract, routine, or manual. This could be a good sign for productivity if such occupations are unclassified because they are novel. It could equally be a bad sign. In any case, there is no evidence for students moving into the more abstract jobs associated with higher productivity growth.

Figure 11.27 shows that students are also not moving away from offshorable jobs. Before, during, and after their online enrollment, they are

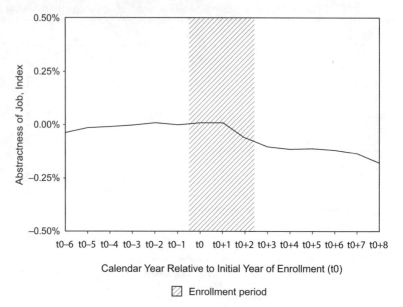

Calendar Year Relative to Initial Year of Enrollment (t0)

☑ Enrollment period

Fig. 11.24   "Abstractness" of job before, during, and after a three-calendar-year episode of postsecondary enrollment that is exclusively online

Notes: Abstractness is measured by the Autor-Dorn index of the skills required in the industry. A standard deviation in this index is about 1 percentage point. This measure changes only when students change the industry of their jobs.

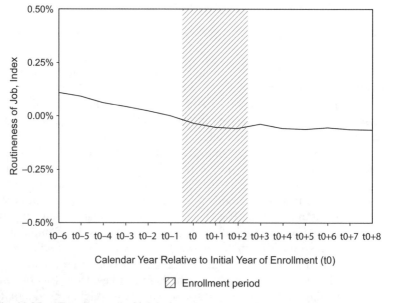

Calendar Year Relative to Initial Year of Enrollment (t0)

☑ Enrollment period

Fig. 11.25   "Routineness" of job before, during, and after a three-calendar-year episode of postsecondary enrollment that is exclusively online

Notes: Routineness is measured by the Autor-Dorn index of the skills required in the industry. A standard deviation in this index is about 1 percentage point. This measure changes only when students change the industry of their jobs.

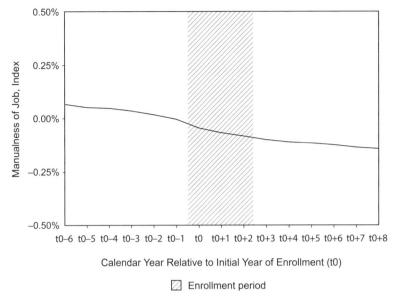

Calendar Year Relative to Initial Year of Enrollment (t0)

▨ Enrollment period

Fig. 11.26 "Manualness" of job before, during, and after a three-calendar-year episode of postsecondary enrollment that is exclusively online

Notes: Manualness is measured using the Autor-Dorn index of the skills required in the industry. A standard deviation in this index is about 1 percentage point. This measure changes only if students change the industry of their jobs.

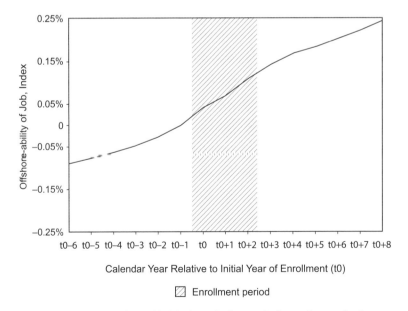

Calendar Year Relative to Initial Year of Enrollment (t0)

▨ Enrollment period

Fig. 11.27 "Offshorability" of job before, during, and after a three-calendar-year episode of postsecondary enrollment that is exclusively online

Notes: Offshorability is measured using the Autor-Dorn index. A standard deviation in this index is about 0.5 percentage points. This measure changes only when students change the industry of their jobs.

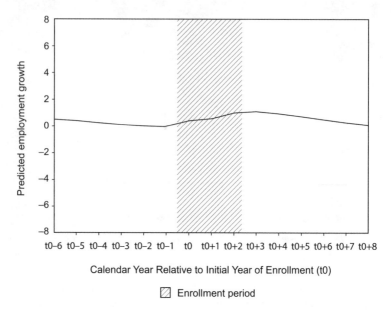

Fig. 11.28   Prospective growth in students' industry before, during, and after a three-calendar-year episode of postsecondary enrollment that is mainly online
Notes: Predicted employment growth in students' industry between 2012 and 2022. A standard deviation in this measure is 16 percentage points. This measure changes only when students change the industry of their jobs.

moving *into* industries that are more offshorable. The trend appears to be unaffected by enrollment.

Mainly online education generates somewhat different patterns. Figure 11.28 shows little indication that mainly online students move into industries with higher projected earnings growth after enrolling. Figure 11.29 suggests that enrollment also does not affect their tendency to join a high-technology industry. They are moving into high-technology industries before, during, and after their episode, but the trend is unaffected. The abstractness of the students' jobs rises quickly *while* they are enrolled: 0.17 standard deviations (0.17 percentage points) in just three years. (See figure 11.30.) However, after the episode, abstractness rises more slowly than it was rising before the episode enrollment. People end up at about the same level of abstractness as their preenrollment trend would have predicted. Finally, figures 11.31, 11.32, and 11.33 show little or no evidence that mainly online enrollment affects the routineness, manualness, or offshorability of a person's job.

Overall, the evidence is slight for the hypothesis that online education shifts people into higher productivity industries. It seems likely that commentators who emphasize that online education generates cutting-edge, high-technology, abstract skills are focusing on programs that and/or students who are highly nonrepresentative. The nonrepresentativeness of their

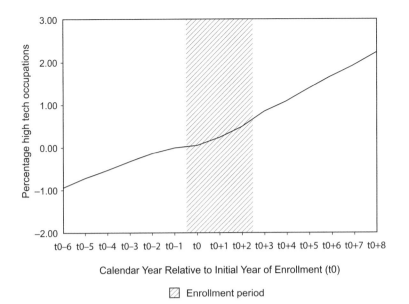

Fig. 11.29 Probability job is in high tech before, during, and after a three-calendar-year episode of postsecondary enrollment that is mainly online

Notes: If percent of high-tech occupations rises by 1 percentage point, this can be interpreted as one in five students taking a job in one of the forty-six four-digit industries that are most high tech in the United States. This measure changes only when students change the industry of their jobs.

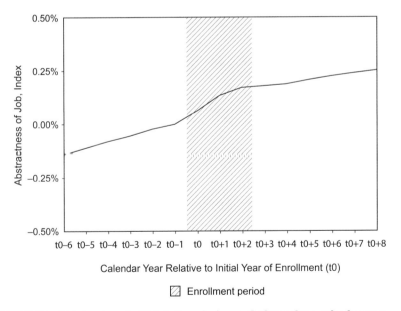

Fig. 11.30 "Abstractness" of job before, during, and after a three-calendar-year episode of postsecondary enrollment that is mainly online

Notes: Abstractness is measured by the Autor-Dorn index of the skills required in the industry. A standard deviation in this index is about 1 percentage point. This measure changes only when students change the industry of their jobs.

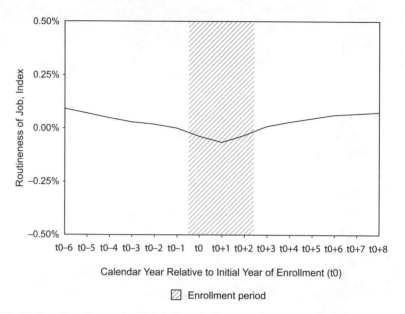

Fig. 11.31 "Routineness" of job before, during, and after a three-calendar-year episode of postsecondary enrollment that is mainly online

Notes: Routineness is measured by the Autor-Dorn index of the skills required in the industry. A standard deviation in this index is about 1 percentage point. This measure changes only when students change the industry of their jobs.

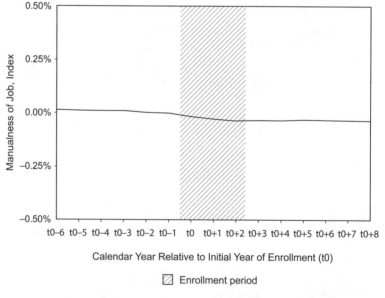

Fig. 11.32 "Manualness" of job before, during, and after a three-calendar-year episode of postsecondary enrollment that is mainly online

Notes: Manualness is measured by the Autor-Dorn index of the skills required in the industry. A standard deviation in this index is about 1 percentage point. This measure changes only when students change the industry of their jobs.

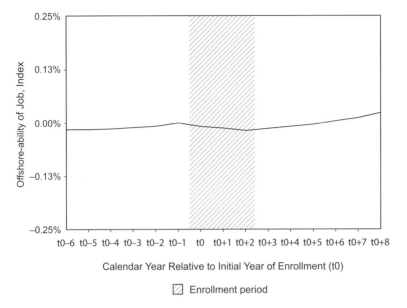

Fig. 11.33   **"Offshorability" of job before, during, and after a three-calendar-year episode of postsecondary enrollment that is mainly online**
Notes: Offshorability is measured by the Autor-Dorn index. A standard deviation in this index is about 0.5 percentage points. This measure changes only when students change the industry of their jobs.

descriptions is confirmed by the studies cited in footnote 6, none of which suggests that advanced or leading-edge courses are prevalent on online education. They suggest that introductory courses dominate.

11.9   Online Education from the Federal Taxpayer's Point of View

As noted in section 11.5, federal taxpayers foot a substantial share of the cost of online postsecondary education. Are they at all likely to recoup their investment? The estimated ROIs suggest not.

For instance, the average social cost of a year of exclusively online post-secondary school is $8,325, of which $3,620 (43.5 percent) is funded up front by federal taxpayers through grants and tax expenditures. Given that social ROIs are below one for exclusively online education and that students never face federal tax rates greater than 43.5 percent, it is not possible that they will repay current federal taxpayers through higher future federal income tax payments. Moreover, because private ROIs are also often well below one for exclusively online education, many students will be unable to repay their federal loans out of higher earnings. Thus, we should not be surprised they are overrepresented among loan defaulters and people who enter income-based repayment schemes through which they are likely to end up repaying only a fraction of what they owe. If students at exclusively online schools repay only

50 percent of their loans, federal taxpayers will have funded 69 percent of the cost of their education with little recoupment through higher future taxes.

Similar calculations can be made for mainly online students. The average social cost of a year of such education is $10,380, of which $3,699 (35.6 percent) is funded up front by federal taxpayers through grants and tax expenditures. Because social ROIs for mainly online schooling are below one and former students are extremely unlikely to pay federal income tax rates above 35.6 percent, current taxpayers cannot recoup their investment in mainly online education through students' higher future tax payments. Moreover, the low private ROIs for most mainly online education suggest that students will fail to repay their loans, which average $5,075 per year, in full. If they repay only 50 percent of their loans, federal taxpayers will have funded 60 percent of the total cost of mainly online education.

11.10   Conclusions

This study attempts a fairly comprehensive examination of online postsecondary education and its effects on students' later earnings and job outcomes. It also calculates ROIs. On the whole, I find little support for optimistic prognostications about online education. It is not substantially less expensive for society than comparably selective in-person education, although of course its costs may fall in the future as online schools gain experience. Students themselves pay *more* for online education than in-person education even though the resources devoted to their instruction are lower. Online enrollment episodes do usually raise students' earnings, but almost never by an amount that covers the social cost of their education. This failure to cover social costs is important for taxpayers, especially for federal taxpayers who are the main funders of online education apart from the students themselves. The failure implies that federal income tax revenues associated with future increased earnings could not come close to repaying current taxpayers.

Most online students' earnings do not rise by an amount that covers even their private costs—the tuition and fees that they themselves, as opposed to governments, pay. This suggests that former online students will struggle to repay their federal loans. Only online students who persist through unusually long enrollment episodes (four or five calendar years) experience earnings increases that usually cover their private costs. We cannot infer from this evidence, however, that students whose online enrollment episodes are short would experience similar benefits if they persisted for more years. Indeed, they may be dropping out precisely because they foresee that their benefits will not be sufficient to justify the costs.

There is only slight evidence that online enrollment moves people toward jobs associated with higher productivity growth. Online enrollment appears to have little or no effect on a person's probability of holding a high-technology job or a job that requires abstract skills.

Overall, the main contribution of this study may be to ground the discussion of online postsecondary education in evidence. Much of the discussion to this point may suffer from undue optimism or pessimism because such evidence has been lacking.

Appendix

Table 11A.1   Median ROI values for online postsecondary education episodes of various lengths

	Exclusively online	Mainly online
Two-calendar-year (one-school-year) episodes		
Social ROI with no Ashenfelter discard	0.64	0.93
Social ROI with two-year Ashenfelter discard	0.39	0.30
Social ROI with three-year Ashenfelter discard	0.39	0.14
Private ROI with no Ashenfelter discard	1.01	1.85
Private ROI with two-year Ashenfelter discard	0.62	0.59
Private ROI with three-year Ashenfelter discard	0.61	0.28
Three-calendar-year (two-school-year) episodes		
Social ROI with no Ashenfelter discard	0.53	0.88
Social ROI with two-year Ashenfelter discard	0.50	0.33
Social ROI with three-year Ashenfelter discard	0.75	0.24
Private ROI with no Ashenfelter discard	0.84	1.75
Private ROI with two-year Ashenfelter discard	0.79	0.65
Private ROI with three-year Ashenfelter discard	1.17	0.47
Four-calendar-year (three-school-year) episodes		
Social ROI with no Ashenfelter discard	0.88	1.03
Social ROI with two-year Ashenfelter discard	0.68	0.47
Social ROI with three-year Ashenfelter discard	0.91	0.51
Private ROI with no Ashenfelter discard	1.39	2.04
Private ROI with two-year Ashenfelter discard	1.06	0.93
Private ROI with three-year Ashenfelter discard	1.43	1.01
Five-calendar-year (four-school-year) episodes		
Social ROI with no Ashenfelter discard	1.11	1.12
Social ROI with two-year Ashenfelter discard	0.79	0.60
Social ROI with three-year Ashenfelter discard	1.01	0.66
Private ROI with no Ashenfelter discard	1.74	2.24
Private ROI with two-year Ashenfelter discard	1.24	1.20
Private ROI with three-year Ashenfelter discard	1.58	1.30
One-calendar-year (part-school-year) episodes[a]		
Social ROI with no Ashenfelter discard	0.96	3.05
Social ROI with two-year Ashenfelter discard	1.00	2.14
Social ROI with three-year Ashenfelter discard	1.18	2.08
Private ROI with no Ashenfelter discard	1.50	6.08
Private ROI with two-year Ashenfelter discard	1.58	4.27
Private ROI with three-year Ashenfelter discard	1.85	4.13

[a]These estimates are not stable. See text.

Table 11A.2   **Median ROI values for online undergraduate postsecondary education episodes of various lengths**

	Exclusively online	Mainly online
Two-calendar-year (one-school-year) episodes		
Social ROI with no Ashenfelter discard	0.71	0.87
Social ROI with two-year Ashenfelter discard	0.43	0.21
Social ROI with three-year Ashenfelter discard	0.39	0.02
Private ROI with no Ashenfelter discard	1.11	1.72
Private ROI with two-year Ashenfelter discard	0.68	0.42
Private ROI with three-year Ashenfelter discard	0.61	0.04
Three-calendar-year (two-school-year) episodes		
Social ROI with 0 no Ashenfelter discard	0.47	0.82
Social ROI with two-year Ashenfelter discard	0.35	0.21
Social ROI with three-year Ashenfelter discard	0.58	0.11
Private ROI with no Ashenfelter discard	0.74	1.63
Private ROI with two-year Ashenfelter discard	0.55	0.42
Private ROI with three-year Ashenfelter discard	0.91	0.21
Four-calendar-year (three-school-year) episodes		
Social ROI with no Ashenfelter discard	0.86	0.98
Social ROI with two-year Ashenfelter discard	0.58	0.39
Social ROI with three-year Ashenfelter discard	0.79	0.41
Private ROI with no Ashenfelter discard	1.35	1.95
Private ROI with two-year Ashenfelter discard	0.91	0.77
Private ROI with three-year Ashenfelter discard	1.24	0.82
Five-calendar-year (four-school-year) episodes		
Social ROI with no Ashenfelter discard	1.09	1.04
Social ROI with two-year Ashenfelter discard	0.68	0.46
Social ROI with three-year Ashenfelter discard	0.90	0.52
Private ROI with no Ashenfelter discard	1.71	2.07
Private ROI with two-year Ashenfelter discard	1.07	0.92
Private ROI with three-year Ashenfelter discard	1.41	1.02
One-calendar-year (part-school-year) episodes[a]		
Social ROI with no Ashenfelter discard	1.07	2.85
Social ROI with two-year Ashenfelter discard	1.07	1.89
Social ROI with three-year Ashenfelter discard	1.30	1.73
Private ROI with no Ashenfelter discard	1.67	5.66
Private ROI with two-year Ashenfelter discard	1.68	3.76
Private ROI with three-year Ashenfelter discard	2.05	3.43

[a]These estimates are not stable. See text.

Table 11A.3   Median ROI values for online graduate postsecondary education episodes of various lengths

	Exclusively online	Mainly online
Two-calendar-year (two-school-year) episodes		
Social ROI with no Ashenfelter discard	−1.36	0.19
Social ROI with two-year Ashenfelter discard	−1.69	−0.28
Social ROI with three-year Ashenfelter discard	−1.73	−0.31
Private ROI with no Ashenfelter discard	−2.13	0.37
Private ROI with two-year Ashenfelter discard	−2.65	−0.56
Private ROI with three-year Ashenfelter discard	−2.72	−0.61
Three-calendar-year (two-school-year) episodes		
Social ROI with no Ashenfelter discard	−0.91	0.51
Social ROI with two-year Ashenfelter discard	−1.26	0.40
Social ROI with three-year Ashenfelter discard	−1.08	0.38
Private ROI with no Ashenfelter discard	−1.43	1.02
Private ROI with two-year Ashenfelter discard	−1.98	0.79
Private ROI with three-year Ashenfelter discard	−1.69	0.76
Four-calendar-year (three-school-year) episodes		
Social ROI with no Ashenfelter discard	−0.06	1.05
Social ROI with two-year Ashenfelter discard	0.03	0.85
Social ROI with three-year Ashenfelter discard	0.20	1.01
Private ROI with no Ashenfelter discard	−0.10	2.09
Private ROI with two-year Ashenfelter discard	0.05	1.69
Private ROI with three-year Ashenfelter discard	0.31	2.02
Five-calendar-year (four-school-year) episodes		
Social ROI with no Ashenfelter discard	0.50	1.67
Social ROI with two-year Ashenfelter discard	0.56	1.71
Social ROI with three-year Ashenfelter discard	0.67	1.82
Private ROI with no Ashenfelter discard	0.79	3.33
Private ROI with two-year Ashenfelter discard	0.88	3.40
Private ROI with three-year Ashenfelter discard	1.05	3.62
One-calendar-year (part-school-year) episodes[a]		
Social ROI with no Ashenfelter discard	−1.91	0.60
Social ROI with two-year Ashenfelter discard	−0.84	−0.02
Social ROI with three-year Ashenfelter discard	−1.52	1.02
Private ROI with no Ashenfelter discard	3.00	1.19
Private ROI with two-year Ashenfelter discard	−1.31	−0.05
Private ROI with three-year Ashenfelter discard	−2.39	2.03

[a]These estimates are not stable. See text.

References

Ashenfelter, Orley, and David Card. 1985. "Using the Longitudinal Structure of Earnings to Estimate the Effect of Training Programs." *Review of Economics and Statistics* 67 (4): 648–60.

Autor, David H., and David Dorn. 2013. "The Growth of Low-Skill Service Jobs and the Polarization of the US Labor Market." *American Economic Review* 103 (5): 1553–97.

Beato, Greg. 2014. "Online Higher Education Retools." *Reason*, Feb. 24. Accessed Jan. 2017. http://reason.com/archives/2014/02/24/online-higher-education-retool /print.

Bettinger, Eric, Lindsay Fox, Susanna Loeb, and Eric Taylor. 2014. "Changing Distributions: How Online College Classes Alter Student and Professor Performance." Unpublished manuscript, Stanford University.

Bowen, William G., Matthew M. Chingos, Kelly A. Lack, and Thomas I. Nygren. 2014. "Interactive Learning Online at Public Universities: Evidence from a Six-Campus Randomized Trial." *Journal of Policy Analysis and Management* 33 (1): 94–111.

Bureau of Labor Statistics. 2017. "Employment Cost Index, Historical Listing Volume III." National Compensation Survey, Jan. 2017 edition. https://www.bls.gov /web/eci/echistrynaics.pdf.

Christensen, Clayton M., and Henry J. Eyring. 2011. *The Innovative University: Changing the DNA of Higher Education from the Inside Out.* Hoboken, NJ: John Wiley & Sons.

Christensen, Clayton M., and Michael B. Horn. 2013. "Innovation Imperative: Change Everything: Online Education as an Agent of Transformation." *New York Times*, Nov. 1. Accessed Jan. 2017. http://www.nytimes.com/2013/11/03/education /edlife/online-education-as-an-agent-of-transformation.html.

Cowen, Tyler, and Alex Tabarrok. 2014. "The Industrial Organization of Online Education." *American Economic Review* 104 (5): 519–22.

Deming, David J., Claudia Goldin, and Lawrence F. Katz. 2012. "The For-Profit Postsecondary School Sector: Nimble Critters or Agile Predators?" *Journal of Economic Perspectives* 26 (1): 139–63.

Deming, David J., Claudia Goldin, Lawrence F. Katz, and Noam Yuchtman. 2015. "Can Online Learning Bend the Higher Education Cost Curve?" *American Economic Review* 105 (5): 496–501.

Deming, David J., Michael Lovenheim, and Richard Patterson. Forthcoming. "The Competitive Effects of Online Education." In *Productivity in Higher Education*, edited by Caroline M. Hoxby and Kevin Stange. Chicago: University of Chicago Press.

Federal Communications Commission. 2009. "High-Speed Services for Internet Access: Status as of June 30, 2008." FCC Report. https://www.fcc.gov/wcb/stats.

Figlio, David, Mark Rush, and Lu Yin. 2013. "Is It Live or Is It Internet? Experimental Estimates of the Effects of Online Instruction on Student Learning." *Journal of Labor Economics* 31 (4): 763–84.

Figueroa, Eric B., and Rose A. Woods. 2007. "Industry Output and Employment Projections to 2016." *Monthly Labor Review* November: 53–85.

Hart, Cassandra, Elizabeth Friedmann, and Michael Hill. 2014. "Online Course-Taking and Student Outcomes in California Community Colleges." Conference paper, Association for Public Policy Analysis and Management.

Hecker, Daniel. 2005. "High-Technology Employment: A NAICS-Based Update." *Monthly Labor Review* July: 57–72.

Ho, Andrew Dean, Justin Reich, Sergiy O. Nesterko, Daniel Thomas Seaton, Tommy Mullaney, Jim Waldo, and Isaac Chuang. 2014. "HarvardX and MITx: The First Year of Open Online Courses, Fall 2012–Summer 2013." HarvardX and MITx Working Paper no. 1. http://dspace.mit.edu/bitstream/handle/1721.1/96649/SSRN -id2381263.pdf;sequence=1.

Hoxby, Caroline M. 2014. "The Economics of Online Postsecondary Education: MOOCs, Nonselective Education, and Highly Selective Education." NBER Working Paper no. 19816, Cambridge, MA.

———. 2015. "Computing the Value-Added of American Postsecondary Institutions." SOI working paper, Statistics of Income, US Internal Revenue Service. Accessed Jan. 2017. https://www.irs.gov/pub/irs-soi/15rpcompvalueaddpostsecondary.pdf.

———. 2018. "Online Postsecondary Education and the Higher Education Tax Benefits: An Analysis with Implications for Tax Administration." *Tax Policy and the Economy* 32 (1): 45–106.

Hoxby, Caroline M., and Christopher Avery. 2013. "The Missing One-Offs: The Hidden Supply of High-Achieving, Low-Income Students." *Brooking Papers on Economic Activity* (Spring 2013): 1–56.

Long, Bridget Terry, and Angela Boatman. 2013. "The Role of Remediation and Developmental Courses in Access and Persistence." In *The State of College Access and Completion: Improving College Success for Students from Underrepresented Groups*, edited by Anthony Jones and Laura Perna. New York: Routledge Books.

Looney, Adam, and Constantine Yannelis. 2015. "A Crisis in Student Loans? How Changes in the Characteristics of Borrowers and in the Institutions They Attended Contributed to Rising Loan Defaults." *Brookings Papers on Economic Policy*, Fall. Accessed Jan. 2017. https://www.brookings.edu/bpea-articles/a-crisis-in-student -loans-how-changes-in-the-characteristics-of-borrowers-and-in-the-institutions -they-attended-contributed-to-rising-loan-defaults/.

McPherson, Michael S., and Lawrence S. Bacow. 2015. "Online Higher Education: Beyond the Hype Cycle." *Journal of Economic Perspectives* 29 (4): 135–53.

Miller, Trey. 2009. "On the Validity of Distance to College as an Instrument for Educational Attainment in Models of the Labor Market Return to College." Working paper, RAND, August.

Pew Charitable Trusts. 2016. "Broadband vs. Dial-Up Adoption Over Time." Accessed Jan. 2017. http://www.pewinternet.org/data-trend/internet-use/connection-type/.

Streich, Francie F. 2014. "Online Education in Community Colleges: Access, School Success, and Labor-Market Outcomes." PhD diss., University of Michigan. https://deepblue.lib.umich.edu/handle/2027.42/108944?show=full.

United States Department of Agriculture, Economic Research Services. 2016. *Commuting Zones and Labor Market Areas*. Online documentation and data. Accessed Jan. 2017. https://www.ers.usda.gov/data-products/commuting-zones-and-labor -market-areas/.

United States General Accountability Office. 2010. *For-Profit Colleges: Undercover Testing Finds Colleges Encouraged Fraud and Engaged in Deceptive and Questionable Marketing Practices*, GAO-10-948T, Aug. Accessed Jan. 2017. http://www .gao.gov/assets/130/125197.pdf.

———. 2011. *For-Profit Schools: Experiences of Undercover Students Enrolled in Online Classes at Selected Colleges*, GAO-12-150, Oct. Accessed Jan. 2017. http:// www.gao.gov/assets/590/586456.pdf.

Waldrop, M. Mitchell. 2013. "Massive Open Online Courses, aka MOOCs, Transform Higher Education and Science." *Nature* magazine, March 13, 1–5.

Wolf, Michael, and Dalton Terrell. 2016. "The High-Tech Industry, What Is It and

Why It Matters to Our Economic Future." *Beyond the Numbers: Employment and Unemployment* 5 (8). US Bureau of Labor Statistics, May. Accessed Jan. 2017. https://www.bls.gov/opub/btn/volume-5/the-high-tech-industry-what-is-it-and -why-it-matters-to-our-economic-future.htm.
Xu, Di, and Shanna Smith Jaggars. 2013. "The Impact of Online Learning on Students' Course Outcomes: Evidence from a Large Community and Technical College System." *Economics of Education Review* 37:46–57.

Comment   Nora Gordon

Hoxby's research makes an important contribution to our understanding of the role online delivery of higher education plays in the earnings dynamics of the students who use it. This online focus is highly policy relevant: as Hoxby details in the chapter, many have hoped that online education would be a low-cost way to avoid the increasing costs of traditional higher education while effectively—and perhaps even more effectively—providing students with skills needed for current labor market success. The current study provides strong evidence that online education is not going to serve as such a panacea.

Hoxby brings the most comprehensive data to date on earnings dynamics following *online* enrollment. The present study speaks to a similar set of policy questions as the literature on returns to brick-and-mortar higher education, some of which have been under scrutiny in recent years particularly as they pertain to for-profit colleges. Do students have access to accurate and timely information about how a degree from a particular institution is likely to pay off? Would they make different choices with access to such information? Do institutions of higher education engage in deceitful marketing? Do some institutions devote too many resources to marketing and recruitment, and too few to student supports and instructional quality? Hoxby's analysis suggests online institutions are far from exempt from these familiar problems.

Finally, a critical contribution of this piece is its focus on *social* returns to online higher education, given the fact that its costs are shared between enrolled students and taxpayers. This is an important point when considering all higher education, including more traditional institutions. While the social costs of student loan default have played prominently in policy discussions, the true full cost to taxpayers includes federal tax expenditures through higher education credits and deductions as well. Hoxby explicitly

Nora Gordon is associate professor at the Georgetown University McCourt School of Public Policy and a research associate of the National Bureau of Economic Research.

For acknowledgments, sources of research support, and disclosure of the author's material financial relationships, if any, please see http://www.nber.org/chapters/c13710.ack.

includes these in her calculations of social return on investment (ROI) to online higher education.

Strengths and Limitations of Data and Methods

Hoxby uses IRS data for all students enrolled in at least partially online enrollment. The IRS data allow linkage of the higher education tax credits or deductions and a student's earnings—before, during, and after the period of enrollment. She links these to IPEDS data on institutional characteristics. She then uses these rich data to describe how earnings dynamics evolve differently postattendance depending on "how online" the at-least-partially-online institution attended is and to estimate ROI to online enrollment with individual fixed effects.

This is necessarily a descriptive study, as there is no experimental or quasi-experimental assignment into online environments. Students self-select into the institutions she observes them attending. The appropriate interpretation of the findings, then, is as presented in the chapter: how have wages been affected by online education—for the types of students who have chosen to use it to date? With relatively few caveats, the data permit a deep exploration of that question and reveal consistent, disappointing patterns about limited wage growth after enrollment in online education.

One significant data limitation Hoxby faces is the need to infer the online nature of a student's coursework. The IRS data link the student to an institution. In the IPEDS data, institutions report how many students are enrolled solely, partially, or not at all in distance education courses separately for non-degree/certificate-seeking undergraduates, degree/certificate-seeking undergraduates, and graduate students. Hoxby then uses these data to calculate student-level probabilities and classify students' experiences as exclusive, mainly, or hardly online.

Given the IPEDS language, Hoxby must make some assumptions about just how online the experience of those students taking "some but not all" distance courses is in order to estimate such probabilities. She interprets it as meaning 50 percent of the student's experience is online. She chooses this share based on the federal "50 percent" rule in place through 2005 requiring institutions to deliver at least half of their instruction in person in order to qualify for federal aid, loans, and tax assistance. She notes that, "many institutions were tightly bound by the 50 percent rule up through 2005," and that "in more recent years, the mainly online category has become, if anything, more online." Fifty percent therefore seems an upper bound of "online-ness" through 2005, and a likely underestimate of the concept in later years (the analysis of online students is weighted toward later years, due to their higher online enrollments). Hoxby acknowledges the imprecise nature of the "mainly online" category throughout the piece. Going forward, it would be useful for the IPEDS (if not already doing so) to collect more precise data based on the number of student-course units that are online versus in person.

The entire universe of students analyzed attended institutions with at least some online offerings. The relevant comparisons are therefore between how individual wages evolve following enrollment for students in more (in Hoxby's terminology, "exclusively" or "mainly") or less ("hardly") online institutions. Among the hardly online institutions, Hoxby notes which are nonselective, to best compare with the (nearly universally nonselective) exclusively or mainly online institutions.

Because of the selection into online education, it is important to understand how online (going forward, shorthand for exclusively or mainly online) students differ from their counterparts attending hardly online nonselective institutions. The online students are older, and earn more before and during their schooling. An unavoidable limitation of this approach is that it does not permit speculation about what would happen in response to particular policy changes, such as an entire state system switching to online only, or a selective institution choosing to offer a greater share of its courses online. It also cannot speak to differences in quality of online instruction. The estimates should be interpreted just as Hoxby does: as impacts of online instruction in the types of institutions that offered it, and on the wages of the types of students who enrolled in it.

Policy Implications

Higher education in the United States is under scrutiny for a host of issues including cost, access, and quality. Online instruction as implemented during the time frame of this study, through 2013, appears an unlikely solution to these problems: wage boosts are small and costs, both private and social, are high. In fact, the most significant policy implications of this work are independent of the online nature of the institutions studied. These findings fit into a broader and entirely consistent literature highlighting low returns to higher education in a subset of institutions, in large part due to low completion rates. For example, Cellini and Chaudhary (2014) find that private returns to for-profit colleges may be too small to warrant their private costs for the average student.

Why would students choose to invest their time and money—and take on debt burden to finance enrollment—in institutions that consistently fail to put their students on a trajectory that yields sufficient income to repay their loans? Recent policy efforts have focused on information problems. Students lacked comprehensive information about how others fared after attending specific institutions, but were offered plenty of persuasive marketing materials from the institutions themselves. This line of reasoning prompted the Obama administration's gainful employment rule for career college programs. In order to comply with this rule, "the estimated annual loan payment of a typical graduate" could not "exceed 20 percent of his or her discretionary income or 8 percent of his or her total earnings." Institutions failing to reach this benchmark would lose access to federal student aid. The rule also requires institution-specific data on debt, graduation rates, and later earnings to be made publicly available. The Trump administration's Education Department is in the process of rewriting this rule.

The gainful employment rule notably applies only to career—vocational—programs. The poor social returns to online education identified here include enrollments in other programs. And many brick-and-mortar institutions, including not only for-profits but also public community colleges, have low attainment rates as well. Efforts to protect student and taxpayer investments should acknowledge that low-quality programs are not confined to online or traditional, or to particular sectors or subject areas: accountability for public dollars should apply across all these realms.

Distance learning is an evolving field, and advances in its pedagogy may well help students learn more online. Low returns on the enrollment dollar, however, are unfortunately far from unique to this context. Improving returns—to both students and taxpayers—on investments in online higher education will require solving the same policy problems that have plagued brick-and-mortar education.

Reference

Cellini, Stephanie R., and Latika Chaudhary. 2014. "The Labor Market Returns to a For-Profit College Education." *Economics of Education Review* 43 (December): 125–40.

12

High-Skilled Immigration and the Rise of STEM Occupations in US Employment

Gordon H. Hanson and Matthew J. Slaughter

12.1   Introduction

US business has long dominated the global technology sector. Among the top ten technology companies in terms of revenues worldwide, six are headquartered in the United States and employ most of their workers in US facilities.[1] The US preeminence in advanced industries is perhaps surprising in light of the perceived weakness of US students in science, technology, engineering, and mathematics (STEM). When it comes to STEM disciplines, US secondary school students tend to underperform their peers in other high-income nations. In the 2012 Program for International Student Assessment (PISA) exam, for instance, US fifteen-year-olds ranked 36th in math and 28th in science out of sixty-five participating countries.[2]

Middling test scores notwithstanding, the US economy has found ways to cope with the labor market demands of the digital age. The country makes up for any shortcomings in "growing its own" STEM talent by importing

Gordon H. Hanson is professor of economics and Pacific Economic Cooperation Chair in International Economic Relations at the School of Global Policy and Strategy at the University of California, San Diego, and a research associate of the National Bureau of Economic Research. Matthew J. Slaughter is the Paul Danos Dean of the Tuck School of Business at Dartmouth, where in addition he is the Earl C. Daum 1924 Professor of International Business. He is also a research associate of the National Bureau of Economic Research.

We thank John Bound, Charles Hulten, and Valerie Ramey for valuable comments and Chen Lui for excellent research assistance. For acknowledgments, sources of research support, and disclosure of the authors' material financial relationships, if any, please see http://www.nber.org/chapters/c13707.ack.

1. These companies (from communications equipment, computers, electronics, internet services, semiconductors, and software and programming) are Apple (US), Samsung (Korea), Hon Hai Precision (Taiwan), Hewlett-Packard (US), IBM (US), Microsoft (US), Hitachi (Japan), Amazon (US), Sony (Japan), and Google (US). See Griffith (2015).

2. See http://www.oecd.org/pisa/.

talent from abroad. Foreign-born workers account for a large fraction of hires in STEM occupations, especially among those with advanced training. Not surprisingly, the tech sector is unified in its support for expanding the number of US visas made available to high-skilled foreign job seekers.[3] Helping maintain US leadership in technology is the country's strength in tertiary education in STEM disciplines, which attracts ambitious foreign students and faculty to US universities. In global rankings of scholarship, US institutions of higher education account for nine of the top ten programs in engineering, for eight of the top ten programs in life and medical sciences, and for seven of the top ten programs in physical sciences.[4]

The United States succeeds in attracting highly trained workers from around the world even though the country's immigration system provides only modest ostensible reward for skill. Family-based immigration absorbs the lion's share of US permanent residence visas. Immediate family members of US citizens, who are eligible for green cards without restriction, accounted for 44.4 percent of admissions of legal permanent residents in 2013 (Office of Immigration Statistics 2014). Additional family members of US citizens and legal residents accounted for another 21.2 percent. Employer-sponsored visas made up only 16.3 percent of the total. These outcomes are consistent with long-standing priorities of US immigration policy. The Immigration Act of 1990, which moderately reformed the landmark Immigration and Nationality Act of 1965, allocated 480,000 visas to family-sponsored categories but just 140,000 visas to employer-sponsored ones.

Despite the pro-family-reunification orientation of US immigration legislation, high-skilled workers find their way into the country and into STEM jobs. The US immigration standards turn out to be more flexible in practice than they appear on paper. A foreign student who succeeds in gaining admission to a US university is likely to garner a student visa. Studying in the United States creates opportunities to make contacts with US employers (Bound, Demirci, et al. 2015) and to meet and to marry a US resident (Jasso et al. 2000), either of which outcome opens a path to obtaining a green card. Although the hurdles involved in securing legal permanent residence can take many years to clear, a foreign citizen with sufficient training and a US job offer is eligible for an H-1B visa, which has come to function as a de facto queue for a green card, at least among those with sought-after skills. These visas, which go primarily to highly educated workers in the tech sector, last for three years and are renewable once. The United States awards 65,000 H-1B visas annually on a first-come, first-served basis, and another 20,000 visas to individuals with a master's or higher degree from a US institution.[5] Other temporary work visas are available to employees of foreign subsid-

---

3. Jordan (2015).
4. See world university rankings by field at http://cwur.org/.
5. Employees of US universities and nonprofit or public research entities are excluded from the H-1B visa cap.

iaries of US multinational companies and to companies headquartered in countries with which the United States has a free trade agreement.

In this chapter, we document the importance of high-skilled immigration for US employment in STEM fields. To begin, we review patterns of US employment in STEM occupations among workers with at least a college degree. These patterns mirror the cycle of boom and bust in the US technology industry (Bound, Braga, et al. 2015). Among young workers with a college education, the share of hours worked in STEM jobs peaked around the year 2000, at the height of the dot-com bubble. The STEM employment shares are just now approaching these previous highs. Next, we consider the importance of immigrant labor to STEM employment. Immigrants account for a disproportionate share of jobs in STEM occupations, in particular among younger workers and among workers with a master's degree or PhD. Foreign-born presence is most pronounced in computer-related occupations, such as software programming. The majority of foreign-born workers in STEM jobs arrived in the United States at age twenty-one or older. Although we do not know the visa history of these individuals, their age at arrival is consistent with the H-1B visa being an important mode of entry for highly trained STEM workers into the US labor market. Finally, we examine wage differences between native- and foreign-born workers. Opposition to high-skilled immigration, and to H-1B visas in particular, is based in part on the notion that foreign-born workers accept lower wages than the native born, thereby depressing earnings in STEM occupations.[6] Whereas foreign-born workers earn substantially less than native-born workers in non-STEM jobs, the native-to-foreign-born earnings difference in STEM is much smaller. Foreign-born workers in STEM fields reach earnings parity with native workers much more quickly than they do in non-STEM fields. In non-STEM jobs, foreign-born workers require twenty years or more in the United States to reach earnings parity with natives; in STEM fields, they achieve parity in less than a decade.

High-skilled immigration has important consequences for US economic development. In modern growth theory, the share of workers specialized in research and development (R&D) plays a role in setting the pace of long-run growth (Jones 2002). Because high-skilled immigrants are drawn to STEM fields, they are likely to be inputs into US innovation. Recent work finds evidence consistent with high-skilled immigration having contributed to advances in US innovation. The US states and localities that attract more high-skilled foreign labor see faster rates of growth in labor productivity (Hunt and Gauthier-Loiselle 2010; Peri 2012). Kerr and Lincoln (2010) find that individuals with ethnic Chinese and Indian names, a large fraction of

---

6. See, for example, the justification provided by Senator Chuck Grassley (R-Iowa) for reforming the H-1B visa program (http://www.grassley.senate.gov/issues-legislation/issues/immigration).

which appear to be foreign born, account for rising shares of US patents in computers, electronics, medical devices, and pharmaceuticals. The US metropolitan areas that historically employed more H-1B workers enjoyed larger bumps in patenting when Congress temporarily expanded the program between 1999 and 2003. Further, the patent bump was concentrated among Chinese and Indian inventors, consistent with the added H-1B visas having expanded the US innovation frontier. Yet, the precise magnitude of the foreign-born contribution to US innovation and productivity growth is hard to pin down. Because the allocation of labor across regional markets responds to myriad economic shocks, establishing a causal relationship between inflows of foreign workers and the local pace of innovation is a challenge. High-skilled immigration may displace some US workers in STEM jobs (Borjas and Doran 2012), possibly attenuating the net impact on US innovation capabilities. How much of aggregate US productivity growth can be attributed to high-skilled labor inflows remains unknown.

When it comes to innovation, there appears to be nothing "special" about foreign-born workers, other than their proclivity for studying STEM disciplines in university. The National Survey of College Graduates shows that foreign-born individuals are far more likely than the native born to obtain a patent, and more likely still to obtain a patent that is commercialized (Hunt 2011). It is also the case that foreign-born students are substantially more likely to major in engineering, math, and the physical sciences, all fields strongly associated with later patenting. Once one controls for the major field of study, the foreign-to-native-born differential in patenting disappears. Consistent with Hunt's (2011) findings, the descriptive results we present suggest that highly educated immigrant workers in the United States have a strong revealed comparative advantage in STEM. The literature has yet to explain the origin of these specialization patterns. It could be that the immigrants the United States attracts are better suited for careers in innovation—due to the relative quality of foreign secondary education in STEM, selection mechanisms implicit in US immigration policy, or the relative magnitude of the US earnings premium for successful inventors—and therefore choose to study the subjects that prepare them for later innovative activity. Alternatively, cultural or language barriers may complicate the path of the foreign born to obtaining good US jobs in non-STEM fields, such as advertising, insurance, or law, pushing them into STEM careers.

In the political debate surrounding H-1B visas, the foreign born are criticized for putting US workers out of jobs due to their willingness to work for low wages (Hira 2010). Critics of the H-1B program portray it as allowing Indian firms in business services, such as Wipro and Infosys, to set up low-wage programming shops in the United States (Matloff 2013). Our results do not support such characterizations. After controlling for observable characteristics, there is little discernible difference in the average earnings of native- and foreign-born workers in STEM occupations. Moreover, the pattern of assimilation among foreign-born STEM workers suggests that immigrants end up in higher-wage and not lower-wage positions. Unknown is how the selection of workers into occupations—or the selective return migration of the foreign born—affect these observed native-immigrant wage differences. If native-born workers with high earnings potential move out of STEM jobs more rapidly over time (into, say, management positions) or if, within STEM occupations, lower-wage immigrants are more likely to return to their home countries, our results may overstate the relative wage trajectory of immigrant workers in STEM jobs.

Section 12.2 presents data used in the analysis, section 12.3 documents the role of STEM in overall US employment, section 12.4 describes the presence of foreign-born workers in STEM occupations, section 12.5 examines earnings differences between native- and foreign-born workers, and section 12.6 concludes.

12.2   Data

The data for the analysis come from the Integrated Public Use Microdata Series (IPUMS) 5 percent samples of the 1980, 1990, and 2000 US population censuses and 1 percent combined samples of the 2010–2012 American Community Surveys (ACS). We also use data from the IPUMS sample of the March Current Population Survey. We define total employment to be total hours worked for individuals in the civilian population not living in group quarters. Because we focus on individuals with a college or advanced degree and who are oriented toward STEM occupations, in much of the analysis we limit the sample to those twenty-five to fifty-four years of age. Excluding those younger than twenty-five drops individuals still in school or still making their schooling decisions. In early sample years, dropping those older than fifty-four excludes the generation of workers who would have made schooling decisions well before the computer revolution. In the census and ACS, hours worked is calculated as weeks worked last year times usual hours worked per week, weighted by sampling weights. Earnings are calculated, alternatively, as average annual earnings, average weekly earnings, or average earnings per usual hours worked.

Our definition of STEM occupations follows that of the Department of Commerce (Langdon et al. 2011), except that we drop the relatively low-skill categories of technicians, computer support staff, and drafters. These excluded categories have a relatively high fraction of workers who have completed no more than a high school degree. The resulting occupations classified as STEM are

- computer-related fields (computer scientists, computer software developers, computer systems analysts, programmers of numerically controlled machine tools);
- engineers (aerospace, chemical, civil, electrical, geological and petroleum, industrial, materials and metallurgical, mechanical);

- life and medical scientists (agricultural and food scientists, biological scientists, conservation and forestry scientists, medical scientists);
- physical scientists (astronomers and physicists, atmospheric and space scientists, chemists, geologists, mathematicians, statisticians); and
- other STEM occupations (surveyors, cartographers, and mapping scientists).

Occupational definitions used by the US Bureau of the Census have expanded over time as a consequence of technological progress (Lin 2011). In order to compare employment patterns from the 1980s to the present, we are obligated to use the 1990 IPUMS occupation categories. This categorization does not include fields that became common only in the later phases of the digital revolution (e.g., information security analysts, web developers, computer network architects). However, these new categories fall almost entirely within the old categories of software developers, computer scientists, and computer systems analysts. Because we work with STEM occupations either as an aggregate or for the broad category of computer-related fields, the proliferation of occupations within information technology does not pose a problem.

12.3   Employment in STEM Occupations

12.3.1   Rising Employment in STEM Fields

To set the stage for discussing the role of foreign-born workers in US employment in science and technology, it is helpful to consider first how national employment in these lines of work has evolved over time. Figure 12.1 uses the March CPS to show the fraction of total work hours in STEM occupations for twenty-five to fifty-four-year-olds across all education categories. This share rises steadily during the 1990s, plateaus after the 2001 dot-com bust, and then rises again in the middle and late first decade of the twenty-first century. When looking at workers in all education categories, STEM jobs still account for a small fraction of total employment, breaking 6 percent only briefly during the sample period.

To put the employment shares in figure 12.1 in context, in table 12.1 we show the total number of full-time equivalent workers in STEM occupations over 2000–2012 and the fractions of these workers with a BA degree and with a BA degree in a STEM discipline. Full-time equivalent workers are calculated as the sum (weighted by survey weights) of usual hours worked per week times weeks worked last year divided by 2000. The STEM workers are, not surprisingly, a relatively highly educated group. Whereas only 34.5 percent of twenty-five- to fifty-four-year-old full-time workers in non-STEM occupations have a BA degree, college education predominates in STEM jobs, ranging from 58.9 percent among network administrators to

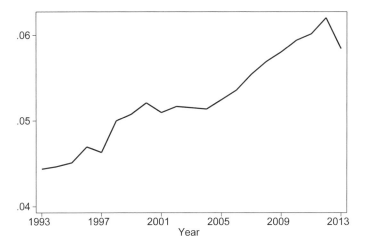

Fig. 12.1   Share of total hours worked in STEM occupations
Source: CPS, 1994–2014.

81.6 percent among engineers and to 91.9 percent among life and physical scientists. In STEM occupations, the majority of those with a BA degree have earned that degree in a STEM field (as seen by taking the ratio of column [3] to column [2] in table 12.1). Consistent with much previous evidence, STEM jobs tend to pay substantially more than non-STEM jobs. Considering just those workers with at least a bachelor's degree, average annual earnings in 2010–2012 for full-time college-educated workers in non-STEM occupations was $78,635, compared with $92,095 for software programmers and $94,297 for engineers. Only earnings for life and physical scientists lag those in non-STEM positions.

Given that STEM jobs tend to require a college education, the upward trend in STEM employment in figure 12.1 may be in part a byproduct of the rising educational attainment of the US labor force. We next examine how employment patterns have changed among workers with at least a BA degree. Figure 12.2 uses the March CPS to show the fraction of total work hours by twenty-five- to fifty-four-year-olds accounted for by STEM occupations in each of three education categories: workers whose highest attainment is a bachelor's degree, workers whose highest attainment is a master's or professional degree, and workers with a PhD. Once we condition on having a college education, employment in the broad science and technology sector has been relatively flat since the late 1990s, ranging from 10–12 percent for college graduates, 9–12 percent for master's and professional degrees, and 14–22 percent for PhDs. (Employment shares among PhDs appear more variable in figure 12.2 due in part to relatively small sample sizes for this subcategory.)

In select lines of work, STEM employment has exploded. Creating soft-

Table 12.1   Characteristics of STEM workers, 2010–2012

| | No. of workers (millions of FTEs) | Share of workers with | | Average income (2012 USD) | | | |
| | | | | All workers | | Workers with BA | |
		BA degree	BA degree in STEM	Annual	Hourly	Annual	Hourly
Non-STEM occupations	88.251	0.345	0.064	53,073	23.9	78,635	34.1
Engineers	1.400	0.816	0.728	89,823	39.7	94,297	41.7
Software programmers	1.325	0.805	0.593	88,300	40.3	92,095	42.1
Network administrators	0.458	0.589	0.333	77,722	35.0	83,966	37.8
Computer scientists	0.392	0.742	0.355	83,378	37.7	88,325	39.9
Physical, life scientists	0.660	0.919	0.649	76,325	34.6	77,528	35.1

Source: Data from ACS 2010–2012.

Notes: Data include all workers twenty-five to fifty-four years old. Values are weighted by annual hours worked/2000 (full-time equivalent units).

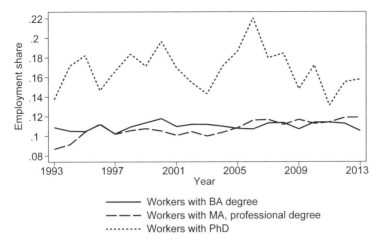

Fig. 12.2   Employment of college-educated males in STEM occupations, share of employment in STEM jobs
Source: CPS, 1994–2014.

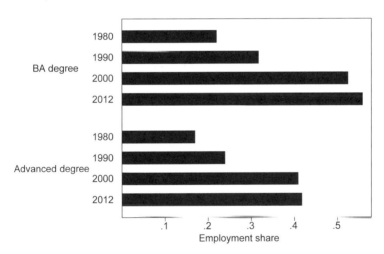

Fig. 12.3   Employment of college-educated males in STEM occupations, share of STEM workers in software, programming
Source: IPUMS census, ACS.

ware, programming computer systems, and managing computer networks were minor occupations in 1980. Today, they are ubiquitous. Computer science is among the most popular majors on many college campuses. The lives of programmers appear in popular culture, inspiring major motion pictures (*The Social Network*, *Steve Jobs: The Man in the Machine*), TV series (*Silicon Valley*), and even contemporary music ("Big Data"). Figure 12.3 shows the share of hours worked in STEM occupations by computer systems analysts and computer scientists, developers of computer software, and programmers of numerically controlled machine tools, where the first two subgroups account for the vast majority of employment in this category. Among bachelor's degree holders, the share of employment in computer-related jobs rises sharply from 22.0 percent in 1980 to 31.7 percent in 1990 before jumping steeply again to 52.5 percent in 2000 and then stabilizing at 55.8 percent for 2010–2012. The STEM employment shares in computer occupations among advanced degree holders (master's degree, professional degree, PhD) show a similar temporal pattern of evolution but are about 10 percentage points lower.

12.3.2   Revealed Comparative Advantage in STEM Occupations

Who gets STEM jobs? Because the rise of information technology is a recent phenomenon, younger workers are those most likely to have chosen a path of study that gives them entry into the STEM labor force. In part because men are more likely to study STEM disciplines in university—especially in computer science and engineering—they are in turn more likely to be employed in STEM occupations once they enter the labor force. To examine occupational sorting by age and gender, we calculate employment shares for five-year age cohorts, separately for men and women. For college graduates, we consider twenty-five- to twenty-nine-year-olds to be the "entry" cohort—that is, the age at which individuals first have stable, full-time work—which allows for the possibility that it may take individuals several years after obtaining their BA to find their professional bearings. Similarly, for those with an advanced degree we discuss results nominally treating thirty- to thirty-four-year-olds as the "entry" cohort.

Figure 12.4 shows the share of hours worked in STEM occupations for males—both native and foreign born—with at least a college education. Consider first panel A, which shows males with a bachelor's degree. Between 1980 and 1990, the share of twenty-five- to twenty-nine-year-olds in STEM jobs climbs from 11.1 percent to 17.5 percent. During the 1980s, which saw the introduction of the Apple Macintosh personal computer, the Microsoft MS-DOS operating system, and the Intel 80386 microprocessor, STEM jobs drew in relatively large numbers of young workers. The STEM employment share for twenty-five- to twenty-nine-year-olds rises again to 19.0 percent in 2000 as the dot-com wave crests, and then declines somewhat to 17.1 percent for the 2010–2012 period, following the Great Recession and the ensuing slow recovery. The shift toward employment in STEM is much lower among individuals who were in their thirties in the 1980s and nonexistent among those forty years old and older in the 1980s.

Turning to hours worked for those with an advanced degree, shown in panel B of figure 12.4, the lure of STEM employment in the 1980s and 1990s is even more pronounced. Among thirty- to thirty-four-year-olds, the share working in STEM rises from 11.6 percent in 1980 to 15.1 percent in 1990

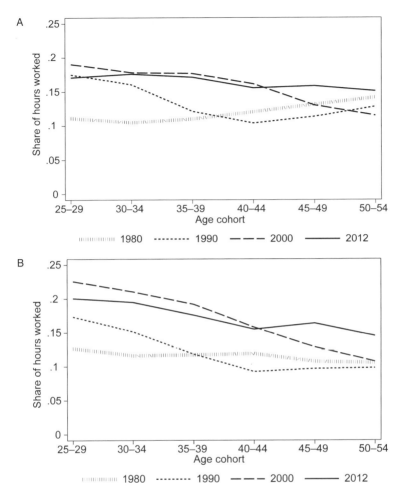

Fig. 12.4   Employment of college-educated males in STEM occupations. *A*, males
with BA degree; *B*, males with advanced degree.
Source: IPUMS census, ACS.

and to 21.0 percent in 2000 before falling to 19.5 percent in 2010–2012. The
higher incidence of STEM employment among the most educated workers
may reflect the need for advanced training in order to perform the job tasks
demanded in science and technology. Alternatively, the disproportionate
share of STEM workers with graduate degrees may reflect an arms race, in
which workers compete via education to improve their chances of obtaining
the high-paying jobs available in information technology industries. Antici-
pating the patterns that we shall see in section 12.4, the arms-race motivation
may be particularly strong among immigrant workers. Those born abroad
may lack access to informal networks through which native-born workers

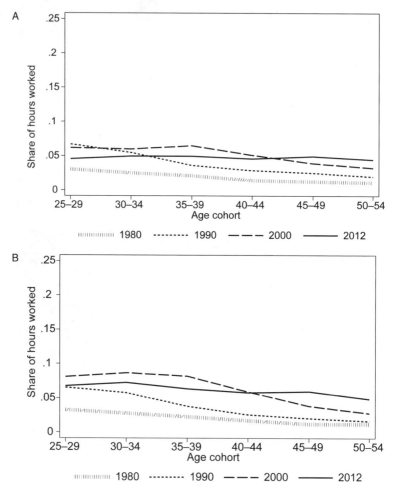

Fig. 12.5   Employment of college-educated females in STEM occupations. *A*, females with BA degree; *B*, females with advanced degree.
Source: IPUMS census, ACS.

obtain information about employment opportunities. Earning an advanced degree provides foreign-born workers with a mechanism for signaling their capabilities, perhaps helping compensate for any lack of informal signaling options.

Silicon Valley is frequently cited in the business press for the lack of professional opportunities that it offers women. The reputation of the tech sector as being male dominated appears to be well founded. Figure 12.5 shows STEM employment shares for females with a bachelor's degree (panel A) and an advanced degree (panel B). Among workers with no more than a bachelor's degree, the share of female employment in STEM occupations is markedly lower than that for males. Among twenty-five- to twenty-nine-year-old women, STEM occupations accounted for only 4.6 percent of employment in 2010–2012 (compared to 17.1 percent for men), a figure that was lower than both 2000 at 6.2 percent (19.0 percent in that year for men) and 1990 at 6.7 percent (17.5 percent in that year for men). For women with an advanced degree (panel B of figure 12.5), specialization in STEM is modestly higher. Among thirty- to thirty-four-year-olds, the share of females in STEM jobs is 7.2 percent in 2010–2012 (19.5 percent in that year for men), down from 8.7 percent in 2000 (21.0 percent in that year for men) and up from 5.8 percent in 1990 (15.1 percent in that year for men). As with men, STEM employment shares are higher among all age cohorts for women with an advanced degree compared to women with no more than a bachelor's degree.

Putting figure 12.5 together with figure 12.4 reveals that the underrepresentation of women in STEM has not improved over time. To see this, we measure occupational specialization using the revealed comparative advantage of males in STEM, given by

[share of male employment in STEM jobs/share of male employment in non-STEM jobs]/[share of female employment in STEM jobs/share of female employment in non-STEM jobs].

Among twenty-five- to twenty-nine-year-olds with a bachelor's degree, revealed comparative advantage for men in STEM rises from 3.0 $(.175/(1 - .175))/(.067/(1 - .067))$ in 1990 to 4.4 $(.171/(1 - .171))/(.045/(1 - .045))$ in 2010–2012. Stated differently, the log odds of a college-educated male being employed in STEM relative to a college-educated female being employed in STEM rises from 1.10 in 1990 to 1.48 in 2010–2012. Among thirty- to thirty-four-year-olds with an advanced degree, revealed comparative advantage for men in STEM rises less sharply from 2.9 $(.152/(1 - .152))/(.058/(1 - .058))$ to 3.1 $(.195/(1 - .195))/(.072/(1 - .072))$, for an increase in the log odds of 1.07 to 1.13. Among the foreign born, more educated women are also underrepresented in STEM jobs when compared to immigrant men. When we turn next to comparing employment patterns for native- and foreign-born workers, will we examine employment for men and women summed together.

12.4   Foreign-Born Workers in STEM Occupations

12.4.1   Immigrant Workers in the US Economy

To provide context for the analysis of specialization patterns by native- and foreign-born workers in STEM occupations, we first examine the share of the foreign born across all occupations. Panel A of figure 12.6 shows the fraction of hours worked accounted for by the foreign born among twenty-five- to fifty-four-year-old workers (males and females combined) with a

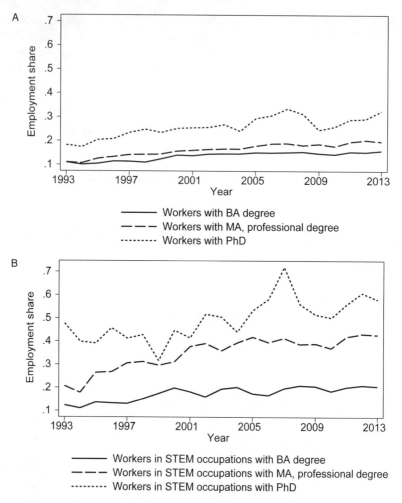

Fig. 12.6   Share of foreign-born workers in employment. *A*, foreign-born share of US employment; *B*, foreign-born share of employment, STEM jobs.
Source: IPUMS census, ACS; CPS, 1994–2014.

bachelor's degree, master's or professional degree, and a PhD. As the literature has documented, the immigrant share of US employment for the more educated is rising steadily over time. Among workers whose highest attainment is a bachelor's degree, the foreign-born employment share reaches 15.2 percent in 2013, up from 10.1 percent in 1993. As is also well known, for workers with at least a college degree immigrant employment shares rise monotonically by education level. In 2013, the foreign born account for 18.1 percent of hours worked among master's and professional degrees and 28.9 percent among PhDs. For comparison, in 2013 the share of the foreign born in the total US civilian labor force was 16.5 percent, up from 9.2 percent in 1990. Immigrants are, then, mildly underrepresented among college graduates, slightly overrepresented among those with master's degrees, and strongly overrepresented among PhDs.

Relative to employment across all occupations, the presence of the foreign born in STEM employment is higher for all education groups, as seen in panel B of figure 12.6, which shows foreign-born employment shares for the same categories as panel A, but now for jobs in STEM. In 2013, the foreign-born share of STEM employment is 19.2 percent among bachelor degrees, higher at 40.7 percent among master's degrees, and higher still at 54.5 percent among PhDs. Since the middle of the first decade of the twenty-first century, immigrants have accounted for the majority of US workers in STEM with doctoral degrees. The majority of advanced degree holders who are foreign born obtained their degrees in the United States (Bound, Turner, and Walsh 2009). Thus, there is a sense in which the United States *is* growing its own STEM talent. Universities in the United States have become a pipeline for advanced degree recipients born abroad to enter the US labor force. These institutions attract foreign students and train them in STEM disciplines before sending them to work for US employers. The large majority of those completing their PhDs in the United States, in particular those from lower- and middle-income countries, intend to stay in the United States after graduation (Grogger and Hanson 2015).

Also apparent in figure 12.6 are differences in the cyclicality of foreign-born employment in STEM by education level. Whereas among college graduates the foreign-born share peaks in 2000 and has been stable since, among master's degree holders the foreign-born share rises by over 10 percentage points in the first decade of the twenty-first century, and among PhDs the foreign-born share rises by a full 25 percentage points between 2001 and 2007, before dipping during the Great Recession.

12.4.2 Revealed Comparative Advantage of Foreign-Born Workers

We have already seen that among the college educated, young workers are relatively likely to select into STEM employment. Since a disproportionate share of the foreign born are workers in their twenties and thirties, it is conceivable that the rising presence of immigrants in US STEM careers is simply a byproduct of differing demographic patterns among natives and immigrants. Evidence on this possibility is seen in panel A of figure 12.7, which shows the share of workers in STEM occupations that are foreign born by five-year age cohorts for those with bachelor's degrees. The foreign-born share among twenty-five- to twenty-nine-year-olds in STEM jobs rises from 5.8 percent in 1980 to 9.1 percent in 1990 and then peaks at 21.1 percent in 2000 before declining to 17.0 percent in 2010–2012. The corresponding shares of non-STEM jobs going to immigrants (for twenty-five- to twenty-nine-year-olds with a bachelor's degree), as shown in panel B of figure 12.7,

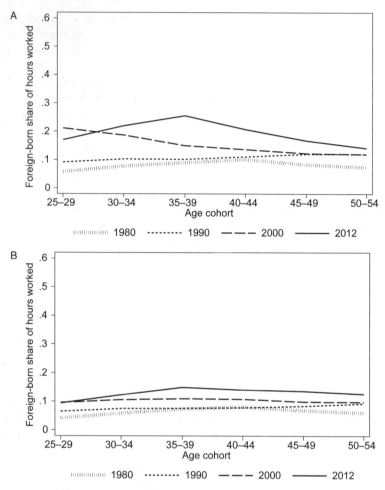

Fig. 12.7   Share of workers who are foreign born bachelor's degree holders.
A, STEM occupations; B, non-STEM occupations.
Source: IPUMS census, ACS.

are 4.2 percent in 1980, 6.5 percent in 1990, 9.5 percent in 2000, and 9.2 percent in 2010–2012. Even controlling for age, the foreign born are strongly overrepresented in STEM employment.

The already substantial presence of immigrants in STEM jobs for a birth cohort at "labor market entry" becomes even larger as the cohort ages. Consider the cohort born between 1971 and 1975, which is the heart of Generation X. Panel A of figure 12.7 shows that by 2010–2012, the share of immigrants among Gen X thirty-five- to thirty-nine-year olds with BA degrees employed in STEM reaches 25.6 percent, up 4.5 percentage points over the level for twenty-five- to twenty-nine-year olds in 2000. This increase

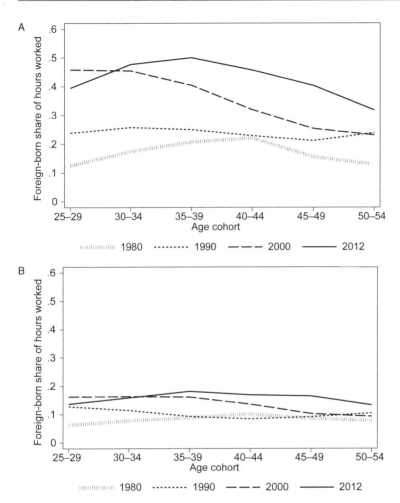

Fig. 12.8   Share of workers who are foreign born advanced degree holders.
A, STEM occupations; *B*, non-STEM occupations.
Source: IPUMS census, ACS.

is accounted for by a combination of immigrants in this birth cohort who arrived during the first decade of the twenty-first century being disproportionately selected into STEM jobs and immigrants in this birth cohort already in the country as of 2000 being relatively unlikely to exit STEM employment. Similar patterns of rising shares of STEM employment going to immigrant workers exist for other birth cohorts, as well.

The relatively strong specialization of immigrant workers in STEM occupations is even more pronounced among for those with advanced degrees, as seen in figure 12.8. For the period 2010–2012, the share of STEM jobs going to the foreign born relative to the share of non-STEM jobs going to the foreign born is 39.4 percent versus 13.6 percent among twenty-five- to twenty-nine-year olds, 47.7 percent versus 15.9 percent among thirty- to thirty-four-year-olds, and 50.0 percent versus 18.2 percent among thirty-five- to thirty-nine-year-olds. Thus, among prime-age workers with an advanced degree, the foreign born now account for one-half of total hours worked in STEM occupations. This fraction is up from one-quarter in the 1990s and from one-fifth in the 1980s. Many of the highly educated workers employed in engineering, science, and technology are at the forefront of US innovation. Foreign-born professionals would seem to have become a vital part of the US R&D labor force. These workers enter STEM employment in their youth and remain in technical occupations after decades of potential labor market experience.

Putting together panels A and B of figure 12.7, and similarly for figure 12.8, the employment of foreign-born workers is consistent with their having a strong revealed comparative advantage in STEM occupations. Among twenty-five- to twenty-nine-year-olds with a bachelor's degree, revealed comparative advantage of foreign-born workers in STEM, which is defined as

[share of foreign-born employment in STEM/share of foreign-born employment in non-STEM]/[share of native-born employment in STEM/ share of native-born employment in non-STEM]

rises from 1.4 (.058/(1 − .058))/(.042/(1 − .042)) in 1980 to 2.0 (.17/(1 − .17))/(.094/(1 − .094)) in 2010–2012. The log odds of a young foreign-born college graduate being employed in STEM relative to a young native-born college graduate being employed in STEM increases from 0.34 to 0.69 over this period. Similar increases are evident among older college-educated workers. The revealed comparative advantage of the foreign born in STEM appears to be even stronger among individuals with advanced degrees. Among thirty- to thirty-four-year-olds with a master's degree, professional degree or PhD, the revealed comparative advantage of the foreign born rises from 2.5 (.174/(1 − .174))/(.077/(1 − .077)) in 1980 to 4.8 (.477/(1 − .477))/ (.159/(1 − .159)) in 2010–2012, for a substantial increase in the log odds of STEM employment for the foreign born relative to the native born of 0.9 to 1.6. Among holders of an advanced degree, the revealed comparative advantage of foreign- over native-born workers in STEM is much larger than that even of male over females workers.

Software development is among the most rapidly growing areas for STEM jobs and among the most hotly contested occupations regarding the allocation of H-1B visas. The revealed comparative advantage of the foreign born in computer-related occupations is manifestly stronger than their comparative advantage in STEM positions overall, as seen in figure 12.9. In this subcategory, 23.0 percent of hours worked among twenty-five- to twenty-nine-year-olds with bachelor's degrees were foreign born in 2010–2012, up from 10.5 percent in 1990 and 60.0 percent of hours worked among thirty- to thirty-four-year-olds with advanced degrees were by the foreign born in

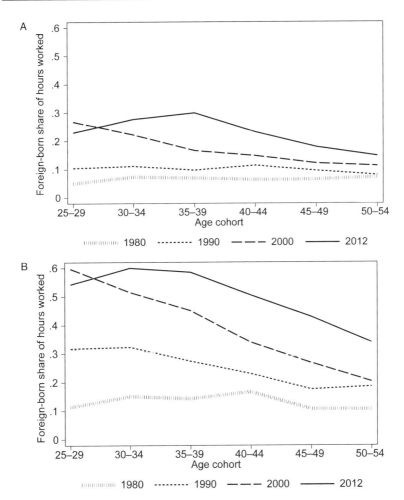

Fig. 12.9   Share of foreign born in computer occupations. *A*, workers with BA degree; *B*, workers with advanced degree.
Source: IPUMS census, ACS.

2010–2012, up from 32.3 percent in 1990. Given that occupational sorting tends to be stable over time for individual birth cohorts, the foreign born would appear to be set to account for a high fraction of US workers who are employed in computer-related jobs for many years to come (unless, for some reason, foreign-born workers currently on H-1B visas fail to gain legal permanent residence at the rates they have in the past).

12.4.3   Age of US Entry by Foreign-Born Workers in STEM Jobs

How do foreign-born STEM workers enter the United States? Although the ACS does not report the types of visas through which an individual first gained entry to the United States or first secured a US job, it does report

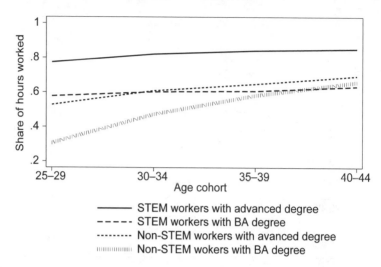

Fig. 12.10   Share of foreign-born workers arriving in the United States at age twenty-one or older, 2012
Source: IPUMS census, ACS.

the age at which an individual first arrived in the United States. The STEM occupations that employ foreign-born workers primarily hire those who arrived in the United States at age twenty-one or older. In figure 12.10, we see that among bachelor's degree holders, those arriving in the United States at age twenty-one or older account for 60.5 percent of immigrant workers with STEM jobs (across all age cohorts in that year), compared to 51.9 percent of immigrant workers in non-STEM jobs. This pattern is even stronger among advanced degree holders. Those arriving in the United States at age twenty-one or older are 82.7 percent of foreign-born workers in STEM with a master's degree, professional degree, or PhD compared to 63.6 percent of similarly educated immigrants in non-STEM jobs. Although we cannot determine the type of visa through which these individuals entered the United States, the pattern of post-age twenty-one entry is consistent with work visas, including the H-1B, being an important admissions channel for STEM-oriented immigrants.

12.4.4   Explanations for Foreign-Born Comparative Advantage in STEM

The preceding results, while consistent with immigrant workers having a comparative advantage in STEM, are silent on the factors behind this outcome. One explanation is that K–12 education in other countries offers stronger training in math and science than is available in the United States. The inferior performance of US fifteen-year-olds in PISA exams is consistent with this possibility. Yet, US students also perform relatively poorly in reading, ranking 24th in this dimension in the 2012 test. Although the ranking for reading is superior to US scores in science (28th) and math (36th), it would not seem to indicate an overriding comparative disadvantage among US high school students in technical fields. Relative to most other high-income countries, US fifteen-year-olds may have an absolute disadvantage in all disciplines and a mild comparative disadvantage in math and science. However, it could be unwise to read too much into the consequences of relatively poor US exam scores, as little is known about the cross-country variation in how individual performance on standardized tests translates into professional success.

A second explanation for immigrant success in STEM is that these jobs are the only positions available to more educated immigrants and that advanced degrees are how one demonstrates competence in technical disciplines. Non-STEM professions in which more educated workers predominate include arts, the media, finance, management, insurance, marketing, medicine, law, and other business services (architecture, consulting, real estate). Some of these fields, such as insurance and marketing, are ones in which the foreign born or nonnative English speakers may have an absolute disadvantage because they lack a nuanced understanding of American culture or because subtleties in face-to-face communication are an important feature of interactions in the marketplace. Others of these fields, such as the law or real estate, may involve an occupational accreditation process that imposes relatively high entry costs on those born abroad.

A third explanation is that US immigration policy has implicit screens that favor more educated immigrants in STEM fields over those in non-STEM fields. The H-1B visas do go in disproportionate numbers to workers in STEM occupations (Kerr and Lincoln 2010). However, there is nothing preordained about this outcome in terms of US immigration policy. The H-1B visas are designated for "specialty occupations," which are defined as those in which (a) a bachelor's or higher degree or its equivalent is normally the minimum entry requirement for the position, (b) the degree requirement is common to the industry in parallel positions among similar organizations, (c) the employer normally requires a degree or its equivalent for the position, or (d) the nature of the specific duties is so specialized and complex that the knowledge required to perform the duties is usually associated with attainment of a bachelor's or higher degree.[7] The H-1B visas are thus available to the more educated in non-STEM lines of work, too. That most H-1B visas are captured by STEM workers may simply be the consequences of strong relative labor demand for STEM labor by US companies.

Are immigrant workers displacing native-born workers in STEM jobs? Rising immigration of more educated workers has not led to an overall

---

7. See http://www.uscis.gov/eir/visa-guide/h-1b-specialty-occupation/understanding-h-1b
-requirements.

expansion in the share of total US employment in STEM occupations. The expansion of labor supply for workers with expertise in technical fields may shift the mix of output toward industries intensive in the use of these skills. Under directed technical change, expanded incentives for innovation emanating from the labor supply shock could provide a further boost to US output in high-tech sectors (Acemoglu 2002). Yet, expanded immigration of highly educated individuals has occurred along with an unchanged share of aggregate employment in STEM occupations, consistent with foreign-born workers having displaced native-born ones in the competition for positions in STEM fields. Of course, many other events occurred in the US labor market in the first decade of the twenty-first century, most notably the bursting of the dot-com bubble and the Great Recession. The magnitude of these shocks makes it difficult to know how employment of US native-born workers in STEM occupations would have fared absent high-skilled immigration.

Evidence on native displacement effects from immigration is mixed. Lewis (2011) and Gandal, Hanson, and Slaughter (2004) find no evidence that immigration inflows shifts the output mix in regional or national economies toward industries intensive in the use of immigrant labor. Borjas and Doran (2012) find that the arrival of Russian mathematicians in the United States induced the exit of incumbent scholars in the subfields of the discipline in which Russia had historically been dominant. Kerr, Kerr, and Lincoln (2015) do not detect evidence of displacement effects of skilled immigrants on native workers, at least inside firms. Within US manufacturing establishments, the arrival of young, high-skilled foreign-born workers is associated with increases and not decreases in the employment of young, high-skilled native-born workers.

12.5   Wage Differences between Native- and Foreign-Born Workers

It is well known that across all occupations, immigrants earn less than natives, even once one controls for age, education, gender, and race. Do similar earnings differences between the native and foreign born materialize when we examine more educated workers and, in particular, those employed in STEM occupations? This issue is of central concern in the public debate about US immigration policy. Concerns have been expressed about foreign-born STEM workers being willing to accept lower earnings than US native-born workers.[8] We aim to provide fresh evidence on the subject.

To begin we compare earnings for native-born and foreign-born workers in STEM occupations. Figure 12.11 shows annual earnings for full-time, full-year male workers twenty-five to forty-four years old who have at least a bachelor's degree. We show earnings by foreign-born status, whether workers have just a bachelor's or an advanced degree, and by year. In 1990, average

8. See, for example, Porter (2013).

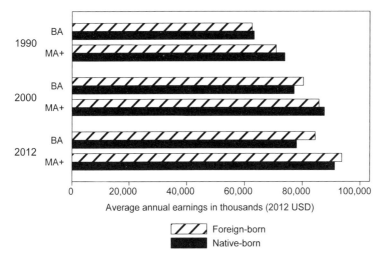

Fig. 12.11   Earnings comparisons, males age twenty-five to forty-four, annual earnings, male full-time STEM workers
Source: Census 1990, 2000; ACS 2010–2012.

annual earnings for natives exceed those for immigrants; in 2000 the picture is mixed, with native-born earnings exceeding those for immigrants among those with an advanced degree but not among those with just a bachelor's degree; and by 2012, the earnings of the foreign born exceed those of the native born in both degree categories. Similar patterns obtain when we examine average weekly wages or average hourly wages. Although the comparison in figure 12.11 is for workers who have selected into STEM jobs, there may be important sources of unobserved heterogeneity between workers. In particular, the foreign born may be relatively likely to work in high-paying occupations. We next perform wage comparisons, while flexibly controlling for individual characteristics.

Pooling data from the 1990 and 2000 population censuses and the 2010–2012 American Communities Surveys, we limit the sample to twenty-five- to fifty-four-year-olds who are full-time (at least thirty-five usual hours worked per week) and full-year (at least forty weeks worked last year) workers with at least a bachelor's degree. We use three measures of earnings: log annual earnings, log weekly earnings (annual earnings divided by weeks worked last year), and log hourly earnings (annual earnings divided by weeks worked last year times usual hours worked per week). All regressions are weighted by annual hours worked (multiplied by the census sampling weight) and include as controls indicators for gender, race, the census geographic region, the year, and a full set of interactions between indicators for education (bachelor's degree, master's degree, professional degree, PhD) and age (five-year age groupings). Later regressions include indicators for the industry of employment.

Table 12.2   Earnings regressions for native born and foreign born

Variable	Log hourly earnings		Log weekly earnings		Log annual earnings	
	(1)	(2)	(3)	(4)	(5)	(6)
STEM = 1	0.191	0.112	0.154	0.069	0.164	0.073
	(0.001)	(0.001)	(0.001)	(0.001)	(0.001)	(0.001)
Foreign born = 1	−0.101	−0.124	−0.120	−0.146	−0.119	−0.149
	(0.001)	(0.001)	(0.001)	(0.001)	(0.001)	(0.001)
STEM × foreign born	0.094	0.095	0.084	0.086	0.079	0.082
	(0.003)	(0.002)	(0.003)	(0.002)	(0.003)	(0.002)
Industry dummies	No	Yes	No	Yes	No	Yes
R^2	0.285	0.327	0.297	0.341	0.296	0.345

Sources: Data from 1990 and 2000 census and 2010–2012 ACS.

Notes: $N = 2,550,537$. Robust standard errors are in parentheses. Sample is full-time, full-year workers twenty-five to fifty-four years old with at least a BA degree. Additional regressors: dummy variables for gender, race, year, census region, and five-year age category interacted with educational degree (BA, MA or prof. degree, PhD). Regressions are weighted by sampling weights.

The regression shown in column (1) of table 12.2 reveals that STEM workers receive hourly earnings that are on average 19.1 log points higher than those of non-STEM workers who have similar demographic characteristics, education, and region of residence. For weekly and annual earnings, shown in columns (3) and (5), the STEM earnings premium is broadly similar at 15.4 log points and 16.4 log points, respectively. Column (2) adds controls for ten one-digit industries, which compresses the STEM hourly earnings premium to 11.2 log points; declines are similar for weekly and annual earnings, shown in columns (4) and (6). Although these findings may seem to suggest that STEM positions are "good jobs" that pay high wages, we should caution that these results are purely descriptive and say nothing about the origin of the STEM earnings differential. This differential may reflect higher-ability workers being disproportionately selected into STEM occupations, such that the coefficient on the STEM earnings dummy picks up the average difference in unobserved ability between STEM and non-STEM positions. Alternatively, the STEM earnings bump may reflect a compensating differential for the higher cost of obtaining the training needed to work in a STEM field (e.g., the extra hours of study required for a computer science or engineering degree). A yet further alternative is that employers that hire relatively large numbers of STEM workers (e.g., Apple, Google, Microsoft) earn rents and share these rents with their employees.

Across all more educated workers, the foreign born in non-STEM occupations earn less than the native born, as shown by the negative and significant coefficient on the indicator for a worker being an immigrant. For hourly earnings, the immigrant wage discount is −10.1 log points (column [1]); for weekly and annual earnings it is comparable at −12.0 log points (column [3]) and −11.9 log points (column [5]), respectively. Immigrant earnings discounts increase modestly when indicators for one-digit industries are added (columns [2], [4], and [6]). These estimated immigrant earnings differentials are also descriptive. They may represent an unobserved-ability differential between similarly educated native- and foreign-born workers or they may capture the limited portability of human capital between countries, such that a degree from, say, China is worth less in the US labor market than is US degree. Earnings differences from either of these sources would be unlikely to diminish over time. A source of *temporary* earnings differences between immigrants and natives is adjustment costs in settling into a new labor market. It may take foreign-born workers a while after arriving in the United States to find employment that is well matched to their particular skills. Assimilation into the US labor market, which we examine in more detail below, may attenuate or even reverse native-immigrant earnings differences.

The earnings discount for foreign-born workers falls considerably when we compare native- and foreign-born individuals employed in STEM occupations. This result is seen in the positive and statistically significant interaction between the STEM indicator and the foreign-born indicator. For hourly earnings in column (1), the immigrant wage discount falls to −0.7 (−10.1 + 9.4) log points; for weekly and annual earnings the immigrant discount falls to −3.6 (−12.0 + 8.4) log points (column [3]) and −4.0 (−11.9 + 7.9) log points (column [5]), respectively. Although all of these differentials are statistically significant, they are far smaller than the earnings differences observed between native and immigrant workers in non-STEM occupations.

Moreover, once we limit the sample to STEM workers—which implicitly allows the returns to education and labor market experience to vary between STEM and non-STEM categories—the immigrant-native earnings difference becomes of indeterminate sign. Unreported results for regressions similar to table 12.2 in which we restrict the sample to workers employed in STEM occupations show that the immigrant earnings differential is positive and significant for hourly earnings (at 1.7 log points without industry controls and 2.6 log points with industry controls), while negative and weakly significant for weekly earnings (−0.3 log points without industry controls and −1.4 log points with industry controls) and negative and strongly significant for annual earnings (−0.7 log points without industry controls and −1.8 log points with industry controls).

Could the immigrant earnings discount be a consequence of adjustment costs that are erased by labor market assimilation? Borjas (2014) finds suggestive evidence that the process of assimilation in immigrant wages—which was evident in earlier decades—has broken down. That is, across all education groups immigrants' earnings appear to be catching up to natives' earnings more slowly than they did in the past. We examine patterns of assimilation for more educated immigrants to see if his findings are repli-

Table 12.3   Year-by-year earnings regressions, non-STEM

	1990 (1)	2000 (2)	2010–2012 (3)
Foreign born, 0–5 years in the United States	−0.289	−0.244	−0.246
	(0.007)	(0.006)	(0.007)
Foreign born, 6–10 years in the United States	−0.222	−0.222	−0.194
	(0.006)	(0.005)	(0.006)
Foreign born, 11–15 years in the United States	−0.104	−0.172	−0.096
	(0.006)	(0.005)	(0.006)
Foreign born, 16–20 years in the United States	−0.034	−0.086	−0.050
	(0.006)	(0.005)	(0.006)
Foreign born, 20+ years in the United States	0.018	0.012	0.003
	(0.004)	(0.004)	(0.004)
R^2	0.165	0.135	0.181
N	692,417	897,896	654,200

Sources: Data from 1990, 2000 census; 2010–2012 ACS.
Notes: Robust standard errors are in parentheses. Sample is full-time, full-year workers twenty-five to fifty-four years old with at least a BA. Additional regressors: dummy variables for gender, race, census region, and five-year age category interacted with ed. degree (BA, MA or prof. degree, PhD). Regressions use sampling weights.

cated among more skilled workers. Because one cannot separately identify wage effects for the birth cohort, the year of immigration, and years since immigration (Borjas 1987), we are unable to decompose the immigrant-native earnings difference into separate effects for the birth cohort (which may reflect time variation in the quality of education), the immigration entry cohort (which may reflect time-varying conditions that shape the pattern of selection into international migration), and years since immigration (which may pick up assimilation effects). Still, it is instructive to examine how earnings for immigrant entry cohorts evolve over time. Tables 12.3 and 12.4 show earnings regressions run separately by year and that include indicators for gender, race, and education-age interactions. The regressions also include indicators for the immigration entry cohort measured as the years a foreign-born individual has resided in the United States (zero to five years, six to ten years, eleven to fifteen years, sixteen to twenty years, twenty or more years) as of a particular year (1990, 2000, 2010–2012), following the structure in Borjas (2014). Table 12.3 shows results for workers employed in non-STEM occupations; table 12.4 shows results for workers employed in STEM occupations.

Looking down column (1) in table 12.3, we see how the immigrant-native earnings difference for recently arrived immigrants (five or fewer years in the United States) compares with that for immigrants who have longer tenure in the country (six to ten years, eleven to fifteen years, sixteen to twenty years, twenty-one or more years). For non-STEM immigrant workers in 2010–

Table 12.4   **Year-by-year earnings regressions, STEM**

	1990 (1)	2000 (2)	2010–2012 (3)
Foreign born, 0–5 years in the United States	−0.173	0.007	−0.057
	(0.012)	(0.007)	(0.008)
Foreign born, 6–10 years in the United States	−0.071	0.043	0.043
	(0.008)	(0.007)	(0.007)
Foreign born, 11–15 years in the United States	0.000	0.045	0.085
	(0.007)	(0.008)	(0.006)
Foreign born, 16–20 years in the United States	0.035	0.059	0.062
	(0.008)	(0.008)	(0.008)
Foreign born, 20+ years in the United States	0.031	0.060	0.041
	(0.007)	(0.006)	(0.006)
R^2	0.184	0.118	0.181
N	85,078	129,497	91,449

Sources: Data from 1990, 2000 census; 2010–2012 ACS.
Notes: Robust standard errors are in parentheses. Sample is full-time, full-year workers twenty-five to fifty-four years old with at least a BA. Additional regressors: dummy variables for gender, race, census region, and five-year age category interacted with ed. degree (BA, MA or prof. degree, PhD). Regressions use sampling weights.

2012 (column [3]), the wage discount relative to natives is −24.6 log points among those with five or fewer years in the United States, −19.4 log points for those with six to ten years in the United States, −9.6 log points for those with eleven to fifteen years in the United States, and −5.0 log points for those with sixteen to twenty years in the United States. Only for the foreign born with twenty-one or more years in the United States does the wage discount relative to the native born disappear. This pattern could be the consequence of assimilation, as immigrants shed their earnings disadvantages relative to the native born over time. It could also be due to selective out-migration of immigrants, if say within any entry cohort those with lower earnings potential in the United States are those most likely to return to their home countries. Or it could be due to decreases over time in the average ability of later immigrant cohorts relative to earlier immigrant cohorts.

Whatever the origin of the entry cohort effect on earnings, it is far different for workers in STEM occupations, as seen in table 12.4. In 2010–2012 (column [3]), recently arrived STEM workers earn 5.7 log points less than their native-born counterparts. This differential becomes positive for those with six or more years in the country, indicating that in less than a decade immigrant STEM workers begin earning more than native-born STEM workers. Again, we cannot say whether or not this pattern reflects assimilation. It could be that lower-wage immigrant workers in STEM are those most likely to be on temporary work visas that either do not get renewed or do not get converted into green cards. Or it could be that native STEM workers are disproportionately likely to get promoted out of STEM jobs into management positions, which may convert them into non-STEM lines of work.

Comparing across columns in tables 12.3 and 12.4, we obtain a sense of how the earnings discount for a particular entry cohort fairs over time. In columns (1) and (2) of table 12.3 for non-STEM workers, we see that the −28.9 log point earnings discount earned by the cohort that entered the United States between 1985 and 1990 (and so had zero to five years in the United States in 1990, column [1]) had fallen to 17.2 log points in 2000 (by which point this entry cohort had eleven to sixteen years in the United States). The corresponding fall in the wage discount for the 1995–2000 entry cohort—from 24.4 log points in 2000 (column [2]) to 9.6 log points in 2010–2012 (column [3])—is even larger. Thus, in contrast to Borjas (2014), we do not see evidence consistent with the assimilation of more educated non-STEM immigrant workers into the US labor market becoming weaker over time. Indeed, if anything, assimilation of more educated non-STEM immigrant workers appears to be accelerating. There is no evidence of a similar acceleration of assimilation for immigrant workers in STEM occupations.

Overall, we observe that the average immigrant earnings discount relative to native-born workers is far smaller in STEM occupations than in non-STEM occupations, that immigrant workers in STEM with six or more years in the United States have earnings parity with natives, and that the process of earnings assimilation for immigration entry cohorts is uneven across time.

12.6 Discussion

The United States has built its strength in high technology in part through its businesses having access to exceptional talent in science and engineering. Although US universities continue to dominate STEM disciplines globally, it is individuals born abroad who increasingly make up the US STEM labor force, particularly among those with advanced degrees. In software development and programming, and other computer-related occupations, the foreign born make up the majority of US workers in STEM jobs with a master's degree or higher. The success of Amazon, Facebook, Google, Microsoft, and other technology standouts thus seems to depend, at least partially, on the ability of the US economy to import talent from abroad. In the press, it is entry-level programmers from abroad admitted under H-1B visas won by foreign outsourcing shops who draw much of the attention. In the data, what catches the eye is the strong and rising presence of foreign-born master's and doctorate degree holders in STEM fields, whose training, occupational status, and earnings put them in the highest rungs of the US skill and wage distributions.

It is little wonder why high-skilled workers from lower-wage countries desire to move to the United States to make their careers. Earnings for technology workers from India rise by a factor of six when individuals succeed in obtaining a US work visa (Clemens 2010). Grogger and Hanson (2011) show that the absolute reward for skill in the US labor market is substantially higher than in other high-income countries (either in pretax or post-tax terms). Although foreign-born workers earn less than their native-born counterparts with similar demographic characteristics and educational attainment, the wage discount for immigrants in STEM jobs is substantially smaller than in non-STEM jobs. Immigrants in STEM occupations with ten or more years of experience in the United States earn equal to or more than native-born workers doing similar tasks. The data thus provide little support for the claim made by critics of US immigration policy that foreign-born workers in STEM jobs accept persistently lower wages than their native-born counterparts.

Our understanding of immigration and its impacts on the US economy is limited by the scarcity of data at the individual level regarding how workers gain entry into the US labor market. We are largely unable to distinguish among workers who arrive on family-based visas, employer-sponsored visas, student visas, or H-1B visas or how these individuals may transition from temporary visa status into permanent residence. These shortcomings in the data impede analysis of how shocks to foreign economies or changes in US immigration policy affect the supply of high-skilled foreign labor in the United States. Relaxing these data constraints is essential for the informed study of how high-skilled immigration affects US economic outcomes, including the pace of productivity growth, the earnings premium commanded by highly skilled labor, and differential wage and employment growth across local labor markets in the United States.

References

Acemoglu, Daron. 2002. "Directed Technical Change." *Review of Economic Studies* 69 (4): 781–809.

Borjas, George. 1987. "Self-Selection and the Earnings of Immigrants." *American Economic Review* 77 (4): 531–53.

———. 2014. *Immigration Economics*. Cambridge, MA: Harvard University Press.

Borjas, George, and Kirk Doran. 2012. "The Collapse of the Soviet Union and the Productivity of American Mathematicians." *Quarterly Journal of Economics* 127 (3): 1143–203.

Bound, John, Breno Braga, Joseph Golden, and Gaurav Khanna. 2015. "Recruitment of Foreigners in the Market for Computer Scientists in the United States." *Journal of Labor Economics* 33 (S1): S187–223.

Bound, John, Murat Demirci, Gaurav Khanna, and Sarah Turner. 2015. "Finishing Degrees and Finding Jobs: U.S. Higher Education and the Flow of Foreign IT Workers." In *Innovation Policy and the Economy*, vol. 15, edited by William Kerr, Josh Lerner, and Scott Stern. Chicago: University of Chicago Press.

Bound, John, Sarah Turner, and Patrick Walsh. 2009. "Internationalization of U.S. Doctorate Education." NBER Working Paper no. 14792, Cambridge, MA.

Clemens, Michael. 2010. "The Roots of Global Wage Gaps: Evidence from Randomized Processing of U.S. Visas." Working paper no. 212, Center for Global Development.

Gandal, Neil, Gordon Hanson, and Matthew Slaughter. 2004. "Technology, Trade, and Adjustment to Immigration in Israel." *European Economic Review* 48 (2): 403–28.

Griffith, Erin. 2015. "The World's Largest Tech Companies: Apple Beats Samsung, Microsoft, Google." *Forbes*, May 11.

Grogger, Jeffrey, and Gordon H. Hanson. 2011. "Income Maximization and the Selection and Sorting of International Migrants." *Journal of Development Economics* 95 (1): 42–57.

———. 2015. "Attracting Talent: Location Choices of Foreign-Born PhDs in the US." NBER Working Paper no. 18780, Cambridge, MA.

Hira, Ron. 2010. "The H-1B and L-1 Visa Programs: Out of Control." Briefing Paper no. 280, Economic Policy Institute.

Hunt, Jennifer. 2011. "Which Immigrants Are Most Innovative and Entrepreneurial? Distinctions by Entry Visa." *Journal of Labor Economics* 29 (3): 417–57.

Hunt, Jennifer, and Marjolaine Gauthier-Loiselle. 2010. "How Much Does Immigration Boost Innovation?" *American Economic Journal: Macroeconomics* 2 (2): 31–56.

Jasso, Guillermina, Douglas S. Massey, Mark R. Rosenzweig, and James P. Smith. 2000. "Assortative Mating among Married New Legal Immigrants to the United States: Evidence from the New Immigrant Survey Pilot." *International Migration Review* 34 (2): 443–59.

Jones, Charles I. 2002. "Sources of U.S. Economic Growth in a World of Ideas." *American Economic Review* 92 (1): 220–39.

Jordan, Miriam. 2015. "U.S. Firms, Workers Try to Beat H-1B Visa Lottery System." *Wall Street Journal*, June 2.

Kerr, Sari Pekkala, William R. Kerr, and William F. Lincoln. 2015. "Skilled Immigration and the Employment Structures of U.S. Firms." *Journal of Labor Economics* 33 (S1): 147–86.

Kerr, William R., and William F. Lincoln. 2010. "The Supply Side of Innovation: H-1B Visa Reforms and U.S. Ethnic Invention." *Journal of Labor Economics* 28 (3): 473–508.

Langdon, David, George McKittrick, David Beede, Beethika Kahn, and Mark Doms. 2011. "STEM: Good Jobs for Now and for the Future." Economics and Statistics Administration no. 03-11, US Department of Commerce.

Lewis, Ethan. 2011. "Immigration, Skill Mix, and Capital-Skill Complementarity." *Quarterly Journal of Economics* 126 (2): 1029–69.

Lin, Jeffrey. 2011. "Technological Adaption, Cities, and New Work." *Review of Economics and Statistics* 93 (2): 554–74.

Matloff, Norman. 2013. "Are Foreign Students the 'Best and the Brightest'? Data and Implications for Immigration Policy." Economic Policy Institute Briefing Paper no. 356, Economic Policy Institute.

Office of Immigration Statistics. 2014. *2013 Yearbook of Immigration Statistics.* Washington, DC: US Department of Homeland Security.

Peri, Giovanni. 2012. "The Effect of Immigration on Productivity: Evidence from U.S. States." *Review of Economics and Statistics* 94 (1): 348–58.

Porter, Eduardo. 2013. "Immigration and the Labor Market." *New York Times*, June 25. http://economix.blogs.nytimes.com/2013/06/25/immigration-and-the-labor-market/?_r=0.

Comment   John Bound

In this chapter Hanson and Slaughter use data from the decennial census, the American Community Survey (ACS), and the Current Population Survey (CPS) to document the rapid growth of the foreign born among US STEM workers. The data used by Hanson and Slaughter do not allow them to identify individuals by visa status. Extending tabulations originally done by Lowell (2000) and Bound et al. (2015) estimate that, as of 2000, close to 500,000 individuals were working in the United States on H-1B visas. The census data Hanson and Slaughter use show 793,000 foreign-born full-time employees working in STEM occupations as of 2000. Since almost all workers on H-1B visas are working in STEM fields, it seems safe to assume that most of the foreign born in Hanson and Slaughter's tabulations are on H-1B visas.

These foreign-born workers appear to be quite productive. Indeed, controlling for education, gender, race, and region, foreign-born STEM workers living in the United States at least six years appear to earn a small premium (roughly 5 percent) over their US-born counterparts. However, those more recently immigrated appear to earn somewhat less than their US counterparts.

As Hanson and Slaughter point out, this pattern of earnings is consistent with a number of very different and not mutually exclusive explanations. First, selection could explain increasing relative earnings among the foreign born. It seems plausible that very productive foreign-born workers are more likely to have employers sponsor them for permanent residency in the United States. If the most productive workers tend to stay, this could explain the observed patterns of earnings. Second, the pattern could simply reflect the acquisition over time by foreign-born workers of skills that are rewarded by the US labor market.

A third explanation for the earnings pattern is found in the cross-employer mobility limitation imposed by the H-1B visa program. Critics of the program say this constraint gives employers some monopsony power over H-1B workers, which could explain their lower relative earnings in the years immediately following immigration to the United States.

While no evidence incontrovertibly demonstrates cost or productivity advantages associated with hiring the foreign born, it seems clear that such advantages must exist. Since the middle of the first decade of the twenty-first century, the H-1B cap has always been reached, often relatively early

John Bound is professor of economics at the University of Michigan and a research associate of the National Bureau of Economic Research.

I would like to thank Gordon Hanson for providing me with data used in the chapter, Gaurav Khanna and Nicolas Morales for comments, and N. E. Barr for editorial assistance. For acknowledgments, sources of research support, and disclosure of the author's material financial relationships, if any, please see http://www.nber.org/chapters/c13708.ack.

in the fiscal year, suggesting the demand for H-1B workers substantially exceeds the quota-determined supply. This excess demand persists despite both pecuniary and nonpecuniary costs associated with hiring foreigners on H-1B visas. For instance, a recent GAO survey found legal and administrative costs to range from \$2,300 to \$7,500 for each H-1B hire (US General Accounting Office 2011).

How Essential to the STEM Workforce Are the Foreign Born?

In their introduction, Hanson and Slaughter seem to suggest that foreign-born scientists are essential to the US world leadership in science and technology—pointing to the poor overall performance of US students in math and science and the US demand for foreign labor.

This story is not as self-evident as it might seem from Hanson and Slaughter's tabulations. The United States has maintained a dominant position in science and technology since the end of World War II, despite having a small foreign-born STEM workforce throughout the 1960s, 1970s, and 1980s, and a public education system that was no better then than it is today.

Understanding the impact that increased high-skilled immigration has had on the US economy ultimately involves evaluating counterfactuals. A very simple, static, partial equilibrium model can illustrate my point. Let β represent the occupational supply elasticity of US nationals to science and engineering, and γ represent the demand elasticity for scientists and engineers. Increases in the availability of foreign talent or changes in the H-1B visa cap can be thought of as exogenous shifts in the supply of foreign-born workers in the US science and technology sector. An exogenous positive shock to the size of the science and engineering workforce in the United States will work to lower wages of scientists and engineers in the United States and, as a result, fewer US nationals will choose these occupations:

$$d \ln(\text{S\&E earnings}) = 1/[\beta + \gamma] \cdot \text{exogenous supply shock}$$

$$d \ln(\text{S\&E employment US nationals}) = \beta/[\beta + \gamma] \cdot \text{exogenous supply shock}.$$

As long as demand curves are downward sloping (finite γ), an exogenous influx of foreign-born scientists and engineers will work to lower wages and employment of US residents in these occupations. How much of the shock will be felt in terms of wages and how much in terms of employment will depend on how elastic the supply of US residents is to these occupations. Although each additional foreign scientist or engineer "crowds out" $\beta/[\beta + \gamma]$ US-born workers from such occupations, the total employment of scientists and engineers working in the United States will grow by a factor of $\gamma/[\beta + \gamma]$.

What do we know about these supply and demand elasticities? Researchers have consistently found that STEM occupational supply elasticities are high (Freeman 1975, 1976; Ryoo and Rosen 2004; Bound et al. 2015). Without some large exogenous supply shift, demand elasticities are harder to gauge, but some evidence indicates that the demand elasticity of STEM workers might be quite high—and trade and endogenous technical change tend to increase demand elasticities. High demand elasticities would imply little crowd-out effect from foreign-born STEM hires.

Some researchers (e.g., Kerr and Lincoln 2010) have used geographic variation in the employment of scientists and engineers on H-1B visas within the United States to directly estimate crowd-out. However, if location is endogenous, such efforts will tend to underestimate crowd-out. Khanna, Morales, and I have worked with calibrated general equilibrium models for workers in the computer science (CS) sector that allow for endogenous technical progress (Bound, Khanna, and Morales 2018). Our calculations produce downward-sloped demand curves, showing that the addition of one foreign-born computer scientist to the CS labor market is associated with an occupational switch out of CS by between 0.33 and 0.61 native computer scientist.

The bottom line: although downward-sloping demand curves indicate crowd-out of native-born by foreign-born workers, a crowd-out effect of around 0.5 suggests that highly skilled immigrants have also significantly increased the size of the STEM workforce in the US economy. The claim that US employers of STEM labor cannot find enough adequately skilled workers within the United States appears to be exaggerated. However, at the same time, it seems very likely that the existence of a pool of skilled foreigners has facilitated the growth of the science and technology sector in the United States.

In addition, the reservoir of foreign talent may act as a buffer, smoothing demand adjustments in the US labor market. One can find suggestions of this kind of effect in Hanson and Slaughter's chapter and in comparisons between how the IT labor market responded to IT booms in the late 1970s and early 1980s versus the boom in the 1990s (Bound et al. 2013).

The simple partial equilibrium model used above, together with most of the literature evaluating the impact of high-skilled immigration on the US economy, do not account for any global effects of US immigration policy—which have likely been significant. As pointed out in the theoretical literature, the US preeminence in advanced technologies benefits the US population (Krugman 1979; Johnson and Stafford 1993; Samuelson 2004). Freeman (2006) has argued that US policies on high-skilled immigration have helped the United States maintain technological leadership in the world. However, he ignores the effects this immigration policy might have had on other countries. The possibility of emigrating to the United States raises the returns to education in technical fields in immigrant-sending countries such as India and China. In addition, many foreign-born STEM workers in the United States eventually emigrate elsewhere, taking their acquired job skills with them. Indeed, Lowell (2000) calculated that roughly half of H-1B visa holders arriving in the United States during the 1990s eventually emigrated.

Both of these potential effects—an increase in returns to STEM education outside the United States and an increase in high-skilled emigration from the United States—imply that US immigration policies allowing foreign-born workers to fill STEM jobs will spur the size and quality of the STEM workforce in sending countries. Khanna and Morales (2017) have tried to quantify these effects, focusing on immigration of computer science workers into the United States from India. Within the context of their model, they find that the H-1B program has indeed spurred CS sector growth in both the United States and in India.

The kind of descriptive evidence that Hanson and Slaughter present in their chapter is important. However, if we are to understand the impact that US policy on high-skilled immigration has had on US workers, consumers, and employers, we need to implicitly or explicitly evaluate counterfactuals. Doing so will require the building and calibration of credible general equilibrium economic models.

A Plea for Data

Hanson and Slaughter end their chapter with a discussion of the need for data to evaluate the impact of high-skilled foreign labor on the US economy. They write: "Relaxing . . . data constraints is essential for the informed study of how high-skilled immigration affects US economic outcomes, including the pace of productivity growth, the earnings premium commanded by highly skilled labor, and differential wage and employment growth across local labor markets in the United States." What I want to emphasize is that, at least in theory, the kind of data that Hanson and Slaughter are talking about exists. Post-9/11 changes in immigration policy should have made tracking immigrants technically possible. What is more, in theory this data could be linked to either Social Security earnings histories or data from the Covered Employment and Wages Program. However, the government has not done these linkages, nor have they given access to this data to researchers. As Hanson and Slaughter emphasize, such data would give us a much more complete picture of the impact that high-skilled immigrants are having on the US economy.

References

Bound, John, Breno Braga, Joseph M. Golden, and Gaurav Khanna. 2015. "Recruitment of Foreigners in the Market for Computer Scientists in the US." *Journal of Labor Economics* 33 (S1): S187–223.
Bound, John, Breno Braga, Joseph M. Golden, and Sarah Turner. 2013. "Pathways to Adjustment: The Case of Information Technology Workers." *American Economic Review, Papers and Proceedings* 103 (3): 203–7.

Bound, John, Gaurav Khanna, and Nicolas Morales. 2018. "Understanding the Economic Impact of the H-1B Program on the United States." In *High-Skilled Migration to the United States and Its Economic Consequences*, edited by Gordon H. Hanson, William R. Kerr, and Sarah Turner, 109–75. Chicago: University of Chicago Press.

Freeman, Richard B. 1975. "Supply and Salary Adjustments to the Changing Science Manpower Market: Physics, 1948–1973." *American Economic Review* 65 (1): 27–39.

———. 1976. "A Cobweb Model of the Supply and Starting Salary of New Engineers." *Industrial and Labor Relations Review* 29 (2): 236–48.

———. 2006. "Does Globalization of the Scientific/Engineering Workforce Threaten U.S. Economic Leadership?" In *Innovation Policy and the Economy*, vol. 6, edited by Adam B. Jaffe, Josh Lerner, and Scott Stern. Cambridge, MA: MIT Press.

Johnson, George E., and Frank P. Stafford. 1993. "International Competition and Real Wages." *American Economic Review* 83 (2): 127–30.

Kerr, William, and William Lincoln. 2010. "The Supply Side of Innovation: H-1B Visa Reforms and U.S. Ethnic Invention." *Journal of Labor Economics* 28 (3): 473–508.

Khanna, Gaurav, and Nicolas Morales. 2017. "The IT Boom and Other Unintended Consequences of Chasing the American Dream." Working paper no. 460, Center for Global Development. https://www.cgdev.org/publication/it-boom-and-other-unintended-consequences-chasing-american-dream.

Krugman, Paul. 1979. "A Model of Innovation, Technology Transfer, and the World Distribution of Income." *Journal of Political Economy* 87 (2): 253–66.

Lowell, B. Lindsay. 2000. "H-1B Temporary Workers: Estimating the Population." Working paper no. 12, University of California, San Diego.

Ryoo, Jaewoo, and Sherwin Rosen. 2004. "The Engineering Labor Market." *Journal of Political Economy* 112 (S1): S110–40.

Samuelson, Paul A. 2004. "Where Ricardo and Mill Rebut and Confirm Arguments of Mainstream Economists Supporting Globalization." *Journal of Economic Perspectives* 18 (3): 135–46.

US General Accounting Office. 2011. "H-1B Visa Program: Reforms Are Needed to Minimize the Risks and Costs of Current Program." Technical Report GAO/HEHS 00-157, US General Accounting Office, Washington, DC.

Contributors

Jaison R. Abel
Federal Reserve Bank of New York
50 Fountain Plaza, Suite 1400
Buffalo, NY 14202

David Autor
Department of Economics, E52-438
Massachusetts Institute of Technology
77 Massachusetts Avenue
Cambridge, MA 02139

Sandy Baum
Urban Institute
2100 M Street, NW
Washington, DC 20037

Canyon Bosler
Department of Economics
University of Michigan
238 Lorch Hall
611 Tappan Avenue
Ann Arbor, MI 48109-1220

John Bound
Department of Economics
University of Michigan
Ann Arbor, MI 48109-1220

Stijn Broecke
Organisation for Economic Co-
    operation and Development
2, rue André Pascal
75775 Paris Cedex 16
France

Mary C. Daly
Federal Reserve Bank of San Francisco
101 Market Street
San Francisco, CA 94105

David J. Deming
Harvard Graduate School of
    Education
Gutman 411
Appian Way
Cambridge, MA 02138

Richard Deitz
Federal Reserve Bank of New York
50 Fountain Plaza, Suite 1400
Buffalo, NY 14202

Douglas W. Elmendorf
Harvard Kennedy School
79 John F. Kennedy Street
Cambridge, MA 02138

John G. Fernald
Professor of Economics
INSEAD
Boulevard de Constance
77305 Fontainebleau, France

Maury Gittleman
US Bureau of Labor Statistics
2 Massachusetts Avenue NE, Suite
   4130
Washington, DC 20212

Grey Gordon
Department of Economics
Wylie Hall, Room 105
Indiana University
100 S Woodlawn Ave
Bloomington, IN 47403

Nora Gordon
McCourt School of Public Policy
Georgetown University
306 Old North
37th and O Streets NW
Washington, DC 20057

Gordon H. Hanson
IR/PS 0519
University of California, San Diego
9500 Gilman Drive
La Jolla, CA 92093-0519

Eric A. Hanushek
Hoover Institution
Stanford University
Stanford, CA 94305-6010

Aaron Hedlund
Department of Economics
University of Missouri
118 Professional Building
909 University Avenue
Columbia, MO 65211

Bart Hobijn
Department of Economics
Arizona State University
PO Box 879801
Tempe, AZ 85287-9801

Caroline M. Hoxby
Department of Economics
Stanford University
Landau Building, 579 Serra Mall
Stanford, CA 94305

Charles R. Hulten
Department of Economics
University of Maryland
Room 3114, Tydings Hall
College Park, MD 20742

Dale W. Jorgenson
Department of Economics
Littauer Center, Room 122
Harvard University
Cambridge, MA 02138

Mun S. Ho
Resources for the Future
1616 P Street, NW
Washington, DC 20036

Frank Levy
Department of Urban Studies and
   Planning
Building 9-517
Massachusetts Institute of Technology
Cambridge, MA 02139

Shelly Lundberg
Department of Economics
University of California, Santa
   Barbara
Santa Barbara, CA 93106-9210

Kristen Monaco
US Bureau of Labor Statistics
2 Massachusetts Avenue NE, Suite
   4130
Washington, DC 20212

Nicole Nestoriak
US Bureau of Labor Statistics
2 Massachusetts Avenue NE, Suite
   4130
Washington, DC 20212

Glenda Quintini
Organisation for Economic Co-
    operation and Development
2, rue André Pascal
75775 Paris Cedex 16
France

Valerie A. Ramey
Department of Economics, 0508
University of California, San Diego
9500 Gilman Drive
La Jolla, CA 92093-0508

Jon D. Samuels
Bureau of Economic Analysis
4600 Silver Hill Road
Washington, DC 20233

Matthew J. Slaughter
Tuck School of Business
Dartmouth College
100 Tuck Hall
Hanover, NH 03755

Robert G. Valletta
Economic Research Department
Federal Reserve Bank of San Francisco
101 Market Street
San Francisco, CA 94105

Marieke Vandeweyer
Organisation for Economic Co-
    operation and Development
2, rue André Pascal
75775 Paris Cedex 16
France

Author Index

Subject Index